D1546467
3 1611 00088 8765

SYMPOSIA OF THE ZOOLOGICAL SOCIETY OF LONDON NO. 67

Ecology, Evolution and Behaviour of Bats

SYMPOSIA OF THE ZOOLOGICAL SOCIETY OF LONDON NO 67

Ecology, Evolution and Behaviour of Bats

The Proceedings of a Symposium held by the Zoological Society of London and the Mammal Society: London, 26th and 27th November 1993

Edited by Paul A. RACEY and Susan M. SWIFT
Department of Zoology, University of Aberdeen

Published for THE ZOOLOGICAL SOCIETY OF LONDON
by CLARENDON PRESS · OXFORD
1995

Oxford University Press, Walton Street, Oxford OX2 6DP

Oxford New York
Athens Auckland Bangkok Bombay
Calcutta Cape Town Dar es Salaam Delhi
Florence Hong Kong Istanbul Karachi
Kuala Lumpur Madras Madrid Melbourne
Mexico City Nairobi Paris Singapore
Taipei Tokyo Toronto
and associated companies in
Berlin Ibadan

Oxford is a trade mark of Oxford University Press

Published in the United States
by Oxford University Press Inc., New York

A catalogue record for this book is available from the British Library

Library of Congress Cataloging in Publication Data
Ecology, evolution, and behaviour of bats : the proceedings of a symposium held by the Zoological Society of London and the Mammal Society : London, 26th and 27th November 1993 / edited by Paul A. Racey and Susan M. Swift.
(Symposia of the Zoological Society of London ; no. 67)
Includes bibliographical references and index.
1. Bats–Congresses. I. Racey, P. A. II. Swift, Susan M.
III. Zoological Society of London. IV. Mammal Society. V. Series.
QL1.Z733 no. 67
[QL737.C5] 591 s–dc20 [599.4] 94–49722
ISBN 0 19 854945 8

Typeset by Light Technology Ltd., Fife

Printed in Great Britain by
Bookcraft (Bath) Ltd
Midsomer Norton, Avon

Preface

After a gestation period of six years the Action Plan on Old World Fruit Bats was published late in 1992 (Mickleburgh, Hutson & Racey 1992).[1] It contained a list of urgently needed research including studies on the role of pteropodids in pollination and seed dispersal in tropical forests. At the same time as the Action Plan was being formulated, The Zoological Society of London and The Mammal Society were planning a Symposium on Recent Advances in the Biology of Bats—the first such international meeting on the second largest order of mammals since that held in Aberdeen with a similar title in 1985 (Fenton, Racey & Rayner 1987).[2] Since the Megachiroptera have received comparatively little attention in the past, the present Symposium provided an opportunity of reviewing the biology of these bats, and almost a day of the two-day programme held at the Zoological Society's meeting rooms in Regent's Park in November 1993 was allocated to papers on Megachiroptera. The selection of aspects of the biology of Microchiroptera for inclusion in the meeting was more difficult. Echolocation was avoided, since symposia on that subject, as well as on Functional Morphology and Conservation Biology, were to be held during the Tenth International Bat Research Conference in Boston in August 1995. The meeting was therefore planned to include a number of other exciting and novel areas of bat biology ranging from ecophysiology through molecular genetics to landscape ecology. Speakers came from nine countries and the meeting was attended by over 200 people.

On advice from the publishers, the title has evolved into the more distinctive *Ecology, evolution and behaviour of bats*. We are grateful to the speakers for submitting high-quality manuscripts, generally on time, to Professor T. H. Kunz of Boston University for helpful suggestions, and particularly to Unity McDonnell for her help throughout all stages of the planning of the meeting and editing of the papers. The spelling of species names follows Corbet & Hill (1991).[3]

P. A. R.
S. M. S.

Aberdeen
September 1994

[1]Mickleburgh, S., Hutson, A. M. & Racey, P. A. (1992). *Action plan for Old World Fruit Bats*. IUCN, Gland, Switzerland.
[2]Fenton, M. B., Racey, P. A. & Rayner, J. M. V. (Eds) (1987). *Recent advances in the study of bats*. Cambridge University Press, Cambridge.
[3]Corbet, G. B. & Hill, J. E. (1991). *A world list of mammalian species* (3rd edn). Oxford University Press, Oxford.

Contents

A review of ecological interactions of fruit bats in Australian ecosystems
G. C. RICHARDS

Part III: Reproductive biology, physiology and energetics

The use of stable isotopes to study the diets of plant-visiting bats
THEODORE H. FLEMING

Part V: Microchiropteran behaviour and ecology

Constraint and flexibility—bats at predators, bats as prey
M. BROCK FENTON

Street lamps and the feeding ecology of insectivorous bats
J. RYDELL & P. A. RACEY

Conservation biology of an endangered species: field studies of greater horseshoe bats
GARETH JONES, P. LAURENT DUVERGÉ & ROGER D. RANSOME

Contributors

Arlettaz, R., Institute of Zoology and Animal Ecology, University of Lausanne, CH-1015 Lausanne, Switzerland.

Barclay, R. M. R., Behavioural Ecology Group, Ecology Division, Department of Biological Sciences, University of Calgary, Calgary, Alberta, Canada T2N 1N4.

Barratt, E. M., Conservation Genetics Group, Institute of Zoology, The Zoological Society of London, Regent's Park, London NW1 4RY, UK.

Bonaccorso, F. J., Department of Zoology and P. K. Yonge Developmental Research School, University of Florida, Gainesville, Florida 32611, USA.

Bruford, M. W., Conservation Genetics Group, Institute of Zoology, The Zoological Society of London, Regent's Park, London NW1 4RY, UK.

Burland, T. M., Conservation Genetics Group, Institute of Zoology, The Zoological Society of London, Regent's Park, London NW1 4RY, UK.

Cox, P. A., Department of Botany & Range Science, Brigham Young University, Provo, UT 84602, USA.

Duvergé, P. L., School of Biological Sciences, University of Bristol, Woodland Road, Bristol BS8 1UG, UK.

Elmqvist, T., Department of Ecological Botany, Umea University, Umea, Sweden 901–87.

Fenton, M. B., Department of Biology, York University, North York, Ontario, Canada M3J 1P3.

Fleming, T. H., Department of Biology, University of Miami, Coral Gables, Florida 33124, USA.

Harris, S., School of Biological Sciences, University of Bristol, Woodland Road, Bristol BS8 1UG, UK.

Heideman, P. D., Institute of Reproductive Biology and Department of Zoology, University of Texas, Austin, Texas 78712, USA; *present address* Department of Biology, PO Box 8795, College of William and Mary, Williamsburg, Virginia 23187–8795, USA.

Hutson, A. M., The Bat Conservation Trust, 45 Shelton Street, London WC2H 9HJ, UK.

Jones, G., School of Biological Sciences, University of Bristol, Woodland Road, Bristol BS8 1UG, UK.

Kalko, E. K. V., University of Tübingen, Animal Physiology, Auf der Morgenstelle 28, D-72076 Tübingen, Germany; *and* Smithsonian Tropical Research Institute, PO Box 2072, Balboa, Republic of Panama.

Kennedy, J. H., Department of Physiology & Pharmacology, University of Queensland, Brisbane, Queensland 4072, Australia.

Kerr, M. H., Hannah Research Institute, Ayr KA6 5HL, Scotland, UK.

Knight, C. H., Hannah Research Institute, Ayr KA6 5HL, Scotland, UK.

Kunz, T. H., Department of Biology, Boston University, 5 Cummington Street, Boston, Massachusetts 02215, USA

Little, L., Department of Physiology & Pharmacology, University of Queensland, Brisbane, Queensland 4072, Australia.

Luckhoff, H. C., Department of Physiology & Pharmacology, University of Queensland, Brisbane, Queensland 4072, Australia.

Martin, L., Department of Physiology & Pharmacology, University of Queensland, Brisbane, Queensland 4072, Australia.

McNab, B. K., Department of Zoology, University of Florida, Gainesville, Florida 32611, USA.

Mayer, F., Institut für Zoologie II, Universität Erlangen-Nürnberg, Staudtstrasse 5, D-91058 Erlangen, Germany.

Neuweiler, G., Zoologisches Institut der Universität München, Luisenstrasse 14, 80333 Munich, Germany.

O'Brien, G. M., Department of Physiology & Pharmacology, University of Queensland, Brisbane, Queensland 4072, Australia.

Pääbo, S., Zoologisches Institut der Universität München, Luisenstrasse 14, 80333 Munich, Germany.

Palmeirim, J. M., Departamento Zoologia, Faculdade de Ciências, Universidade de Lisboa, P-1700 Lisboa, Portugal.

Perrin, N., Ethological Station Hasli, University of Bern, CH–3032 Hinterkappelen, Switzerland.

Petri, B., Zoologisches Institut der Universität München, Luisenstrasse 14, 80333 Munich, Germany.

Pettigrew, J. D., Vision, Touch & Hearing Research Centre, University of Queensland, Brisbane, Queensland 4072, Australia.

Pierson, E. D., Museum of Vertebrate Zoology, University of California, Berkeley, CA 94720, USA.

Pow, C. S. T., Department of Physiology & Pharmacology, University of Queensland, Brisbane, Queensland 4072, Australia.

Racey, P. A., Department of Zoology, University of Aberdeen, Tillydrone Avenue, Aberdeen AB9 2TN, Scotland, UK.

Rainey, W. E., Museum of Vertebrate Zoology, University of California, Berkeley, CA 94720, USA.

Ransome, R. D., School of Biological Sciences, University of Bristol, Woodland Road, Bristol BS8 1UG, UK.

Richards, G. C., Division of Wildlife & Ecology, CSIRO, PO Box 84, Lyneham, ACT 2602, Australia.

Rodrigues, L., Instituto da Conservação da Natureza, Rua Filipe Folque, 46–2, P–1000 Lisboa, Portugal.

RYDELL, J., Department of Zoology, University of Aberdeen, Tillydrone Avenue, Aberdeen AB9 2TN, Scotland, UK.

SIMMONS, N. B., Department of Mammalogy, American Museum of Natural History, Central Park West at 79th Street, New York, NY 10024, USA.

SPEAKMAN, J. R., Department of Zoology, University of Aberdeen, Tillydrone Avenue, Aberdeen AB9 2TN, Scotland, UK.

STERN, A. A., Department of Biology, Boston University, 5 Cummington Street, Boston, Massachusetts 02215, USA.

THOMAS, D. W., Groupe de Recherche en Écologie, Nutrition et Énergétique, Département de Biologie, Université de Sherbrooke, Sherbrooke, Québec, Canada J1K 2RI; *and* Musée du Séminaire de Sherbrooke, Sherbrooke, Québec, Canada J1H 1J9.

TOWERS, P. A., Department of Physiology & Pharmacology, University of Queensland, Brisbane, Queensland 4072, Australia.

UTZURRUM, R. C. B., Center for Tropical Conservation Studies, Silliman University, 6200 Dumaguete City, Philippines; *current address* Department of Biology, Boston University, 5 Cummington Street, Boston, MA 02215, USA.

WALDON, A. K., Department of Physiology & Pharmacology, University of Queensland, Brisbane, Queensland 4072, Australia.

WALSH, A. L., School of Biological Sciences, University of Bristol, Woodland Road, Bristol BS8 1UG, UK.

WANG, D. Y., Department of Physiology & Pharmacology, University of Queensland, Brisbane, Queensland 4072, Australia.

WAYNE, R. K., Conservation Genetics Group, Institute of Zoology, The Zoological Society of London, Regent's Park, London NW1 4RY, UK; *and* Department of Biology, University of California, Los Angeles, CA 90024, USA.

WEBB, P. I., Department of Zoology, University of Aberdeen, Tillydrone Avenue, Aberdeen AB9 2TN, Scotland, UK; *current address* Mammal Research Institute, University of Pretoria, Pretoria 0002, South Africa.

WILDE, C. J., Hannah Research Institute, Ayr KA6 5HL, Scotland, UK.

WILKINSON, G. S., Department of Zoology, University of Maryland, College Park, Maryland 20742, USA.

Organizer of symposium

PROFESSOR P. A. RACEY, Department of Zoology, University of Aberdeen, Tillydrone Avenue, Aberdeen AB9 2TN, Scotland, UK.

Chairmen of sessions

DR R. M. R. BARCLAY, Behavioural Ecology Group, Ecology Division, Department of Biological Sciences, University of Calgary, Calgary, Alberta, Canada T2N 1N4.

PROFESSOR M. B. FENTON, Department of Biology, York University, North York, Ontario, Canada M3J 1P3.

PROFESSOR T. H. KUNZ, Department of Biology, Boston University, 5 Cummington Street, Boston, Massachusetts 02215, USA.

PROFESSOR P. A. RACEY, Department of Zoology, University of Aberdeen, Tillydrone Avenue, Aberdeen AB9 2TN, Scotland, UK.

DR J. R. SPEAKMAN, Department of Zoology, University of Aberdeen, Tillydrone Avenue, Aberdeen AB9 2TN, Scotland, UK.

Part I

Chiropteran monophyly/diphyly

Symp. zool. Soc. Lond. (1995) No. 67: 3–26

Flying primates: crashed, or crashed through?

John D. PETTIGREW

Vision, Touch and Hearing Research Centre
The University of Queensland
4072 Australia

Synopsis

The flying primate hypothesis originated from the finding that megabats shared a number of advanced visual pathway characters with primates that were not found in any other mammalian order, nor in microbats. This hypothesis indicates that primates, colugos and megabats share a common ancestor with each other more recent than any shared with microbats. The hypothesis has found increasing support from other sources of evidence. Examples reviewed here include further derived brain features, both visual and non-visual, immunological studies of serum proteins with monoclonal antibodies and analysis of restriction sites and protein sequences (globins and α-crystallin). DNA sequence data, while supporting the colugo–primate association, have been used to reject a primate–megabat connection, even though the total evidence for a colugo–megabat link is better than the generally accepted evidence for a colugo–primate link, and even though DNA sequence data and protein sequence data on the same genes give conflicting phylogenies. A resolution to this conflict is suggested by a bias in all the published DNA sequence data on bats. The shared substitutions claimed in support of bat monophyly are mostly of adenine (A) or thymine (T), in the same direction as the bias that exists in the overall base composition of DNA from metabolically-active, volant organisms. If the AT content of DNA is taken into account by using the NZ algorithm, the much-vaunted claims for bat monophyly based on DNA sequences are not supported. It is more parsimonious to assume that the AT bias responsible for the claimed association arose independently in the two lines of flying mammals.

Introduction

The 'flying primate' hypothesis arose from the unexpected finding that megabats shared a number of derived brain features with primates that were not shared with other mammals, particularly not with microbats (Pettigrew 1986). Of the many controversial aspects of this hypothesis, perhaps the most contentious is its corollary: that powered mammalian flight has evolved more than once (see the four-part debate on this topic; Pettigrew 1991a, b;

ZOOLOGICAL SYMPOSIUM No. 67
ISBN 0 19 854945 8

Baker, Novacek & Simmons 1991; Simmons, Novacek & Baker 1991). Alternative explanations, such as the convergent evolution of primate-like brain organization in one branch of the bats, have become increasingly unlikely as more primate-like details of neural organization are revealed in different brain systems (Pettigrew, Jamieson, Robson, Hall, McAnally & Cooper 1989; Dann & Buhl 1990; Buhl & Dann 1991; Rosa, Schmid, Krubitzer & Pettigrew 1993; Rosa, Schmid & Pettigrew 1994; Rosa & Schmid 1994).

The hypothesis accounts readily for each of a long list of puzzling differences between the two kinds of flying mammals (Table 1). One example, from the 54 listed, is the absence of laryngeal sonar in megabats despite the favourable energetics and the useful role that would be played by this ability in a nocturnal flier (Speakman 1993). Proponents of the rival hypothesis have provided no satisfactory explanation to account for these wide-ranging differences, which are readily encompassed by the flying primate hypothesis. Supporting molecular evidence for the flying primate hypothesis has also accumulated from protein sequence data on α-crystallin (De Jong, Leunissen & Wistow 1993), from globins (Kleinschmidt, Sgouros, Pettigrew & Braunitzer 1988) and from immunological studies of serum proteins (Schreiber, Bauer & Bauer 1994).

The above evidence, together with support for the flying primate hypothesis from brain research (Calford, Graydon, Huerta, Kaas & Pettigrew 1985; Kennedy 1991; Krubitzer & Calford 1992; Krubitzer, Calford & Schmid 1993) has been completely overshadowed by the results of DNA sequencing studies of bats. Six separate studies were unanimous in rejecting the flying primate hypothesis after sequencing DNA from a variety of mammals. Each study found greater similarity between the DNA sequences of microbats and megabats than between the megabat DNA and primate DNA (Bennett, Alexander, Crozier & Mackinlay 1988; Adkins & Honeycutt 1991; Mindell, Dick & Baker 1991; Ammerman & Hillis 1992; Bailey, Slightom & Goodman 1992; Stanhope, Czelusniak, Si, Nickerson & Goodman 1992). These data have been given wide coverage in the literature, together with reports of the 'crash' of the flying primate hypothesis that provoked the present title. The difficulties of reconciling these new data with the large body of evidence from brain, behaviour and proteins are largely ignored in the reverence that currently attends DNA scripture. Also ignored, by each of these six studies, is a hefty bias towards adenine (A) and thymine (T) in the base composition of the DNA substitutions claimed as evidence of affinity between microbats and megabats.

The increased number of shared AT substitutions in the DNA of microbats and megabats has a more plausible and more parsimonious explanation than the shared flying ancestor proposed by the proponents of monophyly. High AT content of DNA is well-described for bats (Table 2). A new model of DNA evolution, allowing for AT bias, must be incorporated into the analysis before there is justification for ignoring all the morphological and molecular evidence that conflicts with the DNA sequence data.

In the present account I will summarize the evidence in favour of the flying primate hypothesis and give some examples of its heuristic power. I will then attempt to deal with the problem of conflicting evidence, using both the 'total evidence' approach of Kluge (1989) and the consensus approach (Lanyon 1993). Both approaches provide support for the hypothesis, with the extra step required for a second invention of flight outweighed by the many extra steps required by monophyly.

Primate brain features

Since the hypothesis involves a branch from the early evolution of primates, the choice of an appropriately plesiomorphic (i.e. phylogenetically primitive) primate for comparison is paramount. If they arose from the stem of the primate tree as proposed, megabats are not likely to bear comparison with an advanced primate having features not shared with primates in general. For example, the three cone pigments found in anthropoid primates, but not in non-primates, might be thought to provide a clear-cut defining feature of primates, until one discovers that this feature is absent from lemurs and lorises (Jacobs 1993). Megabats have cone photoreceptors, but it would be unreasonable to 'fail' them as primates because they do not measure up to the sophistication of the anthropoid cone photopigments. In fact, megabats, carnivores and prosimians appear to share similar cone types.

Tarsier as a test case

The two most plesiomorphic living primate taxa are *Microcebus* and *Tarsius*. There are three reasons for bringing *Tarsius* into the present discussion of the flying primate hypothesis. (1) Although it is an undisputed primate, *Tarsius* does not occupy an undisputed rank within the primates. The sources of the dispute are instructive and relevant to the debate concerning bats. (2) *Tarsius* appears to represent a basal stem of the primate tree and is therefore a more appropriate taxon for comparison with putative sister taxa than the derived primate taxa commonly used in this context. (3) *Tarsius* shares some characters with the colugo, *Cynocephalus*, that are not found in any other taxa, primate or non-primate.

Lack of consensus over the position of the tarsier

As shown in Fig. 1, there are three possible tree topologies for the relationships of the three primate groups, tarsiers, anthropoids and lemurs+lorises (loosely referred to as prosimians). Each of these three alternatives has received some support from molecular studies. While a recent influential review has endorsed the anthropoid affinity of *Tarsius* (Martin 1993), many workers in the field, particularly palaeontologists, are increasingly uncomfortable with this assignment (Fig. 1b). Accepting *Tarsius* into the anthropoids has the corollary

Table 1. Fifty-four microbat/megabat contrasts. Asterisks mark features shared with primates.

		Microbats	Megabats
1.	Distribution	Worldwide; Neotropical focus	* Palaeotropical
2.	Diet and dentition	Insectivorous with diverse adaptations	Phytophagous
3.	Feeding	Take food on wing	* Land before taking food
4.	Laryngeal, frequency-modulated sonar	Present in all species	Absent
5.	Percussive sonar	Absent	Present in 2 genera
6.	Orientation and navigation	Predominantly acoustic	* Predominantly visual
7.	Defecation/micturition at roost	Many use inverted posture; spine dorsi-flexed	* Upright posture: hanging from pollices
8.	Carrying young	Rare; young parked at roost	* Neonates carried during foraging
9.	Hindlimb	Not used for carrying/grooming	* Reaching, carrying, grooming
10.	Terrestrial locomotion	* Agile, brisk; running, jumping	Awkward, poor alternation
11.	Agonistic display	Acoustic	* Visual
12.	Pollex	Limited use; ulnar deviation in locomotion	* Dexterous, manipulative use in flexion
13.	Spinal cord	Expanded; marginal nucleus	Not expanded; no marginal nucleus
14.	Midbrain	Inferior colliculus >superior colliculus	* Superior colliculus >inferior colliculus
15.	Superior colliculus: retinal input	Primitive pattern	* Advanced, primate-like pattern
16.	Accessory optic system	Prominent medial terminal nucleus	* Medial terminal nucleus reduced
17.	LGN: laminar segregation by eye	Absent	* Present
18.	LGN: magnocellular layers	No magnocellular differentiation	* Paired externally next to optic tract
19.	Rhinal fissure	Visible on lateral surface of forebrain	* Hidden ventrally by neocortical expansion
20.	Hippocampus	Primitive pattern	* Anthropoid pattern in five of seven features
21.	Forebrain/hindbrain ratio	Low	* High
22.	Primary visual cortex	Small fraction of neocortex	* Highest fraction of neocortex in mammals
23.	V2: extrastriate cortex	Small	* Large
24.	V3 and MT: extrastriate cortex	Absent	* Present
25.	Frontal eye fields	Absent	* Present
26.	Somatosensory cortex	2 or 3 somatotopic representations	* 6 somatotopic representations
27.	Motor cortex	Diffuse corticospinal fields	* 4 separate corticospinal areas
28.	Primary auditory cortex (A1)	Low frequencies caudally	* Low frequencies rostrally
29.	Multiple auditory field in addition to A1	Present; up to 7 reported	* Absent; A1 + belt
30.	Cornea: Bowman's membrane	Absent	* Present

Table 1. *Cont.*

		Microbats		Megabats
31.	Retinal nutrition	Choroidal circulation only		Retinal capillary loops + choroidal
32.	Retinal ganglion cell 'horizontal streak'	Below optic nerve head	*	Above optic nerve head
33.	Tapetum lucidum (guanine-crystal type)	Absent	*	Present in many taxa
34.	Visual development	Altricial; eyes open postnatally	*	Precocial; eyes open prenatally
35.	Nycteribiid parasites	Nycteribiinae		Cyclopodiinae; Archinycteribiinae
36.	Fleas	Thaumopsyllinae; diverse monotypics		Ischnopsyllinae
37.	Torpor	General; plesiomorphic		Rare; derived
38.	Spermatozoon: head	Small acrosome and subacrosomal space		Large acrosome and subacrosomal space
39.	Spermatozoon: tail	Coarse fibres originate with fine fibres		Coarse fibres originate behind fine fibres
40.	Penis	Fibrous or bony glans	*	Corpus spongiosum forms glans
41.	Karyotype	Diverse		Uniform
42.	Pinna	Tragus present; incomplete	*	No tragus; completed anteriorly
43.	Middle ear: Paaw's cartilage	Absent	*	Present
44.	Cochlea	Huge; non-allometric; acoustically isolated	*	Allometric
45.	Forelimb metacarpals	Long	*	Short
46.	Hindlimb metatarsals	Long	*	Short
47.	Ankle joint	Flexor tendons separated from gastrocnemius		Tunnel with flexors and gastrocnemius
48.	Cranium: postorbital process	Absent	*	Present
49.	Skin: pilo-erector muscle	Striated	*	Smooth
50.	DNA: AT content	Heterogeneous; slight overall elevation		Highest known for mammals
51.	Serum protein epitopes	Few shared with primates	*	Primate-like pattern
52.	Lens α-crystallin	Indeterminate pattern		Carnivore-ungulate pattern
53.	Globin	Basal pattern	*	Primate pattern
54.	Liver lobulation	Spigelian lobe; separate	*	No Spigelian lobe

The phylogenetic significance of a Table of Differences has justifiably been questioned, on the grounds that polarities are unknown and cladistic analysis is therefore impossible. Note, however, that this problem also attends all molecular sequence data, where outgroup comparison is the only method available to provide character polarity in analysis. Present uncertainties about the mammalian phylogenetic tree make it hazardous, and circular, to choose a eutherian outgroup. If monotremes are used as an outgroup in the above Table, megabats and primates share the derived state for 29 of 54 characters, while microbats share no derived states with primates. If edentates are used as the outgroup, the derived state is shared between megabats and primates for 25 of 54 characters, and for microbats and primates one of the 49 characters is shared derived. The polarity of many character states in the Table can be soundly established by using criteria other than outgroup comparison, such as ontogenetic criteria and character state transformations within rigorously established phylogenies (e.g. the brain data), but the polarity of others flips back and forth as different eutherian outgroups are chosen, like the states of bases and amino acids in molecular sequence data as outgroups are changed.

Table 2. Substitutions supporting the rival hypotheses of bat origins: AT homoplasy in six sets of DNA sequence data

Sequence	kb	Taxa	Monophyly		Flying primates		Reference
			A+T	G+C	A+T	G+C	
CO3mtDNA	0.7	4	38	6	18	24	Bennett *et al.* (1988)
12SmtDNA	0.7	5	28	8	12	26	Mindell *et al.* (1991)
IRBP nDNA	1.2	13	18	4	4	6	Stanhope *et al.* (1992)
ϵ-globin intron	1.2	17	32	7	4	7	Bailey *et al.* (1992)
12SmtDNA	0.3	11	4	1		7	Ammerman & Hillis (1992)
CO3+16S mtDNA	0.8	7	19	7	12	20	Adkins & Honeycutt (1991); P. Tompkins & C. Moritz (unpubl.)
			139	33	50	90	

that living tarsioids are somehow divorced from the well-described, obviously tarsier-like, omomyids, with their extensive ancient fossil record.

The tree topology with *Tarsius* paraphyletic to the other two groups of living primates (Fig. 1a) is the one that is regaining acceptance. The brain data are quite unequivocal in their support of this phylogeny. *Tarsius* has by far the most plesiomorphic brain of any living primate. Its well-developed medial terminal nucleus is unlike any other primate's and quite similar to that found in carnivores. The small cerebellum and corpus callosum also have no parallel in any other living primate. The lateral geniculate nucleus (LGN) is remarkable because on the one hand it strongly affirms that *Tarsius* is a primate (with three pairs of laminae and a prominent pair of external magnocellular laminae), while on the other it reveals a feature that is found in no other primate (reversal of the order of the magnocellular laminae, with the ipsilateral lamina lying externally). This unusual arrangement would be of no use in phylogenetic analysis except for the fact that it is present only in the tarsier and in the colugo, *Cynocephalus*, another putative primate sister taxon. Along with the series of plesiomorphic brain characters, this feature shared with the colugo can be explained only by a very early divergence of *Tarsius* before the diversification of the living primates. There is also growing corroborating evidence that the colugo is a close relative of primates. The unusual

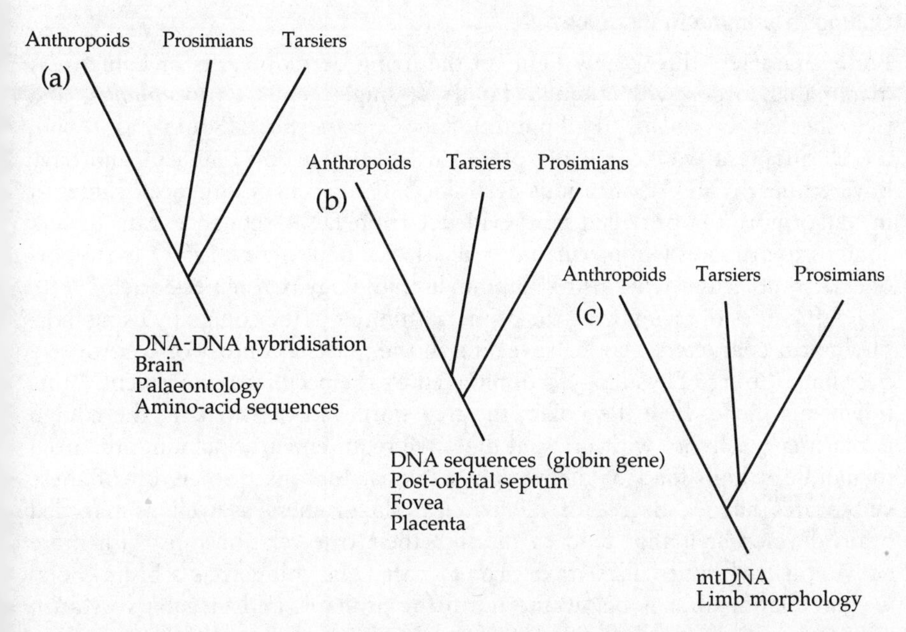

Fig. 1. Three possible topologies for the relationships of tarsiers to anthropoid and lorisiform ('prosimian') primates. Arrangement (b) is widely accepted, but note that there is growing support for arrangement (a) where tarsiers are paraphyletic to all other primates. The present lack of agreement about the position of tarsiers within the primates underscores the importance of the new phylogenetic investigations of primates and their putative sister taxa. The brain data place the tarsier firmly in the position shown in (a).

arrangement shared in the LGN between *Tarsius* and *Cynocephalus* tends to confirm the primate status of the colugo while emphasizing the basal position of *Tarsius* within primates (Rosa, Pettigrew & Cooper in press).

Functional convergence in *Tarsius*

What of the derived morphological characters apparently shared between anthropoids and *Tarsius*? A bony post-orbital septum is found in both *Tarsius* and anthropoids, as has been eloquently emphasized (Cartmill 1972), but this structure is almost certainly independently derived in the two taxa. In *Tarsius* it is a necessary functional acquisition that provides bony protection for the lateral part of the eyeball as it extends far beyond the lateral margin of the skull. This interpretation of the post-orbital septum is supported by developmental work that failed to find homology between the growth patterns of the post-orbital septum in *Tarsius* and in anthropoids (Simons & Rasmussen 1989).

The other characters used to link *Tarsius* and anthropoids, such as a retinal fovea and discoid placentation, have a scattered phylogenetic distribution that reduces their value in making such links.

Colugo: a primate in disguise?

Brain characters throw new light on the living dermopterans or colugos. A mammalian order with a single family, a single genus *Cynocephalus*, and two species, *C. volans* (Philippines) and *C. variegatus* (South-East Asia), the Dermoptera were originally placed with primates by Linnaeus and have hovered nearby in classifications ever since. The recent resurgence of interest in bat origins has provided new evidence from DNA sequence data to link colugos to primates (Ammerman & Hillis 1992; Bailey *et al.* 1992), in support of earlier molecular work using albumin immunology (Cronin & Sarich 1980). The difficulties of recognizing the primate affinities of the colugo by using morphological characters may be revealing in the present context of controversy over bats. Colugo physiology is dominated by the peculiar requirements of its folivorous niche. Like the koala, the tree sloth and the hoatzin, the colugo is a primary folivore with unusual digestive requirements, including the need to handle a high load of dietary phytotoxins. In consequence, all of these vertebrates have a degree of behavioural sluggishness as well as retarded brain development that tend to disguise their true relationships with more active, bigger-brained sister taxa. The fact that the colugo has a brain–body weight relation that is below the norm for primates and megabats (Martin 1993) may not be adequate grounds for its exclusion from a close relationship with primates. The low brain weight is probably a *phenotypic* feature, acquired *in utero* along with hydrocephalus, as a result of high circulating levels of phytotoxins. Folivorous koalas and tree sloths also have enlarged cerebral ventricles and tiny brains (Haight & Nelson 1987).

Colugo–megabat link

The reservations about the colugo's phylogenetic status notwithstanding, a close relationship between colugo and primates is increasingly accepted. The new molecular data hence provide verification for one important prediction of the flying primate hypothesis, which proposed that dermopterans were the early primate branch that gave rise to the megabats. The reasons for proposing a megabat–colugo link are many, but include similarities in the patagium and striking similarities in the brain and behaviour of the two taxa. The parallels between the megabat and colugo patagium are particularly evident when each is transilluminated against a bright sky and the intrinsic humeropatagialis muscle and vasculature are visible. Some patagial characters linking megabat and colugo are also present in microbats, but there are a number of behavioural features, not obviously related to flight, that tie colugo and megabat to the exclusion of microbats and all other mammals. These include the following:

1. Defecation and micturition: accomplished in megabats and the colugo from an erect posture supported by the pollex, the inverse of the normal hanging posture. This contrasts with the microbat posture for excretion, involving some dorsiflexion but no inversion from the normal, inverted hanging posture.
2. Terrestrial locomotion: awkward, with much use of symmetrical limb movements, in contrast to the brisk, well co-ordinated quadrupedal locomotion of microbats.
3. Arboreal locomotion: versatile, with frequent use of an inverted hanging gait where flexor muscles of fore- and hind-limbs are loaded. This inverted, alternating, quadrupedal, arboreal gait is not observed in microbats.
4. Young: carried on forays, to a considerable age, in contrast to the microbat behaviour of carrying the young only for short distances or 'parking' them during foraging.
5. Neck posture: in the inverted, hanging posture, both colugo and megabat usually flex the neck ventrally to change the view in elevation; microbats place the neck in extension and have modified neck vertebrae to suit.
6. Hindlimb: both colugos and megabats make extensive use of the hindlimbs for grooming and, in the case of megabats, for manipulation and for carrying food in flight. The importance of the hindlimb for megabat behaviour is reflected in the greatly increased representation of the hindlimb in the somatosensory cortex. Microbats use the hindlimb only for locomotion and hanging and have a tiny somatosensory representation of the hindlimb.

An ethogram based on these behavioural features strongly links the colugo and megabats. Moreover, the ethogram further supports a primate–colugo–megabat link. Given the stability of Lorenz's ethologically-generated phylogeny of ducks, as well as its abundant support from other studies such as DNA–DNA hybridization (Sibley & Ahlquist 1986), it is difficult to dismiss this link between megabats and colugos, yet this is exactly what has been done as a result of the DNA sequence data. The problem can be put in the following way: the molecular and neural grounds for supporting a colugo–primate link are persuasive, but this case is weak compared with the case for a megabat–colugo link. If colugos are related to primates and megabats are related to colugos, then megabats should be related to primates. Why is it that the DNA data support only the weakest arm of the triad?

This conflict is resolvable if the DNA of bats has evolved along different lines from the DNA of other mammals as a consequence of flight. Convergence of microbat and megabat DNA towards high AT content, as a metabolic consequence of flight, could contribute to a coincidental similarity between them on the one hand, and disguise their true affinities on the other. In support of this idea, when they are not subject to the complicating effects of the high AT content, sequence data link colugo, megabat and primate to the exclusion of microbats. For example, in all of the six DNA data sets claiming monophyly, guanine (G) and cytosine (C) substitutions in isolation from AT substitutions support the rival, flying primate hypothesis! (Table 2). Similarly, protein sequence data do not provide support for monophyly.

Increasing support from other brain characters

There is not space to recount all the new information that has accumulated on the flying fox brain (see Calford *et al.* 1985; Krubitzer *et al.* 1993; Rosa, Schmid, Krubitzer & Pettigrew 1993; Rosa, Schmid & Pettigrew 1994; Rosa & Schmid 1994). It is fair to say that the accumulating data provide increasing support for the flying primate hypothesis. Much of the work has involved exploration of the neocortex. This part of the brain has considerable developmental and phylogenetic plasticity and therefore requires cautious interpretation in the present context. Nevertheless, flying fox neocortical specializations have by far the closest similarity to those of primates amongst all mammals studied, including the intensively-investigated microbat cortex. It is also worth noting that primates are remarkable for their degree of neocortical specialization, so it is appropriate to make some comparisons at this level.

Somatosensory cortex

Flying foxes have six separate representations of the body surface in this cortex, similar to the macaque monkey, the most advanced primate so far studied in this regard. Non-primate mammals have many fewer representations than six (three in microbats, four in cats, three or four in rodents), despite intensive investigation. Moreover, the shape and arrangement of the areas are similar in both primate and megabat.

Extrastriate visual areas

The flying fox has a large number of separate representations of the visual field in the extrastriate visual cortex lying between V1 and somatosensory cortex. These bear a remarkable similarity to extrastriate areas that have been described in primates, and sometimes also in carnivores, but are much less easy to relate to those in other mammals. For example, V2 is very large in area and has a split in the visual field representation, as in primates and carnivores (Rosa, Schmid & Pettigrew 1994). Anterior to V2 there are at least four separate visual areas whose shape and arrangement are like V3, V3A, DM and MT, which have been described only in primates. Because there is still disagreement between laboratories about the exact nature and relationship of primate extrastriate areas (Rosa, Schmid & Pettigrew 1994), it is not possible to be dogmatic about homology between primate and megabat extrastriate cortices. No other mammal apart from the megabat, not even the cat with its well-developed multiple visual cortical areas, comes so close to the primate level of organization in the extrastriate cortex.

Retinal target nuclei

Attention was first drawn to the possible primate affinities of megabats by the finding of the unusual derived pattern of projections from the retina to the visual target nuclei, particularly the reduced, hemidecussate pattern of connections to the visual midbrain and the unusual pattern of lamination in the lateral geniculate nucleus (Pettigrew 1991a, b). Authors critical of these findings have drawn attention to the fact that megabats do not exactly match the primate states for these visual pathways in a number of respects (Thiele, Vogelsang & Hoffmann 1991; Kaas & Preuss 1993). In this regard it is important to bear in mind my remarks in the first section concerning the choice of an appropriate, plesiomorphic primate for comparison. In concluding that their data from the megabat, *Rousettus*, do not conform to the primate pattern because there is some degree of invasion of the ipsilateral hemifield by the retinotectal inputs, Thiele *et al.* (1991) are adopting a criterion from anthropoids that is inappropriate for a basal primate. In all mammals except primates, the ipsilateral invasion is complete. The fact that *Rousettus* has an incomplete invasion immediately places it with the primates rather than with other mammals. Kaas & Preuss (1993) have put forward a criticism that the external, magnocellular layers of the megabat LGN are not homologous with those of primates because they have not been characterized with respect to all of the many techniques that have been applied to the primate LGN over the years. This is an unduly restrictive view, since the megabats share with primates a number of LGN features (large cell layers for each eye lying externally) that are not found in other mammals. In future, if there prove to be differences between these large cell layers in megabats and the corresponding layers in primates, such differences will help to illuminate the phylogeny, particularly with respect to LGN evolution in very early primates and sister taxa. In the meantime, the tools used so far to check LGN characters in megabats have revealed no examples of character states which are in conflict with the hypothesis. For example, CAT 301 antibodies, specific for the magnocellular pathways, reveal a pattern of labelling in the LGN of *Pteropus* that is primate-like, with only the external pair of laminae labelled, in accord with their assignment as magnocellular layers (S. Hockfield & G. Kelly unpubl.).

Motor pathways

In the debate about whether tree shrews were primates, Campbell's (1974) evidence on the pyramidal tract and motor pathways played a crucial role in removing the tree shrews to a more distant relationship with primates. By Campbell's criteria, megabats have a primate-like pyramidal tract. Moreover, they also have a pattern of corticospinal neurons that is found in primates but not in 11 other orders of mammals (Nudo 1985; Kennedy 1991).

Pattern of parvalbumin labelling and neuronal density in the neocortex

Primates are distinct from other mammals in a number of ways relating to the density of cortical neurons that stain the various calcium-binding proteins (Glezer, Hof, Leranth & Morgane 1993). Microbats have a pattern that is typical of basal eutherians and quite unlike the primate pattern, whereas megabats have the primate pattern (P. R. Hof pers. comm.)

Hippocampus

Megabats share with primates five of seven hippocampal features that distinguish them from other mammalian orders (Buhl & Dann 1991). None of these features is found in the hippocampus of the two microbats studied, *Macroderma gigas* (Megadermatidae) and *Mormopterus planiceps* (Molossidae) (E.H. Buhl & J.D. Pettigrew unpubl.).

Support from proteins

Monoclonal antibodies to serum proteins

Using 86 different monoclonal antibodies, Schreiber *et al.* (1994) have looked for epitopes on 26 serum proteins in a variety of mammals, including two microbats and two megabats. The pattern of epitopes lends itself readily to parsimony analysis, with the result shown in Fig. 2. It can be seen that all trees separate microbats from megabats and all trees have the megabats in a close sister-group relation to the single primate representative, the human. Another interesting feature of the trees is the close relation of the carnivores to

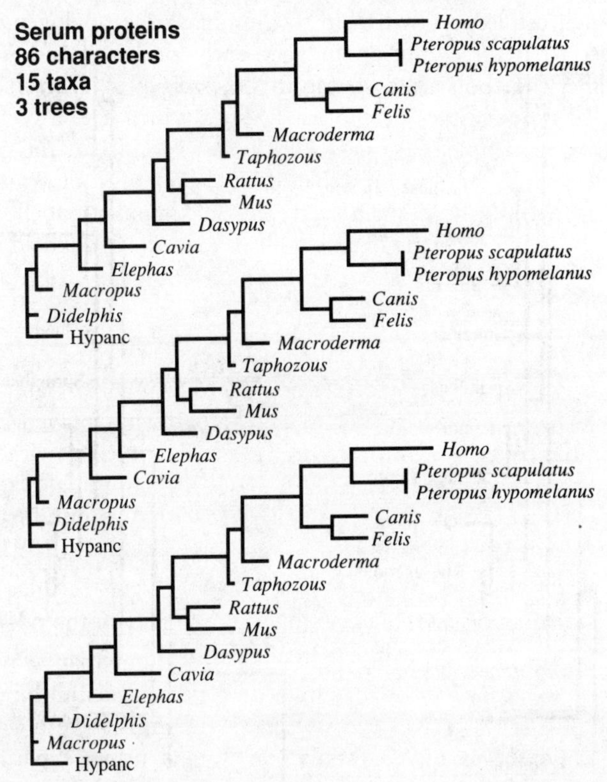

Fig. 2. Cladogram generated from a data matrix of epitopes on serum proteins (Schreiber *et al.* 1994). Thirty equally parsimonious trees were generated, but this was reduced to the three shown if a constraint linking the two carnivores was applied. The human and megabats (*Pteropus*) were joined in all trees, and their nearest sister taxa were the carnivores. The microbats *Taphozous* and *Macroderma* are joined neither with each other nor to the megabats.

the primate–megabat assemblage, a feature that is also seen in the brain data and the data from α-crystallin.

α-Crystallin

Following a protracted study of mammalian relationships using amino acid sequences from α-crystallin, De Jong *et al.* (1993) recently sequenced this protein from a microbat (*Artibeus*) and a megabat (*Pteropus*). As shown in Fig. 3, these two taxa are separated on the most parsimonious trees, with the megabat grouping with carnivores.

Globins

There are considerable amino acid sequence data available from globins (Figs 4, 5). Phylogenetic analysis of these data does not support monophyly of

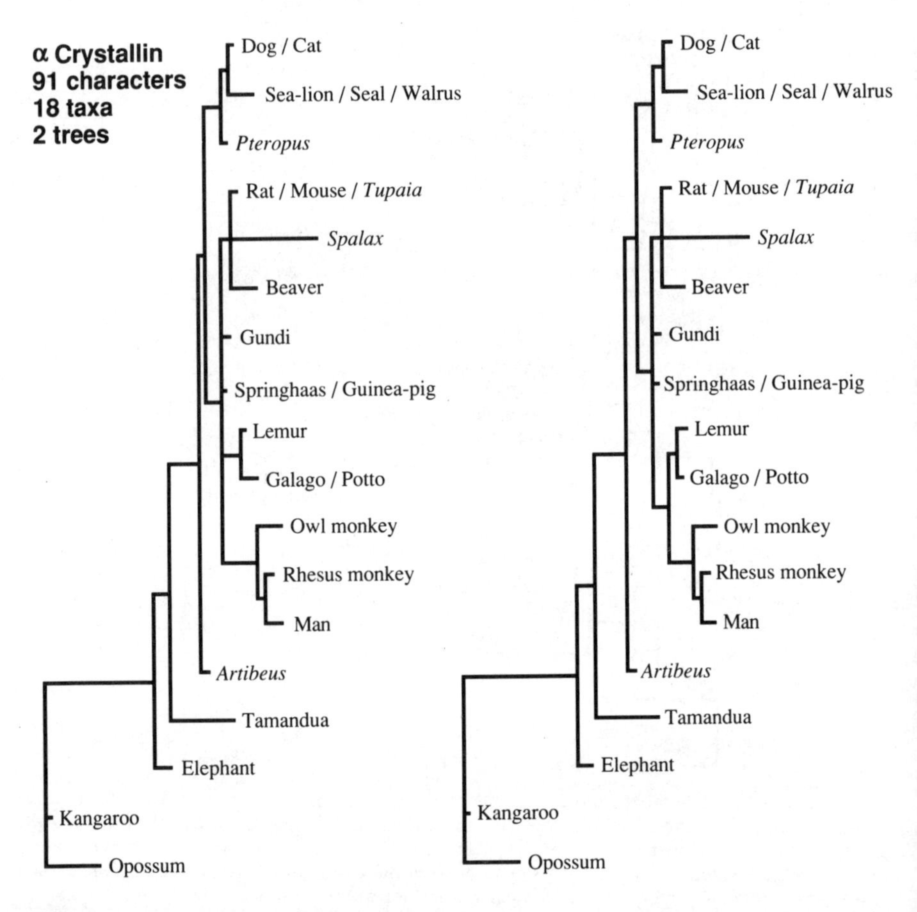

Fig. 3. Cladistic analysis of amino acid sequence data from α-crystallin in selected mammals (data from De Jong *et al.* 1993). Note that the microbat (*Artibeus*) and megabat (*Pteropus*) are well separated on the tree. While these data do not place the megabats with the primates, their position near carnivores is similar to the relationships shown in Fig. 2 where primates, megabats and carnivores are closer to each other than either is to any other mammalian order.

bats, although there is not unequivocal support for the flying primate hypothesis either. While megabats are close to primates in all analyses, the problem is that some microbats tend to cluster in the analysis with megabats while most microbats clearly do not (Pettigrew *et al.* 1989). In view of the AT bias in the DNA of bats and the fact that the globin genes are located in the parts of the genome (L isochores) where this bias is greatest, future analyses could focus on the very small number of amino acid substitutions that are responsible for the splitting of microbats when globin data are used.

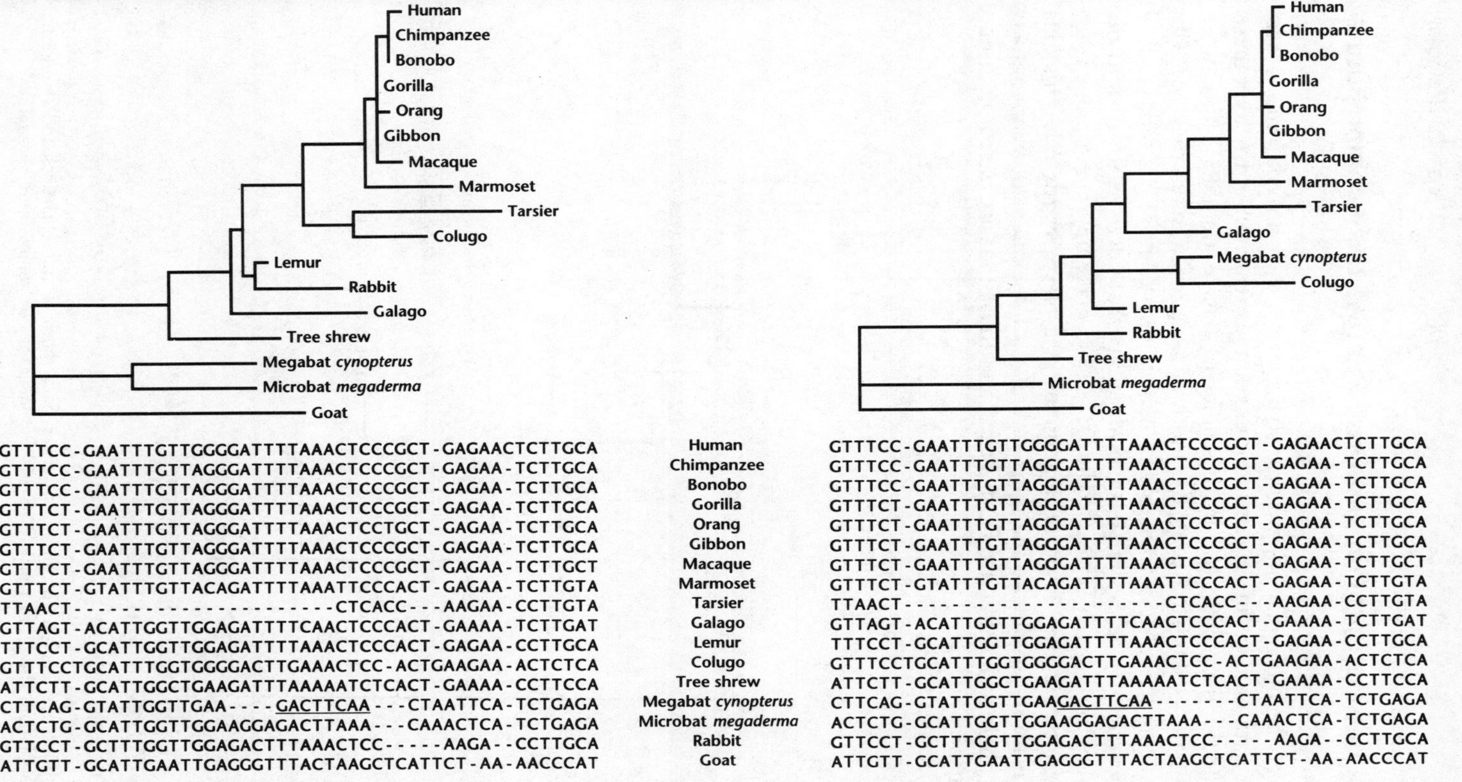

Fig. 4. Problems of DNA sequence alignment in ϵ-globin intron. The dashes indicate gaps that must be introduced to align the sequences, which have different lengths in all of these taxa. Since there are only two positions where the base is conserved across taxa, many alternative alignments are possible. The alignment arrived at subjectively for analysis by Bailey *et al.* (1992) favours monophyly of bats (left), but an equally acceptable alignment, not used in their analysis, favours the rival hypothesis (right). The megabat's fragment *GACTTCAA* lines up with a similar sequence from the microbat in the alignment chosen for analysis, but the same fragment also lines up with similar sequences in the primates and the colugo with the insertion of only one gap.

Conflict between protein and DNA data: AT base compositional bias

In view of the molecular support for flying primates from both amino acid sequence data and serum proteins, the unanimous rejection of the hypothesis by six independent DNA sequence studies of bats is notable. The conflict within the molecular data is all the more notable when taken together with the other conflicting data from brain, behaviour, skeleton, genitalia and all the other systems where primates and megabats share derived features to the exclusion of microbats (Table 1). The DNA data make up a unanimous voice in opposition, but they also are unanimous in a bias towards AT. This bias could explain the conflict in the same way that it has explained other conflicts between phylogenies that are DNA-based and those that are based on amino acid sequences.

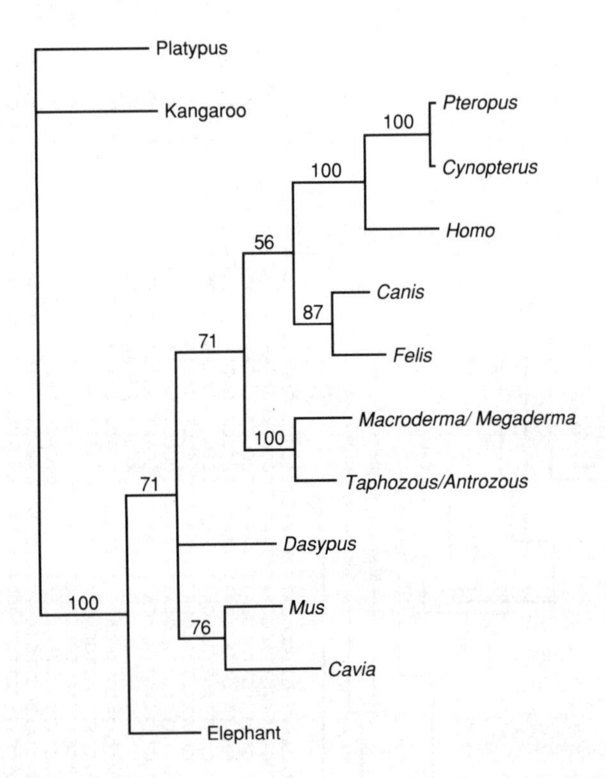

Fig. 5. A bootstrap analysis of globin sequence data (both α and β chains) and the serum protein epitope data shown in Fig. 2. Humans and megabats are joined in all replicates.

AT base compositional bias

All DNA sequence data so far collected on the bat problem have a pronounced AT bias. All six sets of sequences show a 4:1 AT:GC bias in the substitutions that are claimed in support of monophyly. While the total number of substitutions supporting the flying primate hypothesis in the same data set is smaller, they show no AT bias. The number of GC substitutions supporting flying primates is greater, in the same data set, than the number of GC substitutions supporting bat monophyly (Table 2). This bias in the data claimed in support of bat monophyly assumes greater significance alongside the fact that the DNA of bats, particularly megabats, has higher than normal levels of AT. Megabat DNA has the highest proportion of AT known for any vertebrate, in excess of 70% (Arrhigi, Lidicker, Mandel & Bergendahl 1972; Bernardi 1993).

Biases in base composition can lead to preposterous phylogenies when DNA sequence data are used blindly, without allowing explicitly for the bias in the underlying model of DNA evolution assumed in the analysis. A well-documented recent example is the case of the eukaryote, *Dictyostelium*, whose AT-rich DNA causes it to be placed firmly within the prokaryotes by the commonly-used methods of phylogenetic analysis! In contrast, protein sequence data place *Dictyostelium* in a more appropriate place on the tree of life, amongst the eukaryotes (Loomis & Smith 1990). The base compositional bias in the DNA data from bats suggests that a similar distortion may be taking place in these phylogenetic analyses as well (Pettigrew 1994).

None of the six published papers drew attention to the bias, let alone took steps to tackle it. The computer packages used to analyse the DNA data presumably used a model of evolution, such as the Jukes–Cantor model, that made no allowance for the kind of mutational bias responsible for the extreme AT content in megabat DNA and the heterogeneous AT content of microbat DNA. Recent work has provided a method for dealing with data that have base compositional biases, but this currently allows a test on only four taxa at a time (Steel, Lockhart & Penny 1993).

AT homoplasy? Or shared derived high AT?

In the light of the ever-growing list of differences between megabats and microbats (Table 1), it would not be surprising if proponents of monophyly claimed the high AT content of bat DNA as a shared derived feature linking megabats and microbats. As I have pointed out already (Pettigrew 1991b), the convergent pressures operating on organisms with powered flight may be exerted on other systems besides the wings. My suggestion that the evolution of flight might have consequences for DNA has been derided (Gibbons 1992), but appears to be the most parsimonious way to account for the AT biases found so far in all DNA sequence data from bats. The alternative proposal, that high

AT was inherited from a common ancestor of megabats and microbats, can be ruled out on the following grounds:

1. All megabats have an AT content that is higher than is found in any microbat. In view of the normal AT content found in many microbats, compared with the uniformly high AT content of all species of megabats, is it really plausible that high AT was present in a common bat ancestor? One would have to do violence to accepted views about the monophyly of microbats to make a phylogeny unifying all bats that is at the same time compatible with the present distribution of AT content. Such a phylogeny would split the microbats in an attempt to place the megabats on the clade of microbats with moderate AT. This is unacceptable, despite the fact that recent molecular phylogenies of bats have sometimes had this feature. For similar reasons it is not possible to use the reduced cellular content of DNA (C-value) as a shared derived feature of all bats, as shown in Fig. 6.

2. High AT levels are found in the DNA from a variety of unrelated organisms, all of which have unusual metabolism. As can be seen in Fig. 7, high AT is found in the DNA of *Dictyostelium* slime moulds, bees, birds, shrews, microbats and megabats. All of these cases can be explained

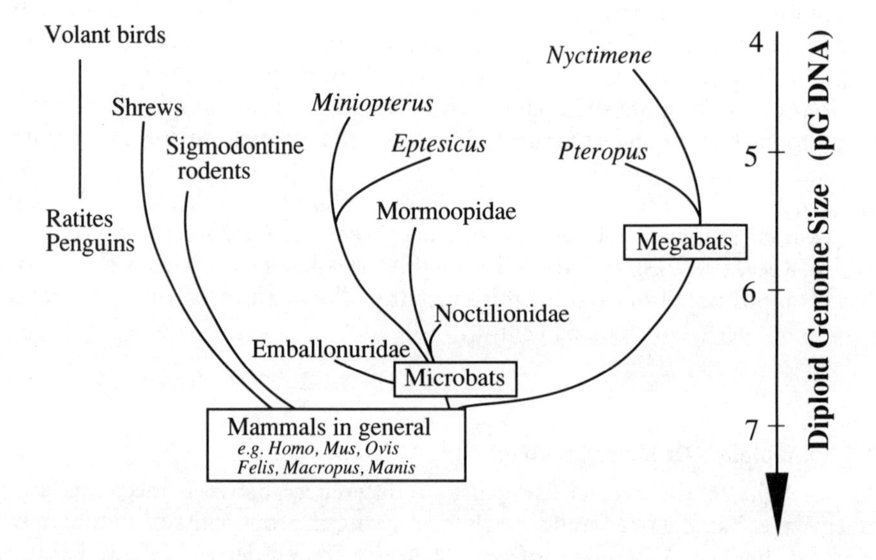

Fig. 6. C-value (diploid cellular DNA content) in homeotherms. The C-value is smallest in volant birds and in megabats. Note wide range of C-value in microbats but uniformly low C-value in megabats. Reduction in C-value reduces nuclear and cellular size and may therefore help to explain the changes in these metabolically-active organisms that would benefit from increased efficiency of transmembrane exchanges. The common occurrence of C-value reductions in unrelated taxa, along with the absence of C-value reduction in some basal microbats such as *Noctilio* and emballonurids, argue that C-value reduction has been acquired independently in the two bat lineages.

in terms of the increased adenine nucleotide levels that would confront DNA replication and repair machinery when compared to control organisms. In most cases the increased adenine nucleotide levels (e.g. ATP) would be the result of increased metabolism, as in the bees, shrews and powered flying vertebrates. Note, however, that while *Dictyostelium* does not have a highly active metabolism, it does have very high levels of the adenine nucleotide, cAMP, used as a secreted signal. DNA from slime moulds using other signals not based on adenine nucleotides, such as *Acrasis* and *Polysphondylium*, do not have high AT levels (Dutta & Mandell 1972). This comparison in the slime moulds supports the interpretation that the increased AT content is related to levels of adenine nucleotide precursor pools. Another instructive comparison, with the same conclusion, is the lower AT content found in ant DNA compared with bee DNA (Bennett *et al.* 1988). These two groups of hymenopterans are distinguished mainly by the reduction in flight abilities in the ants. The high temperatures and high metabolism of bees in flight have been well described.

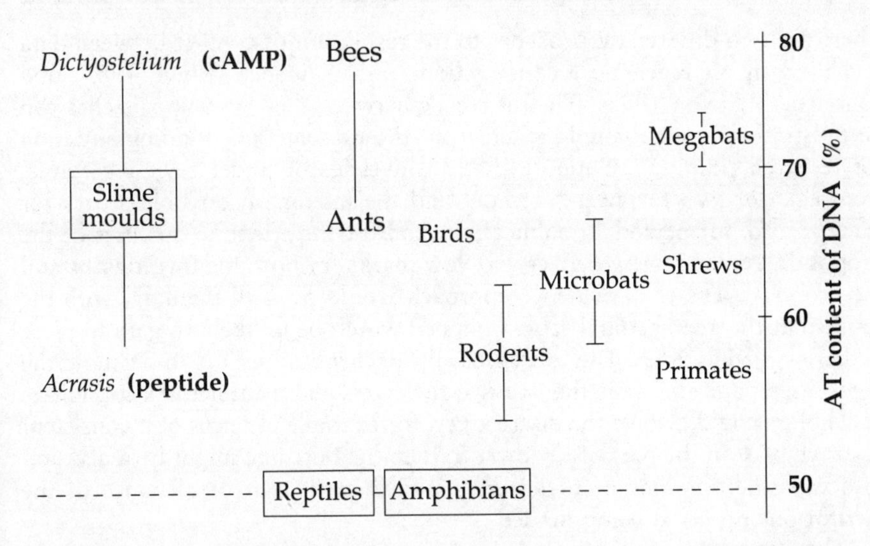

Fig. 7. Mutational biases toward high AT. Whilst reptiles and amphibians have DNA with approximately equal proportions of the different bases, homeotherms have a biased distribution, with high levels of adenine- and thymine-derived nucleotides (high AT). The connection with high metabolism that is suggested by the homeotherm–reptile comparison is further supported by the fact that even higher levels of AT are found in the most metabolically-active mammals, such as shrews and bats. Similarly, bee DNA has a higher AT level than is found in the DNA of the metabolically less-active ant. The highest AT levels in vertebrates are found in the megabats. An exception that may prove this rule is the very high AT content of DNA in *Dictyostelium*.

The occurrence of high AT in DNA from such a wide variety of unrelated taxa suggests that it could have arisen independently in the two kinds of flying mammals. Normal AT values are found in many microbats, and megabat AT values are all outside the microbat range. Neither of these facts can be easily reconciled with the assumption of high AT in a common ancestor of megabats and microbats. The most parsimonious explanation for the AT bias in the substitutions shared by microbats and megabats is therefore that AT bias has been acquired independently in the two lineages, as a metabolic consequence of the demands of flight.

Seen in this light, the extraordinary AT content of bee mtDNA makes some sense, as does the lower value in the somewhat less metabolically active ants. Given these facts about the incidence of high AT, it is hardly more meaningful to group megabats and microbats together on the basis of high AT than it is to group shrews, or birds, or bees in this assemblage.

3. Given that there is a conflict between the DNA sequence data and most other studies except for those that deal with some aspect of the flight apparatus, it is more parsimonious to adopt the hypothesis that flight has independently brought about changes in the DNA of both kinds of bats. By adopting such a hypothesis, supported by the association of high AT with high metabolic activity and/or high concentrations of adenine nucleotide, the DNA data are then brought into line with the consensus view from the numerous other studies that support the flying primate hypothesis.

Resolution of conflicting data sets: total evidence v. consensus

There are two different approaches to the resolution of conflict between data sets used in phylogenetic reconstruction: total evidence (Kluge 1989) and consensus (Lanyon 1993). The difference between these two approaches can be illustrated by an example taken from the present tight funding situation for research proposals. Suppose that an investigator under review receives, from each of two reviewers, acclaim and the maximum possible scores for track record, for personal qualifications and for the scientific qualities of the proposal. A third reviewer gives very low marks for both the investigator and the proposal. The total evidence approach would pool all the data, with the result that the average final score obtained would be unlikely to gain funding in these stringent times. The consensus approach would seek to understand the gross discrepancy between the first two reviewers and the third. If a committee of scholars could resolve the discrepancy, for example in terms of a consistent negative bias on the part of reviewer 3, then the outcome might be a decision to accept the opinions of reviewers 1 and 2, with a greater chance of the meritorious proposal being funded.

Apart from the inherent problem of smearing that is evident from the hypothetical example just given, the total evidence approach has a number of severe limitations when one tries to put it into practice. In the case of the bat controversy, it can be a difficult problem to join DNA sequence data to morphological data in the same data matrix. First, the sizes of the conflicting matrices may be very different, raising the possibility that one data set will swamp the other. Second, when there are uncertainties about alignment, the number of possible data matrices to be tested can be so large as to preclude testing. This is the case for the ϵ-globin intron studied by Bailey *et al.* (1992) and used to reject the flying primate hypothesis. This intron was a different length in every taxon for which a sequence was obtained, with the result that hundreds of gaps had to be inserted to achieve alignment. There were very few conservative sites to aid alignment, with the result that there are many equally acceptable alternative alignments. Alternative alignments can even be found that contradict the major conclusions of the study! (Fig. 4). The positions of most of the gaps can be altered, giving an unmanageable number of alternative possible data matrices to be combined with the morphological data matrix. Pairwise alignment algorithms to choose the 'best' alignments for analysis are not necessarily a solution to the problem in the present case because of the base compositional bias in bat DNA. Since a base compositional bias may increase similarity scores between two unrelated sequences, algorithms using similarity measures, such as Clustal, may not provide a rigorous alignment. Apart from compromising the alignment process, the base compositional biases found in bat DNA suggest that it is unwise to use the raw DNA data without correcting for the effects of the bias. While techniques for dealing with such biases are beginning to appear, such as the four-taxon test developed by Steel *et al.* (1993), they cannot yet be applied to a DNA data matrix that is being joined to a matrix of other data.

A system-by-system consensus approach gives a majority of systems in favour of the flying primate hypothesis, with three systems in opposition (patagium, mtDNA, nDNA). There is strong motivation to put more weight on the DNA data because of the perception that they are superior in their objectivity and their avoidance of problems such as the determination of homology. In the present case, these advantages of DNA data are more apparent than real, since subjectivity played a role in alignment of one data set that appeared in an influential journal and since many of the sequence data are derived from non-coding sequences where homology between sites cannot be as rigorously determined as for coding sequences.

Apart from these reasons for being cautious about the DNA data, there is another consideration with the potential to resolve the conflict completely. The AT bias in all the DNA sets is the only source of conflict. A consideration of GC substitutions from the same data set actually provides support for flying primates. Given the plausible connection between AT bias and the metabolic demands of flight, the consensus approach could resolve all conflicts between the systems. Since the flying primate hypothesis includes a dual origin for flight, it can encompass dual origins of the corollaries of flight such as wing anatomy and AT bias. Seen in this overall context, it is unparsimonious to propose that AT bias and wing anatomy were inherited from a common flying ancestor.

Future investigations

A large proportion of mammalian genes are located in the H3 isochore, a small GC-rich fraction of the genome (Bernardi 1993). Since this isochore is relatively 'protected' from the AT mutational bias that reaches such extremes in megabats, future DNA sequence data could profitably be collected from this part of the genome rather than the L isochores where AT bias is maximal. In addition, DNA data already collected could be reanalysed with the AT bias in mind, along with amino acid sequence data where one could possibly correct for those shared amino acid substitutions that might be coincidentally related to the AT bias.

Megabats, and some microbats, have remarkable changes in genomic organization that are arguably related to flight. Further work on bats could therefore illuminate the currently obscure origins and mechanisms of features of genomic organization such as isochores (that are present in all mammals but dramatically shifted in their distribution in megabats), C-value reduction and AT increase. The flying primate hypothesis therefore has the potential to be a powerful heuristic stimulus for cellular and molecular biology as well as for neurobiology.

Acknowledgements

This work was supported by the Australian Research Council and the National Health and Medical Research Council.

References

Adkins, R. M. & Honeycutt, R. L. (1991). Molecular phylogeny of the superorder Archonta. *Proc. natn. Acad. Sci. USA* **88**: 10317–10321.

Ammerman, L. K. & Hillis, D. M. (1992). A molecular test of bat relationships: monophyly or diphyly? *Syst. Biol.* **41**: 222–232.

Arrhigi, F. E., Lidicker, W. Z. Jr., Mandel, M. & Bergendahl, J. (1972). Heterogeneity in CsCl buoyant densities of chiropteran DNA. *Biochem. Genet.* **6**: 27–39.

Bailey, W. J., Slightom, J. M. & Goodman, M. (1992). Rejection of the 'Flying Primate' hypothesis by phylogenetic evidence from the ϵ-globin gene. *Science* **256**: 86–89.

Baker, R. J., Novacek, M. J. & Simmons, N. B. (1991). On the monophyly of bats. *Syst. Zool.* **40**: 216–231.

Bennett, S., Alexander, L. J., Crozier, R. H. & Mackinlay, A. G. (1988). Are megabats flying primates? Contrary evidence from a mitochondrial DNA sequence. *Aust. J. Biol. Sci.* **41**: 327–332.

Bernardi, G. (1993). The vertebrate genome: isochores and evolution. *Molec. Biol. Evol.* **10**: 186–204.

Buhl, E. H. & Dann, J. F. (1991). Cytoarchitecture, neuronal composition, and entorhinal afferents of the flying fox hippocampus. *Hippocampus* **1**: 131–152.

Calford, M. B., Graydon, M. L., Huerta, M. F., Kaas, J. H. & Pettigrew, J. D. (1985). A variant of the mammalian somatotopic map in a bat. *Nature, Lond.* **313**: 477–479.

Campbell, C. B. G. (1974). On the phyletic relationships of the tree shrews. *Mammal Rev.* **4**: 125–143.

Cartmill, M. (1972). Arboreal adaptations and origin of the order Primates. In *The functional and evolutionary biology of primates:* 97–122. (Ed. Tuttle, R.H.). Aldine Atherton, Chicago.

Cronin, J. E. & Sarich, V. M. (1980). Tupaiid and Archonta phylogeny: the macromolecular evidence. In *Comparative biology and evolutionary relationships of tree shrews:* 293–312. (Ed. Luckett, W.P.). Plenum Press, New York.

Dann, J. F. & Buhl, E. H. (1990). Morphology of retinal ganglion cells in the flying fox (*Pteropus scapulatus*): a Lucifer Yellow investigation. *J. comp. Neurol.* **301**: 401–416.

De Jong, W. W., Leunissen, J. A. M. & Wistow, G. J. (1993). Eye lens crystallins and the phylogeny of placental orders: evidence for a macroscelid-paenungulate clade? In *Mammal phylogeny: placentals*: 5–12. (Eds Szalay, F. S., Novacek, M. J. & McKenna, M. C.). Springer-Verlag, New York.

Dutta, S. K. & Mandell, M. (1972). DNA base composition of some cellular slime molds. *J. Protozool.* **19**: 538–540.

Gibbons, A. (1992). Is 'Flying Primate' hypothesis headed for a crash landing? *Science* **256**: 34.

Glezer, I. I., Hof, P. R., Leranth, C. & Morgane, P. J. (1993). Calcium-binding protein-containing neuronal populations in mammalian visual cortex: a comparative study in whales, insectivores, bats, rodents and primates. *Cereb. Cortex* **3**: 249–272.

Haight, J. R. & Nelson, J. E. (1987). A brain that doesn't fit its skull: a comparative study of the brain and endocranium of the koala, *Phascolarctos cinereus* (Marsupialia: Phascolarctidae). In *Possums and opossums: studies in evolution*: 331–352. (Ed. Archer, M.). Surrey Beatty & Sons and the Royal Zoological Society of New South Wales, Sydney.

Jacobs, G. H. (1993). The distribution and nature of colour vision among the mammals. *Biol. Rev.* **68**: 413–471.

Kaas, J. H. & Preuss, T. M. (1993) Archontan affinities as reflected in the visual system. In *Mammal phylogeny: placentals*: 115–128. (Eds Szalay, F. S., Novacek, M. J. & McKenna, M. C.). Springer-Verlag, New York.

Kennedy, W. (1991). *Origins of the corticospinal tract of the flying fox: correlation with cytoarchitecture and electrophysiology*. MSc thesis: University of Queensland, Brisbane.

Kleinschmidt, T., Sgouros, J. G., Pettigrew, J. D. & Braunitzer, G. (1988). The primary structure of the hemoglobin from the Grey-headed Flying Fox (*Pteropus poliocephalus*) and the Black Flying Fox (*P. alecto*, Megachiroptera). *Biol. Chem. Hoppe-Seyler* **369**: 975–984.

Kluge, A. G. (1989). A concern for evidence and a phylogenetic hypothesis of relationships among *Epicrates* (Boidae, Serpentes). *Syst. Zool.* **38**: 7–25.

Krubitzer, L. A. & Calford, M. B. (1992). Five topographically organized fields in the somatosensory cortex of the flying fox: microelectrode maps, myeloarchitecture, and cortical modules.*J. comp. Neurol.* **317**: 1–30.

Krubitzer, L. A., Calford, M. B. & Schmid, L. M. (1993). Connections of somatosensory cortex in megachiropteran bats: the evolution of cortical fields in mammals. *J. comp. Neurol.* **327**: 473–506.

Lanyon, S. M. (1993). Phylogenetic frameworks: towards a firmer foundation for the comparative approach. *Biol. J. Linn. Soc.* **49**: 45–61.

Loomis, W. F. & Smith, D. W. (1990). Molecular phylogeny of *Dictyostelium discoideum* by protein sequence comparison. *Proc. natn. Acad. Sci. USA* **87**: 9093–9097.

Martin, R. D. (1993). Primate origins: plugging the gaps. *Nature, Lond.* **363**: 223–234.

Mindell, D. P., Dick, C. W. & Baker, R. J. (1991). Phylogenetic relationships among megabats, microbats and primates. *Proc. natn. Acad. Sci. USA* **88**: 10322–10326.

Nudo, R. J. (1985). *A comparative study of cells originating in the corticospinal tract: a comparative morphology in the anthropoid ancestral lineage*. PhD thesis: Florida State University, Tallahassee.

Pettigrew, J. D. (1986). Flying primates? Megabats have the advanced pathway from eye to midbrain. *Science* **231**: 1304–1306.

Pettigrew, J. D. (1991a). Wings or brain? Convergent evolution in the origins of bats. *Syst. Zool.* **40**: 199–216.

Pettigrew, J. D. (1991b). A fruitful wrong hypothesis? Response to Baker, Novacek, and Simmons. *Syst. Zool.* **40**: 231–239.

Pettigrew, J. D. (1994). Flying DNA. *Curr. Biol.* **4**: 1–4.

Pettigrew, J. D. Jamieson, B. G. M., Robson, S. K., Hall, L. S., McAnally, K. I. & Cooper, H. M. (1989). Phylogenetic relations between microbats, megabats and primates (Mammalia: Chiroptera and Primates). *Phil. Trans. R. Soc. (B)* **325**: 489–559.

Rosa, M. P., Pettigrew, J. D. & Cooper, H. M. (In press). Unusual pattern of retinogeniculate projections in the controversial primate, *Tarsius*. *Brain Behav. Evol.*

Rosa, M. P. & Schmid, L. M. (1994). Topography and extent of visual field representation in the superior colliculus of the megachiropteran, *Pteropus*. *Visual Neurosci.* **11**: 1037–1057.

Rosa, M. P., Schmid, L. M., Krubitzer, L. A. & Pettigrew, J. D. (1993). Retinotopic organization of flying foxes (*Pteropus poliocephalus* and *Pteropus scapulatus*). *J. comp. Neurol.* **335**: 55–72.

Rosa, M. P., Schmid, L. M. & Pettigrew, J. D. (1993). Organization of the second visual area in the megachiropteran bat *Pteropus*. *Cereb. Cortex* **4**: 52–68.

Schreiber, A., Bauer, D. & Bauer, K. (1994). Mammalian evolution from serum protein epitopes. *Biol. J. Linn. Soc.* **51**: 359–376.

Sibley, C. G & Ahlquist, J. E. (1986). Reconstructing bird phylogeny by comparing DNA. *Scient. Am.* **254**: 68–78.

Simmons, N. B., Novacek, M. J. & Baker, R. J. (1991). Approaches, methods, and the future of the chiropteran monophyly controversy. *Syst. Zool.* **40**: 239–243.

Simons, E. L. & Rasmussen, D. T. (1989). Cranial morphology of *Aegyptopithecus* and *Tarsius* and the question of the tarsier–anthropoidean clade. *Am. J. phys. Anthrop.* **79**: 1–23.

Speakman, J. R. (1993) The evolution of echolocation for predation. *Symp. zool. Soc. Lond.* No. 65: 39–63.

Stanhope, M. J., Czelusniak, J., Si, J.-S., Nickerson, J. & Goodman, M. (1992). A molecular perspective on mammalian evolution from the gene encoding interphotoreceptor retinoid binding protein, with convincing evidence for bat monophyly. *Molec. phylogenet. Evol.* **1**: 148–160.

Steel, M. A., Lockhart, P. J. & Penny, D. (1993). Confidence in evolutionary trees from biological sequence data. *Nature, Lond.* **364**: 440–442.

Thiele, A. M., Vogelsang, M. & Hoffmann, K.-P. (1991). Pattern of retinotectal projection in the megachiropteran bat *Rousettus aegyptiacus*. *J. comp. Neurol.* **314**: 671–683.

Symp. zool. Soc. Lond. (1995) No. 67: 27–43

Bat relationships and the origin of flight

Nancy B. SIMMONS

Department of Mammalogy
American Museum of Natural History
Central Park West at 79th Street
New York, NY 10024, USA

Synopsis

Interpretation of evolution of any functional system requires a phylogenetic context. In the case of chiropteran flight, two competing hypotheses must be considered: either bats are monophyletic and powered flight evolved only once in mammals, or bats are diphyletic and flight evolved twice. A survey of over 20 molecular and morphological studies indicates that the majority of available data support chiropteran monophyly. Monophyly of Volitantia, a clade containing bats and dermopterans, is supported by morphological data. Interpretation of various traits of the limbs in this context suggests that early volitantians were specialized for 'finger-gliding' and underbranch hanging using both the fore- and hindlimbs. From this ancestry, bats evolved powered flight and hindlimb hanging through modifications of the shoulder, forelimb, hip and ankle. In contrast, living dermopterans have apparently retained the ancestral volitantian body plan and mode of locomotion.

Introduction

Until the 1980s most workers agreed that the common ancestor of all extant bats was a flying mammal, a hypothesis which implies that bats are monophyletic and powered flight evolved only once in mammals. However, bat monophyly and homology of the chiropteran wing have been questioned recently by authors who have suggested that Megachiroptera and Microchiroptera may not be closely related (Jones & Genoways 1970; Smith 1976, 1980; Smith & Madkour 1980; Hill & Smith 1984; Pettigrew 1986, 1991a, b, this volume pp. 3–26; Pettigrew & Jamieson 1987; Pettigrew, Jamieson, Robson, Hall, McAnally & Cooper 1989). There is general agreement that Megachiroptera and Microchiroptera are each monophyletic taxa, but many significant morphological, behavioural and genetic differences exist between the two suborders (Jones & Genoways 1970; Novacek 1987; Pettigrew *et al.* 1989; Sabeur, Macaya, Kadi & Bernardi 1993; Simmons 1994; Pettigrew, this volume pp. 3–26). It was in this context that Jones & Genoways (1970) first raised the possibility that convergent evolution of powered flight

ZOOLOGICAL SYMPOSIUM No. 67
ISBN 0 19 854945 8

might account for the morphological similarities between megachiropterans and microchiropterans. In the 1980s, phylogenetic analyses based principally on characters of the penis and nervous system suggested that bats are diphyletic (e.g. Smith & Madkour 1980; Pettigrew *et al.* 1989), but other studies based on different character sets supported chiropteran monophyly (e.g. Wible & Novacek 1988). These conflicting results led to what has become known as the bat monophyly controversy.

The controversy concerning chiropteran monophyly focuses on relationships of Megachiroptera and Microchiroptera to each other and to other mammalian groups, particularly various 'archontan' mammals including Primates, Dermoptera (gliding lemurs or colugos), Scandentia (tree shrews) and extinct plesiadapids and paromomyids (fossil forms generally considered to be related to either primates or dermopterans). The monophyly hypothesis states that Megachiroptera and Microchiroptera are each other's closest relatives. In contrast, the diphyly hypothesis suggests that Megachiroptera is more closely related to Dermoptera and Primates than to Microchiroptera (Smith & Madkour 1980; Pettigrew 1986, 1991a, b, this volume pp. 3–26; Pettigrew & Jamieson 1987; Pettigrew *et al.* 1989). These alternatives—and different hypotheses concerning the relative phylogenetic positions of non-volant and gliding taxa relative to bats—have significantly different implications concerning the origin of powered flight in mammals.

Understanding the evolutionary origin of any functional system requires a phylogenetic context (Lauder 1981; Brooks & McLennan 1991). If Megachiroptera and Microchiroptera form a monophyletic group, then their shared attributes, including specializations for flight, can be assumed to have been present in their most recent common ancestor. There is no need to hypothesize more than one origin for powered flight if bats are monophyletic. Conversely, if Megachiroptera and Microchiroptera are not close relatives, then anatomical similarities shared by the two groups must be interpreted as the result of either convergent evolution or retention of primitive features. Because wings and powered flight are clearly derived, flight must have evolved twice in mammals if bats are diphyletic.

Most of the characters traditionally used to evaluate phylogenetic relationships among mammals are based on musculoskeletal anatomy. This has caused difficulties in the case of bats because many of the obvious features shared by Megachiroptera and Microchiroptera appear to be specializations for flight. How can we use phylogeny to understand the origins of flight if characters related to flight were used to build the phylogeny in the first place? Circularity seems unavoidable until one considers the methods currently used to draw inferences concerning evolutionary events (Rieppel 1980; Patterson 1982; Brooks & McLennan 1991; Simmons 1993). The first step in the process is construction of a phylogeny. In the case of bats, we can now utilize a wide range of data sources, including molecular data sets, postcranial characters, fetal membranes, etc., in order to form a robust phylogenetic hypothesis of relationships (Simmons 1994). Once this phylogeny has been accepted, then characters of any sort may be mapped upon it for interpretation. Because the phylogenetic hypothesis is based on a diverse array of data, it is not circular to use this phylogeny as a framework for interpreting evolutionary patterns within subsets of the data.

Pettigrew and his colleagues have argued that one should expect homoplasy in any system of obvious adaptive significance, and that phylogenies should therefore be based on characters of no clear functional value (Pettigrew *et al.* 1989). However, distinguishing non-functional from functional characters is a tricky proposition at best, and applying this criterion would be likely to lead us to eliminate virtually all morphological characters from consideration, including Pettigrew's neural characters (Baker, Novacek & Simmons 1991). Simmons (1993) and others have advocated a different approach: rather than eliminating certain classes of data *a priori*, one should consider as many sorts of data as possible when developing phylogenetic hypotheses. When this approach is applied to bats, the hypothesis of chiropteran monophyly is strongly supported.

The case for chiropteran monophyly

Many new morphological and molecular studies were begun following the initial exchange of papers between workers favouring bat diphyly (e.g. Pettigrew 1986, 1991a,b; Pettigrew & Jamieson 1987; Pettigrew *et al.* 1989) and those favouring the monophyly hypothesis (Wible & Novacek 1988; Baker, Novacek, & Simmons 1991; Simmons, Novacek & Baker 1991). The results of most of these phylogenetic studies have now been published, and it is clear that the majority of molecular and morphological evidence supports chiropteran monophyly (Tables 1, 2; for a detailed review see Simmons 1994). Biochemical and molecular studies involving immunological distances, DNA–DNA hybridization, α and β haemoglobin amino acid sequences, nucleotide sequences from genes for ϵ-globin, interphotoreceptor binding protein, cytochrome oxidase subunit II, and 12S rDNA have all provided support for the hypothesis that bats are monophyletic (Table 1). While some analyses have produced inconclusive results, no molecular study provides unambiguous support for the diphyly hypothesis (Simmons 1994).

Morphological data show a somewhat different pattern. Neural and penial characters support diphyly of bats, but other data subsets clearly support bat monophyly (Table 2). When all of the morphological data are considered together, the combined data set strongly supports bat monophyly (Johnson & Kirsch 1993; Simmons 1993, 1994). Over 25 morphological synapomorphies, many of which consist of complex suites of modifications, diagnose the monophyletic order Chiroptera (Table 3; Simmons

Table 1. Results of biochemical and molecular studies[a].

Study	Data	Chiroptera monophyletic?
Cronin & Sarich (1980)	Albumin immunological distances	Yes
Sarich (1993)	Albumin immunological distances	Yes
Kilpatrick & Nunez (1992)	DNA–DNA hybridization	Yes
Pettigrew *et al.* (1989)	β-globin amino acid sequences	No/inconclusive[b]
Pettigrew (1991a)	Haemoglobin amino acid sequences	Inconclusive/Yes[c]
Czelusniak *et al.* (1990); Stanhope, Bailey *et al.* (1993)	α-and β-globin amino acid sequences + data from other proteins[d]	Yes
Stanhope, Bailey *et al.* (1993)	α-and β-globin amino acid sequences	Yes
de Jong, Leunissen & Wistow (1993)	αA-crystallin amino acid sequences	Inconclusive
Baker, Honeycutt & Van Den Bussche (1991)	rDNA restriction sites	Inconclusive
Bailey *et al.* (1992); Stanhope, Bailey *et al.* (1993)	ϵ-globin gene nucleotide sequences	Yes
Stanhope, Czelusniak *et al.* (1992); Stanhope, Bailey *et al.* (1993)	Interphotoreceptor retinoid binding protein gene nucleotide sequences	Yes
Adkins & Honeycutt (1991, 1993)	Cytochrome oxidase subunit II gene (COII) nucleotide sequences	Yes
Mindell, Dick & Baker (1991); Knight & Mindell (1993)	12S rDNA and COI gene nucleotide sequences	Yes[e]
Ammerman & Hillis (1992)	12S rDNA gene nucleotide sequences	Yes
Springer & Kirsch (1993)	12S rDNA gene nucleotide sequences	Yes[f]

[a] For a detailed discussion of these studies and results see Simmons (1994). Unless otherwise indicated, the conclusions indicated are those of the original authors of each study.
[b] Pettigrew *et al.* (1989) claimed that the β-globin data supported bat diphyly, but Baker, Novacek & Simmons (1991) pointed out that diphyly of Microchiroptera and Primates (rather than Chiroptera) was indicated by the β-globin trees. Baker, Novacek & Simmons (1991) concluded that the data employed were inconclusive regarding bat monophyly.
[c] Pettigrew (1991a) indicated that the results of this analysis were inconclusive with respect to bat monophyly. However, Stanhope, Bailey *et al.* (1993) found weak support for bat monophyly when they re-analysed the same data set.
[d] The complete data set used in these studies includes data from a variety of proteins (α- and β-globins, myoglobins, lens αA crystallins, fibrinopeptides, cytochrome c, ribonucleases, and embryonic α- and β-globins), but only α-globin and β-globin data were sampled in both megachiropteran and microchiropteran bats.
[e] Separate analyses of the 12S and COI data indicated that the 12S data support bat monophyly but the COI data are inconclusive. The combined data set supports bat monophyly.
[f] Springer & Kirsch (1993: 149) concluded that their results provided 'mixed support' for bat monophyly.

Table 2. Results of morphological studies[a]

Study	Data	Chiroptera monophyletic?
Smith & Madkour (1980)	Penis + nervous system + skeleton	No
Pettigrew *et al.* (1989); Pettigrew (1991a)	Nervous system	No
Johnson & Kirsch (1993)	Nervous system	No
	Nervous system + diverse characters[b] from Novacek *et al.* (1988)	Yes
Lapoint & Baron (1992)	Volume of brain components	Yes
Luckett (1980, 1993)	Fetal membranes + skeleton	Yes
Wible & Novacek (1988)	Diverse characters[b]	Yes
Thewissen & Babcock (1991, 1993)	Propatagial muscles	Yes
Kay *et al.* (1992)	Cranium	Yes
Beard (1993a)	Cranium + postcranium	Yes
Simmons (1993)	Diverse characters[b]	Yes
Szalay & Lucas (1993)	Postcranium	Yes
Wible & Martin (1993)	Basicranium	Yes

[a] For a detailed discussion of these studies see Simmons (1994). The conclusions indicated here are those of the original authors.
[b] Characters include features of the cranium, postcranial musculoskeletal system, reproductive tract, fetal membranes, vascular system and nervous system.

1994). These characters represent diverse anatomical systems including the dentition, skull, cranial vascular system, postcranial musculoskeletal system, fetal membranes and nervous system. The diversity of this character support, taken together with the biochemical and molecular evidence, effectively refutes Pettigrew's (1991a: 208) suggestion that 'the case for monophyly of bats is based entirely on the wings.' Rather, the case for bat monophyly comprises a broad range of data, most of which indicate that Megachiroptera and Microchiroptera represent a single evolutionary lineage. From this conclusion it follows that powered flight evolved only once in mammals.

Characters related to flight

If chiropteran monophyly is accepted—which seems unavoidable given the evidence now available—we can then examine the nature of the apomorphic features that distinguish bats from other mammals. Because these characters were presumably present in the most recent common ancestor of bats, they provide a starting point from which to investigate the origin of flight.

Of the 26 morphological characters listed in Table 3, six are features of the forelimb that appear to be related to evolution of the flight mechanism. Many of these are unique, including suites of modifications of the scapula (character 8), elbow (9) and wrist and hand (12). Bats also share a unique propatagial

Table 3. Morphological synapomorphies of Chiroptera (Megachiroptera + Microchiroptera)[a]

1.	Deciduous dentition does not resemble adult dentition; deciduous teeth with long, sharp, recurved cusps
2.	Palatal process of premaxilla reduced; left and right incisive foramina fused in midsagittal plane
3.	Postpalatine torus absent
4.	Jugal reduced and jugolacrimal contact lost
5.	Two entotympanic elements in the floor of the middle-ear cavity: a large caudal element and a small rostral element associated with the internal carotid artery
6.	Tegmen tympani tapers to an elongate process that projects into the middle-ear cavity medial to the epitympanic recess
7.	Proximal stapedial artery enters cranial cavity medial to the tegmen tympani; ramus inferior passes anteriorly dorsal to the tegmen tympani
8.	Modification of scapula: reorientation of scapular spine and modification of shape of scapular fossae; reduction in height of spine; presence of a well-developed transverse scapular ligament
9.	Modification of elbow: reduction of olecranon process and humeral articular surface on ulna; presence of ulnar patella; absence of olecranon fossa on humerus
10.	Absence of supinator ridge on humerus
11.	Absence of entepicondylar foramen in humerus
12.	Digits II–V of forelimb elongated with complex carpometacarpal and intermetacarpal joints, lack claws (digits III–V), support enlarged interdigital flight membranes (patagia)
13.	Occipitopollicalis muscle and cephalic vein present in leading edge of propatagium
14.	Modification of hip joint: 90° rotation of hindlimbs effected by reorientation of acetabulum and shaft of femur; neck of femur reduced; ischium tilted dorsolaterally, anterior pubes widely flared and pubic spine present; absence of m. obturator internus
15.	Absence of m. gluteus minimus
16.	Absence of m. sartorius
17.	Vastus muscle complex not differentiated
18.	Modification of ankle joint: reorientation of upper ankle joint facets on calcaneum and astragalus; trochlea of astragalus convex, lacks medial and lateral guiding ridges; tuber of calcaneum projects in plantolateral direction away from ankle and foot; peroneal process absent; sustentacular process of calcaneum reduced, calcaneoastragalar and sustentacular facets on calcaneum and astragalus coalesced; absence of groove on astragalus for tendon of m. flexor digitorum fibularis
19.	Presence of calcar and depressor ossis styliformis muscle
20.	Entocuneiform proximodistally shortened, with flat, triangular distal facet
21.	Elongation of proximal phalanx of digit I of foot
22.	Embryonic disc oriented toward tubo-uterine junction at time of implantation
23.	Differentiation of a free, gland-like yolk sac
24.	Preplacenta and early chorioallantoic placenta diffuse or horseshoe-shaped, with definitive placenta reduced to a more localized discoidal structure
25.	Definitive chorioallantoic placenta endotheliochorial
26.	Cortical somatosensory representation of forelimb reverse of that in other mammals

[a]Morphology and taxonomic distribution of each of these features are discussed in detail in Simmons (1994).

muscle complex (13). Two features of the distal humerus (10 and 11) appear to be synapomorphies of Chiroptera, although they are not unique to bats (both occur in other mammals not believed to be close relatives of Chiroptera; Simmons 1994).

Most of the modifications seen in the scapula of bats probably reflect changes in the roles of different shoulder muscles in locomotion. For example, m. infraspinatus and m. spinodeltoideus are muscles that are important in elevating the wing during the upstroke (Vaughan 1959). Modifications in the shape and size of the infraspinous fossa (where these muscles have their origin) may reflect the increased importance of these muscles in bats as compared to non-volant forms. Other changes in the scapula (e.g. modifications of the scapular spine) probably reflect requirements imposed by differing stresses applied at the scapulohumeral and acromioclavicular joints during flight.

Changes in elbow morphology appear to address other requirements for forelimb-powered flight, including the ability to extend the forearm in a broad arc while simultaneously controlling rotation of the distal end of the limb. Vaughan (1959) suggested that limitations on rotational movements at the humeroradial and radiocarpal joints have permitted the control of adduction and abduction of the wing to be concentrated in centrally located muscles that act on the humerus or proximal end of the radius. Eliminating the need for large distal muscles reduces the mass of the distal wing, thus contributing to energy efficiency during flight.

The wrist and hand modifications seen in bats appear to be related to evolution of the enlarged interdigital patagia that comprise the distal part of the wing. Elongation of digits and loss of claws maximize the area for patagial attachment, and the complex carpometacarpal joints facilitate control of the patagial position and camber. The presence of a propatagial muscle complex, which supports the leading edge of the flight membrane and facilitates generation of lift, appears to be a requirement for both gliding and powered flight in mammals (Pettigrew *et al.* 1989; Thewissen & Babcock 1991, 1993).

In addition to the six forelimb features discussed above, five derived features of the hindlimb of bats may be directly related to flight. These characters encompass several unique features, including reorientation of the hip joint (character 14), lack of differentiation of the vastus muscle complex (17) and evolution of a calcar (19). Two other characters involving loss of thigh muscles (15 and 16) are synapomorphies of bats that apparently evolved independently in some other mammalian groups (Simmons 1994).

The hip modifications seen in bats (Table 3) serve to orient the femur so that it projects laterally rather than ventrally, a change that is clearly significant since the posterior edge of the plagiopatagium inserts on the hindlimb. During flight, the hindlimb projects in approximately the same plane that the forelimb occupies at midstroke (Altenbach 1979), which facilitates a broad upstroke and may contribute to flight efficiency. Changes in relative development of various thigh muscles are probably associated with reorientation of the hindlimb.

Presence of a calcar and m. depressor ossis styliformis (a muscle which controls position of the calcar relative to the ankle and leg) also seems linked to the evolution of flight. The calcar controls the position and tension of the posterior uropatagium, providing a means of changing the shape of this membrane without necessarily moving the hindlimbs relative to the body. As position of the hindlimbs is critical to control of the wing membranes, the presence of a calcar may allow independent control of the wing and uropatagium. Presence of a calcar is clearly not essential for powered flight, however, since several bat lineages have apparently lost the calcar during subsequent evolution (Simmons 1994).

Rather than being directly related to the evolution of flight, some of the hindlimb synapomorphies of bats instead appear to be modifications for hanging from the hindlimbs. These characters include a suite of ankle modifications (character 18) and elongation of the proximal phalanx of digit I of the foot (21). The changes in ankle morphology facilitate full extension of the foot at the ankle, while elongation of the proximal phalanx of digit I serves to bring the claws of all five digits into line (Simmons 1994). These changes may have contributed to the ability of bats to hang freely suspended by their hind feet for extended periods of time.

Given that most of the characters discussed above may be related to the acquisition of flight, why are they listed as separate synapomorphies in Table 3? Two observations account for this decision (Simmons 1994). First of all, a number of these features are not unique to bats, but appear to have evolved independently in other, non-volant mammalian taxa (characters 10, 11, 15, 16). Accordingly, it is possible that these features evolved for reasons not related to flight. Secondly, it seems highly unlikely that all 14 of the postcranial modifications listed in Table 3 evolved simultaneously. In other parts of the mammalian family tree, derived traits that superficially appear to be functionally linked have been demonstrated (often with the help of fossils) to have evolved in a series of steps at different taxonomic levels (Gauthier, Kluge & Rowe 1988; Rowe 1988; Novacek 1992). There is no reason to believe that bat limbs did not evolve in a similar iterative fashion. Because we do not have a clear series of phylogenetic and morphological intermediates between bats and their non-flying relatives[1], we cannot know the relative timing of acquisition of the traits listed in Table 3. However, it seems likely that chiropteran wings and hindlimbs evolved in a series of stages (e.g. Fig. 1; Jepsen 1970; Smith 1977). Modifications in distal structures may have occurred prior to changes in more proximal elements, and changes in the forelimb may have preceded modifications in the hindlimb. Alternatively, the traits listed in Table 3 may have evolved in a more complex mosaic fashion. In the absence of taxa preserving intermediate stages in the process of wing evolution, ontogenetic studies of extant bats may provide the only tests of hypotheses concerning the

[1]The earliest known fossils that clearly belong to the chiropteran lineage include *Icaronycteris* and *Palaeochiropteryx* from the Eocene of North America and Europe, respectively. These taxa were fully volant forms that exhibit virtually all of the skeletal modifications for powered flight that are seen in extant bats (Jepsen 1970; Habersetzer & Storch 1987; Novacek 1987; pers. obs.).

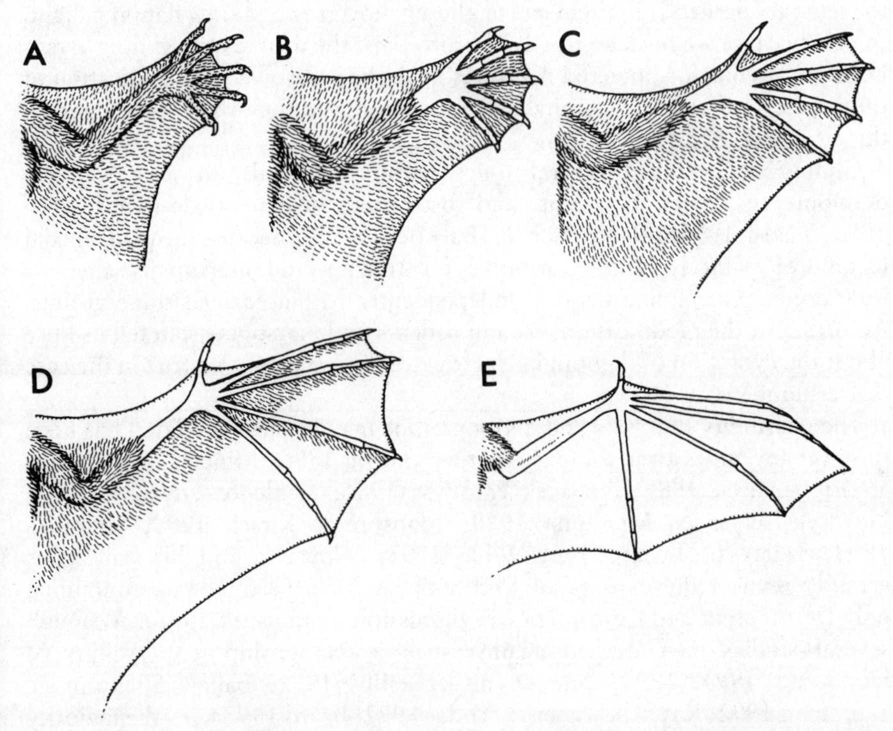

Fig. 1. Progressive stages in the evolution of the bat wing as proposed by Smith (redrawn from Smith 1977: fig. 2). A–C illustrate early stages in which the emphasis was on gliding. D represents an intermediate stage in wing evolution. E shows the wing of a typical extant bat.

relative timing of various evolutionary events. Pending such studies, it seems best to treat the features listed in Table 3 as independent traits.

Gliding and the evolution of powered flight

The evolutionary transformations that led from quadrupedal locomotion to true powered flight in bats have been the source of some debate. Jepsen (1970) suggested that proto-bats captured insects using webbed hands and small skin flaps extending from the arms to the sides of the body, features that presumably evolved for this purpose. With the help of these specializations, proto-bats may have attempted short periods of hovering flight (perhaps when leaping after insects), and selection favouring hovering may have led to the evolution of true powered flight (Jepsen 1970). Pirlot (1977) similarly suggested that early bats passed through a stage in which there was selection for brief periods of hovering flight. In contrast, Smith (1977) and Hill & Smith (1984) have argued that bats evolved from gliding rather than from hovering ancestors (Fig. 1). This view also received support from Clark (1977) based on consideration of the relative energetic requirements of gliding, hovering and slow flapping flight. In part because we have no fossils of proto-bats, the debate over gliding versus hovering origins for powered flight has never been resolved. Consideration of the implications of various phylogenetic hypotheses may shed some light on this problem.

Gliding has evolved several times within mammals, at least once in dermopterans, twice in rodents and three times in marsupials (Thorington 1984; Beard 1993b; Thewissen & Babcock 1993). Because a quadrupedal locomotor habit is clearly primitive for rodents and marsupials, there is little doubt that gliding evolved independently within each of these groups. Accordingly, the modifications seen in rodents and marsupials can tell us little about the evolution of flight in bats. However, this may not be true in the case of Dermoptera.

The possibility that bats and dermopterans may be closely related has been raised many times over the past century (Leche 1886; Miller 1907; Pocock 1926; Novacek 1986; Novacek & Wyss 1986; Wible & Novacek 1988; Novacek, Wyss & McKenna 1988; Johnson & Kirsch 1993; Simmons 1993; Szalay & Lucas 1993; Wible 1993). Novacek and his colleagues recently revived the concept of Volitantia, a clade perceived as containing only Dermoptera and Chiroptera (see discussion in Novacek 1986). Although several studies have argued against such a relationship (e.g. Adkins & Honeycutt 1991, 1993; Ammerman & Hillis 1992; Bailey, Slightom & Goodman 1992; Kay, Thewissen & Yoder 1992; Beard 1993a, b), the majority of the morphological evidence available supports a sister-group relationship between bats and dermopterans (Wible & Novacek 1988; Novacek *et al.* 1988; Thewissen & Babcock 1991, 1993; Johnson & Kirsch 1993; Simmons 1993, unpublished; Szalay & Lucas 1993; Wible 1993). This hypothesis is also supported by results of an analysis that combined morphological data with molecular data from cytochrome oxidase subunit II nucleotide sequences (Novacek 1994). If Dermoptera and Chiroptera are indeed sister taxa, we can learn a great deal about the origin of bat flight by examining the features that dermopterans and bats have in common.

Table 4 lists a series of 17 synapomorphies shared by Chiroptera and Dermoptera, features that were presumably present in their most recent common ancestor. Of these characters, several appear to be directly related to gliding, including features of the upper arm (characters 34 & 35), forearm (34, 37 & 38) and wrist (39). Bats and dermopterans also share the presence of a patagium with two unique features not seen in other volant mammals: a humeropatagialis muscle (36) and interdigital patagia on the manus (40). In this context, it seems likely that primitive volitantians, including the common ancestor of dermopterans and bats, were gliding mammals. This means that the bat lineage passed through a gliding stage prior to evolving the capability of powered flight.

Volitantian gliders (e.g. extinct paromomyids and extant dermopterans) are

Table 4. Morphological synapomorphies of Volitantia (Chiroptera + Dermoptera)[a]

27.	Tooth enamel with horseshoe shaped prisms with associated minor boundary planes (seams)[b]
28.	Fenestra cochlea (round window) faces directly posteriorly[c]
29.	Subarcuate fossa greatly expanded and dorsal semicircular canal clearly separated from endocranial wall of squamosal[c]
30.	Tegmen tympani reduced, tapered to a round process, does not form roof over mallear-incudal articulation or entire ossicle chain[d]
31.	Ramus infraorbitalis of the stapedial artery passes through the cranial cavity dorsal to the alisphenoid[e, f]
32.	Neural spines on cervical vertebrae 3–7 weak or absent[c]
33.	Ribs flattened, especially near vertebral ends[c, g]
34.	Forelimbs markedly elongated[c]
35.	Proximal displacement of the areas of insertion for the pectoral and deltoid muscles; coalesced single proximal humeral torus[g]
36.	Presence of humeropatagialis muscle[c]
37.	Reduction of proximal ulna[c]
38.	Modification of distal radius and ulna: fusion of distal ulna to distal radius; distal radius transversely widened, manus effectively rotated 90°; deep grooves for carpal extensors on dorsal surface of distal radius; disengagement and reduction of the ulna from anterior humeral contact[f,g]
39.	Fusion of scaphoid, centrale and lunate into scaphocentralunate[g]
40.	Patagium continuously attached between digits of manus[c, g]
41.	Elongation of the fourth and fifth pedal rays[g]
42.	Ungual phalanges both proximally and distally deep, compressed mediolaterally[g]
43.	Presence of a tendon locking mechanism on digits of feet[h]

[a]The volitantian synapomorphies listed have been discussed by other authors as noted. Dermoptera is defined here to include extant gliding lemurs (Galeopithecidae = *Cynocephalus*) + extinct Paromomyidae. This grouping is equivalent to Eudermoptera *sensu* Beard (1993a). Several fossil taxa included in Dermoptera by Beard (micromomyids, plesiadapids, carpolestids and saxonellids) are excluded here owing to ambiguity concerning their relationships (Simmons 1993).
[b]Lester, Hand & Vincent (1988)
[c]Wible & Novacek (1988)
[d]Wible & Martin (1993)
[e]Wible (1993)
[f]Simmons (1994)
[g]Szalay & Lucas (1993)
[h]Simmons & Quinn (in press)

unique among mammals in that they are 'finger-gliders' (Beard 1990, 1993 a,b; Szalay & Lucas 1993). The presence of interdigital patagia significantly alters the aerodynamics of gliding in these forms, and voluntary changes in orientation of these membranes—accomplished through movements of the carpus and phalanges—may be an important part of the steering mechanism (Beard 1993b). Finger-gliding in early volitantians may have served as a preadaptation for flight as it evolved in bats. The power-generating portion of the bat wing is the dactylopatagium—the distal part of the wing that comprises the patagia between the second and fifth digits (Hill & Smith 1984). The presence of interdigital patagia in the gliding ancestors of bats may have presented an opportunity for evolution of powered flight that was not open to other mammalian gliders (e.g. rodents and marsupials) because they lacked these membranes.

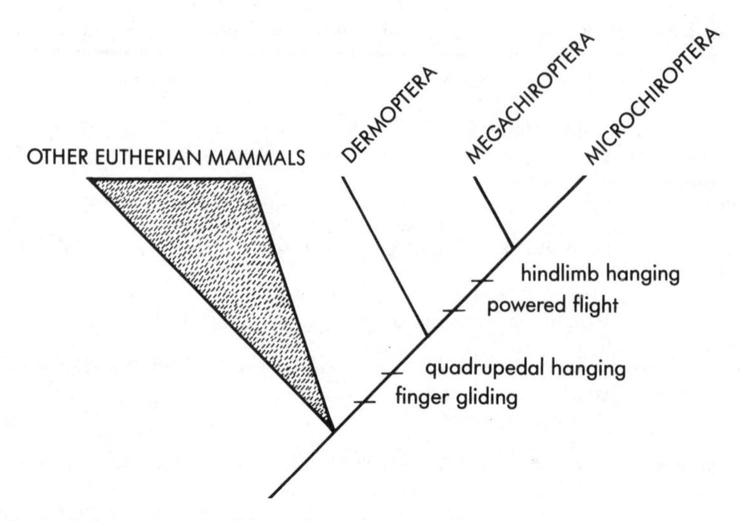

Fig. 2. A phylogenetic tree of bat relationships showing the proposed evolutionary origins of gliding, powered flight and hanging behaviour (see text for discussion). Morphological characters supporting monophyly of Chiroptera (Megachiroptera + Microchiroptera) are given in Table 3; characters supporting Volitantia (Dermoptera + Chiroptera) are given in Table 4.

It seems likely that underbranch hanging is another behaviour that is primitive for dermopterans and bats. Three derived features of the feet are shared by volitantians, including two which appear directly related to hanging behaviour. Szalay & Lucas (1993: 220) noted that dermopterans and bats share elongated fourth and fifth toes on the feet (character 41), and suggested that this condition evolved 'to accommodate habitual, but not obligate, underbranch hanging.' Simmons & Quinn (in press) recently discovered that dermopterans and bats also share the presence of a tendon locking mechanism (TLM) on the digits of the feet (43), a unique feature that permits hanging without continuous contraction of the flexor muscles. In this context, it seems likely that habitual hanging behaviour evolved in the lineage leading to bats before powered flight evolved (Fig. 2). Extant dermopterans hang under branches using all four limbs; ancestral volitantians probably did the same. Habitual hindlimb hanging probably evolved in the bat lineage at some point when flight-related modifications of the wrist and fingers precluded use of the hands for hanging for extended periods. Pettigrew (this volume pp. 3–26) noted that dermopterans and megachiropterans share a number of behavioural traits that are not seen in microchiropterans, including frequent arboreal locomotion using an inverted quadrupedal gait, and defecation and micturition while hanging by the forelimbs. These behaviours, both of which involve use of the forelimbs to support part or all of the body weight, are probably primitive for volitantians. Microchiropteran bats apparently do not use the forelimbs for support while hanging, instead relying exclusively on their hindlimbs[1]. This highly derived behaviour probably evolved from more generalized hanging behaviour like that seen in megachiropterans, which hang principally (though not exclusively) from their hindlimbs.

Conclusions

Consideration of data relevant to phylogenetic relationships among Megachiroptera, Microchiroptera, Dermoptera and other mammalian groups supports the hypothesis that both Chiroptera and Volitantia are monophyletic taxa. Using this phylogeny as a guide, interpretation of various morphological traits of the fore- and hindlimbs suggests that early volitantians were specialized for finger-gliding and quadrupedal underbranch hanging. From this ancestry, bats apparently evolved powered flight and hindlimb hanging, while dermopterans retained the ancestral volitantian body plan and mode of locomotion.

Acknowledgements

I thank P. Racey for his kind invitation to participate in this symposium. K. Koopman, M. Novacek and A. Peffley read earlier versions of this manuscript, and I thank them for their comments. Illustrations were provided by P. Wynne. This research was supported by United States National Science Foundation Grant BSR-9106868.

References

Adkins, R. M. & Honeycutt, R. L. (1991). Molecular phylogeny of the superorder Archonta. *Proc. natn. Acad. Sci. USA* **88**: 10317–10321.

Adkins, R. M. & Honeycutt, R. L. (1993). A molecular examination of archontan and chiropteran monophyly. In *Primates and their relatives in phylogenetic perspective*: 227–249. (Ed. MacPhee, R. D. E). Advances in Primatology Series, Plenum Publishing Co., New York.

Altenbach, J. S. (1979). Locomotor morphology of the vampire bat, *Desmodus rotundus*. *Spec. Publs Am. Soc. Mammal.* No. 6: 1–137.

Ammerman, L. K. & Hillis, D. M. (1992). A molecular test of bat relationships: monophyly or diphyly? *Syst. Biol.* **41**: 222–232.

Bailey, W. J., Slightom, J. L. & Goodman, M. (1992). Rejection of the 'flying primate' hypothesis by phylogenetic evidence from the ϵ-globin gene. *Science* **256**: 86–89.

Baker, R. J., Honeycutt, R. L. & Van Den Bussche, R. A. (1991). Examination of monophyly of bats: restriction map of the ribosomal DNA cistron. *Bull. Am Mus. nat. Hist.* No. 206: 42–53.

[1]Some microchiropteran bats (e.g. emballonurids) employ the forelimbs to brace the body against vertical or slanting surfaces when roosting (Emmons 1990). However, the forelimbs are loaded under compression (rather than tension) when used in this fashion, so this behaviour is not equivalent to the forelimb hanging behaviour seen in megachiropterans.

Baker, R. J., Novacek, M. J. & Simmons, N. B. (1991). On the monophyly of bats. *Syst. Zool.* **40**: 216–231.

Beard, K. C. (1990). Gliding behaviour and palaeoecology of the alleged primate family Paromomyidae (Mammalia, Dermoptera). *Nature, Lond.* **345**: 340–341.

Beard, K. C. (1993a). Phylogenetic systematics of the Primatomorpha, with special reference to Dermoptera. In *Mammal phylogeny: placentals*: 129–150. (Eds Szalay, F. S., Novacek, M. J. & McKenna, M. C.). Springer-Verlag, Berlin.

Beard, K. C. (1993b). Origin and evolution of gliding in Early Cenozoic Dermoptera (Mammalia, Primatomorpha). In *Primates and their relatives in phylogenetic perspective*: 63–90. (Ed. MacPhee, R. D. E.). Advances in Primatology Series, Plenum Publishing Co., New York.

Brooks, D. R. & McLennan, D. A. (1991). *Phylogeny, ecology, and behavior: a research program in comparative biology*. Chicago University Press, Chicago.

Clark, B. D. (1977). Energetics of hovering flight and the origin of bats. In *Major patterns in vertebrate evolution*: 423–425. (Eds Hecht, M. K., Goody, P. C. & Hecht, B. M.). Plenum Press, New York. (*NATO adv. Stud. Inst. Ser. (Life Sci.)* **14**.)

Cronin, J. E. & Sarich, V. M. (1980). Tupaiid and Archonta phylogeny: the macromolecular evidence. In *Comparative biology and evolutionary relationships of tree shrews*: 293–312. (Ed. Luckett, W. P.). Plenum Press, New York & London.

Czelusniak, J., Goodman, M., Koop, B. F., Tagle, D. A., Shoshani, J., Braunitzer, G., Kleinschmidt, T. K., de Jong, W. W. & Matsuda, G. (1990). Perspectives from amino acid and nucleotide sequences on cladistic relationships among higher taxa of Eutheria. *Curr. Mammal.* **2**: 545–572.

de Jong, W. W., Leunissen, J. A. M. & Wistow, G. J. (1993). Eye lens crystallins and the phylogeny of placental orders: evidence for a macroscelid-paenungulate clade? In *Mammal phylogeny: placentals*: 5–12. (Eds Szalay, F. S., Novacek, M. J. & McKenna, M. C.). Springer-Verlag, Berlin.

Emmons, L. H. (1990). *Neotropical rainforest mammals: a field guide*. University of Chicago Press, Chicago & London.

Gauthier, J. A., Kluge, A. G. & Rowe, T. (1988). Amniote phylogeny and the importance of fossils. *Cladistics* **4**: 105–209.

Habersetzer, J. & Storch, G. (1987). Klassifikation und funktionelle Flügelmorphologie paläogener Fledermäuse (Mammalia, Chiroptera). *Cour. Forsch.-Inst. Senckenb.* **91**: 117–150.

Hill, J. E. & Smith, J. D. (1984). *Bats: a natural history*. British Museum (Natural History), London.

Jepsen, G. L. (1970). Bat origins and evolution. In *Biology of bats* **1**: 1–64. (Ed. Wimsatt, W. A.). Academic Press, New York & London.

Johnson, J. I. & Kirsch, J. A. W. (1993). Phylogeny through brain traits: interordinal relationships among mammals including Primates and Chiroptera. In *Primates and their relatives in phylogenetic perspective*: 293–331. (Ed. MacPhee, R. D. E.). Advances in Primatology Series, Plenum Publishing Co., New York.

Jones, J. K. & Genoways, H. H. (1970). Chiropteran systematics. In *About bats: a chiropteran symposium*: 3–21. (Eds Slaughter, R. H. & Walton, D. W.). Southern Methodist University Press, Dallas.

Kay, R. F., Thewissen, J. G. M. & Yoder, A. D. (1992). Cranial anatomy of *Ignacius graybullianus* and the affinities of the Plesiadapiformes. *Am. J. phys. Anthrop.* **89**: 477–498.

Kilpatrick, C. W. & Nunez, P. E. (1992) [1993]. Monophyly of bats inferred from DNA-DNA hybridization. *Bat Res. News* **33**: 62.

Knight, A. & Mindell, D. P. (1993). Substitution bias, weighting of DNA sequence evolution, and the phylogenetic position of Fea's viper. *Syst. Biol.* **42**: 18–31.

Lapoint, F.- J. & Baron, G. (1992) [1993]. Encephalization, adaptation and evolution of bats: statistical evidence for the monophyly of Chiroptera. *Bat Res. News* **33**: 64.

Lauder, G. V. (1981). Form and function: structural analysis in evolutionary morphology. *Paleobiology* **7**: 430–442.

Leche, W. (1886). Uber die Säugethiergattung *Galeopithecus*: eine morphologische Untersuchung. *K. svenska Vetensk-Akad. Handl.* **21** (11): 1–92.

Lester, K. S., Hand, S. J. & Vincent, F. (1988). Adult phyllostomid (bat) enamel by scanning electron microscopy – with a note on dermopteran enamel. *Scanning Microsc.* **2**: 371–383.

Luckett, W. P. (1980). The use of fetal membrane data in assessing chiropteran phylogeny. In *Proceedings fifth international bat research conference*: 245-265. (Eds Wilson, D. E. & Gardner, A. L.). Texas Tech Press, Lubbock.

Luckett, W. P. (1993). Developmental evidence from the fetal membranes for assessing archontan relationships. In *Primates and their relatives in phylogenetic perspective*: 149–186. (Ed. MacPhee, R. D. E.). Advances in Primatology Series, Plenum Publishing Co., New York.

Miller, G. S. (1907). The families and genera of bats. *Bull. U. S. natn. Mus.* No. 57: 1–282.

Mindell, D. P., Dick, C. W. & Baker, R. J. (1991). Phylogenetic relationships among megabats, microbats, and primates. *Proc. natn. Acad. Sci. USA* **88**: 10322–10326.

Novacek, M. J. (1986). The skull of leptictid insectivorans and the higher-level classification of eutherian mammals. *Bull. Am. Mus. nat. Hist.* **183**: 1-111.

Novacek, M. J. (1987). Auditory features and affinities of the Eocene bats *Icaronycteris* and *Palaeochiropteryx* (Microchiroptera, incertae sedis). *Am. Mus. Novit.* No. 2877: 1–18.

Novacek, M. J. (1992). Fossils, topologies, missing data, and the higher level phylogeny of eutherian mammals. *Syst. Biol.* **41**: 58–73.

Novacek, M. J. (1994). Morphological and molecular inroads to phylogeny. In *Interpreting the hierarchy of nature: from systematic patterns to evolutionary process theories*: 85–131. (Eds Grande, L. & Rieppel, O.). Academic Press, New York.

Novacek, M. J. & Wyss, A. R. (1986). Higher level relationships of the Recent eutherian orders: morphological evidence. *Cladistics* **2**: 257–287.

Novacek, M. J., Wyss, A. R. & McKenna, M. C. (1988). The major groups of eutherian mammals. In *The phylogeny and classification of the tetrapods* **2**: 31–71. (Ed. Benton, M. J.). Clarendon Press, Oxford. (*Syst. Ass. spec. Vol.* No. 35B.)

Patterson, C. (1982). Morphological characters and homology. In *Problems in phylogenetic reconstruction*: 21–74. (Eds Joysey, K. A. & Friday, E. A.). (*Syst. Ass. spec. Vol.* No. 21.) Academic Press, New York.

Pettigrew, J. D. (1986). Flying primates? Megabats have the advanced pathway from eye to midbrain. *Science* **231**: 1304–1306.

Pettigrew, J. D. (1991a). Wings or brain? Convergent evolution in the origins of bats. *Syst. Zool.* **40**: 199–216.

Pettigrew, J. D. (1991b). A fruitful, wrong hypothesis? Response to Baker, Novacek, and Simmons. *Syst. Zool.* **40**: 231–239.

Pettigrew, J. D. & Jamieson, B. G. M. (1987). Are flying foxes (Chiroptera: Pteropodidae) really primates? *Aust. Mammal.* **10**: 119–124.

Pettigrew, J. D., Jamieson, B. G. M., Robson, S. K., Hall, L. S., McAnally, K. I. &

Cooper, H. M. (1989). Phylogenetic relations between microbats, megabats and primates (Mammalia: Chiroptera and Primates). *Phil. Trans. R. Soc.* (B) **325**: 489–559.

Pirlot, P. (1977). Wing design and the origin of bats. In *Major patterns in vertebrate evolution*: 375–410. (Eds Hecht, M. K., Goody, P. C. & Hecht, B. M.). Plenum Press, New York. (*NATO adv. Stud. Inst. Ser. (Life Sci.)* **14**.)

Pocock, R. I. (1926). The external characters of the flying lemur (*Galeopterus temminckii*). *Proc. zool. Soc. Lond.* **1926**: 429–444.

Rieppel, O. (1980). Homology, a deductive concept? *Z. zool. Syst. EvolForsch.* **18**: 315–319.

Rowe, T. (1988). Definition, diagnosis, and origin of Mammalia. *J. vert. Paleont.* **8**: 241–264.

Sabeur, G., Macaya, G., Kadi, F. & Bernardi, G. (1993). The isochore patterns of mammalian genomes and their phylogenetic implications. *J. molec. Evol.* **37**: 93–108.

Sarich, V. M. (1993). Some results of twenty-five years with the blood of mammals. In *Mammal phylogeny: placentals*: 103–114. (Eds Szalay, F. S., Novacek, M. J. & McKenna, M. C.). Springer-Verlag, Berlin.

Simmons, N. B. (1993). The importance of methods: archontan phylogeny and cladistic analysis of morphological data. In *Primates and their relatives in phylogenetic perspective*: 1–61. (Ed. MacPhee, R. D. E.). Advances in Primatology Series, Plenum Publishing Co., New York.

Simmons, N. B. (1994). The case for chiropteran monophyly. *Am. Mus. Novit.* No. 3103: 1–54.

Simmons, N. B., Novacek, M. J. & Baker, R. J. (1991). Approaches, methods, and the future of the chiropteran monophyly controversy: a reply to J. D. Pettigrew. *Syst. Zool.* **40**: 239–243.

Simmons, N. B. & Quinn, T. H. (In press). Evolution of the digital tendon locking mechanism in bats and dermopterans: a phylogenetic perspective. *J. mammal. Evol.*

Smith, J. D. (1976). Chiropteran evolution. *Spec. Publs Mus. Texas Tech Univ.* No. 10: 49–69.

Smith, J. D. (1977). Comments on flight and the evolution of bats. In *Major patterns in vertebrate evolution*: 427–438. (Eds Hecht, M. K., Goody, P. C. & Hecht, B. M.). Plenum Press, New York. (*NATO adv. Stud. Inst. Ser. (Life Sci.)* **14**.)

Smith, J. D. (1980). Chiropteran phylogenetics: introduction. In *Proceedings fifth international bat research conference*: 233–244. (Eds Wilson, D. E. & Gardner, A. L.). Texas Tech Press, Lubbock.

Smith, J. D. & Madkour, G. (1980). Penial morphology and the question of chiropteran phylogeny. In *Proceedings fifth international bat research conference*: 347–265. (Eds Wilson, D. E. & Gardner, A. L.). Texas Tech Press, Lubbock.

Springer, M. S. & Kirsch, J. A. W. (1993). A molecular perspective on the phylogeny of placental mammals based on mitochondrial 12S rDNA sequences, with special reference to the problem of the Paenungulata. *J. mammal. Evol.* **1**: 149–166.

Stanhope, M. J., Czelusniak, J., Si, J.-S., Nickerson, J. & Goodman, M. (1992). A molecular perspective on mammalian evolution from the gene encoding interphotoreceptor retinoid binding protein, with convincing evidence for bat monophyly. *Mol. phylogenet. Evol.* **1**: 148–160.

Stanhope, M. J., Bailey, W. J., Czelusniak, J., Goodman, M., Si, J.-S., Nickerson, J., Sgouros, J. G., Singer, G. A. M. & Kleinschmidt, T. K. (1993). A molecular view of primate supraordinal relationships from the analysis of both nucleotide and amino acid sequences. In *Primates and their relatives in phylogenetic perspective*: 251–292. (Ed. MacPhee, R. D. E.). Advances in Primatology Series, Plenum Publishing Co., New York.

Szalay, F. S. & Lucas, S. G. (1993). Cranioskeletal morphology of archontans, and diagnoses of Chiroptera, Volitantia, and Archonta. In *Primates and their relatives in phylogenetic perspective*: 187-226. (Ed. MacPhee, R. D. E.). Advances in Primatology Series, Plenum Publishing Co., New York.

Thewissen, J. G. M. & Babcock, S. K. (1991). Distinctive cranial and cervical innervation of wing muscles: new evidence for bat monophyly. *Science* **251**: 934-936.

Thewissen, J. G. M. & Babcock, S. K. (1993). The implications of the propatagial muscles of flying and gliding mammals for archontan systematics. In *Primates and their relatives in phylogenetic perspective*: 91–109. (Ed. MacPhee, R. D. E.). Advances in Primatology Series, Plenum Publishing Co., New York.

Thorington, R. W. (1984). Flying squirrels are monophyletic. *Science* **225**: 1048-1050.

Vaughan, T. A. (1959). Functional morphology of three bats: *Eumops, Myotis, Macrotus*. *Univ. Kans. Publs Mus. nat. Hist.* **12**: 1–153.

Wible, J. R. (1993). Cranial circulation and relationships of the colugo *Cynocephalus* (Dermoptera, Mammalia). *Am. Mus. Novit.* No. 3072: 1–27.

Wible, J. R. & Martin, J. R. (1993). Ontogeny of the tympanic floor and roof in archontans. In *Primates and their relatives in phylogenetic perspective*: 111-148. (Ed. MacPhee, R. D. E.). Advances in Primatology Series, Plenum Publishing Co., New York.

Wible, J. R. & Novacek, M. J. (1988). Cranial evidence for the monophyletic origin of bats. *Am. Mus. Novit.* No. 2911: 1–19.

Part II:

Fruit bats as keystone species

Symp. zool. Soc. Lond. (1995) No. 67: 47–62

The role of flying foxes (Pteropodidae) in oceanic island ecosystems of the Pacific

William E. RAINEY,
Elizabeth D. PIERSON,

Museum of Vertebrate Zoology
University of California
Berkeley, CA 94720, USA

Thomas ELMQVIST

Department of Ecological Botany
Umea University
Umea, Sweden 901-87

and Paul A. COX

Department of Botany & Range Science
Brigham Young University
Provo, UT 84602, USA

Synopsis

With accumulating evidence that flying foxes (Pteropodidae) are important seed dispersers and pollinators in Palaeotropical ecosystems, we examine their role, focusing on the relatively simple forest communities on remote oceanic islands.

Biogeographical comparisons suggest that plant groups that are pollinated or dispersed by guilds in which flying foxes are important are less diverse in areas that lack flying foxes. Although plant/animal interactions are diffuse in depauperate island biotas, behavioural and morphological traits make it likely that bats play a unique role in longer-distance dispersal of the large-seeded fruits of dominant canopy trees. Anthropogenic extinctions of alternative seed dispersers and pollinators on these islands have greatly increased the importance of flying foxes by truncating the upper end of bird size distribution, thus virtually eliminating other vertebrate dispersers of large-seeded fruits and reducing the guild of pollinators and seed dispersers for small-seeded fruits.

The paradigm of minimal competition among tropical frugivores, which has its roots in Neotropical studies, appears inappropriate for pteropodids on islands, where resource defence and thus the interaction of bat density with local resource abundance may have a fundamental role in structuring patterns of pollination and seed dispersal.

Introduction—keystone species: a rose by any other name?

The concept of keystone species (Paine 1966, 1969) has followed a common trajectory for intuitively appealing ecological constructs, with the initial model of predation structuring rocky intertidal communities being enthusiastically

ZOOLOGICAL SYMPOSIUM No. 67
ISBN 0 19 854945 8

applied to a wide range of interactions (see Simberloff 1991 and Mills, Soulé & Doak 1993 for typologies). Concerns about questionable testability and the malign effects of popular acceptance led Mills *et al.* (1993) to call for the term to be abandoned as unproductively vague and, indeed, dangerous to conservation goals. 'Keystone species' has in fact already vanished from the vocabularies of those attuned to the rapid cycle of innovation and obsolescence in the community ecology vernacular, and been replaced by discussion of interaction strengths among species within communities (e.g. Paine 1992).

Yet the encompassing model of a 'keystone entity' (an ecological analogue to the 'operational taxonomic unit') as a taxon, guild or other assemblage that has a disproportionate role in structuring a community remains a convenient conceptual shorthand. With unequal interaction strengths on some functional axis (not necessarily trophic), there is the potentially testable prediction that severe changes in abundance of a keystone entity would trigger more extensive changes in community structure.

Accumulating evidence indicates that pteropodids are important seed dispersers and pollinators in Palaeotropical ecosystems (Baker & Harris 1957; Crome & Irvine 1986; McCoy 1990; Eby 1991). In what settings, however, they might legitimately be called 'keystone species' remains unclear. Here we examine further the suggestion that on remote, faunally depauperate oceanic islands flying foxes are likely to play a keystone role in structuring forest communities (Cox, Elmqvist, Pierson & Rainey 1991; Elmqvist, Cox, Rainey, & Pierson 1992).

Biogeography: inferences from distribution

On tropical oceanic islands the diversity of pollinator and seed disperser guilds is greatly reduced relative to mainland areas. For example, on Borneo (7.5×10^5 km^2), which has an essentially continental Indomalaysian vertebrate fauna, there are 152 bird and 80 mammal species that are nectarivorous or frugivorous (Smythies 1984; Payne, Francis & Philips 1985). By contrast, on Samoa (3.0×10^3 km^2) these guilds contain 15 indigenous birds (Pratt, Bruner & Berrett 1987) and two mammal species, both pteropodid bats.

The attenuation of the Indomalaysian biota eastward through the South Pacific is consistent with the stochastic model of MacArthur & Wilson (1967). However, an abundant literature (e.g. Whittaker 1992) also indicates that filters other than geographic distance (e.g. morphological constraints, species interactions) shape island species assemblages. Bats have been better colonizers of islands than have other mammals, and for all islands east of the Solomons in the south Pacific and east of the Philippines in the central Pacific are the only native mammals. The bat assemblage on these islands is not a random draw from the source fauna. Microchiropterans are represented by six out of 10 Old World families on the Asian mainland, but only a few emballonurid

Table 1. A comparison of the floras of Samoa and the Marquesas.

	Samoa	Marquesas	
Plant families with species visited by flying foxes			
Anacardiaceae	2	0	
Annonaceae	1	0	
Ebenaceae	4	0	
Guttiferae	3	1	
Meliaceae	6	0	$\chi^2 = 3.93$
Moraceae	5	4	$P < 0.05$
Myrtaceae	5	3	
Rosaceae	2	0	
Sapindaceae	9	5	
Sapotaceae	4	1	
Total no. species	41	14	
Expected no. species	72	49	
Plant families without species visited by flying foxes			
Eleaocarpaceae	4	1	
Euphorbiaceae	22	21	
Lauraceae	5	1	
Loganiaceae	5	3	
Loranthaceae	2	7	$\chi^2 = 1.30$
Oleaceae	3	1	$P > 0.25$
Rhamnaceae	2	3	
Tillaceae	4	2	
Urticaceae	7	11	
Verbenaceae	7	7	
Total species	61	57	
Expected species	72	49	
Angiosperm families	81	68	
Species	586	339	
Species/Family	7.2	4.9	

and molossid species reach remote islands. Among pteropodids, *Pteropus* has undergone an extensive radiation on islands and, on more remote archipelagos, is the only bat taxon present. Of the 174 megachiropteran species, 56 belong to the genus *Pteropus*, and 48 of these (86%) are found only on islands (Rainey & Pierson 1992).

To begin exploring patterns of relationship between flying foxes and plants apparently linked to them, we compared the floras of two archipelagos. Table 1 shows the number of indigenous species (Brown 1931, 1935; Amerson, Whistler & Schwaner 1982; Whistler 1992) in 10 plant families frequently visited by bats (Fujita & Tuttle 1991), and in 10 arbitarily selected plant families not associated with bats, for both Samoa, which has two pteropodid species, and the Marquesas, which have none. To incorporate the eastward attenuation in diversity, expected values (mean number of species/family and thus the number of species in 10 families) were based on the total number of angiosperm species and families in each archipelago. Thus the null hypothesis was that the eastward attenuation of diversity in the bat and non-bat samples would not differ significantly from the proportion in the entire floras. There was no significant difference for non-bat-associated families ($\chi^2=1.3$, $P>0.25$), but in the Marquesas five out of 10 bat-visited families were missing and the total number of species was significantly fewer than expected ($\chi^2=3.93$, $P<0.05$).

We then compared the geographic distribution of 10 forest tree genera[1] which contain species dispersed by flying foxes (Cox *et al.* 1992; Wiles & Fujita 1992) and which are dominant components of the Samoan forest (Amerson *et al.* 1982; Elmqvist, Rainey, Pierson & Cox 1994), with the distribution of 10 arbitrarily chosen genera of primary forest trees[2] which are not known to be dispersed by flying foxes. For each genus we compiled presence or absence data from van Balgooy (1971) for areas with and without flying foxes[3].

A simple summary variable for each area is whether the number of bat-dispersed genera present is greater than, less than, or the same as the number of genera not dispersed by bats. The number of bat-dispersed genera was the same or smaller in six out of the 16 areas with bats and in eight out of the 10 areas lacking bats (P = 0.05, Fisher's Exact Test).

Although these comparisons should be interpreted with caution, since plant distribution data are incomplete, frugivorous bird distribution is ignored, and other variables may simultaneously affect pteropodids and associated plants, both at least suggest a relationship between the presence or absence of flying foxes and plant taxa that have been functionally linked to them.

Ecological evidence: how important are flying foxes as pollinators and seed dispersers?

Relative importance of dispersal by bats and birds

Since birds and bats are the major indigenous vertebrate pollinators and seed dispersers on remote oceanic islands, it is important to evaluate their respective roles. Continental areas show marked niche separation between these two groups (Palmeirim, Gorchov & Stoleson 1989), with little temporal overlap between diurnal plant-visiting birds and nocturnal bats. Additionally, researchers have characterized morphological differences between 'bird' flowers and fruits and 'bat' flowers and fruits (e.g. Marshall 1983).

[1]*Aglaia, Cananga, Diospyros, Dysoxylum, Elaeocarpus, Mammea, Ochrosia, Planchonella, Pometia and Syzygium.*

[2] *Alphitonia, Antidesma, Ascarina, Macaranga, Maesa, Mertya, Myristica, Securinega, Timonius, Weinmannia.*

[3]Flying foxes present: S. E. Asia, Malesia, Philippines, Papua New Guinea, Australia, Bismarck Archipelago, Solomon Is., New Caledonia (including Loyalty Is.), Vanuatu, Mariana Is., E. and W. Caroline Is., Fiji, Samoa, Tonga, Cook Is. Flying foxes absent: Society Is., Tubuais, Rapa, Tuamotus, Marquesas, Hawaii, Easter Is., Juan Fernandez, W. Central and E. Central Pacific (*sensu* van Balgooy).

In less diverse Palaeotropical island communities, however, we observe greater overlap in both foraging time and diet. With reduced predation risk on islands, some *Pteropus* have become partially or primarily diurnal (e.g. Cox 1983; Tidemann 1987), and visit flowering and fruiting trees simultaneously with birds. On Samoa, for example, several of the dominant forest canopy and subcanopy trees produce fruits that are consumed by two species of *Pteropus* and several columbids. Although the bats are generally behaviourally dominant over the birds, it is not unusual to observe bats and doves or pigeons feeding together in the same tree. There remain, however, important differences between birds and bats which help to elucidate their respective roles, particularly as dispersal agents. For example, many bird species, including some of the larger frugivorous columbids, are at least partly seed predators, whereas bats almost never are (Van der Pijl 1957; Howe 1986; Eby 1991).

A qualitative graphical model contrasting the probability of dispersal by birds and pteropodid bats for various sizes of diaspores is presented in Fig. 1. Birds move seeds largely by ingestion and elimination (although some regurgitate seeds or carry fruits short distances), whereas bats move smaller seeds by ingestion and elimination, and larger ones by carrying in their mouths. The upper limit on size of a diaspore dispersed away from the parent tree is set by the largest cross-section the gape can accommodate (Wheelwright 1985). Dispersal distances for ingested seeds are set by the interaction of flight speed, foraging behaviour, passage time and gut load.

The dispersal pattern with small-seeded fruits (e.g. *Ficus*) is similar for both birds and pteropodids in that some seeds are ingested, but handling times for bats are relatively longer because fruits are masticated to extract fluids (some

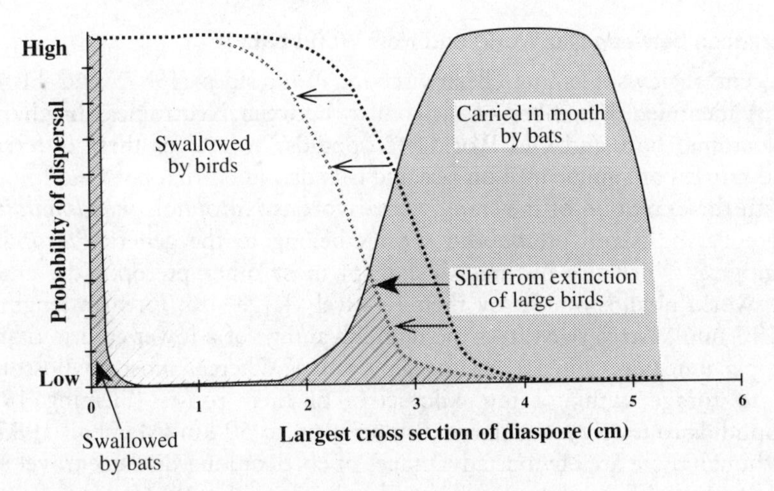

Fig. 1. A graphical model of diaspore size versus the probability of transport away from the parent tree for frugivorous birds and large pteropodid bats.

seeds are discarded with the extracted pulp). For fruits in a middle size range with seeds larger than the maximum actually swallowed by bats, the value of pteropodids as dispersers is probably low. The largest seed observed in *P. conspicillatus* faeces was 3.7 mm, with a probable maximum of about 5.0 mm (G. Richards pers. comm). Such fruits are typically processed in the source tree, with all seeds and fibrous tissue dropped beneath the canopy. Small canopy fruits are rarely carried away even at relatively high forager densities.

For island rainforest canopy trees with larger fruits (and generally larger seeds) two patterns are observed. At low bat densities, fruit is processed in the parent tree and debris, including seeds, is dropped there. As forager densities rise in relation to local fruit resources, however, aggressive interactions increase. Late-arriving bats 'pluck and run', carrying fruits in their mouths some distance away from competitors for processing. Richards (1990) describes a pattern in *P. conspicillatus* of 'residents' (bats that arrive at the resource early in the evening and establish feeding territories) and 'raiders' (late-arriving individuals), and notes that only raiders are responsible for dispersal. Temporal persistence of feeding territories may vary among bat species and plant resources (e.g. local depletion of ripe fruit), but a positive relationship between frequency of aggressive interaction and seed dispersal is likely to be consistent.

Some native island forest fruits are too large for *Pteropus* to transport intact (e.g. *Artocarpus mariannensis*), but pteropodids can carry 200 g fruits ($\approx$ 25% body weight for *P. vampyrus*: Marshall 1983) away from the parent tree. Very large seeds are unlikely to be dispersed by extant island birds, giving *Pteropus* a unique local dispersal role.

Differences between Old World and New World bats

In recent reviews Fleming, Breitwisch & Whitesides (1987) and Fleming (1993) identified a number of differences between Neotropical frugivorous phyllostomid bats and Old World pteropodids. Several of these differences attain particular significance on oceanic islands.

With the exception of the small, nectarivorous *Notopteris macdonaldii*, all remote Pacific island pteropodid species belong to the genera *Pteropus* or *Pteralopex*, which are distinguished from most other pteropodids, and all New World phyllostomids, by their relatively large size (forearm lengths of 86–215 mm). Large size offers the dual advantage of a lower caloric requirement per unit mass and greater foraging range. Whereas most phyllostomids tend to forage within a few kilometres of their roosts (Fleming 1993), pteropodids routinely travel large distances, up to 50 km (Marshall 1985).

Although there are obvious advantages of efficient long-distance travel in an environment with patchy resources and frequent typhoons, large pteropodids have sacrificed manoeuvrability. Whereas many phyllostomid species forage in the understorey, the majority of pteropodids feed in the canopy or on forest edges (Fleming *et al.* 1987). Even in the diverse pteropodid community of mainland Malaysia, Francis (1990) found that only 15–18% of the bat species caught in the understorey were frugivores, as compared with 34–48% reported for Neotropical forests. This may reflect lower understorey fruit availability (Wong 1986) as well as morphological constraint. A related difference is that phyllostomids often forage by hovering at the plant periphery, whereas large pteropodids typically land on trees and forage on fruits or flowers in the interior of the canopy, which would be inaccessible even to a much smaller hovering animal. By this means fruits or flowers seen by investigators as bird-adapted may be exploited by bats.

There are correlated dietary differences between the two groups. Frugivorous phyllostomids consume primarily small-seeded fruits and, as suggested above, for many species diets include or are dominated by understorey or gap plants (Bonaccorso & Humphrey 1984; Fleming 1988). In the Neotropics birds are the primary volant consumers and probable dispersers of large-seeded fleshy fruits although, as in Asia, non-volant mammalian frugivores (rodents, primates, tapir) may be important or essential primary or secondary dispersers for some plants (e.g. Smythe 1989). Although there are some subcanopy pteropodids in areas of high diversity, and some with a diet heavily biased toward *Ficus*, the pattern of heavy exploitation of large-seeded fleshy fruits is a distinctive aspect of the foraging ecology of large pteropodids. Additionally, most phyllostomids, but relatively few pteropodids, roost diurnally in dark settings inimical to seedling survival. Pteropodid day roosts, as well as temporary feeding roosts, may provide sites for seed establishment.

Comparison of Old World and New World plant/animal interactions

Plant/animal interactions in the Palaeotropics tend to be more diffuse than in the New World in the context of both frugivory—e.g. broader dietary overlap among vertebrate frugivore groups (Fleming *et al.* 1987)—and pollination, especially on oceanic islands (Cox 1984). Fleming (1993) suggested that the tendency for Old World taxa to be more generalized in their food habits may arise from the greater spatio-temporal patchiness of resources in Palaeotropical forests. While much remains unknown about regional generalizations on the specificity of plant/pollinator interactions, we can at least point to Old World examples such as the *Freycinetia* species whose pollinators tend to be frugivores (including a wide taxonomic array of vertebrates). On oceanic islands there has been, as well, convergence in reward appearance and sugar composition between inflorescences of some *Freycinetia* and co-occurring canopy fruits consumed by flying foxes (Cox 1984).

Flying foxes act as both pollinators and seed dispersers for a substantial fraction of the plant species they visit. Fourteen of 69 plant genera (20%) used by flying foxes on oceanic Pacific islands (Wiles & Fujita 1992) contain species

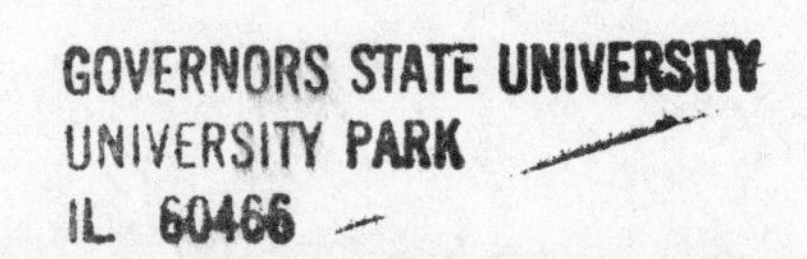

for which flying foxes both disperse the fruits and visit the flowers. By contrast, Heithaus, Fleming & Opler (1975) noted that only one of 25 'bat plants' in a Costa Rican dry forest was both pollinated and dispersed by bats.

From compilations of pteropodid/plant associations (Fujita & Tuttle 1991; Marshall 1985; Wiles & Fujita 1992) it appears likely that flying foxes play a more significant role as seed dispersers than as pollinators. In Marshall's (1985) summary, 37% (n=73) of plant genera were visited for flowers and 73% (n=145) for fruits. In Wiles & Fujita's (1992) analysis, 44% of genera were visited for flowers and 71% for fruits.

Flying foxes as pollinators

Although available data suggest that most pteropodids are primarily frugivorous, it is likely that the frequency of visits to flowers by unspecialized pteropodids has been underestimated. Bats leave conspicuous evidence of fruit consumption and dispersal by discarding pellets of fruit tissue and fruit with distinctive tooth and/or claw marks, but they may leave no distinctive sign of visits to flowers. Relatively little is known outside Australia regarding the diversity of flowers they exploit. Kitchener, Gunnell & Maharadatunkamsi (1991) showed by gut analyses that all 10 pteropodid species examined on Lombok Is., Indonesia, consumed pollen, although the diversity of pollen was inversely correlated with body size.

Visiting flowers does not necessarily mean pollination. While there is a presumption that flying foxes are important pollinators for many tropical plants, the number of plant species for which this has actually been documented is low (e.g. Gould 1978; Cox 1984; Kress 1985; McCoy 1990; Elmqvist, Cox *et al.* 1992). In some cases pollination can reasonably be inferred (Baker & Harris 1957; Start & Marshall 1976; McWilliam 1985–86), but not enough is known of the reproductive biology of many bat-visited plants to evaluate the roles of various flower visitors. Crome & Irvine (1986) showed that even though bird visitors were more numerous, pteropodid bats were the most important pollinators of a major rainforest tree, *Syzygium cormiflorum*, in Australia.

Flying foxes as seed dispersers: impacts on community structure

Recent studies document a wide array of fruits consumed by pteropodids, including those of many dominant forest canopy trees (McWilliam 1985–86; Richards 1990; Cox *et al.* 1991; Eby 1991; Fujita & Tuttle 1991; Wiles & Fujita 1992). Yet the challenging task of investigating the role of pteropodids in dispersal (e.g. scale of movement and probability of establishment) has only just begun. Experiments indicate that viability of seeds ingested or carried by bats is usually high (Richards 1990; Eby 1991). Ingested seeds of both *Ficus* and *Freycinetia* show enhanced germination relative to unmanipulated controls (Utzurrum 1989; P. A. Cox unpubl. obs.).

Table 2. Percentage of fleshy fruits from rainforest tree species dispersed by flying foxes in selected habitats on Savai'i, Western Samoa (July 1988) and Guam (August 1989). Data are from 5 m × 5 m forest floor plots, separated by 20 m along transects (4 plots/transect in Samoa; 5 plots/transect in Guam). All taxa are known to be consumed by bats. Only fruits sufficiently intact to assess the dispersal agent are included in the analysis. *n* = number of transects.

	Lowland (<150 m elev.) *n*=8		Volcanic cone (150–300 m) *n*=8		Montane (>600 m) *n*=8	
Tree species	Total no. fruits	%	Total no. fruits	%	Total no. fruits	%
Samoa						
Inocarpus fagifer	0	–	56	61	0	–
Planchonella torricellensis	5	100	41	0	8	0
Pometia pinnata	12	0	232	28	24	0
Syzygium inophylloides	5	80	401	54	0	–
Terminalia samoensis	0	–	199	88	28	0
			Inland (<150 m) *n*=5			
Guam						
Aglaia mariannensis			151	1		
Ficus sp.			18	0		
Mammea odorata			2	0		
Neisosperma oppositifolia			203	0		
Ochrosia mariannensis			18	0		

Although reported dispersal distances for large mouth-carried fruits are only a few hundred metres, as opposed to estimates of many kilometres for ingested seeds (Van der Pijl 1982; Richards 1990), our experience in Samoa suggests that this apparent difference will decline as data accumulate. We have frequently observed both *Pteropus* species carry 4.0 cm fruit for distances >1.25 km. Earlier we reported bat-dispersed fruit in lowland forest at even longer distances from probable upland source areas (Cox *et al.* 1992). This is consistent with van der Pijl's (1957) model of diplochory in which large-seeded trees reach isolated islands by marine dispersal, but are dispersed to interior areas by bats.

We obtained evidence of seed dispersal by island pteropodids by a comparison between Western Samoa, with two *Pteropus* species, and Guam, where the only extant species is reduced to a remnant population of a few hundred animals (Wiles 1987). In one study, 100 m × 1 m transects were surveyed in different forest communities, and all fleshy fruits located on the forest floor were examined for evidence of bat dispersal. The proportion of fruit dispersed by bats varied from 0–100% per transect in Samoa ($\bar{x}$=37.4%) and between 0 and 1% ($\bar{x}$= 0.2% in Guam (Table 2).

In a second, shorter-term comparison on these islands, 3 m × 3 m debris

Table 3. Estimated mean number of seeds/hectare dispersed by flying vertebrates in Samoa and Guam. Plots (3 m × 3 m plastic sheets) were located in clearings and surveyed daily for 3 days. Number of plots = 7 (Samoa) and 10 (Guam).

	Tafua, W. Samoa (4–6 November 1993)	Anderson AFB, Guam (4–6 August 1989)
Total no. seeds	22	0
Mean no./day/plot (SE)	1.0 (0.3)	0
Estimated no./hectare/day	1111	0

traps were set in cleared areas to estimate the seed rain from volant vertebrates (Table 3). Even though the Samoan plots were run less than two years after a severe typhoon, which had caused marked declines in flying fox populations (Pierson, Elmqvist, Rainey & Cox in press), they still indicated much more vertebrate seed dispersal than did those from Guam, where not a single vertebrate-dispersed seed was identified.

In another approach to inference about their significance in plant community structure, we examined what proportion of the dominant tree species in Samoa were exploited by flying foxes. Eighteen dominant tree species were identified in three lowland forest communities, one in Western Samoa (Elmqvist, Rainey *et al.* 1994) and two in nearby American Samoa (Amerson *et al.* 1982) (Table 4). Flying foxes visit the flowers of at least six (33%) and feed on the fruits of 16 (89%) of these species. Many of these taxa, particularly *Planchonella, Syzygium, Pometia* and *Terminalia*, are highly favoured foods, attracting aggregations of both *Pteropus* species when they are in fruit. Though the links are, at this point, only sketched, it seems likely that pteropodids are playing a significant role in maintaining forest diversity, particularly on islands (Cox *et al.* 1991, 1992; Marshall 1983).

Human effects on island environments

Recent archaeological investigations document widespread faunal extinctions on isolated Pacific islands since the arrival of humans (Steadman 1989, 1993; Steadman & Kirch 1990). Patterns of faunal loss show that extant flying fox populations are a remnant of formerly larger nectarivore/frugivore guilds.

Extinction of vulnerable terrestrial organisms (e.g. large flightless birds) has been a recurrent, nearly inevitable, consequence of human settlement on oceanic islands (see review by Milberg & Tyrberg 1993). The degree of faunal impoverishment roughly tracks the extent of terrestrial resource utilization. Archaeologically, the best-studied region for the fate of island land birds conveniently falls in the Polynesian range of pteropodids, but limited data show comparable extinctions in Micronesia (Steadman 1992a) and the richer faunas of Melanesia (e.g. Balouet & Olson 1989; D. W. Steadman pers. comm.). Steadman (1992b) emphasizes that gaps in the

Table 4. The dominant trees in three forest communities in Samoa, showing records of flying fox visits to flowers or fruits.

	Flowers	Fruits
Tafua lowland forest, Western Samoa**		
Aglaia samoensis		X
Cananga odorata	X	X
Diospyros samoensis		X
Dyzoxylum samoense		X
Garuga floribunda		X
Inocarpus fagifer		X
Mammea glauca	X*	X
Planchonella torricellensis	X	X
Pometia pinnata		X
Syzygium inophylloides	X	X
Coastal forest, American Samoa***		
Diospyros elliptica		X
Diospyros samoensis		X
Syzygium clusiifolium	X	X
Syzygium dealatum	X	X
Lowland forest, American Samoa***		
Calophyllum neo-ebudicum		X
Canarium vitiense		X
Dyzoxylum maota		X
Dyzoxylum samoense		X
Myristica fatua		
Myristica hypargyraea		
Planchonella garberi		X
Planchonella torricellensis	X	X
Pometia pinnata		X
Terminalia richii		X
Syzygium inophylloides	X	X

*Wiles & Fujita (1992). Not observed in Samoa.
**Elmqvist, Rainey *et al.* (1994).
***Amerson *et al.* (1982).

known ranges of both extinct and extant taxa are largely detection artifacts and that avifaunal loss in the pre-European period is certainly more extensive than currently documented.

Extrapolating diet and habits of extinct species from extant forms, it is clear that assemblages of probable avian pollinators, dispersers and, to a lesser extent, seed predators of forest canopy plant taxa have been radically reduced. For example, on 'Eua Is., Tonga, 13 land birds are present today and 23 extinct or extirpated forms were identified in pre-human or post-Polynesian settlement strata (Steadman 1993). The taxa lost include three megapodes of different sizes (forest floor omnivores), a ground dove and three large to very large extinct fruit pigeons (frugivore/seed predators), two small (nectarivore) and one large (frugivore/seed predator) parrot, a honeyeater (nectarivore/frugivore), a thrush (frugivore) and one large white-eye (nectarivore/frugivore).

At the eastern end of the range of *Pteropus*, on Mangaia, Cook Islands, excavation of a human occupation site revealed 12 land birds, eight of which are now extirpated or extinct (Steadman & Kirch 1990). While no pigeons or doves are present in the modern avifauna, there are five in the archaeological assemblage (Steadman 1992a), including *Ducula galeata*, the largest extant pigeon in Polynesia. Two small nectarivorous parrots were also lost on Mangaia, leaving a truly depauperate avifauna with no frugivores or nectarivores. This is an extreme case of *Pteropus* as the last volant vertebrate pollinator and seed disperser in the forest remnants of a heavily altered landscape.

Throughout the Pacific islands, plant and animal introductions have spread into indigenous forests, adding both fruit and floral resources and a few small bird species to the vertebrate pollinator and frugivore guilds, but the net effect of these ongoing alterations on community structure has not been adequately assessed.

Conclusions

While unique evolutionary novelties are an engaging feature of island biotas, species that are recurrent occupants of remote tropical oceanic islands tend to be generalists tolerant of depauperate biotas and patchy fluctuating resources. As elsewhere, frugivorous birds are typically more diverse than pteropodid bats on remote Pacific islands. There is considerable resource overlap within the nectarivore and frugivore guilds but, partly as a consequence of their size and very generalized feeding habits, we suggest that the flying foxes currently play a disproportionate role as seed dispersers and perhaps pollinators by maintaining canopy dominants and thus structuring the plant community. In these anthropogenically altered communities in which most or all large frugivorous birds have been eliminated, pteropodids are the only vertebrates remaining that are capable of dispersing the large-seeded fruits of some dominant canopy trees and thus they may have acquired a residuary keystone role.

Accumulating observations of aggressive resource defence among large pteropodids are an exception to the prevailing view of low levels of competition among frugivores (Fleming 1979). For the plant community, increasing frequency of aggressive interaction should enhance reproductive success by increasing geitonogamous or xenogamous pollination (Elmqvist, Cox *et al.* 1992), as well as increasing typical dispersal distances for diaspores.

From the perspective of conservation biology, we can hypothesize a threshold density of pteropodids in relation to their food resources below which interspecific aggression is low. The geographic scale of pollen and seed dispersal may decline at this level, leading to reductions in local gene flow and recruitment. As noted earlier, resource abundance varies spatially and temporally on several scales, so detecting such an effect may not be simple. Guam, however, offers a grim warning about the endpoint of the sequence.

Acknowledgements

Field work in Samoa was funded by the National Science Foundation, World Wildlife Fund Sweden, and the National Park Service, and in Guam by Bat Conservation International. We thank D. Steadman for a perspective on island extinctions and G. Wiles for discussions and assistance in the field; and the people of Falealupo and Tafua villages for welcoming us into their homes and sharing their wisdom regarding the forest. H. Baker, T. Fleming, F. Pitelka, M. Power, D. Steadman and G. Wiles reviewed earlier versions of the manuscript. We dedicate this paper to the memory of Raymond Fosberg for his friendship and his invaluable contributions to understanding of Pacific island environments.

References

Amerson, A. B., Jr., Whistler, W. A. & Schwaner, T. D. (1982). *Wildlife and wildlife habitat of American Samoa*. II. U. S. Fish and Wildlife Service, Washington, D C.

Baker, H. G. & Harris, B. J. (1957). The pollination of *Parkia* by bats and its attendant evolutionary problems. *Evolution* **11**: 449–460.

Balouet, J.-C., & Olson, S. L. (1989). Fossil birds from Late Quaternary deposits in New Caledonia. *Smithsonian Contr. Zool.* No. 469: 1–38.

Bonaccorso, F. J. & Humphrey, S. R. (1984). Fruit bat niche dynamics: their role in maintaining tropical forest diversity. In *Tropical rain-forest: the Leeds symposium*: 169–183. (Eds Chadwick, A. C. & Sutton, S. L.). Leeds Philosophical and Literary Society, Leeds.

Brown, F. B. H. (1931). Flora of Southeastern Polynesia I. *Bull. Bernice P. Bishop Mus.* **84**: 1–194.

Brown, F. B. H. (1935). Flora of Southeastern Polynesia II. *Bull. Bernice P. Bishop Mus.* **130**: 1–386.

Cox, P. A. (1983). Observations on the natural history of Samoan bats. *Mammalia* **47**: 519–523.

Cox, P. A. (1984). Chiropterophily and ornithophily in *Freycinetia* (Pandanaceae) in Samoa. *Pl. Syst. Evol.* **144**: 277–290.

Cox, P. A., Elmqvist, T., Pierson, E. D. & Rainey, W. E. (1991). Flying foxes as strong interactors in South Pacific island ecosystems: a conservation hypothesis. *Conserv. Biol.* **5**: 448–454.

Cox, P. A., Elmqvist, T., Pierson, E. D. & Rainey, W. E. (1992). Flying foxes as pollinators and seed dispersers in Pacific island ecosystems. *U. S. Fish Wildl. Serv. biol. Rep.* **90**: 18–23.

Crome, F. H. J. & Irvine, A. K. (1986). 'Two bob each way': the pollination and breeding system of the Australian rain forest tree *Syzygium cormiflorum* (Myrtaceae). *Biotropica* **18**: 115–125.

Eby, P. (1991). 'Finger-winged night workers': managing forests to conserve the role of grey-headed flying foxes as pollinators and seed dispersers. In *Conservation of Australia's forest fauna*: 91–100. (Ed. Lunney, D.). Royal Zoological Society, Mosman, NSW.

Elmqvist, T., Cox, P. A., Rainey, W. E. & Pierson, E. D. (1992). Restricted pollination on oceanic islands: pollination of *Ceiba pentandra* by flying foxes in Samoa. *Biotropica* **24**: 15–23.

Elmqvist, T., Rainey, W. E., Pierson, E. D. & Cox, P. A. (1994). Effects of the two tropical cyclones 'Ofa' and 'Val' on the structure of a Samoan lowland rainforest. *Biotropica* **26**: 384–391.

Fleming, T. H. (1979). Do tropical frugivores compete for food? *Am. Zool.* **19**: 1157–1172.

Fleming, T. H. (1988). *The short-tailed fruit bat: a study in plant–animal interactions.* University of Chicago Press, Chicago.

Fleming, T. H. (1993). Plant-visiting bats. *Am. Scient.* **81**: 460–467.

Fleming, T. H., Breitwisch, R. & Whitesides, G. H. (1987). Patterns of tropical vertebrate frugivore diversity. *A. Rev. Ecol. Syst.* **18**: 91–109.

Francis, C. M. (1990). Trophic structure of bat communities in the understorey of lowland dipterocarp rain forest in Malaysia. *J. trop. Ecol.* **6**: 421–431.

Fujita, M. S. & Tuttle, M. D. (1991). Flying foxes (Chiroptera: Pteropodidae): threatened animals of key ecological and economic importance. *Conserv. Biol.* **5**: 455–463.

Gould, E. (1978). Foraging behaviour of Malaysian nectar-feeding bats. *Biotropica* **10**: 184–193.

Heithaus, E. R., Fleming, T. H. & Opler, P. A. (1975). Foraging patterns and resource utilization in seven species of bats in a seasonal tropical forest. *Ecology* **56**: 841–854.

Howe, H. F. (1986). Seed dispersal by fruit-eating birds and mammals. In *Seed dispersal*: 123–189. (Ed. Murray, D. R.). Academic Press, Sydney.

Kitchener, D. J., Gunnell, A. & Maharadatunkamsi (1991). Aspects of the feeding biology of fruit bats (Pteropodidae) on Lombok island, Nusa Tenggara, Indonesia. *Mammalia* **54**: 561–578.

Kress, W. J. (1985). Bat pollination of an Old World *Heliconia*. *Biotropica* **17**: 302–308.

MacArthur, R. H. & Wilson, E. O. (1967). *The theory of island biogeography.* Princeton University Press, Princeton, N. J.

Marshall, A. G. (1983). Bats, flowers and fruit: evolutionary relationships in the Old World. *Biol. J. Linn. Soc.* **20**: 115–135.

Marshall, A. G. (1985). Old World phytophagous bats (Megachiroptera) and their food plants: a survey. *Zool. J. Linn. Soc.* **83**: 351–369.

McCoy, M. (1990). Pollination of eucalypts by flying foxes in northern Australia. In *Flying fox workshop proceedings*: 33–37. (Ed. Slack, J. M.). New South Wales Dept. Agriculture and Fisheries, Sydney.

McWilliam, A. N. (1985–86). The feeding ecology of *Pteropus* in north-eastern New South Wales, Australia. *Myotis* **23–24**: 201–208.

Milberg, P. & Tyrberg, T. (1993). Naive birds and noble savages—a review of man-caused prehistoric extinctions of island birds. *Ecography* **16**: 229–250.

Mills, L. S., Soulé, M. E. & Doak, D. F. (1993). The keystone-species concept in ecology and conservation. *BioScience* **43**: 219–224.

Paine, R. T. (1966). Food web complexity and species diversity. *Am. Nat.* **100**: 65–75.

Paine, R. T. (1969). A note on trophic complexity and community stability. *Am. Nat.* **103**: 91–93.

Paine, R. T. (1992). Food-web analysis through field measurement of per capita interaction strength. *Nature, Lond.* **355**: 73–75.

Palmeirim, J. M., Gorchov, D. L. & Stoleson, S. (1989). Trophic structure of a Neotropical frugivore community; is there competition between birds and bats? *Oecologia* **79**: 403–411.

Payne, J., Francis, C. M. & Phillips, K. (1985). *A field guide to the mammals of Borneo*. Sabah Society, Kota Kinabalu.

Pierson, E. D., Elmqvist, T., Rainey, W. E. & Cox, P. A. (In press.) Effects of tropical cyclonic storms on flying fox populations on the South Pacific islands of Samoa. *Conserv. Biol.*

Pratt, H. D., Bruner, P. L. & Berrett, D. G. (1987). *The birds of Hawaii and the tropical Pacific*. Princeton University Press, Princeton.

Rainey, W. E. & Pierson, E. D. (1992). Distribution of Pacific island flying foxes: implications for conservation. *U. S. Fish Wildl. Serv. biol. Rep.* **90**: 111–122.

Richards, G. C. (1990). The spectacled flying-fox, *Pteropus conspicillatus* (Chiroptera: Pteropodidae), in north Queensland. 2. Diet, seed dispersal and feeding ecology. *Aust. Mammal.* **13**: 25–31.

Simberloff, D. (1991). Keystone species and community effects of biological introductions. In *Assessing ecological risks of biotechnology*: 1–19. (Ed. Ginzberg, L. R.). Butterworth-Heinemann, Stoneham, MA.

Smythe, N. (1989). Seed survival in the palm *Astrocaryum standleyanum*: evidence for dependence on its seed dispersers. *Biotropica* **21**: 50–56.

Smythies, B. E. (1984). *The birds of Borneo*. (4th edn). The Sabah Society, Kota Kinabalu, Sabah and the Malayan Nature Society, Kuala Lumpur.

Start, A. N. & Marshall, A. G. (1976). Nectarivorous bats as pollinators of trees in west Malaysia. In *Tropical trees: variation, breeding, and conservation*: 141–150. (Eds Burley, J. & Styles, B. T.). Academic Press, London.

Steadman, D. W. (1989). Extinction of birds in eastern Polynesia: a review of the record, and comparisons with other Pacific island groups. *J. archaeol. Sci.* **16**: 177–206.

Steadman, D. W. (1992a). Extinct and extirpated birds from Rota, Mariana Islands. *Micronesica* **25**: 71–84.

Steadman, D. W. (1992b). New species of *Gallicolumba* and *Macropygia* (Aves, Columbidae) from archaeological sites in Polynesia. *Los Angeles nat. Hist. Mus., Sci. Ser.* **36**: 329–348.

Steadman, D. W. (1993). Biogeography of Tongan birds before and after human impact. *Proc. natn. Acad. Sci. U.S.A.* **90**: 818–822.

Steadman, D. W. & Kirch, P. V. (1990). Prehistoric extinction of birds on Mangaia, Cook Islands, Polynesia. *Proc. natn. Acad. Sci. U.S.A.* **87**: 9605–9609.

Tidemann, C. R. (1987). Notes on the flying-fox, *Pteropus melanotus* (Chiroptera: Pteropodidae), on Christmas Island, Indian Ocean. *Aust. Mammal.* **10**: 89–91.

Utzurrum, R. C. B. (1989). Fruit consumption and seed dispersal of three fig species by frugivorous bats (Pteropodidae) in a Philippine primary forest. *Bat Res. News* **30**: 80.

van Balgooy, M. M. J. (1971). Plant-geography of the Pacific. *Blumea (Suppl.)*. **6**: 1–222.

Van der Pijl, L. (1957). The dispersal of plants by bats (chiropterochory). *Acta bot. neerl.* **6**: 291–315.

Van der Pijl, L. (1982). *Principles of dispersal in higher plants*. (3rd edn). Springer-Verlag, Berlin.

Wheelwright, N. T. (1985). Fruit size, gape width, and the diets of fruit-eating birds. *Ecology* **66**: 808–818.

Whistler, W. A. (1992). The vegetation of Samoa and Tonga. *Pacif. Sci.* **46**: 159–178.
Whittaker, R. J. (1992). Stochasticism and determinism in island ecology. *J. Biogeogr.* **19**: 587–591.
Wiles, G. J. (1987). The status of fruit bats on Guam. *Pacif. Sci.* **41**: 1–4.
Wiles, G. J. & Fujita, M. S. (1992). Food plants and economic importance of flying foxes on Pacific islands. *U.S. Fish Wildl. Serv. biol. Rep.* **90**: 24–35.
Wong, M. (1986). Trophic organization of understory birds in a Malaysian dipterocarp forest. *Auk* **103**: 100–116.

Symp. zool. Soc. Lond. (1995) No. 67: 63–77

Feeding ecology of Philippine fruit bats: patterns of resource use and seed dispersal

Ruth C. B. UTZURRUM[1]

Center for Tropical Conservation Studies
Silliman University
6200 Dumaguete City
Philippines

Synopsis

Information on diet composition of, feeding patterns in, and seed dispersal by pteropodids was gathered in a primary tropical rain forest area in southern Negros Island, Philippines. Seeds and marked remains of food plants were collected: (1) in 1 m × 1 m seed traps laid out at 3-m intervals along four orthogonal transects around selected fruiting figs, and (2) randomly from the forest floor within a 100 m × 100 m area surrounding these trees. Identity of consumers was determined directly through visual observation and from netting around sampling trees and/or indirectly through analysis of palatal imprints on ejecta. Food habits were recorded for nine species of fruit bats: *Acerodon jubatus, Cynopterus brachyotis, Haplonycteris fischeri, Harpyionycteris whiteheadi, Nyctimene rabori, Ptenochirus jagorii, Pteropus hypomelanus, P. pumilus* and *P. vampyrus*. Figs (F. Moraceae) were found to be a major part of the diet of these bats in the area: at least 11 out of 30 *Ficus* species identified from the area were being eaten in varying amounts. Other components of the diet included fruits of 11 other plant species belonging to seven families. There was some evidence of consumption of leaves and moss. In general, data suggest differential choice of fruit along lines of fruit colour (i.e., bright versus dull-coloured fruits), on the basis of available crop density and on plant height. No size relationships between bats and the fruits consumed were evident.

Analysis of seed shadow patterns strongly indicates that feeding patterns differ between large-sized bat species (> 150 g body mass) and smaller species. Most splats and ejecta generated by large bats were concentrated beneath the crown of the source tree, indicating consumption of fruits on the tree itself. In contrast, smaller bats used feeding roosts in the immediate vicinity of, but not in, a focal feeding tree. Mixed-species seed clumps were found beneath these feeding roosts. This difference in behavioural patterns of fruit consumption causes a more heterogeneous pattern of seed scattering than would be expected of a single-taxon (i.e. fruit bats as a whole)

[1]Current address: Department of Biology, Boston University, 5 Cummington Street, Boston, MA 02215, USA.

ZOOLOGICAL SYMPOSIUM No. 67
ISBN 0 19 854945 8

visitor coterie. The potential importance of these fruit bats as seed dispersers in forests is further supported by seed germination tests; gut-passed seeds of *Ficus chrysolepis* show higher percentages of germination than do seeds from fruits or ejecta.

Introduction

Current data on the feeding ecology of pteropodid bats in the Philippines consist almost entirely of enumeration of food resources in secondary growth areas and cultivated fields and orchards (Guerrero & Alcala 1973; Alcala 1976; Rabor 1977). Except for an assessment of the impact of bat foraging activities on orchards (Guerrero & Alcala 1973), documentation and analysis of patterns of fruit resource use in forest habitats are lacking.

Not surprisingly, virtually nothing is known of the relationships between these frugivores and the plant communities with which they interact; thus, their role in forest maintenance and regeneration in the Philippines is largely unknown. Recent studies in the Philippines, however, are beginning to demonstrate strong relationships between forest habitats and fruit bat diversity and abundance (Heaney, Heideman & Mudar 1981; Heaney, Heideman, Rickart, Utzurrum & Klompen 1989). Most striking is an emerging dichotomy between geographically widespread species that are highly abundant in disturbed and cultivated areas and the Philippine endemics that are largely restricted to forest habitats. Given this relationship, a better understanding and knowledge of food requirements and patterns of resource use in forest areas are of critical importance for at least two reasons. Firstly, shifts in patterns of resource (food and roost) availability may accompany the conversion of forest habitats to agricultural cultivation, and exploitative extraction of forest products could seriously reduce the ability of forest-dwelling species to sustain viable populations through time. Secondly, any decline in numbers and diversity in these bat populations and assemblages could potentially affect the distribution and reproductive biology of those plants that are primarily bat-dispersed or bat-pollinated.

This study documents fruit use by Philippine fruit bats in primary forest habitats. I identified plant species consumed, and examined parameters that characterize patterns of resource use among the species of fruit bats present. I also assessed the impact and potential consequences of fruit consumption on the dispersal of seeds by analysing seed shadow patterns and seed germination tests. Seed dispersal analysis was centred primarily on *Ficus* (fig) species. Although figs are not among the resources listed in published diets of Philippine fruit bats, initial surveys of the forest area studied indicated their importance in the animals' diet.

Data collection and analysis

The study was conducted in a 320-ha forest area surrounding Lake Balinsasayao (9° 21′ N, 123° 10′ E) in the southern tip of Negros Island, Philippines.

Vegetation was typically of a lowland forest type with some montane elements from the level of the lake (850 m) to about 1050 m in elevation. Forest canopy averaged 15 to 20 m in height, although emergents of up to 35 to 40 m were present. The dominant tree species were *Shorea polysperma* and *Syzigium nitidum* (Antone 1983). Small patches of cultivated clearings, typically planted with cash crops (such as *Seachium edule*) and abaca (*Musa textiles*), were present at the edges of, and within, the forest. Drought conditions encompassed the period of the study, with mean monthly rainfall estimated to be 180.1 mm (Heideman & Erickson 1987).

Intensive field data collection was conducted for six months from April to September 1983, with periodic visits made in the months before and after this period over a span of one year. Ten focal sampling trees of the genus *Ficus* were initially selected, primarily on the basis of fruiting condition, i.e. the trees were in early to mid-stages of fruiting, with no apparent evidence of fruit consumption. Fruit samples in various stages of development were obtained and evaluated for size, wet mass, hardness and colour. Herbarium samples of bat-consumed plants were collected and submitted to the Philippine National Museum of Natural History in Manila for identification. Fig identifications follow Corner (1965).

Four 1 m (wide) × 25 m (long) belt transects were established from the base of each sampling tree radiating outwards in four general directions (N, E, S, W). Plastic sheets (1 m × 1 m), or 'seed traps', were laid out at every third metre along the transect length beginning from the tree base. Seed samples were collected from traps daily from the onset to the end of feeding visits. 'Night samples' consisted of those collected prior to 0800, after sheets were cleaned at dusk of the previous day. 'Day samples' consisted of those that accumulated between 0800 and dusk. Samples included partially consumed fruits or non-seed fruit parts and 'seed aggregates', i.e. defecated or spat-out pellets containing one or more intact seeds. Bat faecal material is referred to hereafter as 'splats' (after Janzen 1978) and the masticated pellets as 'ejecta'. All samples were examined for seed content to determine plant source. Additional data on diet composition were recorded from analysis of faecal material from some captured bats. Night samples were quantified and categorized into size classes of large, medium or small, to indicate size class of bat consumer. These categories reflect discrete subsets of species whose palatal/dental characteristics (distance measurements and dental formula) are clustered within subsets but distinct between groups (Utzurrum 1984). These subsets included three large-sized species, three that were medium-sized and two that were small-sized (Table 1). One species was 'intermediate' between large and medium classes.

Consumer identity was determined directly through visual observations (especially for diurnal feeders) and indirectly through sporadic two- to three-hour netting (between 1900 and 2200) around sampling trees and the analysis of dental and/or palatal imprints on ejecta and fruit parts. Information on

Table 1. Species checklist of pteropodid fruit bats by size categories[a]

Species	Average forearm length (mm)	Body mass mass (g)	Size category[b]
Acerodon jubatus[c]	200	1000+	Large
Pteropus hypomelanus	138	450	Large
Pteropus vampyrus	190	1000	Large
Pteropus pumilus[c]	110	180	Intermediate
Harpyionycteris whiteheadi[c]	84	100	Medium
Nyctimene rabori[c]	75	60	Medium
Ptenochirus jagorii[c]	81	80	Medium
Cynopterus brachyotis	63	35	Small
Haplonycteris fischeri[c]	47	20	Small

[a] Excluding 3 principally pollen- and nectar-feeding species: *Eonycteris robusta*[c], *E. spelaea, Macroglossus minimus*, and the typically non-forest species, *Rousettus amplexicaudatus*.
[b] See text (p. 65) for explanation.
[c] Species endemic to the Philippines.

consumers compared across a range of fruit types and characteristics was used to model patterns of fruit choice among the various bat species.

Three descriptive parameters are presented herein to characterize seed dispersal activities of the fruit bats: (1) the density–distance function (i.e. seed density index versus distance from tree source) of the seed shadows around three of the 10 sampling trees; (2) species composition of samples in seed traps; and (3) seed viability based on germination tests. The seed density index is derived from log-transformed counts of seed aggregates. Seed aggregate counts per square metre were first adjusted to correct for proportional oversampling closer to the tree base, as the proportion of area sampled to total area decreases with distance from tree base as a function of circular area. Thus, counts at each trap were multiplied by a factor, X, where $X = [1\ m^2\ (\pi\ r_i^2 - A_p)/(\pi\ r_b^2 - A_t)]$, and where r_b = radius from base of tree to the first square, r_i = radius from base of tree to square to be adjusted, A_t = area of tree base, and A_p= area of sphere from midpoint of tree base to radius preceding square to be adjusted. All adjusted density counts were then log-transformed, $\log_{10}$ (count + 1), because of high variance in counts at each transect (Sokal & Rohlf 1981). Only seeds of the focal sampling tree are included in the seed shadow curves. 'Foreign seed aggregates', i.e. those belonging to plant species other than the species being sampled, are noted separately.

Seed germination tests were done on *F. chrysolepis* following procedures described in Utzurrum (1984) and Utzurrum & Heideman (1991). The trials included 20 replicates of fruit seeds and 16 replicates each of splat and ejecta seeds. Each replicate contained 30 seeds. Mean germination percentages of these three seed types are compared by means of the Kruskal-Wallis test (Sokal & Rohlf 1981).

Results

Fruit consumption patterns

A total of 22 species of food plants were identified from fruit remains, splats and ejecta. These included 11 species of *Ficus* (F. Moraceae; Table 2) and 11 other species belonging to seven plant families (Table 3). These plants ranged from trees to shrubs and included understorey vines (*Piper* spp.) and lianas (*Freycinetia* spp.). Of these 22 fruit resources, 54.5% bore green-coloured fruits, 22.7% had orange to bright red fruits, 18.2% were coloured deep red to dark purple (near black) and the remaining 4.5% (or 1) dull yellow-brown (Tables 2 and 3). Not all bats ate all these resources, but broad overlaps in diets were evident. Figs were found to be the major component of the bats' diets, especially of *A. jubatus, C. brachyotis, H. fischeri, N. rabori* and *P. jagorii.* Fruiting figs were abundant throughout the study period, and consumption of syconia did not vary quantitatively between months when no other resource of similar abundance was available (April to June) and months when fruits of *Syzigium* and *A. simplicifolia* were abundant (July to September). Fig consumption by frugivorous bats has been documented mainly in the Neotropics (Vásquez-Yanes, Orozco, François & Trejo 1975; Fleming, Heithaus & Sawyer 1977; Bonaccorso 1979; Janzen 1979; Morrison 1980; August 1981) and in some parts of the Old World tropics (Marshall 1983, 1985). For Philippine pteropodids these are the first documentations of the species consumed and the species of the consumer.

The 10 fig trees selected for intensive sampling were: *F. chrysolepis* (*n* =4), *F. crassiramea* (*n*=2), *F. pubinervis* (*n*=2), *F. sumatrana* (*n*=1), and *F.* cf. *variegata* var. *sycomoroides* (*n*=1). Two of the trees aborted their crops prior to frugivore visits, including one of the *F. chrysolepis* and one *F. crassiramea.* No evidence of bat visits was ever recorded for *F. sumatrana.* Very few, and only large-sized, seed aggregates of *F. crassiramea* were collected despite a qualitatively large crop, although moderate to heavy consumption of this fig by large bats has previously been noted (P.D. Heideman pers. comm.). Syconia of both *F. crassiramea* and *F. sumatrana* were coloured bright red, were relatively mushy upon ripening and had thin fruit walls (flesh). The other three species studied underwent heavy frugivore visits. Syconia of these species remained green upon ripening and retained a relatively firm consistency upon softening. Two species of fruit doves (*Phapitreron amethystina* and *P. leucotis*) and a hornbill (*Penelopides panini*) were recorded feeding on *F. crassiramea* on two occasions; day samples indicated the occurrence of other avian feeding visits. Flowerpeckers (*Dicaeum* spp.) and *P. amethystina* were seen feeding on *F. sumatrana* syconia. In all recorded instances of avian feeding, consumption took place directly on the tree. Visits lasted no more than 15 min. A group of four macaques (an adult pair and two juveniles) visited one of the fruiting

Table 2. Checklist of bat food resources in the Family Moraceae

Species	Syconium colour (basic)	Size index (L × W) (cm^2)	Tree height category	Crop abundance	Consumer size category[a]
Ficus benjamina	Red to purple	0.5	Canopy	High	Large
F. crassiramea	Bright red	3.8	Canopy	High	Large
F. obscura	Red to purple	0.4	Canopy	Moderate	Large
F. recurva	Red	0.3	Canopy	Moderate	Large
F. aurantiacea var. *parvifolia*	Yellow-brown	63.0	Canopy to emergent	Low	Medium, small
F. chrysolepis	Green	8.1	Subcanopy/canopy	Moderate to high	Large to small
F. congesta	Pale green	4.4	Understorey	Low to moderate	Medium, small
F. nota	Green	1.7	Understorey	Low to moderate	Medium, small
F. pubinervis	Green	3.2	Canopy/emergent	High	Large to small
F. septica	Green	4.8	Understorey	Low to moderate	Medium, small
F. cf. *variegata* var. *sycomoroides*	Green	4.0	Subcanopy/canopy	High	Large to small

[a]See text (p. 65) for explanation of size categories.

Table 3. Checklist of bat food resources: non-fig

Species	Fruit colour (basic)	Size index (L × W) (cm^2)	Tree height category	Crop abundance	Consumer size category[a]
Euphorbiaceae:					
Aparosa simplicifolia	Green	1.8	Understorey	Moderate to high	Medium, small
Melastomaceae:					
Melastoma sp.	Deep purple	1.0	Shrub	Low to moderate	Medium, small
Musaceae:					
Musa sp. (wild)	Green	36.0	Understorey	Low	Medium, small
Myrtaceae:					
Syzigium sp.	Pale green	3.2	Canopy	High	Large to small
Palmae:					
Species 1	Deep purple	1.2	Understorey	Low to moderate	Medium, small
Species 2	Orange to red	10.8	Understorey/subcanopy	Low to moderate	Intermediate, medium
Pandanaceae:					
Freycinetia sp.1	Reddish orange	72.0	Subcanopy/canopy	Low	Intermediate, medium
Freycinetia sp.2	Red	9.0	Understorey/subcanopy	Low	Medium
Piperaceae:					
Species 1	Yellow green	1.6	Understorey	Low	Small
Species 2 and 3	Green	1.5	Understorey	Low	Medium, small

[a]See text (p. 65) for explanation of size categories.

F. chrysolepis for two successive days when syconia were still very immature, i.e. seeds were very poorly developed. Figs were picked at random and all were discarded unconsumed. These two visits lasted 30 and 40 min each. An estimated 15% reduction in crop size per visit was associated with these two bouts.

Seed dispersal patterns

Density–distance curves for four transects of two *Ficus* species were chosen from among seven trees monitored to illustrate the heterogeneous nature of the shadows and the factors that may have influenced these variations (Fig. 1). The west transect of *F. pubinervis* shows a pattern of seed fall typical of seed shadows described by Fleming & Heithaus (1981): the majority of the seed aggregates fell in areas beneath the tree crown (Fig. 1a). Overall, beneath-crown seed density accounted for 75.9% of the total seed aggregates collected in all the four transects of this tree. The east transect of *F. chrysolepis* likewise showed heavy seed fall close to the tree base (Fig. 1b). In this case, however, the area of high seed density was extended over a wider area by the presence of a fruiting conspecific tree whose crown (indicated by dashed lines) abutted that of the sampled individual. The majority of seed aggregates in these two transects were accounted for by large bats, as shown by the partitioned curves (Fig. 1a, b). A *G*-test comparing the fall of large-sized seeds beneath the crown with that beyond the crown in seven of the *Ficus* trees studied showed significantly higher densities beneath the tree crowns (*G* test: $G_{pooled} = 61.29$, $P < 0.05$; Sokal & Rohlf 1981). These indicate that large bats primarily consume fruits on the source tree itself; the finding is in contrast to the generalization, based on Neotropical studies, that fruit bats are 'shuttling' foragers (Heithaus & Fleming 1978; Morrison 1978), a generalization likewise applied to Palaeotropical frugivorous species (Thomas, Cloutier, Provencher & Houle 1988). A second pattern is one in which the density remained relatively high or higher beyond the extent of the tree crown. In the north transect of *F. pubinervis*, this was primarily effected by small and medium-sized bats (Fig. 1c). The spikes in small and medium-sized seed aggregates beyond the tree crown indicate that smaller bats use temporary feeding roosts for processing fruits and thus exhibit a 'shuttling' feeding behaviour as opposed to the 'sit and feed' behaviour of large bats. Beyond-crown seed aggregates for seven trees of these two size classes were significantly more numerous than were those within the extent of the crowns (*G* test: $G_{pooled} = 217.06$, $P < 0.05$; Sokal & Rohlf 1981). Use of temporary night-feeding roosts has been documented for frugivorous species of phyllostomids (Heithaus, Fleming & Opler 1975; Bonaccorso 1979).

The south transect of *F. chrysolepis* followed a steep downhill slope of 30–40°. Seed aggregates falling in areas in the upper slopes closer to the

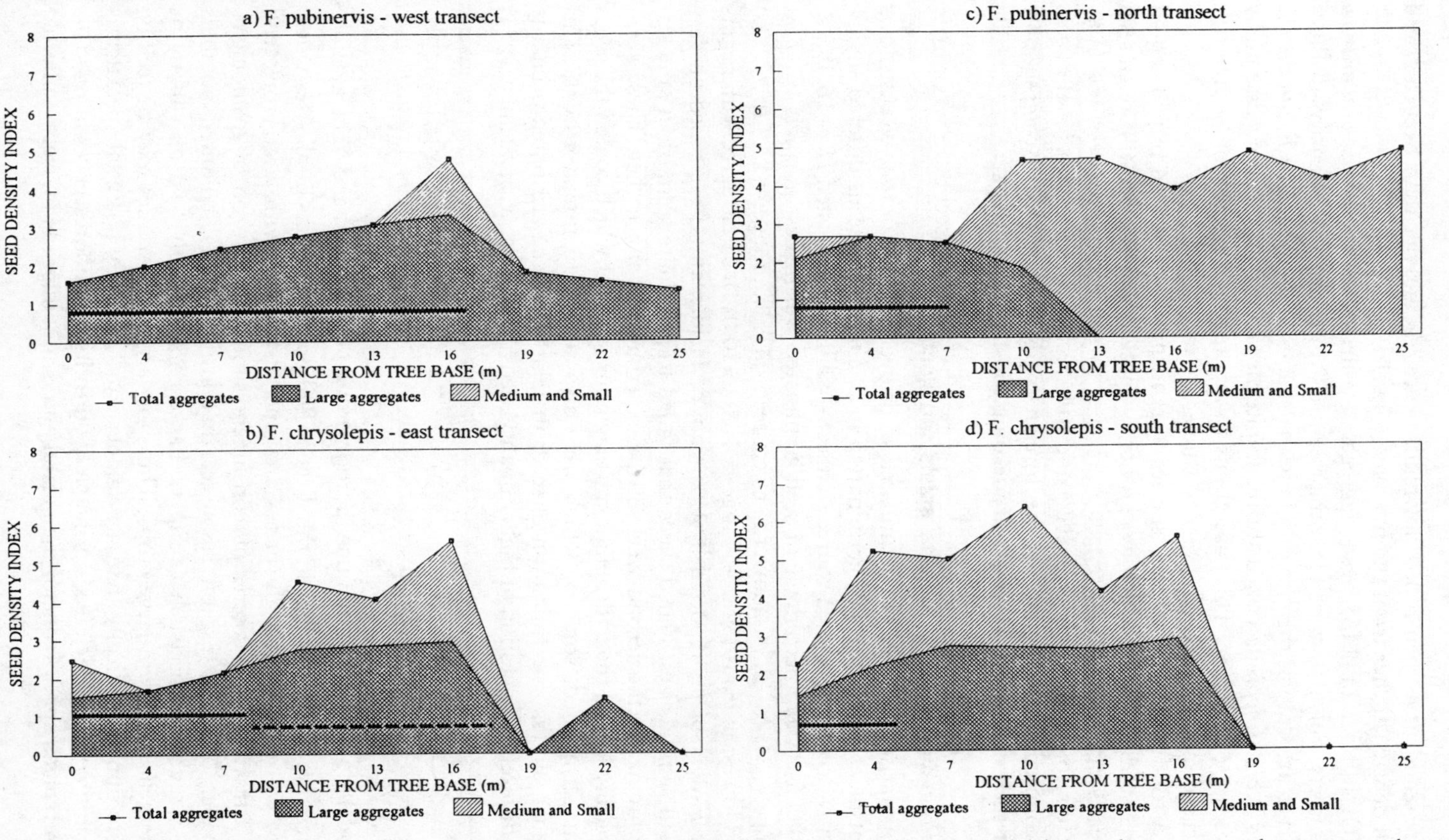

Fig. 1. Seed density–distance curves of four select transects of *Ficus chrysolepis* and *F. pubinervis*. (Solid lines indicate extent of tree crown along transect; dotted line represents crown of fruiting conspecific.)

tree could have then rolled downhill; hence, the heavier than expected seed fall density further from the tree base (Fig. 1d).

Of a total of 1465 seed aggregates collected from the *F. crassiramea* transects, only 1% were from the tree itself. Interestingly, if density–distance curves are drawn, the curves mimic the shapes and patterns of *F. pubinervis* and *F. chrysolepis* described above, despite the fact that the 'shadow' essentially was not of this tree. This case points to a potential weakness of 'source tree focused' analysis in seed dispersal studies.

A total of 13 foreign seed species were identified from seed traps from all trees studied, five of which were other fig species. The number of foreign seed species per tree shadow ranged from zero to ten. Ejecta of leaves and moss were also collected, as were ejecta and fruit peel of wild bananas. Smaller bats accounted for the majority of the 'foreign' loads. This provides strong support for the inferred use of temporary feeding roosts by smaller bats.

Germination of seeds in fruits, splats and ejecta

Comparison of counts of germinated seeds over a period of four weeks shows that mean percentages of germination in *F. chrysolepis* seeds differed significantly according to seed source (Kruskal–Wallis test: adjusted $H = 16.486$, $P < 0.005$). Mean percentage of seed germinated was highest in splats (71.7%; range = 3.3 to 73.3%), lower in ejecta seeds (47.7%; range = 13.3 to 93.3%) and lowest in fruits (32.5%; range = 10.0 to 70.0%). These agree with results for *F. chrysolepis* reported by Utzurrum & Heideman (1991) based on very small sample sizes. Plated seeds that failed to germinate were often attacked by fungus by the second week of the experiment, if not earlier. These seeds typically lacked the hydrophilic mucilaginous material which coated the seeds that germinated, some of which were also subject to fungal growth. This higher success in germination of seeds in splats has been attributed primarily to differential ingestion of higher quality seeds (Utzurrum & Heideman 1991; Janzen 1978).

Discussion

Modelling differential fruit use in frugivorous pteropodids

There are broad overlaps in the diets of Philippine fruit bats of all sizes where they co-occur. Almost all species are generalists. Trees with large crops tend to attract a wide coterie of consumers simultaneously on a given night. Smaller bats also exploit resources that are available in limited numbers on a nightly basis (such as *Piper*). Large bats primarily feed on canopy or emergent trees with large crops. Thus, the large-sized species may be feeding more opportunistically than are smaller bats. The only potential specialist is *H. whiteheadi*, which feeds principally on fruits and inflorescences of pandans (*Freycinetia* spp.). None of the palatal and dental imprints on ejecta collected could be reliably traced to this species. Additionally, the majority of the seeds in faeces retrieved directly from captured individuals in this and other studies (P. Heideman, pers. comm.) are of pandan.

An examination of the basic characteristics of plants and their fruit against the range of bat consumers recorded to feed on these fruits reveals differential patterns of fruit choice among bats of various sizes (see Tables 2 and 3). Plant height and available crop density per tree are traits critical for large bats. Small to medium-sized bats, on the other hand, appear to discriminate primarily on the basis of fruit colour. Fruit size itself is not a consistent indicator of preference, contrary to the relationships between fruit size and bat size in the Neotropics (Heithaus, Fleming & Opler 1975). However, small, bright-coloured fruits are typically absent in the diets of bats less than 50 g in mass. Dull coloration is part of the suite of fruit traits forming the so-called chiropterochorous syndrome (van der Pijl 1957, 1982; Howe 1986). Analyses of the patterns of occurrence of this syndrome are important for understanding the evolutionary relationships between fruit bats and their food plants (Fleming 1979; Marshall 1983). Often overlooked is the need to focus on proximate factors underlying the relationship between preferred intakes of fruits and the possession of chiropterochorous traits. This examination is of equal relevance if we are to understand the importance of specific resources in the maintenance of fruit bat populations, especially forest-dwelling species that do not typically exploit cultivated orchard fruits.

Herein I present a qualitative model of food choice (Fig. 2). Various fruit traits associated with choice are expressed as a universal variable, fruit energy content. Predictions are based on the assumptions that energy requirements are a major factor influencing nightly foraging bouts and that feeding behaviour at fruit sources is indicative of the immediate energy expenditure of the species. Hence, for large bats, the principal energy expenditure is the cost of long flights to and among feeding areas or trees, since they are energetically more conservative at feeding trees (they eat at the tree itself). They are basically constrained by a need to maximize energy intake at each feeding tree or area and to acquire the proportionately greater total amount of energy needed to maintain a large body mass. They are predicted to feed on fruits of low to high energy content provided that the aggregate total of energy available per visit roughly equals or exceeds the cost of the visit to the feeding tree or area (Fig. 2). This requirement will be met by those plant species which provide moderate to large fruit crops per night, provided that such plants possess crowns directly accessible to approaches from open flight spaces (as in canopy and emergent trees). Medium to small-sized bats, however, with their characteristic use of temporary feeding roosts and shorter commuting distances (Heideman & Heaney 1989), are predicted to have higher energy expenditures associated with shuttling from fruit source tree to feeding roosts, even when such movements are combined with occasional commuting between local feeding areas. Thus, these smaller species are predicted to be flexible in their choice along a wide range of crop density and other fruit traits, provided that energy intake per fruit will roughly equal or exceed the cost of shuttling between source and feeding roost (Fig. 2).

Additionally, I predict that the broad-scale dichotomy between bright-coloured figs and dull-coloured figs in the 'preference scale' of bats of all sizes in this study (with the latter apparently preferred over the former subset of available figs) will reflect qualitative and quantitative differences in the amounts of protein, fat and/or minerals present in these fruits. Fruit colour thus, coincidentally, indicates the underlying nutritional quality upon which choice is based and is not the basis of choice itself. Consumption of small fruits of low nightly availability is indicative of this sensitivity to the nutritional quality of fruits.

The types of comparative analyses needed to test these predictions relating energy and nutritional contents of food resources to fruit choice remain largely unexplored, especially in Palaeotropical frugivorous species of bats. Steller (1986) reported estimated daily energy and nitrogen requirements for *Pteropus poliocephalus*, but that estimation may not be truly reflective of diets in the wild. Although fruits rich in carbohydrates are quite common in the tropics, many are low in protein and nitrogen (Snow 1981; Fleming 1988), and available literature indicates that fruits typically consumed by bats are low in fat and protein content (e.g. Thomas 1982; Jordano 1983). I believe that this generalization requires re-examination, given that information on the nutritional quality is heavily biased by Neotropical data the majority of which are on bird diets. It remains to be seen whether the requirement for protein is in fact the factor that limits rates of consumption and, as a corollary, fruit choice, in pteropodid fruit bats, as proposed by Thomas (1984). No estimates

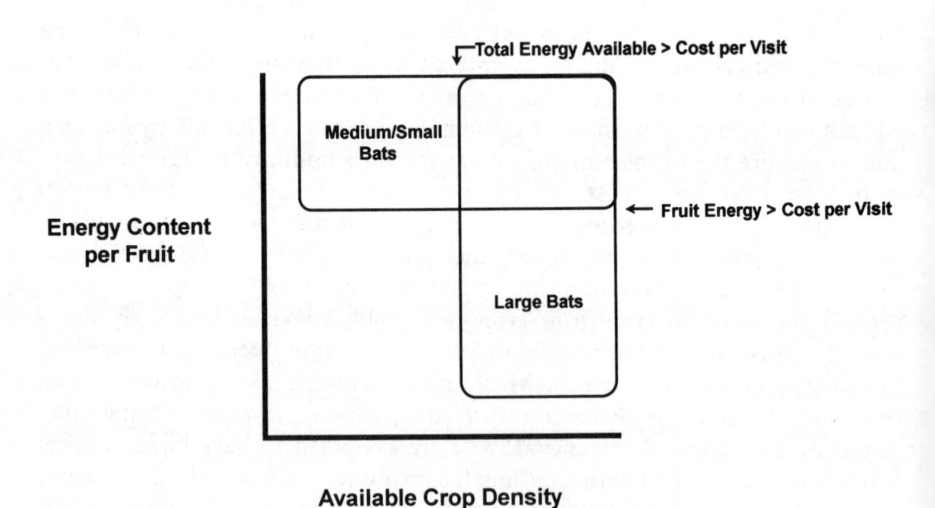

Fig. 2. Conceptual model of fruit choice in frugivorous pteropodids.

of energetic costs of foraging shuttles and commuting trips are available for pteropodids, although estimates for Neotropical species have been reported (e.g., Morrison 1978).

Seed dispersal and its implications for habitat maintenance and regeneration

The critical conclusion that may be inferred from the data on seed shadows and from the germination trials is that bats of different size categories contribute differentially to create a highly heterogeneous pattern of seed scattering. This heterogeneity is reflected in the form of mixed-species seed clumps within primary areas of seed fall from a feeding tree and in the variation in the shapes of seed-density curves. Seed shadows range from those characterized by high seed densities (principally single-peak shadows) beneath tree crowns (Fig. 1a, b), to shadows with extended tails of moderate seed density (Fig. 1c,d) and those with multiple local peaks (Fig. 1b, d). This heterogeneity is, perhaps, unique to the diversity of the bat dispersal coterie, with their attendant differences in feeding behaviour, and has been largely undocumented in Palaeotropical fruit bats prior to this study. Smaller bats, which are principally responsible for generating mixed-species seed plants, could strongly influence local patterns of seedling recruitment and regeneration. The importance of the tails of seed shadows for seedling recruitment and establishment have been discussed in recent literature (Coates-Estrada & Estrada 1988; Portnoy & Wilson 1993). If the observed numbers and types of non-bat tree visitors are indicative of the general disperser activity levels in this forest, especially for the subset of plant species studied, then bats could very well be the principal agent of seed dispersal in this and other sites. Although large bats deposit a majority of seeds in close proximity to source trees, they are probably important for successional patterns over a broader scale as they cover larger areas in their commuting to and from day roosts and between major feeding grounds (Rabor 1977; Heideman & Heaney 1989). The higher quality of seeds that are seen in splats reinforces the conclusion that large bats are important as agents of long-range dispersal, since the few seeds retained following feeding are those most likely to be taken over long distances.

Acknowledgements

P.D. Heideman and L.R. Heaney provided advice and support in the formulation and execution of this project. L. Tag-at provided valuable field assistance. The project was funded largely by a grant from the Faculty Development Committee of Silliman University, Philippines; special thanks to the late L.U. Ausejo. I especially thank Benito Tan and P.D. Heideman for help in resolving problems with fig identifications, J.O. Seamon for assistance in the preparation of this manuscript and P.D. Heideman for reviewing it.

References

Alcala, A. C. (1976). *Philippine land vertebrates: field biology*. New Day Publishers, Quezon City, Philippines.

Antone, M. S. (1983). *Vegetation analysis surrounding Lake Balinsasayao-Lake Danao, Negros Oriental, I. The tree species*. MS thesis: University of the Philippines at Los Banos.

August, P. V. (1981). Fig fruit consumption and seed dispersal by *Artibeus jamaicensis* in the Llanos of Venezuela. *Biotropica* **13** (Suppl.): 70–76.

Bonaccorso, F. J. (1979). Foraging and reproductive ecology in a Panamanian bat community. *Bull. Fla St. Mus. biol. Sci.* **24**: 359–408.

Coates-Estrada, R. & Estrada, A. (1988). Frugivory and seed dispersal in *Cymbopetalum baillonii* (Annonaceae) at Los Tuxtlas, Mexico. *J. trop. Ecol.* **4**: 157–172.

Corner, F. J. H. (1965). Check-list of *Ficus* in Asia and Australia with keys to identification. *Gdns' Bull., Singapore* **21**: 1–186.

Fleming, T. H. (1979). Do tropical frugivores compete for food? *Am. Zool.* **19**: 1157–1172.

Fleming, T. H. (1988). *The short-tailed fruit bat. A study in plant-animal interactions*. The University of Chicago Press, Chicago & London.

Fleming, T. H. & Heithaus, E. R. (1981). Frugivorous bats, seed shadows, and the structure of tropical forests. *Biotropica* **13** (Suppl.): 45–53.

Fleming, T. H., Heithaus, E. R. & Sawyer, W. B. (1977). An experimental analysis of the food location behavior of frugivorous bats. *Ecology* **58**: 619–627.

Guerrero, L. A. & Alcala, A. C. (1973). Feeding habits of pteropodid bats. *Philipp. Biota* **7**: 139–142.

Heaney, L. R., Heideman, P. D. & Mudar, K. M. (1981). Ecological notes on mammals of the Lake Balinsasayao region, Negros Oriental. *Silliman J.* **28**: 122–131.

Heaney, L. R., Heideman, P. D., Rickart, E. A., Utzurrum, R. B. & Klompen, J. S. H. (1989). Elevational zonation of mammals in the central Philippines. *J. trop. Ecol.* **5**: 259–280.

Heideman, P. D. & Erickson, K. R. (1987). The climate and hydrology of the Lake Balinsasayao watershed, Negros Oriental, Philippines. *Silliman J.* **34**: 82–107.

Heideman, P. D. & Heaney, L. R. (1989). Population biology and estimates of abundance of fruit bats (Pteropodidae) in Philippine submontane rainforest. *J. Zool., Lond.* **218**: 565–586.

Heithaus, E. R. & Fleming, T. H. (1978). Foraging movements of a frugivorous bat, *Carollia perspicillata* (Phyllostomatidae). *Ecol. Monogr.* **48**: 127–143.

Heithaus, E. R., Fleming, T. H. & Opler, P. A. (1975). Foraging patterns and resource utilization in seven species of bats in a seasonal tropical forest. *Ecology* **56**: 841–854.

Howe, H. F. (1986). Seed dispersal by fruit-eating birds and mammals. In *Seed dispersal*: 123–189. (Ed. Murray, D. R.). Academic Press, New York.

Janzen, D. H. (1978). A bat-generated fig seed shadow in rainforest. *Biotropica* **10**:121.

Janzen, D. H. (1979). How to be a fig. *A. Rev. Ecol. Syst.* **10**: 13–51.

Jordano, P. (1983). Fig-seed predation and dispersal by birds. *Biotropica* **15**: 38–41.

Marshall, A. G. (1983). Bats, flowers and fruit: evolutionary relationships in the Old World. *Biol. J. Linn. Soc.* **20**: 115–135.

Marshall, A. G. (1985). Old World phytophagous bats (Megachiroptera) and their food plants: a survey. *Zool. J. Linn. Soc.* **83**: 351–369.

Morrison, D. W. (1978). Foraging ecology and energetics of the frugivorous bat *Artibeus jamaicensis*. *Ecology* **59**: 716–723.

Morrison, D. W. (1980). Efficiency of food utilization by fruit bats. *Oecologia* **45**: 270–273.

Portnoy, S. & Wilson, M. F. (1993). Seed dispersal curves: behavior of the tail of the distribution. *Evol. Ecol.* 7: 25–44.

Rabor, D. S. (1977). *Philippine birds and mammals*. University of the Philippines Press, Quezon City.

Snow, D. W. (1981). Tropical frugivorous birds and their food plants: a world survey. *Biotropica* **13**: 1–14.

Sokal, R. R. & Rohlf, F. J. (1981). *Biometry: the principles and practice of statistics in biological research*. (2nd edn). W. H. Freeman and Company, New York.

Steller, D. C. (1986). The dietary energy and nitrogen requirements of the grey-headed flying fox, *Pteropus poliocephalus* (Temminck) (Megachiroptera). *Aust. J. Zool.* **34**: 339–349.

Thomas, D. W. (1982). *Ecology of an African savannah fruit bat community: resource partitioning and role in seed dispersal*. PhD thesis: University of Aberdeen.

Thomas, D. W. (1984). Fruit intake and energy budgets of frugivorous bats. *Physiol. Zool.* **57**: 457–467.

Thomas, D. W., Cloutier, D., Provencher, M. & Houle, C. (1988). The shape of bird- and bat-generated seed shadows around a tropical fruiting tree. *Biotropica* **20**: 347–348.

Utzurrum, R. C. B. (1984). *Fig fruit consumption and seed dispersal by frugivorous bats in the primary tropical rain forest of Lake Balinsasayao, Negros Oriental, Philippines*. MS thesis: Silliman University.

Utzurrum, R. C. B. & Heideman, P. D. (1991). Differential ingestion of viable vs non-viable *Ficus* seeds by fruit bats. *Biotropica* **23**: 311–312.

van der Pijl, L. (1957). The dispersal of plants by bats (chiropterochory). *Acta bot. neerl.* **6**: 291–315.

van der Pijl, L. (1982). *Principles of dispersal in higher plants*. Springer-Verlag, Berlin.

Vázquez-Yanes, C., Orozco, A., François, G. & Trejo, L. (1975). Observations on seed dispersal by bats in a tropical humid region in Vera Cruz, Mexico. *Biotropica* **7**: 73–76.

Symp. zool. Soc. Lond. (1995) No. 67: 79–96

A review of ecological interactions of fruit bats in Australian ecosystems

G. C. RICHARDS

Division of Wildlife and Ecology
CSIRO
Canberra, Australia

Synopsis

The 13 species of fruit bats in Australia can be separated into five ecological groups: large-sized (>300 g) and small-sized (<60 g) specialist frugivores and nectarivores, and large-sized generalists. Each group contains only one abundant species and one or more rare species. Large nectarivores have a wide distribution which is related to the extent and species diversity of eucalypt forests and woodlands. The frugivorous species have restricted distributions that correspond to the similarly restricted and diminished distribution of rainforest. All identifiable ecological niches available in forest in Australia appear to be filled by at least one frugivorous and nectarivorous species.

Specialized frugivores locate food visually, and for one (*Pteropus conspicillatus*), the dominant dietary components are light-coloured fruits that are prominent against the dark background of the upper canopy of rainforest. The distribution of colony sites of this species is either within or adjoining tropical rainforest. The small tube-nosed bat *Nyctimene robinsoni* appears to forage only in the sub-canopy zone of rainforest and does not roost in colonies, opting rather to roost by day camouflaged within or near the last food tree used the previous night.

Reduced molariform dentition that reflects the absence of mastication of food has evolved in specialized nectarivores such as *Pteropus scapulatus* and two species of macroglossine bats. Large and highly mobile nectarivores locate food (primarily *Eucalyptus* blossom) using olfaction, with broad-scale movements over hundreds of kilometres being related to mass flowering. Eucalypt flowering patterns in north Queensland show that droughts can cause reduced nectar production or flowering failure. For *P. scapulatus* this explains occasional natural mortality coincident with migration to coastal regions. Since flowering failure occurs after 3–4 consecutive months of below-average rainfall, broad-scale migrations of this species can be predicted.

The small amount of research that has been conducted to date shows that, although a broad group of pollinators other than bats is available for them, a large suite of *Eucalyptus* species are dependent upon flying-foxes as their major

ZOOLOGICAL SYMPOSIUM No. 67
ISBN 0 19 854945 8

source of outcrossed pollen. Similarly, the successful regeneration of many rainforest trees is contingent upon seed dispersal by large frugivorous bats; these trees are predominantly those having light-coloured fruits. Successful seed dispersal relates to feeding territoriality; a 'raiders versus residents' concept indicates that the invasion of feeding territories results in the carriage of large propagules away from their parent tree and over long distances. This is especially important in the transmission of genes between different tracts of fragmented rainforest. It is suggested that in the long term the proportion of trees in Australia's tropical rainforest having light-coloured fruits may gradually decline. Ecological mutualism is shown to exist between pteropodid bats and the native forests of Australia.

Introduction

Bats are thought to have colonized the Australian land mass during the Pleistocene when it was connected to New Guinea and ocean barriers with Asia were shorter than they are today (Holloway & Jardine 1968; Hand 1984; Hall 1984). However, Archer, Hand & Godthelp (1991) consider that the Pteropodidae may not have entered Australia until the Quaternary, and support the theory that the primary source is undoubtedly New Guinea. The present day pteropodid fauna reflects this ancestry.

The Family Pteropodidae is represented in Australia by 13 species of fruit bats in five genera, with all but four species having an extralimital distribution into New Guinea. Of the 43 genera and 173 species of Pteropodidae in the Old World, only 12% and 7% respectively are found in Australia. There are only four endemic pteropodids in Australia (*Pteropus poliocephalus, P. brunneus, Pteropus* sp. nov. and *Nyctimene robinsoni*), less than 3% of the world's total.

Although Australia is a large continent, only a small proportion of its 7.6 million square kilometres is inhabited by fruit bats. Suitable habitat is generally found only along eastern and northern coastal regions, in subtropical and tropical environments. Much of this habitat is dominated by myrtaceous eucalypt forest and woodland, but rainforest is also present in scattered fragments. The relationship between the distribution patterns of fruit bats and these types of forests leads to the hypothesis that ecological mutualism exists between the two, a hypothesis to be tested in this review.

Species of fruit bats in Australia

The 13 species of fruit bats recorded in Australia are shown in Table 1. One of these, *P. brunneus*, is considered to be extinct. Eight species (in the genera *Pteropus* and *Dobsonia*) are large and exceed 300 g as adults, and five species (*Nyctimene, Syconycteris* and *Macroglossus*) are small, weighing less

Table 1. Species of fruit bats recorded from Australia and their IUCN conservation status, as determined by Richards & Hall (in press).

Species		Status
Large fruit bats (Flying-foxes)		
Pteropus poliocephalus	Grey-headed flying-fox	**Vulnerable**
Pteropus scapulatus	Little red flying-fox	–
Pteropus alecto	Black flying-fox	–
Pteropus conspicillatus	Spectacled flying-fox	**Vulnerable**
Pteropus macrotis	Large-eared flying-fox	**Rare**
Pteropus brunneus	Dusky flying-fox	**Extinct**
Pteropus sp. nov.	Torresian flying-fox	**Rare**
Dobsonia moluccense	Bare-backed fruit-bat	**Rare**
Small fruit bats		
Nyctimene robinsoni	Eastern tube-nosed bat	–
Nyctimene vizcaccia	Torresian tube-nosed bat	**Rare**
Nyctimene (cf. *cephalotes*)	Cape York tube-nosed bat	**Rare**
Blossom bats		
Syconycteris australis	Eastern blossom-bat	–
Macroglossus minimus	Northern blossom-bat	–

than 100 g. The two groups have distinctly different roles and interactions with forest ecosystems in Australia. Richards (1987, 1990a) concluded that *P. conspicillatus* has a specialist association with rainforest, whereas other species such as *P. scapulatus* specialize on blossom and nectar (Hall 1983a). Tube-nosed bats (*Nyctimene*) and blossom-bats (Macroglossinae) are principally sub-canopy fruit and nectar feeders respectively (Nelson 1964; Richards 1983a, 1986).

On the basis of dental characters and known diet, Australian fruit bats can be allocated to broad categories that reflect their general ecology (Table 2). Specialist species are those that have approximately 90% of their food as one type; generalists have a broader diet. Within each cell in Table 2, only one of the specialists has a large geographical distribution, and the remaining species in each cell have quite restricted distributions. Table 2 shows that six of the 12 extant fruit bats are apparently specialist frugivores, four are specialist nectarivores and the remaining two are dietary generalists, and include fruit, nectar and pollen in their diet.

The two extant dietary generalists, *P. alecto* and *P. poliocephalus*, appear to be competing where they are sympatric. Comparison of the distribution of *P. alecto* in eastern Australia in the 1920s (Ratcliffe 1932) with the present-day range reveals that this more aggressive species is extending southward. At the same time, both the northern and southern limits of *P. poliocephalus* have shifted southwards. These subtle changes in distribution patterns may be a reflection of the short evolutionary history of fruit bats in Australia (Holloway & Jardine 1968; McKean 1970; Hall 1981) as outlined above,

Table 2. Classification of Australian fruit bats in an ecological framework, based on dental morphology and diet. Specialists are considered to be those species with approximately 90% of their food as one type; generalists have a broader diet. Species that are abundant and widespread are indicated by an asterisk. *Pteropus brunneus* is considered to be extinct.

Frugivorous specialists	Nectarivorous specialists	Dietary generalists
Large species (weight > 300 g)		
*P. conspicillatus**	*P. scapulatus**	*P. alecto**
Pteropus sp. nov.	*P. macrotis*	*P. poliocephalus*
D. moluccense		*P. brunneus*
Small species (weight < 60 g)		
*N. robinsoni**	*S. australis**	
N. vizcaccia	*M. minimus*	
N. (cf. cephalotes)		

and may indicate that *P. alecto* arrived in Australia much later than did *P. poliocephalus*.

Diet of Australian fruit bats

Formal dietary studies such as those of Richards (1990b), Parry-Jones (1987), Parry-Jones & Augee (1991) and Law (1992a,b, 1993) have been supplemented by a plethora of natural history observations, many of which are summarized by Marshall (1985).

Table 2 shows that *P. poliocephalus* has been defined as a dietary generalist, a definition supported by the studies of several authors (Nelson 1965; Hall & Richards 1979; Richards 1983b; McWilliam 1986; Parry-Jones 1987; Eby 1991a). In fact, Eby (1991b) provided evidence that colonies of this species have some individuals that primarily select blossom and others that primarily select fruit. Twenty individuals from roosts in northern New South Wales were radio-tracked during nightly foraging throughout a seasonal cycle. Half of them fed primarily on subtropical rainforest fruit and had annual movements between roosts of approximately 50 km, whereas the others fed primarily on *Eucalyptus* flowers and moved up to 800 km from their origin. These annual differences in movement patterns were correlated with peaks in the fruiting and flowering phenology of forests in eastern Australia (Eby 1991b).

Fruit diets

A primarily frugivorous diet has been recorded for three flying-fox species: *P. conspicillatus* (Richards 1990a), *Pteropus* sp. nov. which is resident only in Torres Strait (L. S. Hall & G. C. Richards unpubl.), *D. moluccense*

(Dwyer 1975; Hall 1983b; Hyde, Pernetta & Senabe 1984), and one small (40 g) fruit bat *N. robinsoni* (Richards 1986). *Pteropus conspicillatus* is the largest frugivorous specialist pteropodid in Australia and has the most restricted distribution of Australia's four species of *Pteropus*. It is found only in coastal north-eastern Queensland and shows a close association with tracts of rainforest, roosting either within rainforest or never further than 7 km away (Richards 1990a).

Data on the foods eaten by *P. conspicillatus* were obtained by Richards (1990b) over a five-year period. The fruits of 26 native species were incorporated and all were rainforest canopy species. The flowers of 10 tree species were also eaten, and a single record of foraging on foliage was also reported (Richards & Prociv 1984). The overall ratio of fruit species to flower species was close to 4:1.

Van der Pijl (1957) suggested that the colour and mode of presentation of fruit may be a pattern that influences choice by fruit-eating bats. Of the sample of natural fruits in the diet of *P. conspicillatus*, 22 out of 25 were in light-coloured categories, and the majority were presented on the periphery of the tree canopy (Richards 1990b), where they were highly visible. The neurophysiological emphasis on vision and olfaction in *Pteropus* (Calford, Graydon, Huerta, Kaas & Pettigrew 1985) explains this relationship with food colour and its visibility. Möhres & Kulzer (1956) and Neuweiler (1962) also showed that the eyes of *Pteropus* are highly adapted for nocturnal vision, being particularly suited to recognizing light colours (Neuweiler 1968).

Tube-nosed bats (*Nyctimene*) forage mainly on fruit obtained from the sub-canopy stratum of rainforests, with emphasis upon the fruits of understorey shrubs, sub-canopy trees and cauliflorous upper canopy trees such as *Ficus nodosa* and *Syzigium cormiflorum* (Richards 1986).

Flower diets

The major proportion of the flower diet of Megachiroptera in Australia comes from the Family Myrtaceae, and in particular the genera *Eucalyptus*, *Melaleuca*, *Angophora* and *Syncarpia* (Ratcliffe 1932; Hall & Richards 1979; Marshall 1985; McWilliam 1986; Eby 1991a,b; Mickleburgh, Hutson & Racey 1992; McCoy 1993). *Banksia* (Family Proteaceae) are also highly important for small blossom-bats (Law 1992a,b). These plant genera have widespread distributions in Australia's forests and woodlands (Pryor 1976; Boland *et al.* 1984; Clemson 1985), have a geographic range that closely matches that of the Megachiroptera and are amongst the best nectar-producing genera in the northern tropics (Taylor & Dunlop 1985).

McCoy (1989, 1990, 1993) noted a wide taxonomic diversity in the flower diet of Megachiroptera in the monsoonal tropics of northern Australia, and showed that it comprised 52 species in 19 genera from 11 families, which is similar to the diversity of this diet elsewhere in the Old World (Marshall 1983). The two *Pteropus* in this study were the 600 g black flying-fox (*P. alecto*) and the 300 g little red flying-fox (*P. scapulatus*). These bats had similar diets, with an 85% overlap in the plant species that they utilized. The diet of the sympatric 15 g northern blossom-bat (*Macroglossus minimus*) was quite different, being explained by the gross difference in body size and foraging mode between this and the *Pteropus* species, overlapping by 36% when compared with *P. alecto* (a dietary generalist), and by 46% when compared with *P. scapulatus* which is also a specialist nectarivore.

Foliage diets

The ingestion of foliage was first reported for Australian fruit bats by Ratcliffe (1932) and later by other authors (Richards & Prociv 1984; Parry-Jones & Augee 1991). The former authors observed *Pteropus* feeding on the leguminous *Albizia procera*. Marshall (1985) showed that folivory, although rarely reported, is quite widespread in the Pteropodidae. Lowry (1989) also observed folivory (upon *A. lebbek*) in *P. alecto* and calculated that the liquid fraction extracted from chewed leaves contained about 51% of the crude protein of the leaves, and was itself 36% protein (on a dry-matter basis). As a nutritional strategy, Lowry concluded that the potential attraction of tree legumes appeared to be their relatively high protein levels and lack of toxic secondary compounds.

Foraging ecology

Frugivores

As a result of studies of nocturnal territorial behaviour of *P. conspicillatus*, Richards (1990b) proposed a 'raiders versus residents' model for long-distance seed dispersal in *Pteropus*. Individuals that raid the feeding territories established early at night are inevitably evicted. They take a fruit before escaping from the aggression within the food tree and in so doing move propagules away from the parent tree. Janzen (1983) showed that seedlings growing at a distance from the parent tree had a greater chance of survival to maturity than did those growing near the parent. The 'raiders versus residents' model leads to the conclusion that *P. conspicillatus* have an integral role in the seed dispersal and regeneration of a suite of rainforest trees, particularly those having light-coloured fruit (Richards 1990c).

The carriage of fruits may be the primary dispersal role for Australian flying-foxes, but internal dispersal also has its place. The larger *Pteropus* in Australia have an oesophageal lumen distendable to 4–5 mm (G. C. Richards unpubl.), which places a limit on the size of seeds that can be transported internally. Seeds contained in *P. conspicillatus* faecal material grew to seedlings in a glasshouse, verifying that they were still viable after passing through the digestive tract (Richards 1990b). Germination trials by Eby (1991a) of both ingested and ejected seeds showed that all but one of the species incorporated in the diet of *P. poliocephalus* were viable after dispersal by these bats.

Nectarivores

McCoy (1989, 1990, 1993) developed an '*index of advertisement*' for flowers of trees, based upon eight of the main characteristics established by van der Pijl (1957) and Faegri & van der Pijl (1971) that describe bat-flowers. Those species visited by Megachiroptera had at least five of these characters; those not visited had a lower index. A review of the *index of advertisement* for 240 species of *Eucalyptus* in south-eastern Australia listed by Brooker & Kleinig (1983) showed that over 90% have the potential to be bat-flowers that could be utilized by *Pteropus*. By developing a '*honey index*' Cocks & Dennis (1978) were able to assess the potential for viable honey production of 83 eucalypt and rainforest communities on the south coast of New South Wales. At least half of these communities had a high value for honey production and could therefore be expected to be of value as food for *Pteropus*, in particular the *P. poliocephalus* that seasonally visit this region. If, however, it can be shown that *Pteropus* have a significant role in the reproductive biology of these eucalypt communities, then there is a high proportion of coastal forests that may be dependent upon these bats.

Functional ecology of Australian fruit bats

Pollination

Most woody plants in Australian forests are obligate outbreeders, especially members of the Myrtaceae and Proteaceae (Johnson & Briggs 1975; Crome & Irvine 1986; Lamont, Collins & Cowling 1987). Outbreeding is the primary advantage that any plant species has in a co-dependent association with bats (Janzen 1970; Gould 1978; Lack 1978; Lemke 1985). Gene flow is particularly important in the genus *Eucalyptus*, which is characterized by high levels of outcrossing and reduced viability in seeds from self-pollinated flowers (Moran & Bell 1983).

Several Australian studies have identified bats as major pollen vectors. Radio-collared *P. poliocephalus* feeding on *Melaleuca quinquenervia* and blossoms of various *Eucalyptus* were highly mobile, feeding through the night on several trees within a stand as well as moving between stands of flowering conspecifics that were several kilometres apart (Eby 1991a). McCoy (1989, 1990, 1993) provided evidence that thousands of viable pollen grains collect each night on the bodies of foraging pteropodids (*P. alecto*, *P. scapulatus* and *M. minimus*). Movements of animals from plant to plant indicated that more pollen was moved a greater distance by bats than by other vertebrate flower visitors in McCoy's study area in the tropics (McCoy 1989, 1990, 1993).

Most tree species in the diet of pteropodids grow in multi-species communities and it is rare that conspecific trees grow side by side (Boland *et al.* 1984). Consequently, efficient pollinators must move 50 m or more to transport pollen to the nearest flowering conspecific. McCoy (1993) observed that 95% of movements by *P. scapulatus* from flowering tree to flowering tree were 50 m or more, but for nectarivorous birds such as honeyeaters (Family Meliphagidae), only 20% of movements between flowering plants are of the order of 50 m (Keast 1968; Hopper & Moran 1981). Although not as yet determined with necessary experimental rigidity, this evidence indicates strongly that pteropodid flower visitors are the primary donors of outcrossed pollen in Australian eucalypt forests and woodlands.

Although it is highly likely that bats have the potential to be a major pollinator in eucalypt dominated ecosystems, the question remains as to whether *viable* pollen can be transported. McCoy (1993) established that 78% of pollen grains removed from two *Pteropus* spp. and *M. minimus* were of high enough quality to ensure fertilization. It was also shown that pteropodid bats carry significantly greater pollen loads than those published for other bats and for all other Australian flower-visiting vertebrates. In an experiment with 568 tagged and covered flowers, McCoy also demonstrated that (for *Pteropus*) the physical act of landing and foraging upon the large inflorescences of eucalypts does in fact, ensure pollination and successful seed set by these animals.

Crome & Irvine (1986) showed that 12-g eastern blossom-bats (*Syconycteris australis*) contributed more to seed production in the myrtaceous rainforest tree *Syzigium cormiflorum* than did other visitors (Table 3). Although they were the major pollinator for this tree species, these bats spent less time foraging at the tree than did other groups that were observed (birds, possums, moths). Visiting for only a few hours after dusk and briefly before dawn, blossom bats were therefore likely to be more efficient than other pollinators.

Table 3. Pollinator effectiveness in three cauliflorous *Syzigium cormiflorum* (Myrtaceae) in night and day caging experiments. Entries are the percentage of total buds pollinated by that source (after Crome & Irvine 1986).

Tree no.	Fertilized by bats	Fertilized by birds	Fertilized by insects	Total fertilized
1	48.1	16.6	17.0	81.7
2	34.2	22.0	23.0	79.2
3	46.0	18.0	18.0	71.7
Mean	42.8	18.9	19.3	77.5

Seed dispersal

Wide-ranging seed dispersal encourages genetic exchange between fragments of forest or isolated populations of particular species, decreasing the amount of genetic subdivision of taxa (Loveless & Hamrick 1984). Since the success of self-regeneration for many tropical trees improves if their propagules are moved away from the parent tree (Baker, Bawa, Frankie & Opler 1983; Janzen 1983), those tree species that encourage visits by pteropodids ensure that such a process occurs. Crome (1975) showed that rainforest pigeons preferred to feed upon fruits that were brightly coloured, exactly the opposite colours to those selected by *P. conspicillatus*. This suggests that within Australian rainforest ecosystems, where *Pteropus* and pigeons are sympatric, it is possible that rainforest canopy tree communities may partition the suite of dispersal agents that are available. Species of pteropodids in Australia may be the only dispersal agent for many rainforest trees, and would consequently play an important role in the long-term survival of light-coloured fruiting taxa (Richards 1990b, 1991).

Furthermore, it is possible that some tree species dispersed only by pteropodids may be *pivotal* taxa (keystone mutualists), in accordance with the concept of Howe (1984): 'Although most species of trees produce when other fruits are readily available in the forest, others . . . [species listed] . . . bear fruits during annual periods of fruit scarcity, and consequently maintain species of fruit-eating birds and mammals which are critical for the dispersal and ultimate recruitment of many tree species at other times of the year.' Howe & Westley (1988) expand further on this concept.

Broad-scale migrations: influences and effects

Several studies have shown that the movement patterns of flying-foxes (*Pteropus*) are related to the local availability of food; a decline in food supply causes migration out of one area to another with an abundant food source (Nelson 1965, 1989; McWilliam 1986; Eby 1991a,b; G. C. Richards unpubl.).

Availability of food resources

What little is known about the broad-scale availability of native fruit as food for bats in Australia indicates that nowhere are flowers or fruit in continuous supply, year after year. Nelson (1965) noted that the amount of blossom produced by any one species varies, observing that in some years most of the trees of a species in an area flower at the same time, in other years most of the trees flower individually at different times over a period of 4–5 months, and in other years only a few trees may flower. This study was the first to identify the unpredictable nature of the food supply available to pteropodid bats.

Phenological studies of fruit production in rainforests indicate high seasonality by most species, with the greatest number of species fruiting in summer (Holmes 1987; Innis 1989) The latter author also showed that winter was a period of low fruit availability. Some genera such as *Ficus* and *Solanum* have fruit throughout the year (Eby 1991a; G. C. Richards unpubl.). In the rainforests of north Queensland, many tree species have a sequential fruit set and ripening related to altitude, where the fruit in coastal populations ripens earlier than that in the cooler uplands (Crome 1975). Thus, frugivorous pteropodids may be faced with a food supply that is highly predictable in its seasonal availability.

Although the majority of *Eucalyptus* species have extensive distributions in Australia (Boland *et al.* 1984), their flowering patterns show some seasonality and often occur sequentially along latitudinal and altitudinal gradients (Pryor 1976). In many species, individual trees or populations do not flower annually, and flowering is patchy throughout the range of some tree species (Clemson 1985; Eby 1991a,b). Therefore, nectarivorous pteropodids may differ from frugivorous ones by having a food supply that is highly variable in its availability, and it can be shown that migration patterns differ between the different food specialists (Eby 1991b). Eby also classified the intensity of flowering by Myrtaceae and Proteaceae communities to allow regional mapping of food resources and correlation with movements and migrations of *P. poliocephalus* in New South Wales.

Many eucalypts in tropical Queensland flower synchronously, but the month of first flowering and the period of flowering vary, so that a range of food is provided throughout the year when environmental conditions are good (G. C. Richards unpubl.) In the monsoonal tropics of the Northern Territory, the flowering period for food trees averaged five months, but in tropical Queensland it averaged only three months. The Queensland tree species could be grouped into winter-flowering or summer-flowering species and had flowering times averaging 3.9 and 2.8 months.

Food availability appeared to be virtually continuous and therefore more reliable for nectarivorous bats in the Northern Territory than in Queensland. This is reflected in the migratory patterns of *Pteropus* when the two regions are compared: McCoy (1993) noted that in the Northern Territory tropics *P. alecto* and *P. scapulatus* were always present and occupied colony sites for long periods, and the movements that occurred were small and local. Conversely, G.C. Richards (unpubl.) recorded the occupation time of 22 colony sites in Queensland's tropical Gulf of Carpentaria region as ranging from two weeks to two months, and 21 sites were occupied only once in 410 inspections. During six years of study, three large-scale migrations of *P. scapulatus* colonies were recorded. These migrations could be related to the failure of winter-flowering food trees during droughts. *Eucalyptus* trees will have poor flowering when rainfall is lower than normal (Pryor 1976; Landsberg 1986) or, if in flower, may have reduced nectar production during droughts (Clemson 1985). Species such as *E. maculata* that hold buds for long periods may abort them during periods of climatic stress, and other species such as *E. albens* may still flower but with reduced amounts of nectar (Clemson 1985). The circumstances surrounding the migrations by *P. scapulatus* in north Queensland were related to variations in rainfall and food supply, and were correlated with approximately four consecutive months of below-average rainfall.

Migrations by frugivorous species

Studies of *P. conspicillatus* showed that this species is a specialized frugivore (Richards 1990b), distributed only in north-eastern Queensland (Hall & Richards 1979), and roosting either within or adjacent to tropical rainforest.

The occupancy of traditional roost sites by *Pteropus conspicillatus* in Queensland was either a continuous or a seasonal residence. Colonies resident along the coastal section of the study area were most likely to occupy roost sites continuously, whereas traditional sites on the Atherton Tableland were only occupied on a winter or summer basis, without continuity. The winter versus summer occupancy of traditional *Pteropus* roost sites has also been noted by many authors (Ratcliffe 1932; Nelson 1965; Hall & Richards 1991; Eby 1991b).

Differences in fruiting seasons and in types of rainforest (Tracey 1982; Tracey & Webb 1975) have some influence on the pattern of seasonal movements by *P. conspicillatus*. The altitudinal difference between the north Queensland coast and adjacent uplands (over 700 m) accounts for differences in rainforest species composition, as well as for differences in the fruiting seasons of species that are common to both regions (Crome 1975; Tracey 1982). It is possible that some colonies of *P. conspicillatus* utilizing a particular rainforest type may have to accommodate a greater variability in food supply, whereas others do not, especially during the young-rearing phase when energetic demands are high. In lowland areas, and possibly upland areas as well, Crome (1975) observed that the late dry-season months (August and September) constituted the period of overall peak fruiting in rainforest. The birth period of *P. conspicillatus* lags behind this fruiting peak by one month. Crome also observed that the late wet-season months (March to May) were the period of least fruit abundance, corresponding with the period when colonies have left their traditional maternity roosts on the Atherton Tableland.

Migrations influenced by differential ripening of fruit also occur in the Torres Strait region of north Australia, an area located between Cape York and the New Guinea mainland and containing approximately 40 islands and atolls. L. S. Hall & G. C. Richards (unpubl.) gathered evidence that *P. alecto* leave New Guinea in September each year and progressively infiltrate islands southward, following the progressive ripening of native fruits and others such as mangoes. By the following March these animals have disappeared from the area and are presumed to have returned to New Guinea.

Radio-tracking studies of *Nyctimene robinsoni* show that these bats reduce the effort of searching for food at dusk by roosting during the day in the food tree last visited (C. Tidemann, J. E. Nelson & G. C. Richards unpubl.). For an animal that has an apparently high metabolic rate (body water turnover has been measured at over 1000 ml^{-1} kg^{-1} d^{-1}—G. C. Richards unpubl.—and is one of the highest for any mammal), this strategy provides immediate savings in the energetic cost of commuting to and from permanent roost sites and foraging grounds.

Migrations by nectarivorous species

The migrations by *P. scapulatus* mentioned above ended at the Atherton Tableland, a region where they were not normally distributed (G.C. Richards, unpubl.), and were correlated with food availability (the failure of inland eucalypt flowering caused by below-average rainfall). During one migration (in 1986), many individuals were emaciated and dying.

Migrations by dietary generalists

Long-distance migrations by *P. poliocephalus* were suggested in the early research of Ratcliffe (1932), whereas the studies of Nelson (1965) and McWilliam (1986) indicated that localized movement patterns were perhaps more typical of this species. Recent work by Eby (1991a,b) demonstrates that both long and short migrations occur during the annual cycle of this species. However, L. S. Hall (unpubl.) has gathered evidence that during the last decade, populations of some *Pteropus* species have become reduced in size, large colonies have disappeared completely or have fragmented into smaller groups and migration pathways have been broken. The implications of these problems are as yet not well defined, but they may have a major significance due to the apparent dependence of a myriad of tree species upon fruit bats for their successful reproduction.

The studies of McWilliam (1986) in north-eastern New South Wales revealed a seasonal migration of *P. poliocephalus* from the coast to inland areas. This was attributed to both climate (temperature) and food availability, but the influence of the weather is an as yet unknown influence on the ecology of this species. There is strong taxonomic evidence indicating that two races may exist, separable by fur length and density: a southern (cold-adapted) form and a northern (subtropical) form. McWilliam's (1986) study may have included the northern form, affected to an unknown degree by temperature when at the southern limit of its range. However, there is stronger evidence that food availability is the driving force for movements and migrations of dietary generalist species.

As outlined above, Eby (1991a) provided strong evidence from radio-tracking studies that colonies of *P. poliocephalus* have individuals that primarily select blossom and others that primarily select fruit. Those feeding primarily upon fruit had annual movements between roosts of approximately 50 km, whereas those that fed primarily on *Eucalyptus* flowers moved up to 800 km during the annual cycle, following the mass flowering of eucalypt communities. One individual in the latter category moved 210 km in three nights to a new feeding area. The sequential flowering southward of *E. maculata* was the predominant reason for migration (Eby 1991a).

The mutualism hypothesis

Megachiropteran bats have shared a long association with angiosperms, these plants probably evolving in the South-East Asian region around 130 million years ago and achieving world-wide dominance over gymnosperms about 90 million years ago. Furthermore, the first recognizable rainforest formations were in existence 60 million years ago, and megachiropteran bats have been in existence for 35 million years (Mickleburgh *et al.* 1992). Frugivory in this suborder arose before nectarivory (Marshall 1983). Although it would stand to reason that, given the period of evolution, a strong association between bats and plants would exist, a review of evidence is necessary before this conclusion can be attained for the Australian Pteropodidae. Evidence supporting the concept of a mutualistic relationship between Australian forest trees and megachiropteran bats includes:

- the correlation between the distribution patterns of fruit bats and myrtaceous forest, and the fact that Myrtaceae blossom is the primary food of many pteropodid taxa;
- 10 of the 12 extant pteropodid species are dietary specialists and all ecological niches in forest in Australia available to pteropodids are filled by at least one common specialist and several less abundant specialists;
- that eucalypt flowers have most of the characteristics that support the bat-pollination syndrome of Faegri & van der Pijl (1971), particularly their odour and colour;
- that the Australian Pteropodidae have innumerable sensory, morphological and anatomical adaptations for locating or utilizing plants;
- the pattern of mass flowering by many of the Myrtaceae taxa, which draws large populations of *Pteropus* as potential pollen vectors from long distances;
- that several common and widespread pteropodids have been shown to be competent pollen vectors and satisfy the requirements of the Myrtaceae for outcrossing; a large proportion of pollen is carried undamaged and pollen transferred by pteropodids will cause fertilization to occur;
- that most of the fruits known to be eaten by pteropodids fit the bat–plant

syndrome of van der Pijl (1957), particularly their light colour and presentation on the periphery of the tree;
- that pteropodids in Australia may be the only dispersal agent for many rainforest trees and may therefore play an important role in the long-term survival of some taxa;
- the 'raiders versus residents' model for the dispersal of large fruits and seeds by *Pteropus*, and a paucity of other dispersal agents, both suggest that some trees are totally dependent upon pteropodids for their regeneration;
- and that germination trials of both ingested and ejected seeds of approximately 60 species showed that all but one was viable after dispersal by pteropodids.

The evidence that a mutually advantageous system is in existence appears to be conclusive. The Myrtaceae appear to be highly dependent upon pteropodids in Australia for outcrossed pollination; a large suite of rainforest tree species are dependent upon pteropodids for seed dispersal, yet without these food sources this family of bats would not exist. Until further research is conducted, the effect upon ecosystem function of current conservation problems is unknown. One of these problems, the culling of at least 240 000 individuals in the period 1986–1992 (D.E. Wahl unpubl.) by fruit-growers in north coastal NSW alone, must be having a major effect.

Acknowledgements

This review would not have been possible without the efforts of many researchers studying fruit bats, including Peggy Eby, Leslie Hall, Maria McCoy, John Nelson, Kerryn Parry-Jones and Chris Tidemann, but particularly Francis Ratcliffe, whose work in the 1920s stands as baseline information for studies today. Much of the research was done without appropriate funding, a situation that it is vital to correct if our knowledge base and current guesswork is to be improved.

References

Archer, M., Hand, S. J. & Godthelp, H. (1991). *Riversleigh. The story of animals in ancient rainforests of inland Australia*. Reed Books, Sydney.

Baker, H. G., Bawa, K. S., Frankie, G. W. & Opler, P. A. (1983). Reproductive biology of plants in tropical forests. In *Ecosystems of the world*. **14A**. *Tropical forest ecosystems. Structure and function*: 183–215. (Ed. Golley, F. B.). Elsevier Scientific Publishing Co., New York

Boland, D. J., Brooker, M. I. H., Chippendale, G. M., Hall, M., Hyland, B.P.M., Johnston, R. D., Kleinig, D. A. & Turner, J. D. (1984). *Forest trees of Australia*. Thomas Nelson, Melbourne.

Brooker, M. I. H. & Kleinig, D. A. (1983). *Field guide to the eucalypts of south-eastern Australia*. Inkata Press, Melbourne.

Calford, M. B., Graydon, M. L., Huerta, M. F., Kaas, J. H. & Pettigrew, J. D.

(1985). A variant of the mammalian somatotopic map in a bat. *Nature, Lond.* **313**: 477–479.
Clemson, A. (1985). *Honey and pollen flora.* Inkata Press, Melbourne.
Cocks, K. D. & Dennis, E. (1978). Apiculture. In *Landuse on the south coast of New South Wales.* **4**. *Land function studies*: 23–30. (Gen. Eds Austin, M. P. & Cocks, K. D.). CSIRO, Melbourne.
Crome, F. H. J. (1975). *The ecology of fruit pigeons in tropical northern Queensland. Aust. Wildl. Res.* **2**: 155–185.
Crome, F. H. J. & Irvine, A. K. (1986). 'Two bob each way': the pollination and breeding system of the Australian rainforest tree *Syzigium cormiflorum* (Myrtaceae). *Biotropica* **18**: 115–125.
Dwyer, P. D. (1975). Notes on *Dobsonia moluccensis* (Chiroptera) in the New Guinea Highlands. *Mammalia* **39**: 113–118.
Eby, P. (1991a). 'Finger-winged night workers': managing forests to conserve the role of grey-headed flying foxes as pollinators and seed dispersers. In *Conservation of Australia's forest fauna*: 91–100. (Ed. Lunney, D.). Royal Zoological Society of NSW, Mosman.
Eby, P. (1991b). Seasonal movements of grey-headed flying-foxes *Pteropus poliocephalus* (Chiroptera: Pteropodidae), from two maternity camps in northern New South Wales. *Aust. Wildl. Res.* **18**: 547–559.
Faegri, K. & van der Pijl, L. (1971). *The principles of pollination ecology.* Pergamon Press, New York.
Gould, E. (1978). Foraging behaviour of Malaysian nectar-feeding bats. *Biotropica* **10**: 184–193.
Hall, L. S. (1981). The biogeography of Australian bats. *Monogr. biol.* **41**: 1557–1583.
Hall, L. S. (1983a). Little red flying-fox *Pteropus scapulatus.* In *The Australian Museum complete book of Australian mammals*: 277–279. (Ed. Strahan, R.). Angus and Robertson, Sydney.
Hall, L. S. (1983b). Bare-backed fruit-bat *Dobsonia moluccense.* In *The Australian Museum complete book of Australian mammals*: 284–285. (Ed. Strahan, R.). Angus and Robertson, Sydney.
Hall, L. S. (1984). And then there were bats. In *Vertebrate zoogeography and evolution in Australasia*: 837–852. (Eds Archer, M. & Clayton, G.). Hesperian Press, Perth.
Hall, L. S. & Richards, G. C. (1979). Bats of eastern Australia. *Q. Mus. Bklt* No. 12: 1–66.
Hall, L. S. & Richards, G. C. (1991). Flying fox camps. *Wildl. Aust.* **28**: 19–22.
Hand, S. (1984). Bat beginnings and biogeography: a southern perspective. In *Vertebrate zoogeography and evolution in Australasia*: 853–904. (Eds Archer, M. & Clayton, G.). Hesperian Press, Perth.
Holloway, J. D. & Jardine, N. (1968). Two approaches to zoogeography: a study based on the distribution of butterflies, birds and bats in the Indo-Australian area. *Proc. Linn. Soc. Lond.* **179**: 153–188.
Holmes, G. (1987). *Avifauna of the Big Scrub Region.* Unpublished report to National Parks and Wildlife Service of New South Wales.
Hopper, S. D. & Moran, G. F. (1981). Bird pollination and the mating system of *Eucalyptus stoatei. Aust. J. Bot.* **29**: 625–638.
Howe, H. F. (1984). Implications of seed dispersal by animals for tropical reserve management. *Biol. Conserv.* **30**: 261–281.
Howe, H. F. & Westley, L. C. (1988). *Ecological relationships of plants and animals.* Oxford University Press, New York & Oxford.

Hyde, R. L., Pernetta, J. C. & Senabe, T. (1984). Exploitation of wild animals. In *The research report of the Simbu Land Use Project.* **4**. *South Simbu: studies in demography, nutrition, and subsistence*: 291–380. I.A.E.S.R., Port Moresby.

Innis, G. J. (1989). Feeding ecology of fruit pigeons in sub-tropical rainforests of south-eastern Queensland. *Aust. Wildl. Res.* **16**: 365–394.

Janzen, D. H. (1970). Herbivores and the number of trees in tropical forests. *Am. Nat.* **104**: 501–528.

Janzen, D. H. (1983). Food webs : who eats what, why, how, and with what effects in a tropical forest? In *Ecosystems of the world.* **14A**. *Tropical rainforest ecosystems. Structure and function*: 167–182. (Ed. Golley, F. B.). Elsevier Scientific Publishing Co, New York.

Johnson, L. A. S. & Briggs, B. G. (1975). On the Proteaceae – the evolution and classification of a southern family. *Bot. J. Linn. Soc.* **70**: 83–182.

Keast, A. (1968). Seasonal movements of the Australian honeyeaters (Meliphagidae) and their ecological significance. *Emu* **67**: 159–209.

Lack, A. (1978). The ecology of flowers of the savannah tree *Maranthes polyandra* and their visitors, with particular reference to bats. *J. Ecol.* **66**: 287–295.

Lamont, B. B., Collins, B. G. & Cowling, R. M. (1987). Reproductive biology of the Proteaceae in Australia and South Africa. *Proc. ecol. Soc. Aust.* **14**: 213–224.

Landsberg, J. J. (1986). *Physiological ecology of forest production.* Academic Press, London.

Law, B. S. (1992a). Physiological factors affecting pollen use by Queensland blossom bats (*Syconycteris australis*). *Funct. Ecol.* **6**: 257–264.

Law, B. S. (1992b). The maintenance nitrogen requirements of the Queensland blossom bat (*Syconycteris australis*) on a sugar/pollen diet: is nitrogen a limiting resource? *Physiol. Zool.* **65**: 634–648.

Law, B. (1993). Sugar preferences of the Queensland blossom bat *Syconycteris australis*: a pilot study. *Aust. Mammal.* **16**: 17–21.

Lemke, T. O. (1985). Pollen carrying by the nectar-feeding bat *Glossophaga soricina* in a suburban environment. *Biotropica* **17**: 107–111.

Loveless, M. D. & Hamrick, J. L. (1984). Ecological determinants of genetic structure in plant populations. *A. Rev. Ecol. Syst.* **15**: 65–96.

Lowry, F. B. (1989). Green-leaf fractionation by fruit bats: is this feeding behaviour a unique nutritional strategy for herbivores? *Aust. Wildl. Res.* **16**: 203–206.

Marshall, A. G. (1983). Bats, flowers and fruit: evolutionary relationships in the Old World. *Biol. J. Linn. Soc.* **20**: 115–135.

Marshall, A. G. (1985). Old World phytophagous bats (Megachiroptera) and their food plants: a survey. *Zool. J. Linn. Soc.* **83**: 351–369.

McCoy, M. (1989). *Pollination of two eucalypt species by flying-foxes.* Unpublished abstracts of the Eighth International Bat Research Conference, Sydney, 1989.

McCoy, M. (1990). Pollination of eucalypts by flying foxes in northern Australia. In *Flying fox workshop proceedings*: 33–37. (Ed. Slack, J. M.). New South Wales Dept. Agriculture and Fisheries, Sydney.

McCoy, M. (1993). *Feeding ecology of Flying-foxes (Pteropodidae) in Northern Australia.* PhD thesis: Australian National University, Canberra.

McKean, J. L. (1970). Geographical relationships of New Guinean bats (Chiroptera). *Search* **1**: 244–245.

McWilliam, A. N. (1986). The feeding ecology of *Pteropus* in north-eastern New South Wales, Australia. *Myotis* **23–24**: 201–208.

Mickleburgh, S. P., Hutson, A. M. & Racey, P. A. (1992). *Old World fruit bats: an action plan for their conservation*. Chiroptera Specialist Group, Species Survival Commission, IUCN, Gland, Switzerland.

Möhres, F. P. & Kulzer, E. (1956). Uber die Orientierung der Flughunde (Chiroptera—Pteropodidae). *Z. vergl. Physiol.* **38**: 1–29.

Moran, G. F. & Bell, J. C. (1983). *Eucalyptus*. In *Isozymes in plant genetics and breeding*: 423–441. (Eds Tanksley, S.D. & Orton, T. J.). Elsevier, Amsterdam.

Nelson, J. E. (1964). Notes on *Syconycteris australis* Peters, 1867 (Megachiroptera). *Mammalia* **28**: 429–432.

Nelson, J. E. (1965). Movements of Australian flying foxes (Pteropodidae: Megachiroptera). *Aust. J. Zool.* **13**: 53–73.

Nelson, J. E. (1989). Pteropodidae. In *Fauna of Australia*. **1B**. *Mammalia*: 836–844. (Eds Walton, D. W. & Richardson, B. J.). Aust. Govt. Publ. Service, Canberra.

Neuweiler, G. (1962). Bau und Liestung des Flughundauges (*Pteropus giganteus giganteus* Brünn.). *Z. vergl. Physiol.* **46**: 13–56.

Neuweiler, G. (1968). Verhaltensbeobachtungen an einer indischen Flughundkolonie (*Pteropus g. giganteus* Brünn.). *Z. Tierpsychol.* **26**: 166–199.

Parry-Jones, K. (1987). *Pteropus poliocephalus* (Chiroptera: Pteropodidae) in New South Wales. *Aust. Mammal.* **10**: 81–85.

Parry-Jones, K. A. & Augee, M. L. (1991). Food selection by grey-headed flying-foxes (*Pteropus poliocephalus*) occupying a summer colony site near Gosford, New South Wales. *Aust. Wildl. Res.* **18**: 111–124.

Pryor, L. D. (1976). *The biology of the eucalypts*. Edward Arnold, London.

Ratcliffe, F. N. (1932). Notes on the fruit bats (*Pteropus* spp.) of Australia. *J. Anim. Ecol.* **1**: 32–57.

Richards, G. C. (1983a). Queensland blossom-bat *Syconycteris australis*. In *The Australian Museum complete book of Australian mammals*: 288–289. (Ed. Strahan, R.). Angus and Robertson, Sydney.

Richards, G. C. (1983b). Grey-headed flying-fox *Pteropus poliocephalus*. In *The Australian Museum complete book of Australian mammals*: 275–276. (Ed. Strahan, R.). Angus and Robertson, Sydney.

Richards, G. C. (1986). Notes on the natural history of the Queensland tube-nosed bat, *Nyctimene robinsoni*. *Macroderma* **2**: 64–67.

Richards, G. C. (1987). Aspects of the ecology of spectacled flying-foxes, *Pteropus conspicillatus* (Chiroptera: Pteropodidae) in tropical Queensland. *Aust. Mammal.* **10**: 87–88.

Richards, G. C. (1990a). The Spectacled flying-fox, *Pteropus conspicillatus* (Chiroptera: Pteropodidae), in north Queensland. 1. Roost sites and distribution patterns. *Aust. Mammal.* **13**: 17–24.

Richards, G. C. (1990b). The Spectacled flying-fox, *Pteropus conspicillatus* (Chiroptera: Pteropodidae), in north Queensland. 2. Diet, seed dispersal and feeding ecology. *Aust. Mammal.* **13**: 25–31.

Richards, G. C. (1990c). Rainforest bat conservation: unique problems in a unique environment. *Aust. Zool.* **26**: 44–46.

Richards, G. C. (1991). The conservation of forest bats in Australia: do we really know the problems and solutions? In *Conservation of Australia's forest fauna*: 81–90. (Ed. Lunney, D). Royal Zool. Soc. NSW, Mosman.

Richards, G. C. & Hall, L. S. (In press) *An action plan for bat conservation in Australia*. Report to Australian Nature Conservation Agency, Canberra, Australia.

Richards, G. C. & Prociv, P. (1984). Folivory in *Pteropus*. *Aust. Bat Res. News* No. **20**: 13–14.

Taylor, J. A. & Dunlop, C. R. (1985). Plant communities of the wet-dry tropics of Australia: the Alligator Rivers region, Northern Territory. *Proc. ecol. Soc. Aust.* **13**: 83–127.

Tracey, J. G. (1982). *The vegetation of the humid tropical region of north Queensland.* CSIRO, Melbourne,

Tracey, J. G. & Webb, L. J. (1975). *Key to the vegetation of the humid tropical region of north Queensland plus 15 maps at 1:100,000 scale.* CSIRO Divn. Plant Industry, Canberra.

van der Pijl, L. (1957). The dispersal of plants by bats (Chiropterochory). *Acta bot. neerl.* **6**: 291–315.

Part III

Reproductive biology, physiology and energetics

Symp. zool. Soc. Lond. (1995) No. 67: 99–110

The use of stable isotopes to study the diets of plant-visiting bats

Theodore H. FLEMING
Department of Biology
University of Miami
Coral Gables
Florida 33124, USA

Synopsis

Dietary analysis in bats has traditionally involved examination of stomach contents or faecal materials or the identification of plant materials (pollen, seeds) on the fur of animals or under feeding roosts. While they provide detailed information about what kinds of 'prey' bats are eating, these methods provide only a 'snapshot' of an animal's diet and provide no long-term information about the general diets and trophic positions of plant-visiting bats. In this paper, I describe the application of stable isotope techniques to the study of the diets of plant-visiting bats. My colleagues and I have used carbon isotope analyses to determine whether CAM or C_3 plants are the more important in the diets of Sonoran desert flower-visiting bats. Our results indicate that the nectarivorous bat *Leptonycteris curasoae* (Phyllostomidae) is a seasonal CAM specialist in mainland Mexico and a year-round CAM specialist in Baja California. Using this technique, we have discovered that the carnivorous bat *Antrozous pallidus* (Vespertilionidae) is also a legitimate pollinator of cactus and agave flowers in the deserts of the south-western United States. We have now begun to use carbon and nitrogen stable isotopes to study the relative importance of plant and insect tissue as sources of carbon and nitrogen in plant-visiting bats. We predict that, because of their ability to echolocate, phyllostomid bats obtain more of their nitrogen from non-plant sources than do non-echolocating pteropodid bats.

Introduction

Dietary information is important for understanding the ecology and behaviour of all species of animals because diet has a profound influence on many, if not most, aspects of a species' biology (Elton 1927). In plant-visiting birds and mammals, for example, spatial and temporal variation in the availability of fruit and flower resources influences not only diets but also reproductive patterns, overall abundance, population fluctuations, daily and seasonal movement patterns, intra- and interspecific social interactions, and mating patterns (Fleming 1992).

ZOOLOGICAL SYMPOSIUM No. 67
ISBN 0 19 854945 8

Dietary analysis in bats has traditionally involved the examination of stomach contents or faecal materials or the identification of plant materials (pollen, seeds) on the fur of animals or deposited in day roosts or under night feeding roosts (Whitaker 1988; Thomas 1988). While they provide important detailed information about what kinds of 'prey' bats are eating, these methods often provide only a 'snapshot' of an animal's diet and may not provide long-term information about general diets and trophic positions of bats, particularly in comparing bats that feed in different habitats and on different classes of food (e.g. insect-eaters with fruit-eaters or with fish-eaters). Different techniques are needed to examine these broader aspects of bat diets.

In the last two decades, plant physiologists, plant and animal ecologists, historical ecologists and geochemists have used stable isotopes of carbon, nitrogen and sulphur to study a wide variety of biological and physical processes, including long-term fluctuations in global air temperature, global patterns of CO_2 and water cycling, and the movement of energy and nutrients through ecosystems (Ehleringer, Rundel & Nagy 1986; Peterson & Fry 1987; Tieszen & Boutton 1989). Early studies using stable carbon isotopes examined the relative amounts of marine and terrestrial food in the diets of prehistoric humans (Chisholm, Nelson & Schwarcz 1982), the relative importance of corn in the diets of North American Indians through time (van der Merwe 1982) and the relative importance of C_3 and C_4 plant tissues in the diets of mammalian herbivores in East Africa (Tieszen, Hein, Qvortrup, Troughton & Imbamba 1979). More recent studies have examined the relative contributions of marine and of freshwater or terrestrial foods in the diets of polar bears, owls and alcid birds (Hobson 1990; Ramsay & Hobson 1991; Hobson & Sealy 1991).

The use of stable isotopes to study the diets of bats is in its infancy. Results of studies to date, however, indicate that this technique holds considerable promise for providing new insights into the diets of bats. After briefly reviewing methods involved in stable isotope analysis, I will describe the application of this technique to the study of the diets of bats with special emphasis on plant-visiting species.

Stable isotope analysis

Two stable isotopes exist for carbon (^{12}C and ^{13}C) and nitrogen (^{14}N and ^{15}N). The heavier isotope of each of these elements is much less common in the atmosphere than is the lighter isotope (1.1% and 0.4% for ^{13}C and ^{15}N, respectively). Plants incorporate these isotopes during photosynthesis and nitrogen fixation (in legumes) or uptake, and animals obtain these isotopes from plants either directly (herbivores) or indirectly (carnivores).

Stable isotope analysis involves determining the ratio of two isotopes in plant or animal tissue as expressed by a δ value or deviation per thousand (‰) from an arbitrary standard using the formula:

$$\delta X = \frac{R_{Sample} - R_{Standard}}{R_{Standard}} \times 1000$$

where X is ^{13}C or ^{15}N and R is $^{13}C/^{12}C$ or $^{15}N/^{14}N$. The standard reference materials for carbon and nitrogen are PeeDee limestone and atmospheric N_2, respectively. Purified samples of CO_2 and/or N_2 are obtained from biological material by placing small amounts of freeze-dried tissue into Vycor ampoules containing cupric oxide, copper and small amounts of silver foil, and burning it at 800°C for several hours (Buchanan & Corcoran 1959). The two gases are then cryogenically purified in a vacuum system, and their isotopic composition is determined using a mass spectrometer.

Early studies of plant tissues revealed that there is a bimodal distribution of δ ^{13}C values relative to that of the atmosphere (δ ^{13}C = −7.7‰). Plants using the Calvin or C_3 photosynthetic pathway have δ values of about −27‰, whereas plants using the Hatch-Slack or C_4 pathway have δ values of about −12‰; succulent plants using the Crassulacean Acid Metabolism or CAM pathway also have δ values of about −12‰ (Smith & Epstein 1971). The δ ^{15}N value of the atmosphere is 0‰ and typical values for the leaves of non-nitrogen-fixing plants are − 8 to +3‰. Values for nitrogen-fixing plants range from −2 to +2‰.

Fractionation, or the differential uptake of the heavy isotope, commonly occurs in both plants and animals and is the basis for determining the trophic position of animals eating foods of known isotopic composition. C_3 plants, for example, are depleted in ^{13}C relative to the atmosphere owing to discrimination (fractionation) during the primary photosynthetic carboxylation reaction catalysed by ribulose-1, 5-bis-phosphate carboxylase. The δ values of herbivorous animals closely resemble those of their plant food but are typically 'enriched' in ^{13}C by about 1‰ (i.e. they are about 1‰ less negative). Further enrichment occurs in primary and secondary carnivores by about 1‰ per trophic transfer. Trophic enrichment is somewhat higher (by 3–5‰ per transfer) in ^{15}N owing to nitrogen excretion in animals.

Considerable variation in isotopic composition occurs among different tissues and metabolites in animals, and care must be taken in comparative studies to examine the same kinds of tissues. For example, studies of the composition of rodents fed isotopically known diets indicate that the protein collagen in bone is enriched by 2 to 6‰ in ^{13}C compared to the diet, whereas fat (lipids) is depleted in ^{13}C by 2 to 8‰ (Tieszen, Boutton, Tesdahl & Slade 1983). Similar variation also occurs in plant tissues, so that care must be taken to determine the δ values of the specific tissues (e.g. seeds, leaves, fruits, nectar) eaten by animals.

Turnover rates of carbon and nitrogen vary among animal tissues owing to differences in rates of metabolism (Tieszen, Boutton *et al.* 1983), and these differences can provide different levels of temporal resolution in dietary analyses. For example, rates of carbon turnover are very slow in bone collagen, and hence isotopic analysis of this tissue reflects the composition of the average diet over the entire lifetime of an animal. In contrast, carbon turnover rates are relatively high (of the order of a few weeks) in muscle, analysis of which thus provides information on diet during the preceding month or two. Finally, analysis of rumen, stomach, and faecal material indicates present dietary composition.

The relative contribution of different classes of food (e.g. C_3 and CAM plants, marine and freshwater or terrestrial foods) to the diets of animals can be calculated by using the following mass balance equation:

$$\delta_D = \alpha_I \delta_I + (1 - \alpha_I)\, \delta_{II}$$

where δ_D is the δ value of the animal tissue being analysed, α_I is the fraction of the diet composed of food type I with a value of δ_I and the rest of the diet comes from food type II with a value of δ_{II}. According to this equation, the diet of an animal whose muscle tissue is $\delta\ ^{13}C = -16.0‰$ consists of 75.4% CAM carbon ($\delta\ ^{13}C = -12.6‰$) and 24.6% C_3 carbon ($\delta\ ^{13}C = -26.4‰$).

The analysis of bat diets using stable isotopes

To date, only three studies of bat diets using stable isotopes have been published: Des Marais, Mitchell, Meinschein & Hayes (1980); Fleming, Nunez & Sternberg (1993) and Herrera, Fleming & Findley (1993). Each of these studies was based on carbon stable isotopes.

Des Marais *et al.* (1980) examined the ^{13}C contents of individual hydrocarbons in bat guano to determine plant–herbivore–bat relationships in the Carlsbad region of New Mexico (Table 1). They collected guano deposited over the past 40 years from a cave containing large numbers of *Tadarida brasiliensis* (Molossidae) and *Myotis velifer* (Vespertilionidae). These bats feed primarily over the Pecos River Valley, an area whose natural vegetation

Table 1. Delta ^{13}C values of samples of plant tissues and bat guano from Eddy County, New Mexico. Data are from table 5 in Des Marais *et al.* (1980).

Material analysed	*n*	Mean $\delta\ ^{13}C$ (1 SD)
Native C_3 plant species	27	−26.0 (2.2)
Native C_4 and CAM plant species	31	−13.1 (1.1)
Summer crops	4	−20.0 (7.9)
Bat guano		−20.1 (0.4)

contains a mixture of C_3 and C_4 plants and four agricultural crops of C_3 plants; the dominant crops are cotton and alfalfa. Analysis of the alkanes present in the guano indicated the existence of two major frequency classes: a lighter class whose δ ^{13}C averaged −20.7‰ and a heavier class whose δ ^{13}C averaged −23.2‰. Des Marais *et al.* (1980) speculated that the lighter alkanes were derived from agricultural crops and that the heavier alkanes were derived from native C_3 vegetation. They suggested that the bats roosting in this cave appear to be eating two classes of insects in nearly equal proportions: insects feeding on C_3 natives and insects eating C_3 crops. If this is true, then the bats apparently derive about 50% of their carbon from economically important crop insects. These results have obvious importance for the conservation of bats in this ecosystem (and probably other agricultural ecosystems; see Whitaker 1993). The authors concluded by noting that the study of older guano deposits might reveal temporal changes in the composition of vegetation (e.g. the time of introduction of exotic crops and rates of change in the composition of the native vegetation).

My colleagues and I have used carbon stable isotope analysis to study the diets of several species of nectar-feeding bats in the arid regions of the south-west United States and Mexico (Fleming *et al.* 1993; Herrera *et al.* 1993). The objective of our first study was to determine whether CAM (Cactaceae and Agavaceae) or C_3 plants were the more important in the diets of the nectar-feeding phyllostomid bats *Leptonycteris curasoae, L. nivalis, Choeronycteris mexicana* and *Glossophaga soricina.* The first three species migrate seasonally into the Sonoran and Chihuahuan deserts from other parts of Mexico, whereas the latter species is a non-migratory resident of tropical and subtropical parts of Mexico and Central and South America. Nectar-feeding bats in central and southern Mexico are known to visit flowers of C_3 plants (mainly in the Bombacaceae and Convolvulaceae) and CAM plants, whereas they visit only CAM plants in the north (Alvarez & Gonzalez 1970; Quiroz, Xelhuantzi & Zamora 1986; Cockrum 1991).

Our source of tissue for this study was the muscle from one or two toes per individual (< 4 mg of tissue), mostly from museum specimens collected at different times and localities. We analysed nectar and flower material from three species of bat-pollinated columnar cacti from the Sonoran desert and flower tissue from 18 species of *Agave* and 10 species of Mexican bat-pollinated C_3 plants. We also collected flowers and the insects visiting them, breast muscle samples from six species of birds and toes from three species of bats at Bahia Kino in the central coastal region of Sonora, Mexico.

Our data from Bahia Kino (Fig. 1) illustrate the kinds of information that can be gained from carbon stable isotope analyses. First, we confirmed the large difference between the δ ^{13}C values of C_3 and CAM plants. Second, the conspicuous insects visiting both C_3 plants (bees and butterflies) and CAM plants (bees) were mostly C_3 in carbon composition. Although honeybees are common visitors to cactus flowers, they appear to obtain far more of their carbon from C_3 shrubs and trees than from cacti. Third, except for certain doves and a woodpecker, the birds and *Macrotus californicus* bats (Phyllostomidae) that we sampled had obtained $\geq$ 61% of their carbon from C_3 plants (calculations based on a 1‰ correction to the δ ^{13}C of animal tissues to account for trophic enrichment; this is a conservative correction for insectivorous birds and bats because they are two trophic levels removed from plants). Fourth, the bat *Antrozous pallidus* (Vespertilionidae), though normally an insectivorous or carnivorous bat, contained substantial amounts of CAM carbon (see below). Finally, *Leptonycteris* bats contained more CAM carbon in June than in April.

The major results of our geographic survey are summarized in Fig. 2, in which I have recalculated our original δ ^{13}C data as percentage CAM contribution. As expected, the non-migratory *Glossophaga soricina* from southern Mexico is strongly C_3 in carbon composition year-round. In contrast, the carbon composition of individuals of *L. curasoae* from the Mexican mainland changes seasonally. Bats contain a mixture of C_3 and CAM carbon during the late fall and winter months (which is the tropical dry season when many C_3

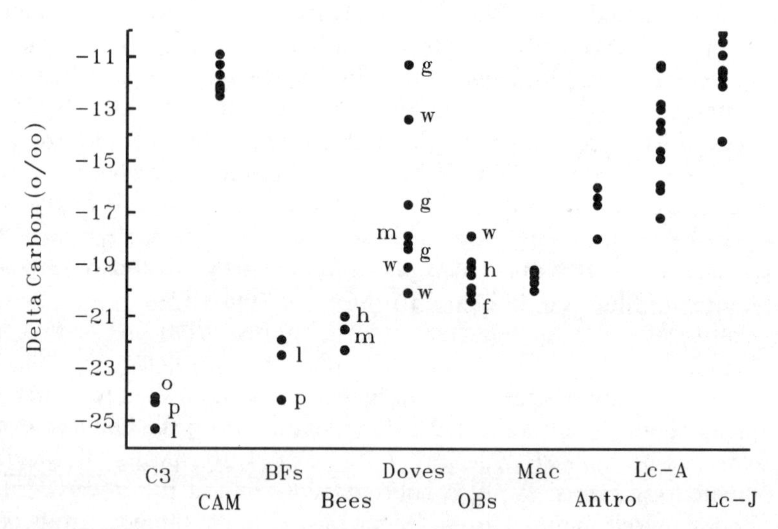

Fig. 1. Delta ^{13}C values for tissues collected in April through June at Bahia Kino, Sonora, Mexico. Plant tissues include three C_3 species (l = *Lycium* sp., o = *Olneya tesota*, p = *Prosopis glandulosa*) and three CAM species (*Carnegia gigantea, Pachycereus pringlei* and *Stenocereus thurberi* (Cactaceae). Insect tissues include butterflies (BFs, l = lycaenids, p = pierids) and bees (h = honeybees, m = megachilids). Bird tissues include doves (g = common ground dove, m = mourning dove, w = white-winged dove) and other birds (OBs, f = ashy-throated flycatcher, h = house finch, w = ladder-backed woodpecker). Bat tissues include *Macrotus californicus* (Mac), *Antrozous pallidus* (Antro) and *Leptonycteris curasoae* from April (Lc–A) and June (Lc–J).

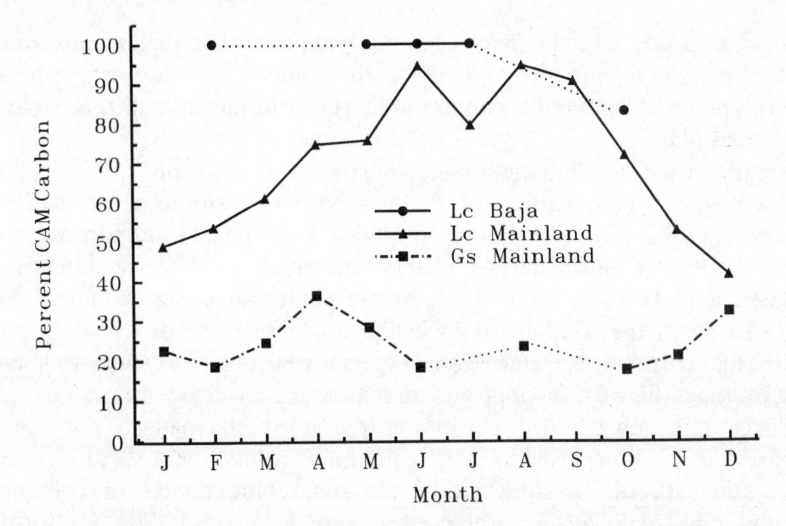

Fig. 2. Percentage of CAM carbon in muscle tissue from two species of nectar-feeding phyllostomid bats: Gs = *Glossophaga soricina*, Lc = *Leptonycteris curasoae*. Tissues come from either mainland Mexico or Baja California. Data recalculated from Fleming *et al.* (1993) with a 1 ‰ correction factor for trophic level effects (i.e., 1 ‰ subtracted from the original monthly mean values of $\delta\ ^{13}C$).

plants are flowering) but are nearly 100% CAM in composition during the summer months. Our limited data for *L. nivalis* and *Choeronycteris mexicana* show a similar pattern. Surprisingly, individuals of *L. curasoae* from Baja California contain nearly 100% CAM carbon year-round.

These results indicate that the degree of specialization on CAM plants varies seasonally and geographically in *L. curasoae* and probably also in *L. nivalis* and *C. mexicanus*. *L. curasoae* appears to be a CAM specialist in two situations: when it migrates from central Mexico to the Sonoran desert and in Baja California. On the basis of these results, we have hypothesized that in the spring migrants fly north along a nectar corridor consisting of several species of columnar cacti located in the coastal lowlands of western Mexico; they fly south in the fall along a corridor consisting of several species of paniculate agaves located in the Sierra Madre Occidental mountains. Our results also suggest that the bats living in Baja California do not migrate to the mainland during the fall and winter.

During our work with *L. curasoae* at Bahia Kino, we captured 20 individuals of the pallid bat, *Antrozous pallidus*, whose faces were heavily covered with cactus pollen. Since other workers have reported capturing this species, whose normal diet consists of large ground-dwelling insects and arthropods (including scorpions), at *Agave* inflorescences, we initiated a carbon stable isotope study to test the following hypothesis: *A. pallidus* is a regular visitor to flowers of certain bat-pollinated cacti and agaves in the Sonoran and Chihuahuan deserts. We predicted that pallid bats would contain substantial amounts of CAM carbon during the flowering periods of those plants but would contain only C_3 carbon in habitats lacking bat-adapted plants.

We again used toe muscle tissue, mostly from museum specimens, to conduct a geographic survey of the carbon composition of *A. pallidus* (Herrera *et al.* 1993). Our results provided only partial support for our hypothesis. In our entire dataset of 110 individuals, δ ^{13}C was negatively correlated with latitude, indicating that the pallid bat contains more CAM or C_4 carbon in the southern than in the northern parts of its geographic range. This correlation, which was especially strong for specimens collected in June (Fig. 3), is opposite to that in *L. curasoae*, in which individuals contain more CAM carbon in the northern than in the southern parts of its range. Pallid bats contained $\geqslant$ 69% CAM or C_4 carbon at only six of 18 sites within the geographic ranges of columnar cacti and paniculate agaves; these sites were located in Baja California, Sonora and Arizona. Carbon composition of individuals at the other 12 sites was either pure C_3 or mostly C_3. Pallid bats also contained substantial amounts of non-C_3 carbon at one site far outside the cactus-agave range (Union County, New Mexico). Comparison of the δ ^{13}C values of the pallid bat with those of three other species of insectivorous bats (*Macrotus californicus, Eptesicus fuscus* and *Tadarida brasiliensis*) caught at the same time in the same places indicated that the pallid bat almost always contained more CAM or C_4 carbon than did the other bats (Fig. 1).

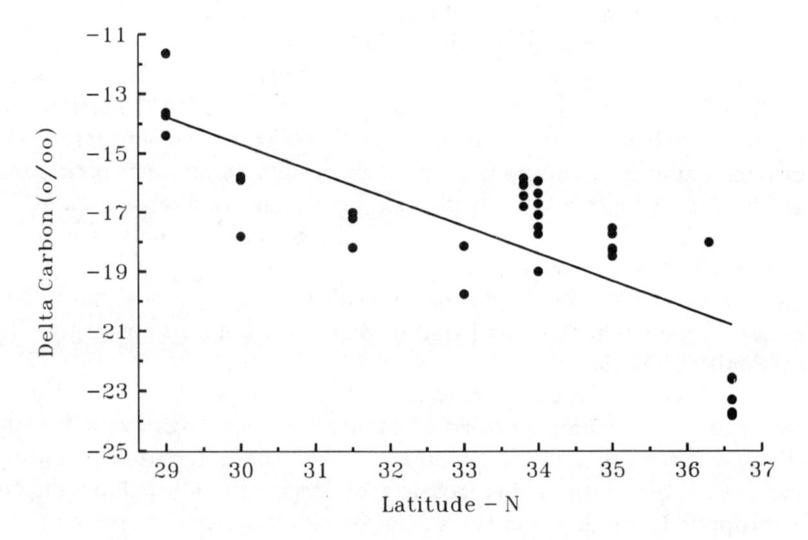

Fig. 3. The relationship between δ ^{13}C and latitude in muscle tissue from the pallid bat, *Antrozous pallidus*, samples from June. Each point represents a different bat.

Our stable isotope results, plus observations of captive pallid bats, indicate that this species visits cactus flowers to eat beetles that enter the flowers soon after they open, not to ingest nectar. While eating insects, however, this bat picks up large amounts of pollen on its face and contacts the cactus stigma, making it a legitimate pollinator of cactus, and probably of agave, flowers. At some locations outside the ranges of cacti and agaves, this bat appears to eat insects feeding on C_4 plants, in contrast to other sympatric insectivorous bats which appear to eat insects feeding primarily on C_3 plants. Through the use of stable isotopes, therefore, we have been able to confirm that *A. pallidus* obtains substantial amounts of CAM carbon from eating insects associated with desert succulent flowers and obtains substantial amounts of C_4 carbon through eating insects different from those consumed by other common insectivorous bats.

Current stable isotope work in our laboratory is aimed at determining the relative importance of plant and insect tissue as sources of carbon and nitrogen in plant-visiting bats. This study was inspired by the work of Don Thomas, who suggested (Thomas 1984) that a basic dichotomy exists in these bats. Phyllostomid plant-visitors can echolocate and are known to supplement their fruit and nectar/pollen diets with insects and hence can be considered to be facultative plant-visitors. In contrast, because they cannot echolocate, pteropodid bats apparently obtain all of their energy and protein from plant products and should thus be obligate plant-visitors. On the basis of this dichotomy, Thomas predicted that fruit consumption in phyllostomids should be driven by the energy, not the protein, content of their fruit. Consumption of fruits by pteropodids should be driven by the protein, not the energy, content of their fruit. Since the protein content of most fruits eaten by bats is very low, Thomas predicted that pteropodid bats over-ingest energy to meet their daily protein requirements. Since phyllostomids can feed on insects, a richer source of protein than fruit, they should be less likely than pteropodids to over-ingest energy. One of the important implications of this proposed dichotomy is that pteropodid bats might be more coevolved with their food plants than are phyllostomids.

Testing this proposed dichotomy by using carbon and nitrogen stable isotopes is straightforward in theory. If this dichotomy is correct, then phyllostomid tissues should be significantly more enriched in ^{13}C and ^{15}N than those of pteropodids (Fig. 4). Pteropodids should occupy the 'herbivore' trophic level, whereas phyllostomids should be intermediate between herbivores and primary carnivores (i.e. insectivores). In practice, testing this prediction involves collecting bat muscle tissue (in larger amounts than is found in the muscle from a single toe because of the lower density of nitrogen compared with carbon in muscle and other tissues), fruit and/or flower tissue and insect tissue. Muscle tissue from sympatric plant-visiting *and*

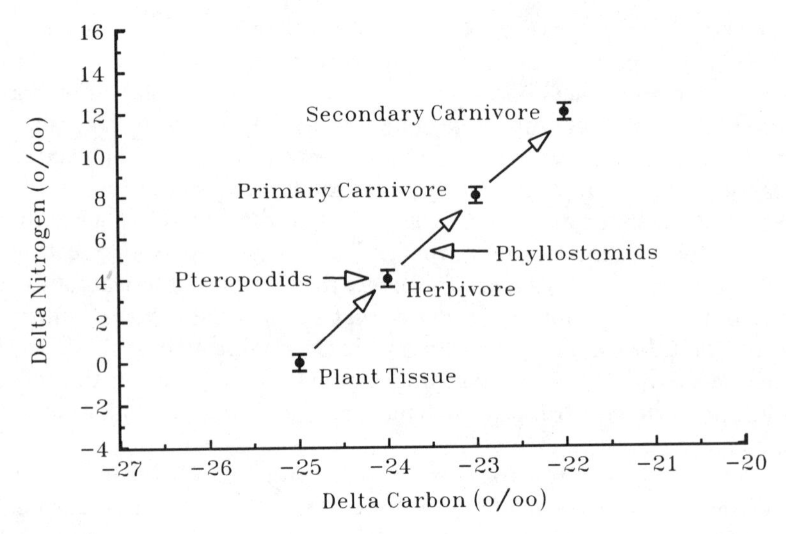

Fig. 4. Hypothetical example of trophic enrichment in $\delta\ ^{13}C$ and $\delta^{15}N$ in a terrestrial food chain and the predicted trophic positions of plant-visiting pteropodids and phyllostomids according to Thomas (1984).

insectivorous bats needs to be collected for a proper statistical comparison. The hypothesized dichotomy predicts that phyllostomid tissue should be roughly intermediate in its $\delta\ ^{13}C$ and $\delta\ ^{15}N$ values between the values of pteropodid and insectivorous bats.

In conclusion, stable isotope analysis is a powerful tool for studying the diets of bats. It provides a range of temporal scales for looking at diets, it can discriminate between different potential sources of food and it can provide information about which microhabitats (e.g. open or closed forest, in areas containing both C_3 and C_4 plants; C_3 canopy trees or shrubs, or CAM epiphytes; terrestrial or aquatic sites) contribute carbon and nitrogen to the diets of bats. Bat ecologists should seriously consider supplementing the more traditional methods of collecting dietary data with stable isotope analyses in future studies.

Acknowledgements

My work with nectar-feeding bats has been supported by the National Geographic Society, the U. S. National Fish and Wildlife Foundation, the U.S. National Science Foundation and the University of Miami General Research Fund. I thank my colleague Leo Sternberg for introducing me to stable isotope techniques, for generous access to his laboratory and for helpful discussions. My graduate student Gerardo Herrera has kindly conducted isotope analyses for me at a moment's notice.

References

Alvarez, T. & Gonzalez, Q. L. (1970). Analisis polinico del contenido gastrico de murcielagos Glossophaginae de Mexico. *An. Esc. nac. Cienc. Biol. Méx.* **18**: 137–165.

Buchanan, D. & Corcoran, B. (1959). Sealed tube combustions for the determination of ^{14}C and total carbon. *Analyt. Chem.* **31**: 1635–1638.

Chisholm, B. S., Nelson, D. E. & Schwarcz, H. P. (1982). Stable-carbon isotope ratios as a measure of marine versus terrestrial protein in ancient diets. *Science* **216**: 1131–1132.

Cockrum, E. L. (1991). Seasonal distribution of northwestern populations of the long-nosed bats, *Leptonycteris sanborni* Family Phyllostomidae. *An. Inst. Biol. Univ. nac. Aut. Mex.* (*Ser. Zool.*) **62**: 181–202.

Des Marais, D. J., Mitchell, J. M., Meinschein, W. G. & Hayes, J. M. (1980). The carbon isotope biogeochemistry of the individual hydrocarbons in bat guano and the ecology of the insectivorous bats in the region of Carlsbad, New Mexico. *Geochim. cosmochim. Acta* **44**: 2075–2086.

Ehleringer, J. R., Rundel, P. W. & Nagy, K. A. (1986). Stable isotopes in physiological ecology and food web research. *Trends Ecol. Evol.* **1**: 42–45.

Elton, C. S. (1927). *Animal ecology*. Sidgwick & Jackson Ltd., London.

Fleming, T. H. (1992). How do fruit- and nectar-feeding birds and mammals track their food resources? In *Effects of resource distribution on animal–plant interactions*: 355–391. (Eds Hunter, M.D., Ohgushi, T. & Price, P. W.). Academic Press, San Diego.

Fleming, T. H., Nunez, R. A. & Sternberg, L. (1993). Seasonal changes in the diets of migrant and non-migrant nectarivorous bats as revealed by carbon stable isotope analysis. *Oecologia* **94**: 72–75.

Herrera, L. G, Fleming, T. H. & Findley, J. S. (1993). Geographic variation in carbon composition of the pallid bat, *Antrozous pallidus,* and its dietary implications. *J. Mammal.* **74**: 601–606.

Hobson, K. A. (1990). Stable isotopic determinations of the trophic relationships of seabirds: preliminary investigations of alcids from coastal British Columbia. *Occ. Pap. Can. Wildl. Serv.* No. 68: 16–20.

Hobson, K. A. & Sealy, S. G. (1991). Marine protein contributions to the diet of northern saw-whet owls on the Queen Charlotte Islands: a stable-isotope approach. *Auk* **108**: 437–440.

Peterson, B. J. & Fry, B. (1987). Stable isotopes in ecosystem studies. *A. Rev. Ecol. Syst.* **18**: 293–320.

Quiroz, D. L., Xelhuantzi, M. S. & Zamora, M. C. (1986). Analisis palinologico del contenido gastrointestinal de los murcielagos de Juxtlahuaca, Guerrero. *Inst. nac. Antrop. Hist. Ser. Prehist.* **1986**: 1–62.

Ramsay, M. A. & Hobson, K. A. (1991). Polar bears make little use of terrestrial food webs: evidence from stable-carbon isotope analysis. *Oecologia* **86**: 598–600.

Smith, B. N. & Epstein, S. (1971). Two categories of $^{13}C/^{12}C$ ratios for higher plants. *Pl. Physiol., Rockville* **47**: 380–384.

Thomas, D. W. (1984). Fruit intake and energy budgets of frugivorous bats. *Physiol. Zool.* 57: 457–467.

Thomas, D. W. (1988). Analysis of diets in plant-visiting bats. In *Ecological and behavioral methods for the study of bats*: 211–220. (Ed. Kunz, T. H.). Smithsonian Institution Press, Washington.

Tieszen, L. L. & Boutton, T. W. (1989). Stable carbon isotopes in terrestrial ecosystem research. In *Stable carbon isotopes in ecological research*: 166–195. (Eds Rundel, P. W., Ehleringer, J. R. & Nagy K. A.). Springer-Verlag, New York. (*Ecol. Stud. Anal. Synth.* **68.**)

Tieszen, L. L., Boutton, T. W., Tesdahl, K. G. & Slade, N. A. (1983). Fractionation and turnover of stable carbon isotopes in animal tissues: implications for δ ^{13}C analysis of diet. *Oecologia* **57**: 32–37.

Tieszen, L. L., Hein, D., Qvortrup, S. A., Troughton, J. H. & Imbamba, S. K. (1979). Use of δ ^{13}C values to determine vegetation selectivity in East African herbivores. *Oecologia* **37**: 351–359.

van der Merwe, N. J. (1982). Carbon isotopes, photosynthesis, and archaeology. *Am. Scient.* **70**: 596–606.

Whitaker, J. O., Jr. (1988). Food habit analysis of insectivorous bats. In *Ecological and behavioral methods for the study of bats*: 171–189. (Ed. Kunz, T. H.). Smithsonian Institution Press, Washington.

Whitaker, J. O., Jr. (1993). Bats, beetles, and bugs. *Bats* **11** (1) : 23.

Symp. zool. Soc. Lond. (1995) No. 67: 111–122

The energetics of pteropodid bats

Brian K. McNAB — *Department of Zoology, University of Florida, Gainesville, Florida 32611, USA*

and Frank J. BONACCORSO — *Department of Zoology, and P. K. Younge Developmental Research School, University of Florida, Gainesville, Florida 32611, USA*

Synopsis

The temperature regulation and energetics of pteropodid bats are highly variable. Their thermoregulation is effective in species that weigh more than 60 g, but smaller species often exhibit imprecise regulation at cool environmental temperatures. The smallest nectarivores usually maintain a small temperature differential with the environment, especially at low altitudes. Nectarivores either regulate body temperature at high altitudes or are limited in distribution to lowlands. Basal rate in pteropodids varies with body size, with the propensity to enter torpor and with the size of the island on which pteropodids are found. The low rates of metabolism found in small-island *Pteropus*, produced directly or by a reduction in body mass, reduce resource use in an environment characterized by limited resources. A lower rate of metabolism is also found in species of *Dobsonia* endemic to small islands, compared with mainland *Dobsonia* of the same size, and in the genus *Melonycteris*, which is endemic to small islands, compared with the equally sized *Eonycteris* from Borneo. Large *Pteropus* reduce heat loss at low environmental temperatures, probably through peripheral vasoconstriction, thereby permitting rate of metabolism to fall without compromising the maintenance of a high core temperature. We conclude that the evolution of specialized nectarivory, which appears to have occurred independently at least five times in the Pteropodidae, is usually associated with a reduction in both mass and rate of metabolism, a combination that leads to a variable body temperature.

Introduction

The concept that monophyletic clades of closely related organisms are physiologically, behaviourally and ecologically uniform is often facilitated by the paucity of information. For example, published data on the energetics of pteropodids suggest that these bats are effective thermoregulators and have rather high basal rates of metabolism (McNab 1989). The only exceptions

ZOOLOGICAL SYMPOSIUM No. 67
ISBN 0 19 854945 8

previously known were members of the subfamily Nyctimeninae, which appeared to be imprecise thermoregulators at low environmental temperatures (Bartholomew, Dawson & Lasiewski 1970), and the African nectarivore *Megaloglossus woermanni*, which sometimes entered a torpid state at low ambient temperatures (Kulzer & Storf 1980). This perspective on pteropodid energetics, however, suffered from a shortage of data because data were available on only 10 in a family of 173 extant species.

Our recent work in Papua New Guinea and on species of the genus *Pteropus* held in captivity at the Lubee Foundation (Gainesville, Florida) has permitted us to expand knowledge of the energetics of this family and suborder. This information paints a much more diverse pattern in their energetics, an observation that is noteworthy in light of the recent analysis of the phylogeny of the Pteropodidae by Kirsch and co-workers (J.A.W. Kirsch unpubl.) based on DNA hybridization. These workers, for example, argue that morphologically specialized nectarivory has evolved independently in this family at least five times.

Here we survey the standard energetics of pteropodid bats, the details of which will appear elsewhere, and show the highly flexible nature of physiological adaptation to ecological conditions in this family. We then attempt to place these observations in a phylogenetic context. In spite of similarities in the energy expenditure of pteropodids and phyllostomids, the New World family that includes frugivores and nectarivores, some aspects of the energetics of pteropodids are not anticipated from our knowledge of the energetics of phyllostomids.

Temperature regulation of pteropodids

Most frugivorous bats, including those that belong to the Pteropodidae, regulate body temperature with some precision at ambient temperatures down, at least, to 10 °C (Fig.1). This precision is found in species belonging, at least, to the genera *Pteropus, Dobsonia* and *Rousettus*. The principal exceptions in this family are bats that weigh less than 60 g, namely the nectarivores *Syconycteris, Macroglossus, Megaloglossus, Eonycteris* and *Melonycteris*, and the small frugivores *Paranyctimene* and *Nyctimene*.

The response to low temperature in *Syconycteris australis* and *Macroglossus minimus* deserves special attention. In lowland tropical forests (elevation <200 m) these nectarivores show little temperature regulation when exposed to ambient temperatures below 20–25 °C. Rather, they maintain a temperature differential with the environment, at least down to an ambient temperature of 10–15 °C (Fig. 2). This differential is 2–6 °C in *Syconycteris* and 2–10 °C in *Macroglossus*. As a consequence, rates of metabolism fall with environmental temperature (Fig. 3). These species have a thermal behaviour that is common in insectivorous bats but unknown in phyllostomid nectarivores (McNab 1969, 1989). The torpor in *Syconycteris* and *Macroglossus* is facultative

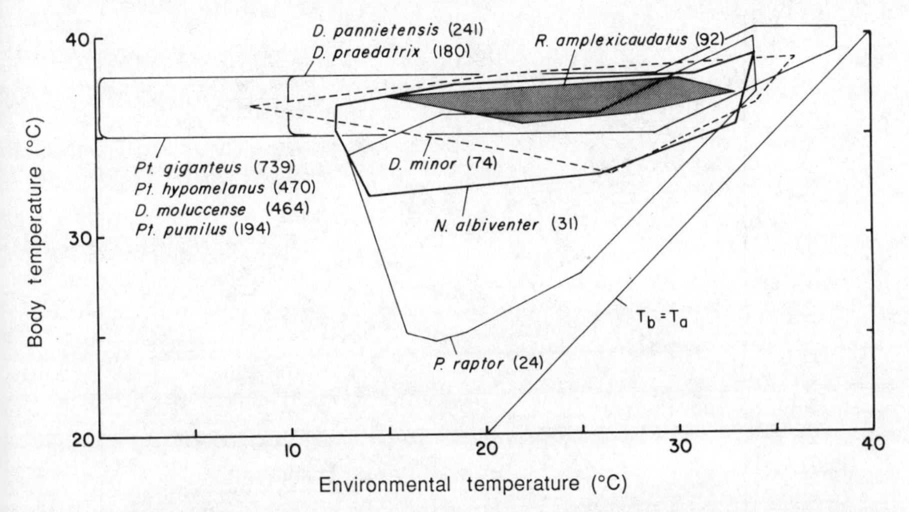

Fig. 1. Body temperature (°C) as a function of environmental temperature (°C) in nyctimenine and pteropodine bats. Body mass in grams is indicated in parentheses. Note that smaller species have more variable body temperatures.

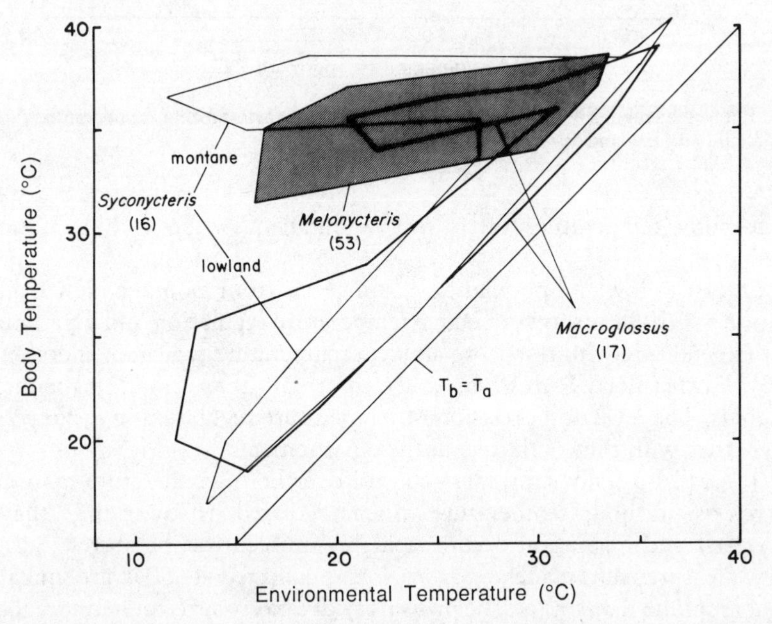

Fig. 2. Body temperature (°C) as a function of environmental temperature (°C) in macroglossine bats. Body mass in grams is indicated in parentheses. Note that temperature regulation is most highly developed in larger species and in highland populations.

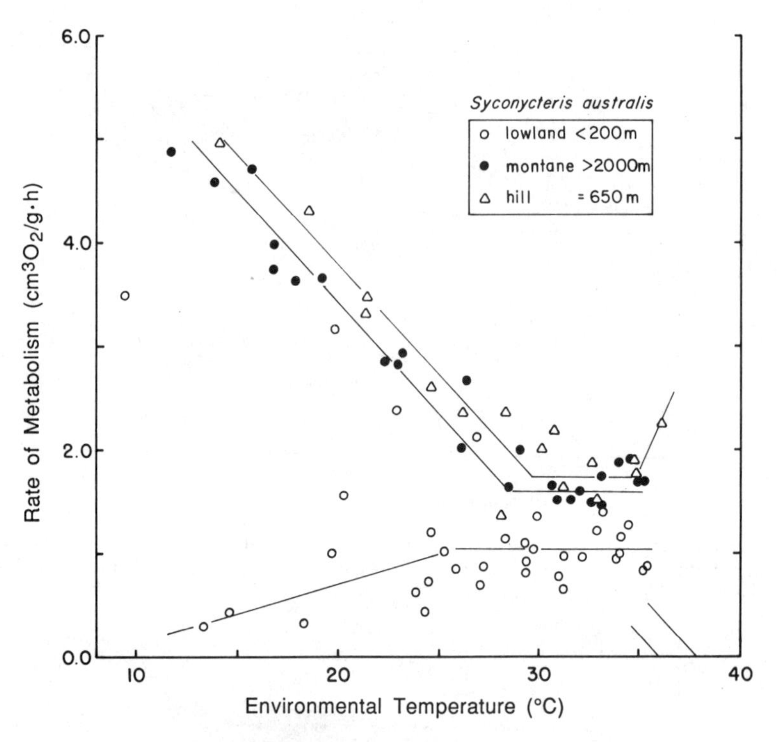

Fig. 3. Rate of metabolism (cm^3O_2/g.h) as a function of environmental temperature (°C) in highland, hill, and lowland *Syconycteris australis*.

in that body temperature often rises spontaneously or with a stimulus to the bat.

In contrast to lowland populations, *S. australis* from montane New Guinea (elevation = 2100 m) shows effective temperature regulation and a pattern of energy expenditure with respect to ambient temperature typical of endotherms (Fig. 3). Furthermore, *S. australis* collected at 650 m are similar to montane individuals. The level and precision of temperature regulation in *Syconycteris* clearly varies with the conditions in the environment.

Macroglossus usually shows less control over body temperature than does *Syconycteris*: its body temperature, when regulated, is lower than that of *Syconycteris* and regulation occurs at higher ambient temperatures (Fig. 2). The few measurements made on *Macroglossus* collected at 650 m are similar to those of montane *S. australis*. The propensity of *Syconycteris* to thermoregulate more effectively than *Macroglossus* is compatible with their distribution in New Guinea: *M. minimus* is found only to 1200 m, whereas *S. australis* is found at elevations of up to 3000 m (Bonaccorso in press). The limited temperature regulation of *M. minimus* is also compatible with the suggestion of Hood & Smith (1989) that females of this species store spermatozoa.

The nectarivores *Megaloglossus woermanni* from West Africa (Kulzer & Storf 1980), *Eonycteris spelaea* from Borneo (McNab 1989) and *Melonycteris melanops* from New Britain, all of which are usually classified as belonging to the Macroglossinae (but see below), are variable thermoregulators. *Megaloglossus*, which weighs 12–20 g, sometimes permits body temperature to fall to near ambient levels (Kulzer & Storf 1980). *Eonycteris* and *Melonycteris*, presumably in relation to their larger mean body masses (51.6 and 53.3 g, respectively), do not let body temperature fall as much as do small nectarivores.

The thermoregulation of *Paranyctimene raptor* (24 g) and *Nyctimene albiventer* (31 g) is much less precise than that of similarly sized phyllostomids, such as *Sturnira lilium* (22 g) and *Vampyrops lineatus* (22 g), but is similar to that of small (*c.* 10 g) tropical phyllostomids like *Ametrida minor*, *Artibeus cinereus* and *Rhinophylla pumilio* (McNab 1969): body temperature (Fig. 1) and rate of metabolism are highly variable at low T_as. In New Guinea *P. raptor* and *N. albiventer* are limited in distribution to elevations of less than 1200 and 1700 m, respectively (Bonaccorso in press). Notice that the larger of these two species has more effective temperature regulation and ascends to higher elevations.

Basal rate of metabolism in pteropodids

The interspecific variation in the basal rate of pteropodids is appreciable (Fig. 4). Compared with a curve describing the mean relationship in mammals between basal rate and body mass (McNab 1988), nine pteropodids have low basal rates, two have high basal rates and nine have basal rates approximately equal to the average of mammals. Six of the nine species with low basal rates weigh less than 60 g, usually in association with poor temperature regulation. Indeed, all pteropodids that show marginal temperature regulation and/or entrance into facultative torpor have basal rates that fall below the curve that describes the minimal cost of continuous endothermy (the so-called boundary curve; McNab 1983). At greater masses pteropodids may have high, standard or low basal rates. This differentiation occurs between and within genera. Thus, *Pteropus* has either standard or low basal rates and *Dobsonia* has either standard or high basal rates (Fig. 4). Corrected for mass, the pteropodid with the highest basal rate is *D. moluccense*. In this species, subadults with a mass of 208 g have a basal rate equal to 145% of the expected value, adult females at 345 g have a basal rate equal to 149% and adult males, with a mass of 464 g, have a basal rate that is 143%. The high basal rate in *D. moluccense* is therefore independent of age, sex and body mass.

What factor other than body size accounts for this variation in basal rate? Mass alone accounts for 96.6% of the variation in total basal rate in the 19 species of pteropodids. The regression of total basal rate as a power function of mass actually had 20 entries because *S. australis* was entered twice, once each for the lowland and the montane populations. This can be justified because these populations are as different as many species pairs. Nevertheless, much residual variation in basal rate remains when mass alone is considered (Fig. 4).

An analysis of covariance was used to determine whether basal rate is significantly correlated with factors other than mass (Table 1). The model used was $\log_{10}$ basal rate = $f(\log_{10}$ mass, subfamily affiliation, body temperature, food habits, island size), where both basal rate and mass were entered as logarithms to convert a power function to a linear equation. Body temperature was represented as regulated ($n = 16$), marginally regulated ($n = 2$), or not regulated ($n = 2$); food was fruit ($n = 14$) or nectar ($n = 6$); island size indicated whether a species was limited in distribution to small islands ($n = 6$) or was found on large islands and continents ($n = 14$). The final analysis included only those factors that retained significance.

The results were that subfamily affiliation, as described by Andersen (1912), and food habits were statistically insignificant and eliminated, but that $\log_{10}$ basal rate was significantly correlated with $\log_{10}$ mass ($P \leqslant 0.0001$), temperature regulation ($P = 0.0051$) and island size ($P = 0.0106$).

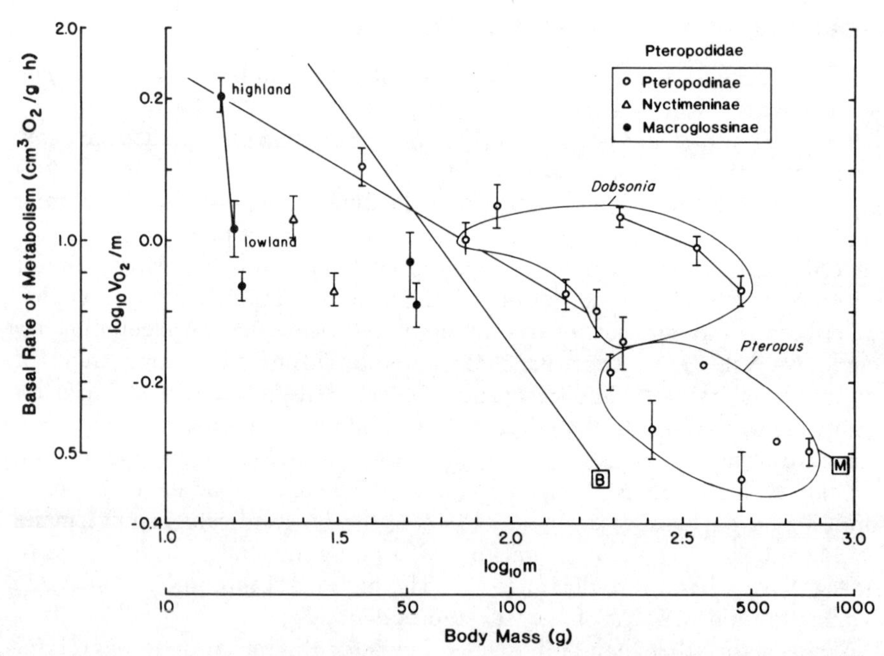

Fig. 4. Log_{10} basal rate of metabolism (cm^3O_2/g.h) as a function of $\log_{10}$ body mass (g) in pteropodid bats. The mean relationship in mammals (*M*, McNab 1988) and the curve that separates continuous from discontinuous endothermy (*B*, McNab 1983) are indicated. Measurements on a species are connected.

Table 1. Basal rates of metabolism in pteropodid bats.

Species	Mass (g)	Basal rate of metabolism $cm^3 O_2/h$	T_b[a]	Food habits[b]	Island size[c]	Reference
Nyctimeninae						
Paranyctimene raptor	23.6	35.7	I	F	L	This study
Nyctimene albiventer	30.9	26.3	I	F	L	This study
Macroglossinae						
Syconycteris australis (highland)	14.5	23.9	R	F/N	L	This study
Syconycteris australis (lowland)	16.5	17.2	N	F/N	L	This study
Macroglossus minimus	16.4	1.44	N	N	L	This study
Eonycteris spelaea	51.6	48.0	R	N	L	McNab (1989)
Melonycteris melanops	53.3	43.2	R	N	S	This study
Pteropodinae						
Cynopterus brachyotis	37.4	47.5	R	F	L	McNab (1989)
Dobsonia minor	73.7	74.4	R	F	L	This study
Rousettus amplexicaudatus	91.5	102.5	R	F	L	This study
Rousettus aegyptiacus	146.0	122.6	R	F	L	Noll (1979)
Dobsonia praedatrix	179.5	141.8	R	F	S	This study
Pteropus pumilus	194.2	126.2	R	F	S	This study
Dobsonia pannietensis	241.4	173.8	R	F	S	This study
Pteropus rodricensis	254.5	134.9	R	F	S	This study
Pteropus scapulatus	362.0	242.5	R	N	L	Bartholomew, Leitner *et al.* (1964)
Dobsonia moluccense	463.8	394.2	R	F	L	This study
Pteropus hypomelanus	470.1	216.2	R	F	S	This study
Pteropus poliocephalus	598.0	316.2	R	F	L	Bartholomew, Leitner *et al.* (1964)
Pteropus giganteus	739.2	377.0	R	F	L	This study

[a]Body temperature: regulated (R), intermediate (I), or not regulated (N)
[b]Food habits: fruit (F) or nectar (N)
[c]Island size: small (S) or large (L)

This model accounts for 98.6% of the variation in $\log_{10}$ basal rate. Precise thermoregulators had basal rates 1.60 times those of nonregulators, and large island and continental species had basal rates 1.24 times those of species restricted to small islands. 'Small' islands were those with an area of less than 40 000 km^2, although additional data indicate that this is a continuous function (see McNab 1994).

As argued elsewhere (McNab 1994), the reduction of basal rate in species limited to small islands appears to reflect a limitation in available resources. This response is also found in rodents and fruit-eating pigeons living on small islands. Other means of reducing resource requirements on oceanic islands include the evolution of flightlessness in some birds and the evolution of a small body size (which is notable in small-island *Pteropus*, as well as in rodents and flightless rails living on small islands). The consumption of fruit, an abundant food on many tropical islands, may also facilitate long-term survival on oceanic islands.

A response to low temperature in large pteropodids

Recent measurements on the energetics of some tropical mammals weighing between 2 and 15 kg have shown that rate of metabolism at low environmental temperatures may equal, or be less than, the basal rate without compromising the maintenance of high core temperatures. This pattern has been seen in a tree-kangaroo (*Dendrolagus matschiei*) and the red panda (*Ailurus fulgens*) (McNab 1988) and in tropical, arboreal viverrids (McNab 1995) and lemurs (B.K. McNab pers. obs.). The decrease in rate of metabolism reflects a decrease in heat loss produced by peripheral vasoconstriction. At these times, body mass is divided into a core, in which body temperature is regulated, and a periphery (or shell), in which temperature is permitted to fall to near ambient levels (McNab 1988). This behaviour saves appreciable amounts of energy, but requires inactivity. It occurs, as far as we know, only in tropical, arboreal, solitary endotherms that are not exposed to cold temperatures for extended periods.

We now report this behaviour in intermediate to large species of the genus *Pteropus*. It is found in *Pt. giganteus* (739 g; Fig. 5) and *Pt. hypomelanus* (470 g), but not in *Pt. pumilus* (194 g) and *Pt. rodricensis* (255 g). Preliminary information indicates that this behaviour is found only in lemurs that weigh more than a kilogram. Why it occurs at smaller masses in pteropodids than in lemurs is unclear, unless small bats are more effective in separating a core from a shell by cutting off circulation to their wing membranes. *Pteropus* in the mass range from 255 to 470 g should be examined to determine the mass where this change in behaviour occurs. We note that Bartholomew, Leitner & Nelson (1964) reported no evidence of this behaviour in *Pt. scapulatus* (362 g) and that we found no evidence of its presence in *Dobsonia moluccense*, either in adult females (345 g) or in adult males (464 g). Its absence in *D. moluccense* might reflect body size or that this species has an unusually high basal rate.

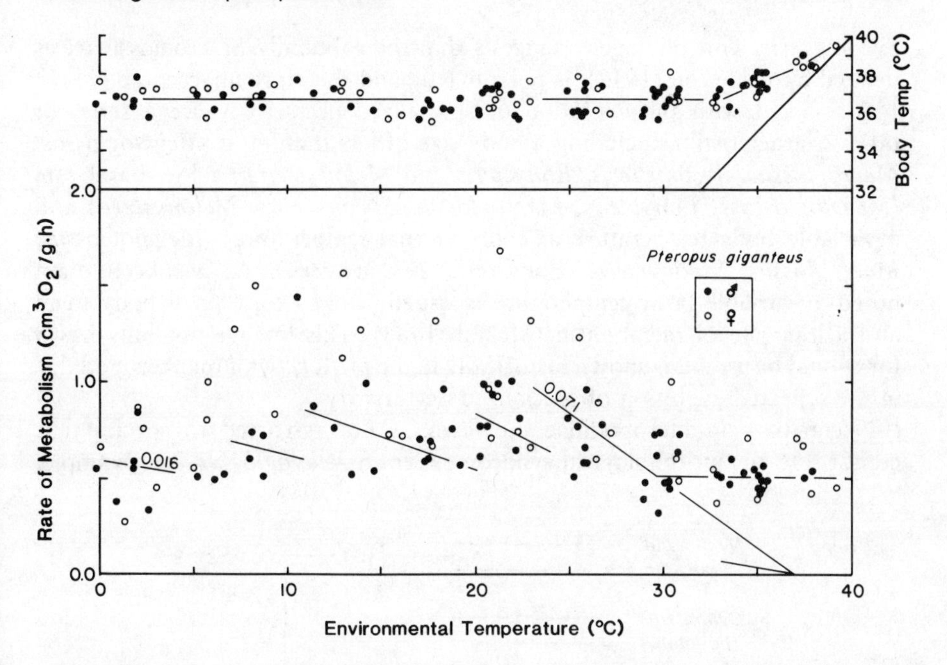

Fig. 5. Body temperature (°C) and rate of metabolism (cm^3O_2/g.h) in *Pteropus giganteus* as a function of environmental temperature (°C).

Species that show this behaviour not only are larger, but also are characterized by a low basal rate, suggesting that it is another means of reducing energy expenditure.

Energetics in a phylogenetic context

Physiological characters like temperature regulation and basal rate of metabolism, and ecological characters like fruit- or nectar-eating, can be placed in a phylogenetic context. We can then ask whether the evolutions of these physiological and ecological characteristics are connected. Such conclusions, of course, depend on the validity of the phylogeny used, and we urge caution because phylogenies change as new techniques and methodologies appear. The recent development of molecularly based phylogenies, such as those of Sibley & Ahlquist (1990) for birds, have had a radical impact on the perceived relationships of the organisms studied. A similar impact is expected as these techniques are applied to mammals.

With this caution in mind, we examine Kirsch and co-workers' molecular phylogeny of the Pteropodidae from a physiological and ecological viewpoint. Their phylogeny (Fig. 6) hypothesizes that specialized nectarivory in Megachiroptera possibly evolved five times, i.e. in *Megaloglossus*, in *Eonycteris*, in *Notopteris*, in *Melonycteris* and once in *Macroglossus* and

Syconycteris. This phylogeny suggests that the subfamily Macroglossinae, as defined by Andersen (1912), is polyphyletic and therefore unacceptable.

Associated with the evolution of specialized nectarivory are a series of other characteristics, including a body size of less than 60 g (*Megaloglossus, Macroglossus, Syconycteris, Eonycteris* and *Melonycteris*), a low basal rate (*Megaloglossus?, Eonycteris, Macroglossus, Syconycteris, Melonycteris*) and a variable body temperature at cool external temperatures (*Megaloglossus, Macroglossus, Syconycteris, Eonycteris, Melonycteris*). As has been often noted, a variable body temperature is usually linked to a small body mass and a low rate of metabolism (McNab 1983). This linkage not only has a functional basis, but is shown historically in the phylogeny of the Pteropodidae in the repeated evolution of specialized nectarivory.

Nectarivory in Pteropodidae, however, is not restricted to species that conform to the morphological syndrome seen in *Macroglossus*. For example,

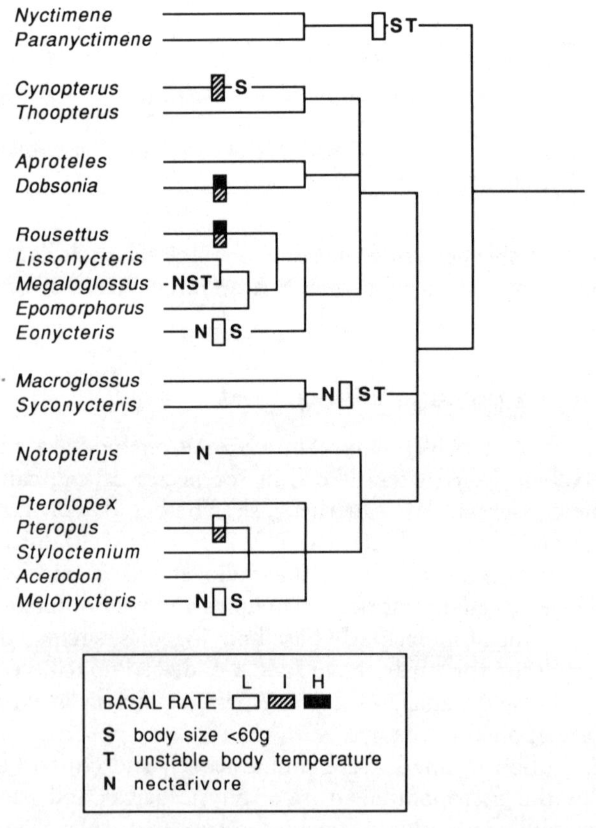

Fig. 6. A cladogram of the relationships within the Pteropodidae based on DNA hybridization by Kirsch and co-workers. Mapped on the cladogram are nectarivory (N), small body size (< 60 g, S), and rate of metabolism (low, L; intermediate, I; high, H).

Pteropus scapulatus (G. Richards pers. comm.) and *Pt. tonganus* (Wodzicki & Felten 1975) feed extensively on pollen and nectar without the development of a highly specialized morphology and a small size. Some *Rousettus* and *Cynopterus* may feed opportunistically on pollen and nectar, also without much morphological specialization. These species usually have an intermediate size and basal rate and a precisely regulated body temperature. Furthermore, the specialized nectarivores belonging to Glossophaginae, a subfamily of the Phyllostomidae, have both high basal rates and good temperature regulation (McNab 1969, 1989), coupled with a small body mass.

This analysis demonstrates that the connections existing amongst nectarivory, body size, rate of metabolism and temperature regulation are complicated, even if some elements of these relationships are understood. Thus, why do lowland nectarivorous pteropodids permit body temperature to fall to near-ambient levels, whereas this behaviour is unknown in nectarivorous phyllostomids of similar or smaller body sizes? An answer that this difference reflects phylogeny is far too simplistic.

Another distinctive group of pteropodids is the subfamily Nyctimeninae, the sister group to all other pteropodids (Fig. 6). The two species of this group that have been studied are characterized by rather small masses (<31 g), low basal rates and variable body temperatures. Again, these characters are connected both functionally and phylogenetically. Because the nyctimenines are the sister group to all other living pteropodids, these characteristics may be similar to those that were found in the 'protopteropodids' from which all pteropodids evolved. Nyctimenine bats are solitary, foliage-roosting, sedentary and cryptically coloured, characteristics that may be associated with low basal rates.

Most other pteropodids tend to be large (for bats), and have intermediate basal rates and generally highly effective temperature regulation. One exception is that subadult *Rousettus amplexicaudatus* from New Guinea showed marginal thermoregulation, even though they weighed 50–60 g. This failure may reflect an ontogenetic stage in the development of precise endothermy in this species, or it might suggest that the clade that includes *Rousettus, Lissonycteris, Megaloglossus, Epomophorus* and *Eonycteris* (Fig. 6) had a variable form of temperature regulation at its base and that this variability can be overcome only with the establishment of a sufficiently large body mass.

We conclude that the diversity in energy expenditure and temperature regulation is great in pteropodid bats. Much of this diversity is complexly associated with the variation in, and interactions among, body size, food habits and phylogeny.

Acknowledgements

We thank the many people that have made this work possible, especially the Lubee Foundation, Gainesville, Florida, and the Christensen Research

Institute, Madang, Papua New Guinea. We thank Mr John Sayaget, Lubee Foundation, and the directors of CRI, Drs Matthew Jebb and Larry Orsak. Dr J. A. W. Kirsch, University of Wisconsin, kindly permitted us to use his new phylogeny of the Pteropodidae. Dr Jon Reiskind, University of Florida, helped us with the intricacies of cladograms. We thank Monica Armstrong for providing us with the measurements on *Pteropus rodricensis*. The Department of Zoology, University of Florida, was helpful in many ways. This is publication number 114 in the CRI series and number 6 from the Lubee Foundation.

References

Andersen, K. (1912). *Catalogue of the Chiroptera in the collection of the British Museum.* **1.** *Megachiroptera.* (2nd edn). British Museum (Natural History), London.

Bartholomew, G. A., Dawson, W. R. & Lasiewski, R. C. (1970). Thermoregulation and heterothermy in some of the smaller flying foxes (Megachiroptera) of New Guinea. *Z. vergl. Physiol.* **70**: 196–209.

Bartholomew, G. A., Leitner, P. & Nelson, J. E. (1964). Body temperature, oxygen consumption, and heart rate in three species of Australian flying foxes. *Physiol. Zool.* **37**: 179–198.

Bonaccorso, F. J. (In press). *The bats of Papua New Guinea.* Christensen Research Institute, Madang, Papua New Guinea.

Hood, C. S. & Smith, J. D. (1989). Sperm storage in a tropical nectar-feeding bat, *Macroglossus minimus* (Pteropodidae). *J. Mammal.* **70**: 404–406.

Kulzer, E. & Storf, R. (1980). Schlaf-Lethargie bei dem africanischen Langzungenflughund *Megaloglossus woermanni* Pagenstecher, 1885. *Z. Säugetierk.* **45**: 23–29.

McNab, B. K. (1969). The economics of temperature regulation in neotropical bats. *Comp. Biochem. Physiol.* **31**: 227–268.

McNab, B. K. (1983). Energetics, body size, and the limits to endothermy. *J. Zool., Lond.* **199**: 1–29.

McNab, B. K. (1988). Complications inherent in scaling the basal rate of metabolism in mammals. *Q. Rev. Biol.* **63**: 25–54.

McNab, B. K. (1989). Temperature regulation and rate of metabolism in three Bornean bats. *J. Mammal.* **70**: 153–161.

McNab, B. K. (1994). Resource use and the occurrence of terrestrial and freshwater vertebrates on oceanic islands. *Am. Nat.* **144**: 643–660.

McNab, B. K. (1995). Energy expenditure and conservation in mixed-diet and frugivorous carnivorans. *J. Mammal.* **76**: 206–222.

Noll, U. G. (1979). Body temperature, oxygen consumption, noradrenaline response and cardiovascular adaptations in the flying fox, *Rousettus aegyptiacus. Comp. Biochem. Physiol.* (*A*) **63**: 79–88.

Sibley, C. G. & Ahlquist, J. E. (1990). *Phylogeny and classification of birds. A study in molecular evolution.* Yale University Press, New Haven.

Wodzicki, K. & Felten, H. (1975). The peka, or fruit bat (*Pteropus tonganus tonganus*) (Mammalia: Chiroptera), of Nuie Island, South Pacific. *Pacific Sci.* **29**: 131–138.

Symp. zool. Soc. Lond. (1995) No. 67: 123–138

Maternal investment and post-natal growth in bats

Thomas H. KUNZ
and April A. STERN

Department of Biology
Boston University
Boston, Massachusetts 02215, USA

Synopsis

We analysed post-natal growth data for body mass from 33 species of free-ranging and captive bats, using the logistic growth equation. When these data were examined by means of linear regression and covariance analysis, we found that growth rates decreased linearly with increasing asymptotic body mass. When we removed the effect of body mass, growth rates showed no significant differences with respect to diet (insect or fruit), taxonomic affiliation (Megachiroptera or Microchiroptera), growth condition (captive or free-ranging), or basal metabolic rate. Climate (tropical or temperate) was the only variable that had a significant effect on post-natal growth rates, with temperate bats growing faster than tropical species. This climatic effect was also evident when insectivorous bats were examined separately. While post-natal growth rates may provide a valuable index of maternal investment, milk energy output of females during lactation should provide the most direct link between the environment and growth of pups. Milk composition and milk-energy output as indices of maternal investment have been investigated in only a few species of bats. From the limited data available it appears that the milk of insectivorous species contains a higher percentage of dry matter, fat and protein than does that of frugivorous species.

Introduction

Although post-natal growth is an important life-history trait in vertebrates (Case 1978; Ricklefs 1979) and an important index of maternal investment (Oftedal 1984; Kirkwood 1985; Costa, Le Boeuf, Huntley & Ortiz 1986; Gittleman & Oftedal 1987; Oftedal & Gittleman 1988; McLaren 1993), little attention has been given to the proximate and evolutionary forces influencing post-natal growth rates in mammals. Moreover, few life-history analyses of mammals have included representatives of the Chiroptera, the second largest order of mammals in number of species. Previous analyses of life-history variation in this group have either grossly under-represented bats (e.g. Wootton 1987; Harvey & Read 1987; Read & Harvey 1989;

ZOOLOGICAL SYMPOSIUM No. 67
ISBN 0 19 854945 8

Promislow & Harvey 1990) or excluded them entirely (e.g. Millar 1981; Western & Ssemakula 1982; Martin 1984; Martin & MacLarnon 1985). Proximate factors known to influence growth rates include food supply, climate, habitat, maternal factors and social environment. In the following account, we examine interspecific patterns of post-natal growth in bats, and consider factors that may be important selective forces in moulding these patterns of growth.

Previous reviews of post-natal growth in bats (Orr 1970; Tuttle & Stevenson 1982) have been largely qualitative. Case's (1978) allometric analysis of post-natal growth in vertebrates included 17 species of bats, but his conclusions are limited because he analysed only the early linear period of growth and did not remove the effects of body size. Moreover, the growth data on bats that were available at the time of Case's analysis were strongly biased towards small insectivorous species (<30 g as adults), most of which were members of a single family (Vespertilionidae). From the various mammal species that Case (1978) analysed, he considered bats to have intermediate growth rates, although he gave no explanation for this pattern.

The objectives of the present review are to compare patterns of post-natal growth in the Chiroptera using data representing a wide range of asymptotic body masses (3.6 g to 557 g), phylogeny (Megachiroptera and Microchiroptera), diet (insects or fruit), climate (temperate or tropical), and study conditions (free-ranging or captive). We also review the available data on milk composition in bats. While our analyses of post-natal growth and milk composition are still limited to relatively few species, they offer some insight into factors that may affect maternal investment patterns in the Chiroptera.

Bats are unique among mammals in that females provide their young with milk until they achieve at least 90% of adult wing dimensions and at least 70% of adult (postpartum) body mass (Fig. 1). Generally, bat pups are not capable of sustained flight and foraging until they achieve adult wing dimensions, and thus remain nutritionally dependent on their mothers for longer periods of time than do the young of most terrestrial mammals (Barclay this volume pp. 245–258). Perhaps because of this constraint, bats have smaller litter sizes than terrestrial mammals, although litter masses at birth are comparable for the two groups (Kurta & Kunz 1987). Thus, the energy and nutrients that female bats invest in their pups should play a crucial role in the pattern of post-natal growth before pups achieve flight (Barclay 1994). Although individuals of some species begin to fly and feed on solid food well before they are fully weaned, patterns of post-natal growth should be strongly influenced by the quality and quantity of milk allocated to, and assimilated by, pups during the pre-weaning period.

Methods

Data on post-natal growth were analysed for 33 species of bats (Table 1), representing four of 17 families and both suborders (Megachiroptera and Microchiroptera). Because our analysis is based on less than 4% of the total number of species currently recognized (Koopman 1993), our conclusions must be considered preliminary. We evaluated published and unpublished data on post-natal growth (body mass, g) and used the logistic growth model for our analysis, because previous studies (Kunz & Robson in press) indicated that this model, when compared with von Bertalanffy and Gompertz models (Zullinger, Ricklefs, Redford & Mace 1984) best fits the empirical data for bats (also see Hughes, Rayner & Jones in press). Although forearm length is an important variable in growth studies (Kunz 1987), we chose body mass because it appears to be more sensitive to environmental variation (K. M. Hoying & T.H. Kunz unpubl). We used the Marquardt-Levenberg algorithm (Marquardt 1963) to derive growth parameters from the logistic equation:

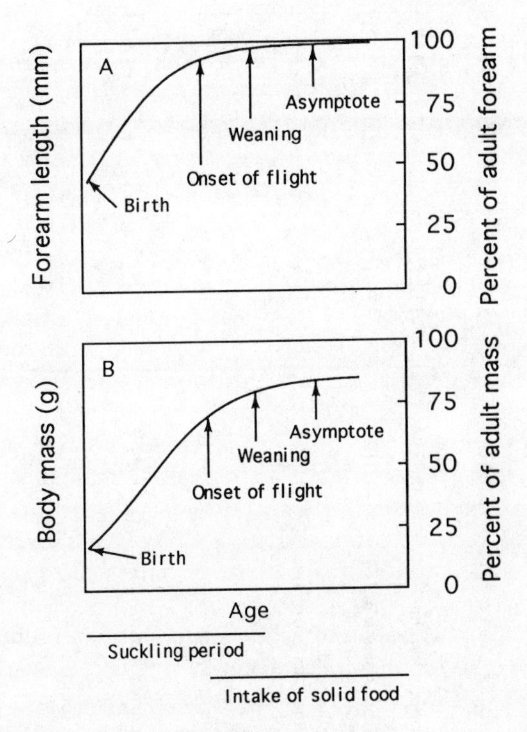

Fig. 1. Generalized model for post-natal growth in bats, showing the relationships between growth of (A) forearm length and (B) body mass, and the timing of birth, suckling, onset of flight, weaning and intake of solid food.

Table 1. Logistic growth constants of bats and their physiological and ecological correlates.

Taxonomic affiliation	Logistic growth parameters		Physiological and ecological correlates					Source
	Growth constant (K)	Asymptotic mass (A)	Adult mass (g)	BMR[a]	Diet[b]	Climate[c]	Condition[d]	
Megachiroptera								
Pteropodidae								
Hypsignathus monstrosus	0.0072	463.0	—	—	fr	tr	ca	Bradbury (1977)
Pteropus hypomelanus	0.0124	469.7	470.1	211.6	fr	tr	ca	T. H. Kunz (unpubl.); B. K. McNab (pers. comm.)
Pteropus poliocephalus	0.0238	557.0	598.0	316.9	fr	te	ca	T. H. Kunz (unpubl.); Bartholomew, Leitner & Nelson (1964)
Pteropus pumilus	0.0217	132.7	194.2	127.2	fr	tr	ca	T. H. Kunz (unpubl.); B. K. McNab (pers. comm.)
Pteropus scapulatus	0.0170	379.1	362.0	242.5	fr	tr	ca	G. O'Brien (pers. comm.)
Microchiroptera								
Emballonuridae								
Taphozous georgianus	0.0506	25.33	—	—	in	tr	fl	Jolly (1990)
Taphozous longimanus	0.0686	20.54	—	—	in	tr	fl	Krishna & Dominic (1983)
Molossidae								
Molossus molossus	0.1010	11.47	15.6	17.2	in	tr	ca	Häussler, Möller & Schmidt (1981); McNab (1969)
Tadarida brasiliensis	0.1680	10.63	10.4	20.8	in	te	fl	Kunz & Robson (in press); Herreid & Schmidt-Nielsen (1966)
Noctilionidae								
Noctilio albiventris	0.0573	25.37	27.0	23.8	in	tr	ca	Brown, Brown & Grinnell (1983); McNab (1989)
Phyllostomidae								
Artibeus jamaicensis	0.0340	52.89	45.0	56.3	fr	tr	ca	Taft & Handley (1991); McNab (1989)
Carollia perspicillata	0.0629	17.30	14.9	31.4	fr	tr	ca	Kleiman & Davis (1979); McNab (1989)
Desmodus rotundus	0.0109	26.90	29.0	26.4	bl	tr	ca	Schmidt & Manske (1973); McNab (1989)

Taxonomic affiliation	Logistic growth parameters		Physiological and ecological correlates					Source
	Growth constant (*K*)	Asymptotic mass (*A*)	Adult mass (g)	BMR[a]	Diet[b]	Climate[c]	Condition[d]	
Phyllostomus discolor	0.0530	42.41	34.0	35.02	fr	tr	ca	Rother & Schmidt (1985); McNab (1989)
Phyllostomus hastatus	0.0620	77.72	84.0	70.56	fr,in	tr	fl	A. Stern & T. H. Kunz (unpubl.); McNab (1989)
Rhinolophidae								
Rhinolophus ferrumequinum	0.1250	15.64	—	—	in	te	ca	R. Ransome *et al.*(unpubl.)
Vespertilionidae								
Antrozous pallidus	0.0955	22.27	00.0	00.00	in	te	ca	Brown (1976)
Eptesicus fuscus	0.1470	12.80	16.9	20.28	in	te	fl	Burnett & Kunz (1982); Herreid & Schmidt-Nielsen (1966)
Eptesicus serotinus	0.0404	28.84	—	—	in	te	ca	Kleiman (1969)
Miniopterus schreibersii	0.0744	15.47	—	—	in	te	fl	Dwyer (1963)
Myotis daubentonii	0.0690	9.75	—	—	in	te	fl	Krátký (1981)
Myotis lucifugus	0.2242	6.70	6.5	9.3	in	te	fl	Kunz & Anthony (1982); Hock (1951)
Myotis myotis	0.1500	20.91	—	—	in	te	fl	Krátký (1970)
Myotis velifer	0.1670	8.94	11.9	7.7	in	te	fl	T. H. Kunz & S. K. Robson (unpubl.); Riedesel & Williams (1976)
Nyctalus lasiopterus	0.1220	29.66	—	—	in	te	ca	Maeda (1972)
Nyctalus noctula	0.1310	19.77	—	—	in	te	ca	Kleiman (1969)
Pipistrellus mimus	0.1103	3.36	—	—	in	tr	fl	S. Isaac & G. Marimuthu (pers. comm.)
Pipistrellus pipistrellus	0.1198	3.95	—	—	in	te	fl	Rakhmatulina (1971)
Pipistrellus savii	0.1870	5.48	—	—	in	te	fl	Tiunov (1992)
Pipistrellus subflavus	0.1340	4.98	—	—	in	te	fl	K. M. Hoying & T. H. Kunz (unpubl.)
Plecotus auritus	0.2500	5.87	—	—	in	te	ca	De Fanis & Jones (1995)
Scotophilus heathii	0.1033	28.57	—	—	in	tr	fl	Krishna & Dominic (1983)
Vespertilio superans	0.1680	11.66	—	—	in	te	fl	Tiunov (1989)

[a]Basal metabolic rate. [b]fr = fruit; in = insects; bl = blood. [c]tr = tropical; te = temperate. [d]fl = field; ca = captivity

$$M(t) = A\{e^{-K(t-I)} + 1\}^{-1}$$

where M = mass (g), A = asymptotic size, K = growth constant (days^{-1}), and I = inflection point.

In theory asymptotic body mass is achieved by bats when the post-natal growth rate becomes zero, usually in the season of birth. Asymptotic mass of young bats is usually less than adult mass, because it does not include accretionary growth after the first year, nor does it include the deposition of fat in autumn which is characteristic of most temperate species in the post-weaning period. When possible, growth parameters were derived directly from the analysis of original growth data. However, because few published studies on post-natal growth included original data, either graphically or in tabular form, we derived growth curves (and growth parameters) from published graphs or from plotted data points using the program DigiMatic (Version 2.0.1 for Macintosh). This software made it possible to represent data points and line graphs digitally from published and unpublished growth curves that otherwise would not be available for comparison. Graphs were plotted using SigmaPlot (Version 4.11 for Macintosh).

Our allometric analysis is based on data for post-natal growth rates for both free-ranging and captive bats. In the case of captive bats, only the growth data from young suckled by their mothers were used for analysis—data on hand-fed bats were excluded. We did not include data for some species because the original data or curves did not allow us to compute each of the parameters for the logistic growth equation. Thus, we could not use data on post-natal growth that only included the early linear phase of post-natal growth (e.g. O'Farrell & Studier 1973; Tuttle 1975; McWilliam 1987). Moreover, we did not include data on growth curves derived from cross-sectional (grab) samples of wild populations (e.g. Short 1961; Medway 1972; Pagels & Jones 1974; Yokoyama, Ohtsu & Uchida 1979; Thomas & Marshall 1984), or when we judged the asymptotic body masses to deviate markedly from those of free-ranging populations (e.g. Jones 1967; Noll 1979). When growth data were available for both sexes in sexually dimorphic species, we only used data for females. When growth rates did not differ between sexes, growth parameters were based on combined data. Ideally, we would have preferred to use growth data only for free-ranging bats, but obtaining post-natal growth data on some free-ranging species is impractical, especially for the large pteropodids and small solitary species.

For allometric analyses, we used $\log_{10}$ transformations on continuous variables, including the logistic growth constant, asymptotic body mass and basal metabolic rate, to improve the symmetry of data for analysis (Hoaglin, Mosteller & Tukey 1983). We chose least-squares regression analysis instead of reduced major axis analysis to maintain a constant dimensionality of slopes across different data sets. We used regression analysis to test for significant allometric relationships, and analysis of covariance (ANCOVA) to test for differences in slopes and elevations of these regressions. We first tested each data set for homogeneity of variance. If the slopes of regressions were the same, we then tested for differences in intercepts.

The $\log_{10}$ of the logistic growth constant (K) for pup mass was used as the dependent variable; independent variables included $\log_{10}$ asymptotic mass and BMR (continuous variables). Categorical variables included taxonomic affiliation (Megachiroptera and Microchiroptera), diet (insects and fruit), climate (temperate and tropical) and growth condition (captive and free-ranging). We chose the species as the primary unit for allometric analysis, although there are criticisms of this approach (see Harvey & Mace 1982). Suborder designations were used to test for taxonomic affiliation because sample sizes for most taxonomic groups (e.g., genus or family) were too small for meaningful analysis. Climate designations (temperate and tropical) probably are not independent of diet, because frugivorous species are known almost exclusively from tropical climates (except *Pteropus poliocephalus*); however, because obligate insectivory is found in both temperate and tropical regions (six tropical and 17 temperate species in the present study), we used these data to test for climatic effects.

Results and discussion

Interspecific patterns of post-natal growth

Among the 33 species analysed, we found a significant negative correlation between post-natal growth rate and asymptotic body mass ($r^2 = 67.0\%$, $F = 65.87$, $P < 0.0005$) (Fig. 2). To examine the effects of diet, we omitted the one omnivore (*Phyllostomus hastatus*) and sanguivore (*Desmodus rotundus*) from the data set. With these two species removed the correlation of the allometric relationship increased by 10.8% ($r^2 = 77.8\%$, $F = 106.05$, $P < 0.0005$). When the effects of body mass were removed, the additional increase (+ 1.9%) due to diet (insects and fruit) was not significant ($F = 3.68$, $P = 0.065$). However, the coefficient of determination for frugivorous species ($r^2 = 70.6\%$, $n = 8$) was greater than that for insectivorous species ($r^2 = 26.5\%$, $n = 23$). This suggests that fruit-eating bats may have greater nutritional constraints and feed on a narrower range of food quality. We found no significant phylogenetic effect (Megachiroptera vs. Microchiroptera) on post-natal growth ($F = 0.2601$, $P = 0.614$) after removing the effect of body mass (Fig. 3a). However, the absence of a significant phylogenetic effect is not surprising, given the small sample of megachiropterans ($n = 5$) available for analysis. Moreover, we found no significant effect of growth condition (captive vs. field) on post-natal growth ($F = 1.493$, $P = 0.231$) after removing the effect of body mass (Fig. 3b). However, because the variance in captive versus field data for log growth rates differs significantly ($P < 0.05$, captive, $r^2 = 62.0\%$, and free-ranging, $r^2 = 25.5\%$), the statistical validity of this comparison remains questionable.

After the effect of body mass was removed, climate (temperate vs. tropical) was the only extrinsic variable that significantly affected post-natal growth rates (F = 9.63, P = 0.004), indicating that temperate species have higher growth rates than their tropical counterparts (Fig. 3c). Because frugivorous species are known primarily from tropical regions (except *Pteropus poliocephalus*) and insectivorous species are known from both climatic regions, we examined the effects of climate by comparing growth rates between temperate and tropical insectivorous species. From this analysis, the negative allometric relationship (r^2 = 26.6%) for all insectivorous bats was significant (F = 8.93, P = 0.007) and post-natal growth rates of temperate insectivorous species were significantly greater (F = 4.72, P = 0.042) than those of insectivorous bats from tropical regions (Fig. 3d). The effects of climate increased (+ 11%) the overall allometric relationship among insectivorous species to 37.6% but the separate coefficients of determination for tropical (21.0%, n = 6) and temperate insectivores (21.7%, n = 17) were lower. Asymptotic body mass had no significant effect on post-natal growth for tropical insectivorous species

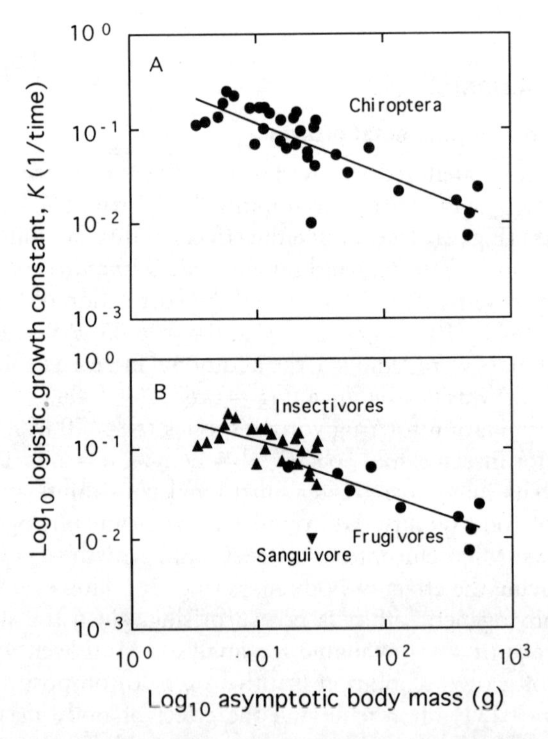

Fig. 2. (A) Allometric relationship between post-natal growth rates and asymptotic body mass in the Chiroptera. (B) After removing the effect of asymptotic body mass, there was no significant effect of diet (insectivory or frugivory) on post-natal growth. The single sanguivore was not included in this analysis, but it appears to have a relatively slow growth rate for its body mass.

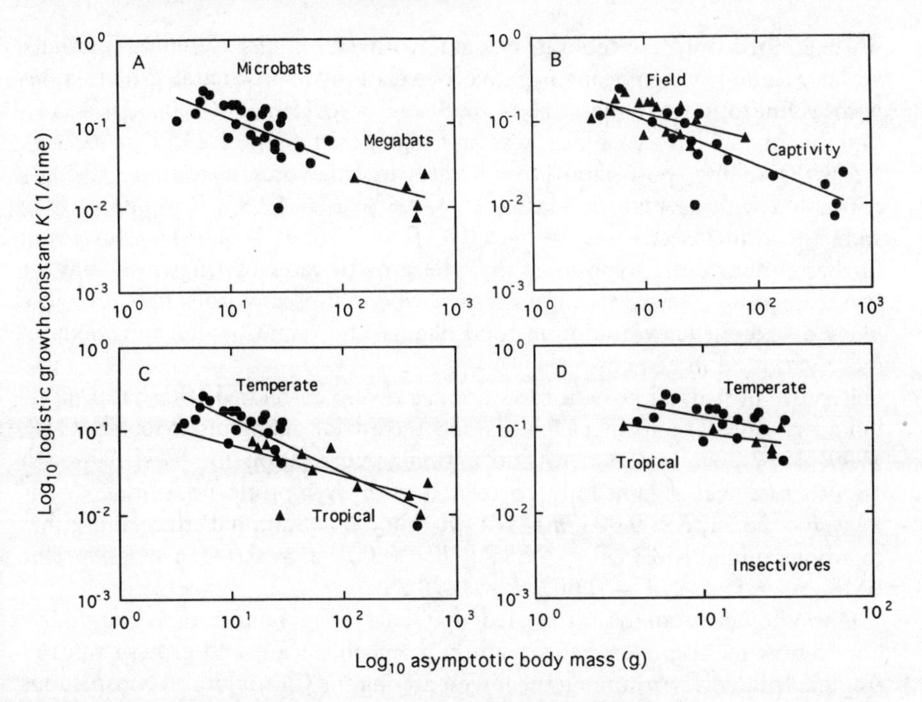

Fig. 3. (A) Allometric relationship between post-natal growth rate, asymptotic body mass and phylogeny (megabats or microbats). When the effect of body mass was removed there was no significant effect of phylogeny on post-natal growth rate. (B) Allometric relationship between post-natal growth rates, asymptotic body mass and growth condition (captive or free-ranging). When the effect of body mass was removed, there was no significant effect of growth condition on post-natal growth rate. (C) Allometric relationship between post-natal growth rate, asymptotic body mass and climate (temperate or tropical). When the effect of body mass was removed there was a significant effect of climate. (D) Allometric relationship between post-natal growth rates and asymptotic body mass for temperate and tropical insectivorous bats. When the effect of body mass was removed there was a significant difference in growth rates between these two groups.

($P = 0.201$), perhaps because of small sample size. Thus, tropical species appear to have growth rates that are independent of asymptotic body mass, whereas temperate species are negatively correlated. When we examined the effects of study condition on post-natal growth rates of insectivorous species, we found no significant differences ($F = 1.29$, $P = 0.89$), although the coefficient of determination for laboratory-reared bats ($r^2 = 52.0\%$, $n = 7$) was 3.6 times greater than that for free-ranging bats ($r^2 = 14.2\%$, $n = 16$).

The higher coefficient of determination for bats reared in captivity as opposed to free-ranging bats suggests that extrinsic factors may have a more important influence on post-natal growth rates in captive than in free-ranging bats. This observation suggests that the nutritional conditions in captivity may push post-natal growth rates to their maximum, whereas growth conditions experienced by free-ranging bats appear to be more variable. When we examined only the laboratory data (with *Desmodus rotundus* omitted), we also found a significant negative correlation of post-natal growth rates with asymptotic body mass ($r^2 = 82.6\%$, $F = 67.54$, $P < 0.005$, $n = 15$), but the overall effect of diet was not significant ($F = 1.23$, $P = 0.289$). Notwithstanding, post-natal growth rates in frugivores were more strongly correlated with asymptotic mass ($r^2 = 70.6\%$, $F = 17.85$, $P = 0.006$) than were those in insectivores ($r^2 = 52\%$, $F = 7.50$, $P = 0.041$). This result further supports the hypothesis that the growth rates of frugivores may be constrained by a relatively uniform diet, whereas insectivorous bats are more likely to experience variation in food quality and quantity and, thus, exhibit more variation in growth rates.

Because post-natal growth rates are negatively correlated with basal metabolic rate (BMR) and diet in birds and terrestrial mammals (McNab 1988, 1989, 1992), we expected to find a similar correlation for bats. Although growth rate was significantly correlated with asymptotic body mass ($r^2 = 62.7$, $F = 26.21$, $P < 0.005$, $n = 16$), we found no additional effects due either to whole animal BMR ($r^2 = 59.8\%$), $F = 0.005$, $P = 0.94$) or mass-specific BMR ($r^2 = 59.8\%$, $F = 0.002$, $P < 0.96$).

Our allometric analysis revealed that bats from temperate regions have higher post-natal growth rates than do tropical species, and growth rates of tropical frugivores are among the lowest among the Chiroptera. A conspicuous exception to this generalization is the sanguivorous vampire bat, *Desmodus rotundus*, which has the slowest growth rate ($K = 0.01$) among all the species that we examined (Fig. 2, Table 1). Unfortunately, data on post-natal growth rates for free-ranging frugivorous bats do not exist, partly because many fruit-eating species are solitary (or occupy inaccessible foliage roosts) and they are less amenable to study than are gregarious species which live in caves or buildings. If frugivorous bats in captivity are being pushed to their maximum growth limits owing to high planes of nutrition, we would expect free-ranging species overall to have lower rates than those observed in captivity. We suggest that frugivorous species that roost in caves and other enclosed roosts should be the focus of future studies, to help clarify these effects of diet and growth conditions on post-natal growth rates in bats.

The majority of species included in our analysis were from temperate regions and feed primarily on insects and other arthropods. No data on temperate frugivorous or omnivorous species exist, so the only valid comparison we could make was between temperate and tropical insectivorous taxa. Our analysis shows that temperate and tropical insectivorous species differ in growth rates, suggesting a strong selection pressure for faster growth rates in cooler climates. One hypothesis to account for higher growth rates at more northern latitudes is that bats may be selected in response to a shorter growing season. This would allow individuals to achieve maximum somatic growth and to deposit important fat reserves in preparation for winter hibernation or migration. Our interpretation is consistent with Boyce's (1979) hypothesis that accelerated growth should be found among mammals living in highly seasonal environments.

Methodological considerations

Observed post-natal growth rates in bats may be affected by the methods used to collect the empirical data as well as by the methods used in their analysis. Acceptable methods for collecting empirical growth data for bats were reviewed by Kunz (1987). These include marking and measuring individual pups at birth and monitoring changes in their linear size and their body mass on subsequent days. This method is equally appropriate for free-ranging and captive situations. In field situations, however, successful recaptures of pups will vary, depending upon the size of the colony, the fidelity that mothers and their pups show to the roost and the access that investigators have to the pups. Growth rates derived from measurements of pups captured on different dates (so-called cross-sectional samples) are less desirable and are usually biased towards measurements of smaller and younger pups, because they are usually easier to capture than older, more elusive individuals. Thus, growth rates based on cross-sectional samples are usually lower than those determined from capture–recapture data (Kunz 1987). In studies where cross-sectional samples have been used in growth analysis, corrections based on known-age individuals were necessary to compensate for this potential bias (see Tuttle 1975).

Milk composition

Reports of milk composition for a wide range of terrestrial and marine mammals show considerable interspecific variation in both quality and quantity produced (Oftedal 1984, 1985; Ortiz, Le Boeuf & Costa 1984). Much of this variation can be attributed to differences in maternal diet. For example, carnivorous and insectivorous species typically produce milk higher in protein and fat than do those which feed on mixed diets of plants and animals or exclusively on plant parts. Even among animal-eating mammals, marked differences in milk composition reflect the general composition of their diets (Gittleman & Oftedal 1987). Thus, we would predict that bats which feed on fruit and nectar should have slower post-natal growth rates than those which feed on insects and vertebrates. Why sanguivorous bats should have such slow growth rates remains puzzling, especially given the high protein content of blood available to lactating females.

Although milk composition has been investigated for several species of bats (Huibregtse 1966; Jenness & Studier 1976; Kunz, Stack & Jenness 1983; Quicke, Sowler, Berry & Geddes 1984), few data are available for meaningful interspecific comparisons, either because the stage of lactation was not reported or because sample sizes were too small to be representative. Moreover, few reliable data are available for making comparisons across dietary categories. From our analyses of milk samples of four species of insectivorous bats, *Eptesicus fuscus, Myotis lucifugus, Myotis velifer* and *Tadarida brasiliensis* (Kunz, Stack *et al.* 1983; Kunz, Oftedal, Robson, Kretzmann & Kirk in press), one omnivorous species (*Phyllostomus hastatus*), and three frugivorous species (*Pteropus hypomelanus, P. rodricensis*, and *P. vampyrus*) (T. H. Kunz, A. A. Stern & O. T. Oftedal unpubl.), significant interspecific differences are evident. Milks of insectivorous bats generally have higher dry matter, fat and protein content than do those of omnivorous or frugivorous species. The relatively low fat, protein and dry matter content of milk from frugivores is consistent with the relatively low fat and protein content of fruit eaten by most plant-visiting bats (Morrison 1980; Fleming 1988; Kunz & Diaz in press). Although we found no significant effect of adult diet on post-natal growth in bats based on our allometric analysis ($P = 0.065$), we expect that milk composition and milk-energy output will prove to be important factors influencing post-natal growth rates in bats.

Conclusions

Previous comparative studies on post-natal growth of bats have been mostly qualitative. Results of our comparative analysis, showing an inverse relationship between post-natal growth rates and asymptotic body mass, are consistent with findings reported for other mammalian taxa. The significant negative correlation between post-natal growth rates in bats and climate (temperate or tropical) is consistent with life-history models that predict higher post-natal growth rates in highly seasonal environments. Although basal metabolic rate (BMR) was highly correlated with body mass and growth rates in other mammalian taxa, we found no significant effect of BMR on growth rates in bats that could not be explained by asymptotic body mass. Although post-natal growth rates in bats and other mammals can provide valuable indices of maternal investment, milk-energy output of females during lactation should offer the most direct link between environmental effects and pup growth. We suggest that future studies on post-natal growth in bats include estimates of food availability, maternal diet, milk composition and milk-energy output of mothers, time of first solid food, duration of lactation, and the maternal pup environment, to help clarify factors that influence post-natal growth rates.

Acknowledgements

We are grateful to A. Field, K. Atkinson, J. Seyjagat and F. Bonaccorso for their assistance in collecting growth data on bats housed at the Lubee Foundation, B.K. McNab for providing unpublished estimates of BMR, P. Hughes, G. Jones, S. Isaac, G. O'Brien, R. Ransome and M. Tiunov, for making original data on post-natal growth available for our analysis. We also thank D. Jones who helped to digitize graphical data, and V. Hayssen and

S. Robson for assistance on statistical matters. V. Hayssen also kindly reviewed and made helpful suggestions on the manuscript. This paper was made possible by grants to T.H.K. from the National Science Foundation (BSR 87–00585) and the Lubee Foundation, Inc.

References

Barclay, R. M. R. (1994). Constraints on reproduction by flying vertebrates: energy and calcium. *Am. Nat.* **144**: 1021–1031.

Bartholomew, G. A., Leitner, P. & Nelson, J. E. (1964). Body temperature, oxygen consumption, and heart rate in three species of Australian flying foxes. *Physiol. Zool.* **37**: 179–198.

Boyce, M. S. (1979). Seasonality and patterns of natural selection for life histories. *Am. Nat.* **114**: 569–583.

Bradbury, J. W. (1977). Lek mating behavior in the hammer-headed bat. *Z. Tierpsychol.* **45**: 225–255.

Brown, P. (1976). Vocal communication in the pallid bat, *Antrozous pallidus*. *Z. Tierpsychol.* **41**: 34–54.

Brown, P. E., Brown, T. W. & Grinnell, A. D. (1983). Echolocation, development, and vocal communication in the lesser bulldog bat, *Noctilio albiventris*. *Behav. Ecol. Sociobiol.* **13**: 287–298.

Burnett, C. D. & Kunz, T. H. (1982). Growth rates and age estimation in *Eptesicus fuscus* and comparison with *Myotis lucifugus*. *J. Mammal.* **63**: 33–41.

Case, T. J. (1978). On the evolution and adaptive significance of postnatal growth rates in the terrestrial vertebrates. *Q. Rev. Biol.* **53**: 243–282.

Costa, D. P., Le Boeuf, B. J., Huntley, A. C. & Ortiz, C. L. (1986). The energetics of lactation in the Northern elephant seal, *Mirounga angustirostris*. *J. Zool., Lond.(A)* **209**: 21–33.

De Fanis, E. & Jones, G. (1995). Post-natal growth, mother-infant interactions and development of vocalizations in the vespertilionid bat, *Plecotus auritus*. *J. Zool., Lond.* **235**: 85–97.

Dwyer, P. D. (1963). The breeding biology of *Miniopterus schreibersi blepotis* (Temminck) (Chiroptera) in north-eastern New South Wales. *Aust. J. Zool.* **11**: 219–240.

Fleming, T. H. (1988). *The short-tailed fruit bat. A study in plant-animal interactions.* University of Chicago Press, Chicago.

Gittleman, J. L. & Oftedal, O. T. (1987). Comparative growth and lactation energetics in carnivores. *Symp. zool. Soc. Lond.* No. **57**: 41–77.

Harvey, P. H. & Mace, G. M. (1982). Comparison between taxa and adaptive trends: problems of methodology. In *Current problems in sociobiology*: 343–361. (Eds Kings' College Sociobiology Group). Cambridge University Press, Cambridge.

Harvey, P. H. & Read, A. F. (1987). How and why do mammalian life histories vary? In *Evolution and life histories of mammals: theory and pattern:* 213–232. (Ed. Boyce, M. S.). Yale University Press, New Haven.

Häussler, U., Möller, E. & Schmidt, U. (1981). Zur Haltung und Jugendentwicklung von *Molossus molossus*. *Z. Säugetierk.* **46**: 337–351.

Herreid, C. F., II & Schmidt-Nielsen, K. (1966). Oxygen consumption, temperature, and water loss in bats from different environments. *Am. J. Physiol.* **211**: 1108–1112.

Hoaglin, D. C., Mosteller, F. & Tukey, J. F. (1983). *Understanding robust and exploratory data analysis.* John Wiley and Sons, New York.

Hock, R. J. (1951). The metabolic rates and body temperatures of bats. *Biol. Bull. mar. biol. Lab., Woods Hole* **101**: 289–299.

Hughes, P., Rayner, J. M. V. & Jones, G. (In press). Ontogeny of 'true' flight and other aspects of growth in the bat *Pipistrellus pipistrellus. J. Zool., Lond.*

Huibregtse, W. H. (1966). Some chemical and physical properties of bat milk. *J. Mammal.* **47**: 551–554.

Jenness, R. & Studier, E. H. (1976). Lactation and milk. *Spec. Publs Mus. Texas Tech Univ.* No. 10: 201–218.

Jolly, S. E. (1990). The biology of the common sheath-tail bat, *Taphozous georgianus* (Chiroptera: Emballonuridae), in central Queensland. *Aust. J. Zool.* **38**: 65–77.

Jones, C. (1967). Growth, development, and wing loading in the evening bat, *Nycticeius humeralis* (Rafinesque). *J. Mammal.* **48**: 1–19.

Kirkwood, J. K. (1985). Patterns of growth in primates. *J. Zool., Lond. (A)* **205**: 123–136.

Kleiman, D. G. (1969). Maternal care, growth rate and development in the Noctule (*Nyctalus noctula*), Pipistrelle (*Pipistrellus pipistrellus*) and Serotine (*Eptesicus serotinus*) bats. *J. Zool., Lond.* **157**: 187–211.

Kleiman, D. G. & Davis, T. M. (1979). Ontogeny and maternal care. *Spec. Publs Mus. Texas Tech Univ.* No. 16: 387–402.

Koopman, K. F. (1993). Chiroptera. In *Recent mammals of the world*: 137–242. (Eds Wilson, D. E. & Reeder, D. M.). Smithsonian Institution Press, Washington, D.C.

Krátký, J. (1970). Postnatale Entwicklung des Grossmausohrs, *Myotis myotis* (Borkhausen, 1797). *Acta Soc. zool. bohemoslov.* **34**: 202–218.

Krátký, J. (1981). Postnatale Entwicklung der Wasserfledermaus, *Myotis daubentoni* Kuhl, 1819 und bisherige Kenntnis dieser Problematik in Rahmen der Unterordnung Microchiroptera (Mammalia: Chiroptera). *Folia Mus. Rerum nat. Bohemiae occident.* **16**: 3–34.

Krishna, A. & Dominic, C. J. (1983). Growth of young and sexual maturity in three species of Indian bats. *J. Anim. Morph. Physiol.* **30**: 162–168.

Kunz, T. H. (1987). Post-natal growth and energetics of suckling bats. In *Recent advances in the study of bats*: 395–420. (Eds Fenton, M. B., Racey, P. & Rayner, J. M. V.). Cambridge University Press, Cambridge.

Kunz, T. H. & Anthony, E. L. P. (1982). Age estimation and post-natal growth in the bat *Myotis lucifugus. J. Mammal.* **63**: 23–32.

Kunz, T. H. & Diaz, C. A. (In press). Folivory in fruit-eating bats, with new evidence from *Artibeus jamaicensis* (Chiroptera: Phyllostomidae). *Biotropica.*

Kunz, T. H., Oftedal, O. T., Robson, S. K., Kretzmann, M. B. & Kirk, C. (In press). Changes in milk composition during lactation in three species of insectivorous bats. *J. comp. Physiol.*

Kunz, T. H. & Robson, S. K. (In press). Post-natal growth and development of the Mexican free-tailed bat, *Tadarida brasiliensis*: birth size, growth rates and age estimation. *J. Mammal.*

Kunz, T. H., Stack, M. H. & Jenness, R. (1983). A comparison of milk composition in *Myotis lucifugus* and *Eptesicus fuscus* (Chiroptera: Vespertilionidae). *Biol. Reprod.* **28**: 229–234.

Kurta, A. & Kunz, T. H. (1987). Size of bats at birth and maternal investment during pregnancy. *Symp. zool. Soc. Lond.* No. **57**: 79–106.

McLaren, I. A. (1993). Growth in pinnipeds. *Biol. Rev.* **68**: 1–79.

McNab, B. K. (1969). The economics of temperature regulation in neotropical bats. *Comp. Biochem. Physiol.* **31**: 227–268.
McNab, B. K. (1988). Food habits and the basal rate of metabolism in birds. *Oecologia* 77: 343–349.
McNab, B. K. (1989). Complications inherent in scaling the basal rate of metabolism in mammals. *Q. Rev. Biol.* **63**: 25–54.
McNab, B. K. (1992). A statistical analysis of mammalian rates of metabolism. *Funct. Ecol.* **6**: 672–679.
McWilliam, A. N. (1987). The reproductive and social biology of *Coleura afra* in a seasonal environment. In *Recent advances in the study of bats*: 324–350. (Eds Fenton, M. B., Racey, P. & Rayner, J. M. V.). Cambridge University Press, Cambridge.
Maeda, K. (1972). Growth and development of large noctule, *Nyctalus lasiopterus* Schreber. *Mammalia* **36**: 269–278.
Marquardt, D. W. (1963). An algorithm for least squares estimation of non-linear parameter. *J. indust. appl. Math.* **11**: 431–441.
Martin, R. D. (1984). Scaling effects and adaptive strategies in mammalian lactation. *Symp. zool. Soc. Lond.* No. 51: 87–117.
Martin, R. D. & MacLarnon, A. M. (1985). Gestation period, neonatal size and maternal investment in placental mammals. *Nature, Lond.* **313**: 220–223.
Medway, Lord (1972). Reproductive cycles of the flat-headed bats *Tylonycteris pachypus* and *T. robustula* (Chiroptera: Vespertilioninae) in a humid equatorial environment. *Zool. J. Linn. Soc.* **51**: 33–61.
Millar, J. S. (1981). Pre-partum reproductive characteristics of eutherian mammals. *Evolution* **35**: 1149–1163.
Morrison, D. W. (1980). Efficiency of food utilization by fruit bats. *Oecologia* **45**: 270–273.
Noll, U. G. (1979). Postnatal growth and development of thermogenesis in *Rousettus aegyptiacus. Comp. Biochem. Physiol.* (*A*) **63**: 89–93.
O'Farrell, M. J. & Studier, E. H. (1973). Reproduction, growth, and development in *Myotis thysanodes* and *M. lucifugus* (Chiroptera: Vespertilionidae). *Ecology* **54**: 18–30.
Oftedal, O. T. (1984). Milk composition, milk yield and energy output at peak lactation: a comparative review. *Symp. zool. Soc. Lond.* No. 51: 33–85.
Oftedal, O. T. (1985). Pregnancy and lactation. In *The bioenergetics of wild herbivores*: 215–238. (Eds Hudson, R. J. & White, R. G.). CRC Press, Boca Raton, Florida.
Oftedal, O. T. & Gittleman, J. L. (1988). Patterns of energy output during reproduction in carnivores. In *Carnivore behavior, ecology, and evolution*: 355–378. (Ed. Gittleman, J. L.). Chapman & Hall, London, and Cornell University Press, Ithaca.
Orr, R. T. (1970). Development: prenatal and postnatal. In *Biology of bats* **1**: 217–231. (Ed. Wimsatt, W. A.). Academic Press, New York.
Ortiz, C. L., Le Boeuf, B. J. & Costa, D. P. (1984). Milk intake of elephant seal pups: an index of parental investment. *Am. Nat.* **124**: 416–422.
Pagels, J. F. & Jones, C. (1974). Growth and development of the free-tailed bat, *Tadarida brasiliensis cynocephala* (Le Conte). *SWest. Nat.* **19**: 267–276.
Promislow, D. E. L. & Harvey, P. H. (1990). Living fast and dying young: a comparative analysis of life-history variation among mammals. *J. Zool., Lond.* **220**: 417–437.
Quicke, G. V., Sowler, S., Berry, R. K. & Geddes, A. M. (1984). Composition of mammary secretion from the epauletted fruit bat, *Epomophorus wahlbergi. S. Afr. J. Sci.* **80**: 481– 482.

Rakhmatulina, I. K. (1971). The breeding, growth, and development of pipistrelles in Azerbaidzhan. *Soviet J. Ecol.* **2**: 131–136. (English translation.)

Read, A. F. & Harvey, P. H. (1989). Life history differences among the eutherian radiations. *J. Zool., Lond.* **219**: 329–353.

Riedesel, M. L. & Williams, B. A. (1976). Continuous 24-hr oxygen consumption studies of *Myotis velifer*. *Comp. Biochem. Physiol. (A)* **54**: 95–99.

Ricklefs, R. W. (1979). Adaptation, constraint, and compromise in avian postnatal development. *Biol. Rev.* **54**: 269–290.

Rother, G. von & Schmidt, U. (1985). Die ontogenetische Entwicklung der Vokalisation bei *Phyllostomus discolor* (Chiroptera). *Z. Säugetierk.* **50**: 17–26.

Schmidt, U. & Manske, U. (1973). Die Jugendentwicklung der Vampirfledermäuse (*Desmodus rotundus*). *Z. Säugetierk.* **38**: 14–33.

Short, H. L. (1961). Growth and development of Mexican free-tailed bats. *SWest. Nat.* **6**: 156–163.

Taft, L. K. & Handley, C. O. Jr. (1991). Reproduction in a captive colony. *Smithson. Contr. Zool.* No. 511: 19–41.

Thomas, D. W. & Marshall, A. G. (1984). Reproduction and growth in three species of West African fruit bats. *J. Zool., Lond.* **202**: 265–281.

Tiunov, M. P. (1989). The postnatal growth and development of *Vespertilio superans* (Chiroptera). *Zool. Zh.* **68**: 156–160. (In Russian.)

Tiunov, M. P. (1992). Parturition, postnatal growth and development of youngs in *Hypsugo savii* (Chiroptera). *Zool. Zh.* **71**: 91–95. (In Russian.)

Tuttle, M. D. (1975). Population ecology of the gray bat (*Myotis grisescens*): factors influencing early growth and development. *Occ. Pap. Mus. nat. Hist. Univ. Kansas* No. 36: 1–24.

Tuttle, M. D. & Stevenson, D. (1982). Growth and survival of bats. In *Ecology of bats*: 105–150. (Ed. Kunz, T.H.). Plenum Press, New York.

Western, D. & Ssemakula, J. (1982). Life history patterns in birds and mammals and their evolutionary interpretation. *Oecologia* **54**: 281–290.

Wootton, J. T. (1987). The effects of body mass, phylogeny, habitat, and trophic level on mammalian age at first reproduction. *Evolution* **41**: 732–749.

Yokoyama, K., Ohtsu, R. & Uchida, T. A. (1979). Growth and LDH isozyme patterns in the pectoral and cardiac muscles of the Japanese Lesser horseshoe bat, *Rhinolophus cornutus cornutus* from the standpoint of adaptation for flight. *J. Zool., Lond.* **187**: 85–96.

Zullinger, E. M., Ricklefs, R. E., Redford, K. H. & Mace, G. M. (1984). Fitting sigmoidal equations to mammalian growth curves. *J. Mammal.* **65**: 607–636.

Symp. zool. Soc. Lond. (1995) No. 67: 139–149

Lactation in vespertilionid bats

Colin J. WILDE,
Marian A. KERR,
Christopher H. KNIGHT
and Paul A. RACEY

Hannah Research Institute
Ayr KA6 5HL
Scotland, UK

Department of Zoology
University of Aberdeen
Aberdeen AB9 2TN, Scotland, UK

Synopsis

Lactating pipistrelle bats (*Pipistrellus pipistrellus*) feed one, occasionally two, young from two thoracic mammary glands for about four weeks. Throughout lactation, suckling is interrupted each night as the bats forage for food, sometimes over considerable distances. The energetic costs of milk secretion and aerial foraging are considerable and, particularly at times of food shortage, are accommodated by a reduction in body temperature and torpor in the lactating animal. The mammary gland is also subject to this fall in temperature, so that milk secretion is likely to undergo pronounced diurnal variation, a situation similar to that in rodents, where it is dependent on the daily feeding pattern. Milk secretion is tailored to the intermittent suckling pattern by local regulatory mechanisms within the mammary gland, which respond to the frequency or completeness of milk removal. This mechanism is shared by other species but appears to be a predominant influence in bat mammary tissue, so much so that the developmental changes it induces obscure those changes usually associated with stage of lactation.

Introduction

Parental investment in lactation is not wholly altruistic. Investment to maximize offspring survival is tempered by the overall energetic demand imposed on the mother (Peaker 1989). On the one hand, in some rodent species the lactating mother is often pregnant and so may respond to unfavourable conditions by abandoning the lactation, choosing instead to invest in the next litter (Trivers 1985). Species which have few or only a single offspring each year would, on the other hand, be expected to make a much greater investment in each lactation. For example, red deer are often in such poor condition at the end of a season's lactation that they fail to conceive at the rut, and so breed only in alternate years (Loudon & Kay 1984). Lactating bats are likely to fall into this second category. The

ZOOLOGICAL SYMPOSIUM No. 67
ISBN 0 19 854945 8

pipistrelle (*Pipistrellus pipistrellus*), for instance, gives birth to a singleton, occasionally twins, which depend on the mother's milk for up to four weeks *post partum*.

In many species, investment in lactation takes the form of mobilization of body reserves accumulated during pregnancy (Bauman & Elliot 1983; Weiner 1987), greatly increased food intake (Williamson 1980; Vernon 1988), and homeorhetic channelling of nutrients to the mammary gland (reviewed by Vernon 1989). Deposition of body reserves during pregnancy is not, however, an attractive strategy in the lactating bat. The energetic cost of flight for aerial foraging makes any increase in body weight or foraging time disadvantageous. For the same reason, hyperphagia is unlikely to contribute significantly to the energetic cost of lactation, not least because the lactating bat would need to accommodate the considerable additional energetic cost of prolonged aerial foraging. Accordingly, there is no evidence for hyperphagia in *P. pipistrellus* (Swift 1980; Racey & Swift 1985). Clearly, these factors limit the bat's scope for metabolic adaptation of the sort employed by other species to fuel milk secretion.

These deductions are supported by studies of two British species, the brown long-eared bat *Plecotus auritus* and the pipistrelle, which concluded that the additional metabolic load of lactation was not met to any significant degree by mobilization of body stores accumulated during pregnancy (Racey & Speakman 1987). Similar conclusions came from a New England study of the little brown bat *Myotis lucifugus* (Kurta, Bell, Nagy & Kunz 1989). In this species, body reserves stored during pregnancy accounted for only 2% of total energy intake, and food intake increased by only 25% during lactation. During lactation, this bat allocated most of its metabolic energy (>60%) to aerial foraging. Nevertheless, milk output accounted for 80% of the remainder, emphasizing the strong parental commitment to a sustained lactation. Therefore, it appeared that mechanisms for energy conservation must play a critical role in maintaining lactation.

The principal mechanism of energy conservation during lactation, as at other times, appears to be the bat's ability to enter torpor, reducing its body temperature to just a few degrees above ambient temperature and its energy utilization to just 5–10% of the endothermic rate (Racey & Speakman 1987). A disadvantage of this method of energy conservation is its non-specificity. Mammary metabolism should be reduced in proportion to that in other tissues, with a consequent decrease in milk secretion, unless the gland is protected in some way such that milk output for the offspring is maintained. We wished to determine if this was the case, or if torpor in the lactating bat was associated with species- and tissue-specific adaptations in the development or function of the lactating mammary gland; such adaptations might act to optimize the animal's investment in lactation.

Methods

Animals

Mammary development and function were studied in lactating pipistrelle bats from a captive colony maintained in the Department of Zoology, University of Aberdeen, under licence from Scottish Natural Heritage, formerly the Nature Conservancy Council. Bats were caught during pregnancy, trained to feed on mealworms in captivity (Racey 1971) and inspected daily for signs of birth. Animals in lactation for up to 18 days were killed by cervical dislocation, and the thoracic mammary glands were rapidly removed. Tissue from the two glands was pooled, except when one gland had been suckled immediately before tissue collection. Portions of tissue were stored frozen in liquid nitrogen for enzyme assay, or cultured immediately for assay of metabolic activity.

Tissue culture

Two groups of 20 explants were prepared rapidly (within 10 min) from pooled or single-gland mammary tissue, and cultured on stainless steel grids in Medium 199 containing insulin (5 μg/ml), hydrocortisone (0.1 μg/ml) and prolactin (1 μg/ml). After a 5 min equilibration period, this was replaced by fresh medium containing hormones and either L-[4,5-^{3}H] leucine (10 μCi/ml) for measurement of protein synthesis, or [U-^{14}C] glucose (1 μCi/ml) and [6-^{3}H] thymidine (2 μCi/ml) for measurement of lactose and DNA synthesis respectively. Protein synthesis over 15 min – 1h periods in culture was assayed as trichloroacetic acid-precipitable radioactivity in explant homogenates and culture medium. In some cases, [^{3}H]leucine was replaced by [L-^{35}S] methionine (100 μCi/ml), and radiolabelled proteins were resolved by SDS-polyacrylamide gel electrophoresis and fluorography. Total radioactivity incorporated in DNA was measured by precipitation with trichloroacetic acid, and [^{14}C] lactose was extracted from explant homogenate and culture medium by selective precipitation (Kuhn & White 1975).

Temperature dependence of tissue protein synthesis was tested by equilibrating and culturing groups of explants from a single gland at 37 °C or ambient temperature (22 °C) in the presence of L-[4,5-^{3}H] leucine (15 μCi/ml) or L-[^{35}S] methionine (100 μCi/ml) for 1 h. Explants cultured at 37 °C were maintained at this temperature throughout the preparation procedure.

Enzyme assay

Enzyme activities were assayed in a particle-free supernatant prepared by centrifugation (10 000 g, 4 °C, 1 min) of tissue homogenates (glass/PTFE homogenizer, 10 strokes) in Tris-sucrose buffer (30 mM Tris-HCI pH 7.4 containing 1 mM EDTA, 1 mM glutathione and 2 mM phenylmethanesulphonyl fluoride). Acetyl-CoA carboxylase, fatty acid synthetase and galactosyltransferase activities were measured under conditions designed to realize maximum activity, and where activity was linearly related to the amount of sample and incubation time. DNA was assayed by a fluorimetric method (Labarca & Paigen 1980). Details of culture media and assay reagents are described in Wilde, Kerr, Knight, Racey & Burnett (1992).

Results and discussion

Effect of torpor

In several species, synthetic activity during short-term culture of mammary explants at 37 °C reflects the performance of the tissue from which they were prepared (Wilde, Henderson & Knight 1986; Shipman, Docherty, Knight & Wilde 1987; Knight, Hillerton, Kerr, Teverson, Turvey & Wilde 1992). On this basis, we considered that comparison of synthetic activity at 37 °C and ambient temperature (22 °C) should reflect the ability of bat mammary tissue *in vivo* to tolerate the fall in body temperature associated with torpor. When cultured under such conditions, the rate of protein synthesis in freshly-prepared explants was significantly lower at ambient temperatures. [^{3}H] leucine incorporation was reduced by 49% compared with explants cultured at 37 °C ($P < 0.01$), a decrease similar to that observed in lactating mouse mammary explants cultured under the same conditions. This suggested that bat mammary tissue does not possess a homeostatic mechanism to protect milk protein synthesis against a general decrease in body temperature. Therefore, daytime periods of torpor are likely to produce a pronounced diurnal variation in the rate of milk secretion.

A similar situation occurs in small terrestrial mammals which feed mainly nocturnally (Carrick & Kuhn 1978), even during lactation when food intake increases dramatically (Wilde & Kuhn 1979). In lactating rats, lactose synthesis, the principal determinant of milk volume, was closely linked to food intake, being lowest between late morning and early evening (Carrick & Kuhn 1978). Indeed, when food intake was restricted by scheduled feeding, diurnal variation in the rate of lactose synthesis was more pronounced. Food intake in the early morning was followed by a maximum rate of lactose synthesis 6 h later, and a subsequent decrease to 20% of this value immediately before the next meal (Wilde & Kuhn 1979). Thus, the effect of torpor may be to induce diurnal variation in milk secretion in lactating pipistrelles similar to that imposed by food intake in other mammals.

Effect of stage of lactation

Short-term rates of protein and lactose synthesis in mammary explants from goats (Wilde, Henderson & Knight 1986), mice (Shipman *et al.* 1987) and cows (Knight, Hillerton *et al.* 1992) vary with stage of pregnancy or lactation. In mice, for example, mammary protein synthesis rates remained low until parturition and then more than doubled by day 15 of lactation, when peak milk yield was attained (Shipman *et al.* 1987). In contrast, neither protein nor lactose synthesis rates in freshly-prepared bat mammary explants changed significantly with stage of lactation (protein synthesis, $P = 0.171$; lactose synthesis, $P = 0.351$; Table 1). This suggested that secretory cell differentiation was not an important feature of bat mammary development, during a period when milk yield, estimated by $^{3}H_2O$ transfer from mother to young, increased approximately fourfold (0.37 ml/day on day 5–7 compared with 1.37 ml/day on day 11 of lactation; C. H. Knight, P. A. Racey & C. J. Wilde unpubl. work).

This was supported by measurements of mammary enzyme activities. The activities of three key enzyme markers of differentiation, acetyl-CoA carboxylase, fatty acid synthetase and galactoyltransferase also did not change significantly with stage of lactation (Fig. 1). A weak trend towards higher activities of acetyl-CoA carboxylase and fatty acid synthetase in later lactation was heavily influenced by two high values on days 16 and 17 of lactation. The absence of a relationship cannot be attributed to increasingly inefficient tissue homogenization or fractionation in later lactation; the activities of lactate dehydrogenase and aryl esterase, two 'house-keeping' enzymes unrelated to the differentiated state of secretory cells, were similar in early- and late-lactation tissue samples (results not shown). By comparison, acetyl-CoA carboxylase, fatty acid synthetase and galactosyltransferase activities were multiplied by 1.9, 3.3 and 4.0 respectively in mouse mammary tissue between parturition and day 15 of lactation (Shipman *et al.* 1987), indicating that cellular differentiation was an important determinant of the increasing rate of milk secretion during the period (Knight, Maltz & Docherty 1986).

In the absence of a significant increase in cell differentiation, pipistrelle mammary development could, alternatively, arise predominantly from an increase in secretory cell number. However, DNA synthesis, an index of cell proliferation in tissue explants, did not change significantly with stage of lactation ($P = 0.283$), and was consistently lower than the rate measured in comparable explant cultures from growing mouse mammary glands (Shipman *et al.* 1987). We concluded that bat mammary development, unlike that in

Table 1. Rates of protein and lactose synthesis in bat mammary tissue explants freshly prepared at stages during lactation

Stage of lactation	Day 0–10	Day 10–20
Protein synthesis (DPM min^{-1} μg DNA^{-1})	15±3	18±3
Lactose synthesis (nmol h^{-1} μg DNA−1)	0.37±0.06	0.25±0.03

Values are the mean ± SEM for five determinations. Statistical significance was assessed by Student's *t*-test.

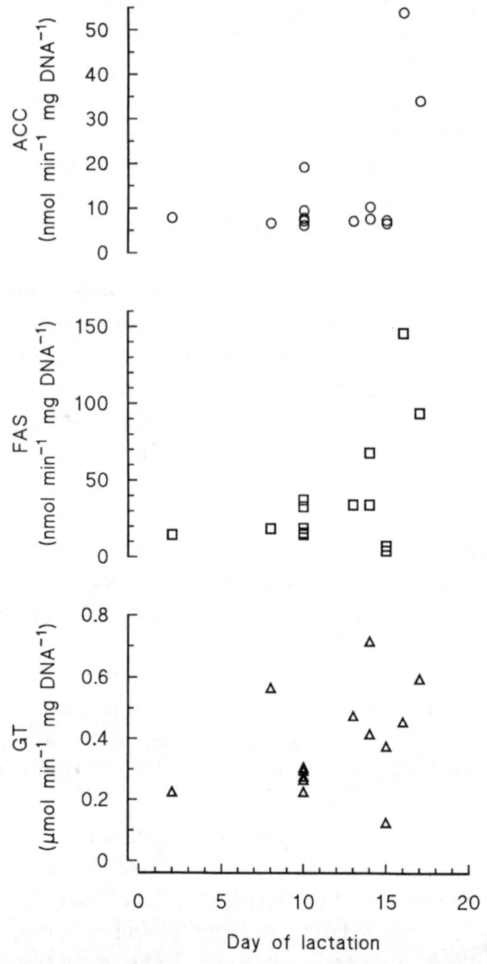

Fig. 1. Key enzyme activities in bat mammary tissue during lactation. Activities of acetyl-CoA carboxylase (ACC), fatty acid synthetase (FAS) and galactosyltransferase (GT) were measured in tissue pooled from both thoracic glands, as described in the Methods section.

terrestrial mammals, is governed by factors which presumably, in adapting the gland to meet demand by the offspring, override and obscure any effect of stage of lactation.

Effect of milk accumulation

A number of bats in the study had their offspring attached to one gland immediately before tissue collection. In these animals, the suckled gland was apparently devoid of milk, an observation supported by the absence

Table 2. Key mammary enzyme activities in bat mammary tissue from suckled and unsuckled glands. Values are the mean ± SEM for six determinations (four determinations in the case of galactosyl transferase) on day 6–10 of lactation.

Enzyme	Suckled gland	Unsuckled gland
Acetyl-CoA carboxylase (nmol min^{-1} μg DNA^{-1})	15.2±3.5	8.3±3.3*
Fatty acid synthetase (μmol min^{-1} μg DNA^{-1})	37.3±6.7	14.3±3.6*
Galactosyltransferase (μmol min^{-1} μg DNA^{-1})	0.32±0.03	0.27±0.12

*$P < 0.05$ compared with suckled glands (Student's paired *t*-test).

of lactose in the tissue homogenate. In contrast, the other 'unsuckled' gland was engorged with milk. Comparison of enzyme activities and explant synthesis rates in suckled and unsuckled glands indicated that secretory cell differentiation was greatly influenced by the extent to which the gland was filled with milk. Two of the three key enzyme markers of cell differentiation, acetyl-CoA carboxylase and fatty acid synthetase, were significantly higher in the suckled gland (Table 2), as was the rate of total protein synthesis in freshly prepared mammary explants. This local regulation of mammary development by milk accumulation may account for the absence of any progressive change with stage of lactation. The rate of DNA synthesis was also higher in explants from suckled glands, suggesting that cell proliferation was also regulated locally by milk accumulation and removal.

Local regulation of mammary development by mechanisms sensitive to frequency or completeness of milk removal has also been observed in ruminants. In comparison to twice-daily milking, frequent milking of goats stimulated milk secretion acutely, increased cellular differentiation after 10 days and ultimately, after several months, resulted in a net increase in mammary cell number (Wilde, Henderson, Knight, Blatchford, Faulkner & Vernon 1988). Conversely, once-daily or incomplete milking decreased milk yield within hours and mammary cell differentiation after days or weeks of treatment (Wilde, Blatchford, Knight & Peaker 1989; Wilde & Knight 1990). In these cases, changes in secretory cell differentiation represented, chronologically, a secondary response to changes in the milking regimen. In the lactating bat, however, the differences in cell activity between suckled and unsuckled glands appear to constitute a relatively acute response to milk accumulation. Certainly the abundance of milk in the unsuckled gland showed that the difference was not one between active and quiescent glands. On the other hand, the presence of a large volume of milk in the unsuckled gland could indicate that it was being suckled only intermittently. This could occur if the offspring were permanently attached to a teat whilst the mother was in the roost and if only one gland were suckled on each occasion. Then, assuming there is an extended interval whilst the mother forages, it is conceivable that there would be pronounced independent cycles of milk accumulation and removal in each gland. However, pipistrelle offspring show no consistent preference for one gland (P. M. Hughes pers. comm.), unlike pigs (Hartman, Ludwick & Wilson 1962) and marsupials (Nicholas 1988), which show a consistent teat order. Moreover, the lactating evening bat *Nycticeius humeralis* is reported to participate in communal nursing in the roost (Wilkinson 1992), a habit which would tend to randomize suckling patterns. In any case, suckling of each mammary gland is likely to become more intermittent as the young develop and begin to move around the roost. All these factors suggest that each bat mammary gland is suckled frequently, if not sequentially, and that changes in mammary secretory cell differentiation appear to represent a relatively acute response in this species, rather than a longer-term adaptation as observed in lactating goats and cows.

Milk accumulation may regulate cellular differentiation in bat mammary tissue through the action of an inhibitory milk protein similar to those identified in goats (Wilde, Henderson, Knight *et al.* 1988; Wilde & Peaker 1990) and cows (C. J. Wilde & C. V. P. Addey unpubl. work). Indeed, since there is preliminary evidence for structurally-related proteins in human (Prentice, Addey & Wilde 1989) and tammar wallaby milk (Hendry, Wilde, Nicholas & Bird 1992), this may be a mechanism shared by most, if not all, mammals. The caprine factor acts in an autocrine manner, blocking milk protein secretion at the endoplasmic reticulum–Golgi level (Rennison, Kerr, Addey, Turner, Wilde & Burgoyne 1993). This action on membrane trafficking may account for the down-regulation of prolactin receptors (and probably of those for other hormones) observed when milking frequency is altered (McKinnon, Knight, Flint & Wilde 1988), and when mammary secretory cells are treated *in vivo* (C. N. Bennett, C. H. Knight & C. J. Wilde unpubl. work) or *in vitro* (Bennett, Knight & Wilde 1990) with the inhibitory protein. Since changes in mammary hormone receptor distribution *in vivo* are associated with up- or down-regulation of secretory cell differentiation, and the inhibitory protein affects cellular differentiation directly in cell cultures, it is likely that regulation of mammary differentiation, exemplified by changes in key enzyme activities and tissue explant synthesis rates in the present study, are a consequence of the autocrine inhibitor's primary effect on membrane trafficking and subcellular hormone receptor distribution.

The reason why changes in cell differentiation should apparently constitute an acute response in the lactating bat is a matter for conjecture. It may be that the tissue in this animal is particularly sensitive to autocrine regulation, either by dint of rapid variation in inhibitory protein levels or through intracellular amplification of the inhibitory signal. Whatever the reason, the presence of such a mechanism renders the mammary gland's capacity for milk production exquisitely sensitive to milk accumulation and provides an efficient method for matching milk production to demand by the offspring. Autocrine inhibition may, in addition, serve to slow milk accumulation progressively during extended aerial foraging, an important consideration if, as the study suggests, maximal metabolic rate during flight initially coincides with maximal synthetic capacity of the mammary gland. Rapid milk accumulation during flight may be one reason why bats return to the roost and suckle during the night (Racey, Speakman & Swift 1987). A period of nocturnal suckling would also relieve autocrine feedback and allow maximal milk secretion during the next period of foraging, at dawn (Racey & Swift 1985). On the other hand, the pipistrelle's bimodal flight pattern may simply reflect the dusk and dawn peaks of aerial insect activity (Swift 1980). Since the energy cost of commuting back to the roost is a relatively trivial part of the daily energy budget (Speakman, Racey, Catto, Webb, Swift & Burnett 1991), a period inside the roost represents, in either case, a worthwhile economy, particularly when energy demand is increased by lactation. Elucidation of the relative importance of intra-mammary and whole-body factors in regulating pipistrelle lactation will require further investigation, particularly with respect to the temporal relationship between foraging, suckling (including milk removal and autocrine inhibition) and torpor. Frequent observation of torpor in lactating pipistrelles (J. R. Speakman & P. A. Racey unpubl.) suggests that strategic control is exerted predominantly by heterothermic or systemic mechanisms sensitive to food intake. However, autocrine mechanisms sensitive to milk removal may provide an additional tactical level of control, which modulates lactation according to the demands of the offspring.

Conclusion

The high energetic cost of milk production and aerial foraging in lactating pipistrelle bats is accommodated by their ability to enter torpor and by close matching of mammary synthetic capacity to the demand for milk by the offspring. In the first case, regulation is exerted at a whole body level and milk secretion is not protected against the general decrease in metabolic activity associated with a fall in body temperature. In the latter, regulation is exerted within the mammary gland, by a mechanism acutely sensitive to the extent of milk accumulation.

Acknowledgements

This work was funded by the Scottish Office Agriculture and Fisheries Department.

References

Bauman, D. E. & Elliot, J. M. (1983). Control of nutrient partitioning in lactating ruminants. In *Biochemistry of lactation*: 437–468. (Ed. Mepham, T. B.). Elsevier Press, Amsterdam.

Bennett, C. N., Knight, C. H. & Wilde, C. J. (1990). Regulation of mammary prolactin binding by secreted milk proteins. *J. Endocr.* **127** (Suppl.): 141.

Carrick, D. J. & Kuhn, N. J. (1978). Diurnal variation and response to food withdrawal of lactose synthesis in lactating rats. *Biochem. J.* **174**: 319–324.

Hartman, D. A., Ludwick, T. M. & Wilson, R. F. (1962). Certain aspects of lactation performance in sows. *J. Anim. Sci.* **21**: 883–886.

Hendry, K. A. K., Wilde, C. J., Nicholas, K. R. & Bird, P. H. (1992). Evidence for an inhibitor of milk secretion in milk of the tammar wallaby. *Proc. Aust. Soc. Biochem. molec. Biol.* **24**: POS-2–3.

Knight, C. H., Hillerton, J. E., Kerr, M. A., Teverson, R. M., Turvey, A. & Wilde, C. J. (1992). Separate and additive stimulation of bovine milk yield by the local and systemic galactopoietic stimuli of frequent milking and growth hormone. *J. Dairy Res.* **59**: 243–252.

Knight, C. H., Maltz, E. & Docherty, A. H. (1986). Milk yield and composition in mice: effect of litter size and lactation number. *Comp. Biochem. Physiol. (A)* **84**: 127–133.

Kuhn, N. J. & White, A. (1975). The topography of lactose synthesis. *Biochem. J.* **148**: 77–84.

Kurta, A., Bell, G. P., Nagy, K. A. & Kunz, T. H. (1989). Energetics of pregnancy and lactation in free-ranging little brown bats *(Myotis lucifugus)*. *Physiol. Zool.* **62**: 804–818.

Labarca, O. & Paigen, K. (1980). A simple, rapid and sensitive DNA assay procedure. *Analyt. Biochem.* **102**: 344–352.

Loudon, A. S. I. & Kay, R. N. B. (1984). Lactational constraints on a seasonally breeding mammal: the red deer. *Symp. zool. Soc. Lond.* No. 51: 233–252.

McKinnon, J., Knight, C. H., Flint, D. J. & Wilde, C. J. (1988). Effect of milking frequency and efficiency on goat mammary prolactin receptor number. *J. Endocr.* **119** (Suppl.): 167.

Nicholas, K. R. (1988). Control of milk protein synthesis in the marsupial *Macropus eugenii*: a model system in which to study prolactin-dependent development. In *The developing marsupial*: 41–54. (Eds Tyndale-Biscoe, C. H. & Janssens, P. A.). Springer-Verlag, Berlin.

Peaker, M. (1989). Evolutionary strategies in lactation: nutritional implications. *Proc. Nutr. Soc.* **48**: 53–57.

Prentice, A., Addey, C. V. P. & Wilde, C. J. (1989). Evidence for local feedback control of human milk secretion. *Biochem. Soc. Trans.* **15**: 122.

Racey, P. A. (1971). The breeding, care and management of vespertilionid bats in the laboratory. *Lab. Anim., Lond.* **4**: 171–183.

Racey, P. A. & Speakman, J. R. (1987). The energy costs of pregnancy and lactation in heterothermic bats. *Symp. zool. Soc. Lond.* No. 57: 107–125.

Racey, P. A., Speakman, J. R. & Swift, S. M. (1987). Reproductive adaptations of heterothermic bats at the northern borders of their distribution. *S. Afr. J. Sci.* **83**: 635–638.

Racey, P. A. & Swift, S. M. (1985). Feeding ecology of *Pipistrellus pipistrellus* (Chiroptera: Vespertilionidae) during pregnancy and lactation. I. Foraging behaviour. *J. Anim. Ecol.* **54**: 205–215.

Rennison, M. E., Kerr, M. A., Addey, C. V. P., Turner, M. D., Wilde, C. J. & Burgoyne, R. D. (1993). Inhibition of constitutive protein secretion from lactating mouse mammary epithelial cells by FIL (feedback inhibitor of lactation), a secreted milk protein. *J. Cell Sci.* **106**: 641–648.

Shipman, L. J., Docherty, A. H., Knight, C. H. & Wilde, C. J. (1987). Metabolic adaptations in mouse mammary gland during a normal lactation cycle and in extended lactation. *Q. Jl. exp. Physiol.* **72**: 303–311.

Speakman, J. R., Racey, P. A., Catto, C. M. C., Webb, P. I., Swift, S. M. & Burnett, A. M. (1991). Minimum summer populations and densities of bats in N. E. Scotland, near the northern borders of their distributions. *J. Zool., Lond.* **225**: 327–345.

Swift, S. M. (1980). Activity patterns of Pipistrelle bats *(Pipistrellus pipistrellus)* in north-east Scotland. *J. Zool., Lond.* **190**: 285–295.

Trivers, R. H. (1985). *Social evolution.* Benjamin/Cummings, Menlo Park, California.

Vernon, R. G. (1988). The partition of nutrients during the lactation cycle. In *Nutrition and lactation in the dairy cow*: 32–52. (Ed. Garnsworthy, P. C.). Butterworths, London.

Vernon, R. G. (1989). Endocrine control of metabolic adaptation during lactation. *Proc. Nutr. Soc.* **48**: 23–32.

Weiner, J. (1987). Limits to energy budget and tactics in energy investments during reproduction in the Djungarian hamster *(Phodopus sungorus sungorus* Pallas 1770). *Symp. zool. Soc. Lond.* No. 57: 167–187.

Wilde, C. J., Blatchford, D. R., Knight, C. H. & Peaker, M. (1989). Metabolic adaptations in goat mammary tissue during long-term incomplete milking. *J. Dairy Res.* **56**: 7–15.

Wilde, C. J., Henderson, A. J. & Knight, C. H. (1986). Metabolic adaptations in goat mammary tissue during pregnancy and lactation. *J. Reprod. Fert.* **76**: 289–298.

Wilde, C. J., Henderson, A. J., Knight, C. H., Blatchford, D. R., Faulkner, A. & Vernon, R. G. (1988). Effect of long-term thrice-daily milking on mammary enzyme activity, cell population and milk yield in the goat. *J. Anim. Sci.* **64**: 533–539.

Wilde, C. J., Kerr, M. A., Knight, C. H., Racey, P. A. & Burnett, A. (1992). Effect of stage of lactation and milk accumulation on mammary cell differentiation in lactating bats. *Exp. Physiol.* 77: 873–879.

Wilde, C. J. & Knight, C. H. (1990). Milk yield and mammary function in goats during and after once daily milking. *J. Dairy Res.* 57: 441–447.

Wilde, C. J. & Kuhn, N. J. (1979). Lactose synthesis in the rat, and the effects of litter size and malnutrition. *Biochem. J.* **182**: 287–294.

Wilde, C. J. & Peaker, M. (1990). Autocrine control in milk secretion. *J. agric. Sci.* **114**: 235–238.

Wilkinson, G. S. (1992). Communal nursing in the evening bat *Nycticeius humeralis*. *Behav. Ecol. Sociobiol.* **31**: 225–235.

Williamson, D. H. (1980). Integration of metabolism in tissues of the lactating rat. *FEBS Lett.* **117**(Suppl.): K93–K105.

Symp. zool. Soc. Lond. (1995) No. 67: 151–165

Synchrony and seasonality of reproduction in tropical bats

Paul D. HEIDEMAN[1]

Institute of Reproductive Biology
and Department of Zoology
University of Texas
Austin, Texas 78712, USA

Synopsis

Many tropical bat populations reproduce synchronously (the temporal clustering of reproduction) and seasonally (clustering of reproductive events at approximately the same period in two or more years), but statistical testing of synchrony and seasonality has been difficult. A randomization technique using circular statistics was used to test seasonal peaks and year-to-year variation in reproductive timing over four years for five species of fruit bats on Negros Island in the Central Philippines. Reproduction was significantly synchronous in all five species, with single annual peaks in births in *Haplonycteris fischeri* and *Nyctimene rabori*, and two annual peaks in births in *Cynopterus brachyotis*, *Harpyionycteris whiteheadi* and *Ptenochirus jagorii*. For all five species, there were significant peaks at similar times in most years, demonstrating that reproduction was seasonal as well as synchronous. The exact timing of peaks varied significantly among years, by up to eight weeks, for *C. brachyotis*, *H. fischeri* and *P. jagorii*. A literature review reveals little additional information on variation in seasonal timing in the field or on the regulation of seasonal reproduction in tropical bats. New field and laboratory studies that examine variation in reproductive timing and regulation of reproductive timing are necessary to understand the evolution of reproductive synchrony and seasonality in the tropics.

Introduction

Reproductive synchrony and seasonality appear to be typical of most species of bats in the tropics (Racey 1982). Reproductive synchrony and seasonality will be favoured under two conditions. First, the environment must have predictable seasonal variation and second, individuals must be able to maximize fitness by timing a critical reproductive stage to coincide with the season of optimal conditions for that stage. Because the terms synchrony and seasonality

[1]Present address: Department of Biology, P.O. Box 8795, College of William and Mary, Williamsburg, Virginia 23187–8795, USA

ZOOLOGICAL SYMPOSIUM No. 67
ISBN 0 19 854945 8

can be used in several ways, I will state my definitions here. Synchrony is any significant temporal clustering of an event among individuals. Seasonality is any significant temporal clustering of an event among individuals that is repeated at approximately the same time in each year. Most mammals that reproduce synchronously also reproduce seasonally, but work by Higgins (1993) shows that precise reproductive synchrony can occur in the absence of seasonality. Thus, data from a single year can be used to establish synchrony of reproduction, while two or more years of data are necessary, technically, to assess seasonality.

There has been a lack of unbiased and objective methods to define seasonal peaks, assess their statistical significance and compare peaks among years statistically. A major problem has been that field data sets often violate one or more assumptions of standard parametric and nonparametric statistical tests. Those statistical tests for which assumptions can be met often ignore some of the data. For example, *G*-tests or Chi-square tests using the percentage of individuals that are pregnant lose information on embryo size, and they are generally limited to establishing the presence of synchrony and seasonality without identifying specific peaks and valleys. Thus, few studies have been able to describe variation in reproductive timing and then discuss its possible causes and significance.

The purpose of this paper is to discuss our understanding of the timing of seasonal reproduction and variation in timing in tropical bats. I will use data from a community of fruit bats in tropical rainforest in the Philippines to (1) assess seasonality statistically and (2) test for differences in timing among years. I will discuss the results in the context of a review of laboratory studies on the environmental and physiological factors that may affect reproductive timing in bats.

Methods

Reproductive timing was assessed for five species of frugivorous bats (*Haplonycteris fischeri, Nyctimene rabori, Cynopterus brachyotis, Harpyionycteris whiteheadi* and *Ptenochirus jagorii*) from Negros Island, Philippines (9° N, 121° E). Bats were captured between June 1982 and June 1983, February to July 1984 and February to September 1987. Most of the captures were from a slightly disturbed primary submontane forest site at 800–1200 m elevation, surrounding Lake Balinsasayao, but some *Cynopterus brachyotis* were captured 17 km to the south-east in a lowland orchard in Dumaguete City. The capture sites and methods have been discussed in detail elsewhere (Heideman 1988, 1989a, b; Heaney, Heideman, Rickart, Utzurrum & Klompen 1989; Heideman & Heaney 1989). Mean annual rainfall over southern Negros ranges from about 1200 to 3100 mm, with the highest totals occurring in upland areas. The drier period in most years is from February to April, with January and May as transitional months; the timing of rainfall is classed as seasonal according to Moh's *Q* index

(Heideman & Erickson 1987). During the dry period, average monthly rainfall is above 50 mm at many sites for all months and probably averages 100 mm or more in every month at some upland sites.

For each captured female, records were made of the presence or absence of an embryo, the greatest diameter of the embryo (including surrounding uterine tissue), lactational status and age category. Approximately 25–50% of the individuals captured for each species were kept as museum specimens after reproductive autopsy; the remainder were released after assessment of reproductive status (Heideman & Heaney 1989). Accuracy of palpation of embryos and estimation of embryo size of intact females was similar to, or higher than, that reported elsewhere for *Haplonycteris fischeri* (Heideman 1988): approximately 95% accuracy for presence/absence of the smallest embryos (2–4 mm uterine swellings) and ± 20% for embryo size. Because the timing of reproduction in primiparous females can differ from that of parous adults (Heideman 1988, 1989a), only data from parous adult females (defined as those with fused phalangeal epiphyses and having enlarged teats) were used in the analyses. A small number of adult females were captured more than once. Because multiple records of an individual recaptured during a single pregnancy (or between two pregnancies) are not independent, only a single capture record during each pregnancy or each interval between pregnancies was included in the analyses.

Field data that have unequal samples of females among sampling periods and that use embryo sizes to estimate birth dates violate assumptions of standard parametric and non-parametric statistical tests. Therefore, I used randomization methods (Manly 1991) to test for the significance of peaks within years, and to compare the timing of peaks among years. The algorithms will be described in more detail elsewhere, but a brief description is included here.

Both algorithms estimated birth dates using an approximation of embryonic growth rate (maximum diameter of conceptus just before birth divided by gestation length, see Table 1), calculated a statistic, and then calculated the same statistic for a series of randomized data sets. One algorithm calculated the circular variance (Batschelet 1981) of birth dates in a data set as a measure of clustering. It randomly reassigned capture information (embyro presence and size and lactational status) to the capture dates and compared the circular variance of the original data set with that of the randomized data set for 10 000 iterations. Peaks in births were considered statistically significant if the circular variance of the randomized data sets was higher than that of the original data set in more than 95% of the iterations.

A second algorithm calculated mean vector angle of births (Batschelet 1981; equivalent to the mean birth date) for birth dates for two years under comparison. Then it combined the two data sets and used the combined data set to reconstruct each data set by random selection without replacement from the combined data sets. Mean vector angles for birth dates were calculated from randomized data sets 10 000 times. Birth peaks were considered

Table 1. The approximate duration of gestation and lactation in four species of frugivorous pteropodids on Southern Negros Island. Estimates of gestation are based upon the time between records of the first pregnant females and first lactating females within particular peaks. Estimates of lactation are based upon the time between records of the last births and last females still lactating following a particular peak.

Species	Duration of gestation	Duration of lactation
C. brachyotis	120	50
H. whiteheadi	135	120
N. rabori	130	120
P. jagorii	120	90

significantly different if the difference between the mean vector angles of the two original data sets was larger than the differences between the randomized data sets in more than 95% of the iterations.

In some species, there were two birth peaks in each year. There is no unbiased method to identify and analyse double peaks, and an *ad hoc* method following Batschelet (1981) was chosen. Data sets were divided at a point midway between potential peaks, and each half was analysed separately.

The results of both statistical tests should be robust to small errors in the measurement of embryos and in the estimation of embryonic growth rate. Unbiased measurement error should have the effect of flattening peaks in births, making the tests more conservative. Inaccuracies in the estimation of embryonic growth rate should have the same effect whenever sampling is evenly distributed throughout the period of pregnancy. However, because the actual date of each projected birth depends upon the estimated growth rate, an inaccurate estimate will provide inaccurate birth dates. The result could be birth peaks that have been shifted away from their true position, although a conclusion of significant synchrony or of significant differences between two samples might still be correct in many cases. Assessing robustness by repeating statistical tests using a range of possible values can reduce the risk of accepting a result that is dependent upon estimates that may be incorrect.

In order to test the robustness of the results, randomization tests were repeated over a range of embryonic growth rates and embryo sizes (± 10% or more) and, for species with double peaks, a range of division points (± 2 months). In only one case were the results not robust to variation in these parameters. The significance of the August/September birth peak in *C. brachyotis* in 1982 was above 0.05 for some division points.

Because parturition dates can be estimated for lactating females, albeit imprecisely, data on lactation were also considered in tests for robustness of results and in one test for synchrony in which results using only pregnant females were equivocal (*C. brachyotis* in August/September 1982). In these cases it was assumed that lactating females were at the midpoint of lactation; thus, parturition dates for lactating females were estimated by backdating a number of days equal to half the duration of lactation (Table 1).

Results

Five species (*Cynopterus brachyotis, Haplonycteris fischeri, Harpyionycteris whiteheadi, Nyctimene rabori* and *Ptenochirus jagorii*) had statistically significant synchrony of births and seasonality of births (Tables 2 & 3, Figs 1–5). There were significant differences in the timing of birth peaks among years for three of these species (*C. brachyotis, H. fischeri* and *P. jagorii*: Table 3). For each of these species, one or two birth peaks occurred three to seven weeks earlier in 1983 than it did in the corresponding parts of one or more other years (Figs 1, 2 & 5). In addition, the August birth peak of *P. jagorii* in 1982 was significantly earlier than this peak in 1984 and the August/September birth peak of *P. jagorii* in 1984 was significantly later than this peak in 1987.

Sample sizes were too small or temporally narrow for statistical assessment in five frugivorous species. Three of these species apparently produce young synchronously and, probably, seasonally in a single annual birth peak on Negros (*Acerodon jubatus*: April–June; *Pteropus hypomelanus*: March and May; *Pteropus vampyrus*: April and May), and data from several other islands in the Philippines suggest that reproductive timing is similar on other islands, although perhaps not identical on all (Rabor 1977; Mudar & Allen 1986; Heideman 1987; Rickart, Heaney, Heideman & Utzurrum 1993; P.D. Heideman & L.R. Heaney unpubl. data). For a fourth species, *Dobsonia chapmani*, which is probably now extinct (Heaney & Heideman 1987), there exists a single reproductive record from a one- to two-week-old juvenile captured with an adult female on 20 May, 1948 (specimen 67859 in the Field Museum of Natural History). The data for the last species, *Pteropus pumilus*, are difficult to interpret. Examination of data from macroscopic examination of reproductive tracts and from released females suggested that this species may produce young at any time of year (Heideman 1987). However, histological examination of one specimen that was apparently pregnant with a small, 5 mm, uterine swelling revealed no signs of embryo or ovum (unpubl. data), suggesting that small uterine swellings may not reliably indicate pregnancy in *P. pumilus*. If other records of small uterine swellings are disregarded, the remaining data suggest that this species may produce young seasonally on Negros, with a single peak of births around March and April. However, more records are necessary to test this hypothesis.

In the frugivorous bat community as a whole, births were centred on the dry season (January/February to April/May) and first half of the wet season (May/June to August/September). Those species with a single annual peak

Table 2. The attained level of significance of clustering of births and, in parentheses, the midpoint of each birth peak for five species of frugivorous pteropodids on Southern Negros Island.

Species	Birth peak	Significance and date of peak in			
		1982	1983	1984	1987
C. brachyotis	Mar.–Apr.	—	0.0402 (Mar. 18)	0.0683 (Apr. 25)	0.0068 (Apr. 1)
	Aug.	0.0481 (Sept. 13)	0.0054 (July 17)	0.0365 (Aug. 13)	NS: 0.24 (Aug. 29)
H. fischeri	June–July	—	0.0000 (June 15)	0.0000 (July 5)	0.0000 (July 3)
H. whiteheadi	Jan.–Feb.	—	0.0000 (Jan. 24)	—	—
	July–Aug.	—	0.0414 (July 26)	0.0057 (Aug. 2)	NS: 0.13 (July 27)
N. rabori	Mar.–Apr.	—	0.0000 (Apr. 3)	0.0000 (Apr. 24)	—
P. jagorii	Mar.–May	—	0.0000 (Mar. 29)	0.0006 (Apr. 14)	—
	Aug.–Sept.	0.0000 (Aug. 16)	0.0037 (Aug. 8)	0.0008 (Sept. 2)	0.0000 (Aug. 15)

NS= $P < 0.10$
—, no data.

Table 3. The attained level of significance for comparisons of the timing of birth peaks among years in frugivorous pteropodids on Southern Negros Island.

Species	Birth peak	Significance of differences in timing in					
		1982 vs. 1983	1982 vs. 1984	1982 vs. 1987	1983 vs. 1984	1983 vs. 1987	1984 vs. 1987
C. brachyotis	Mar.–Apr.	—	—	—	0.0073	NS	(0.0867)
	Aug.	0.0124	NS	NS	NS	0.0162	NS
H. fischeri	June–July	—	—	—	0.0001	0.0000	NS
H. whiteheadi	Jan.–Feb.	—	—	—	—	—	—
	July–Aug.	—	—	—	NS	NS	NS
N. rabori	Mar.–Apr.	—	—	—	(0.10)	—	—
P. jagorii	Mar.–May	—	—	—	0.0000	—	—
	Aug.–Sept.	0.0016	0.0299	NS	0.0002	(0.0992)	0.0255

NS $= P > 0.10$; values between 0.05 and 0.10 are in parentheses.
—, missing data.

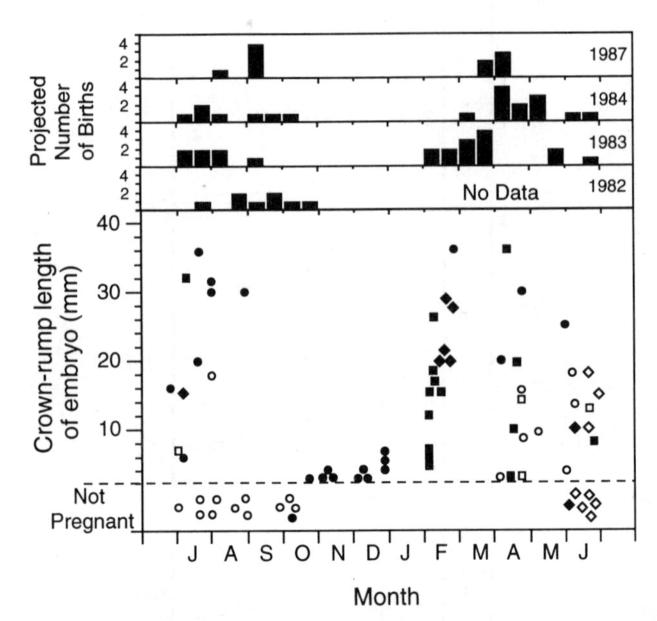

Fig. 1. Reproductive records and estimates of births from pregnant adult *Cynopterus brachyotis* on Southern Negros Island in 1982/83 (circles), 1984 (squares), and 1987 (diamonds). Open symbols indicate lactation. At the top of the figure are frequency distributions of births for 1982, 1983, 1984 and 1987.

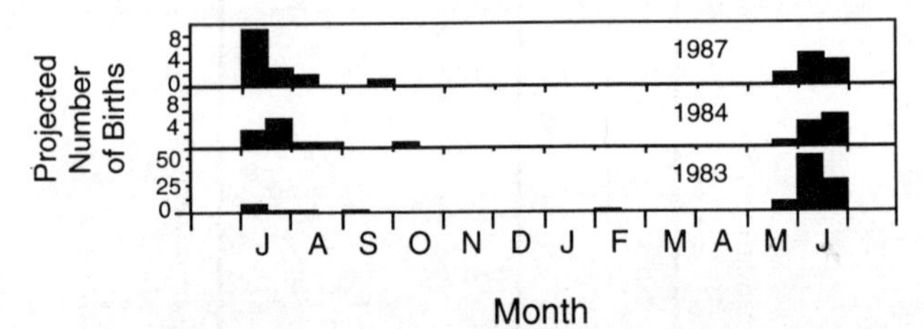

Fig. 2. Frequency distributions of births from pregnant adult *Haplonycteris fischeri* on Southern Negros Island in 1983 (lower panel), 1984 (middle panel), and 1987 (upper panel).

produced young between the mid dry season and early wet season. Those species with two annual peaks in births generally produced peaks that flanked the mid dry season and early wet season. Females in these latter species typically produced one young just before or during the first half of the dry season, and a second young between the beginning and middle of the dry season.

Discussion

In all five species tested statistically, reproduction was both synchronous and seasonal, suggesting that there is, or has been, predictable seasonal variation in

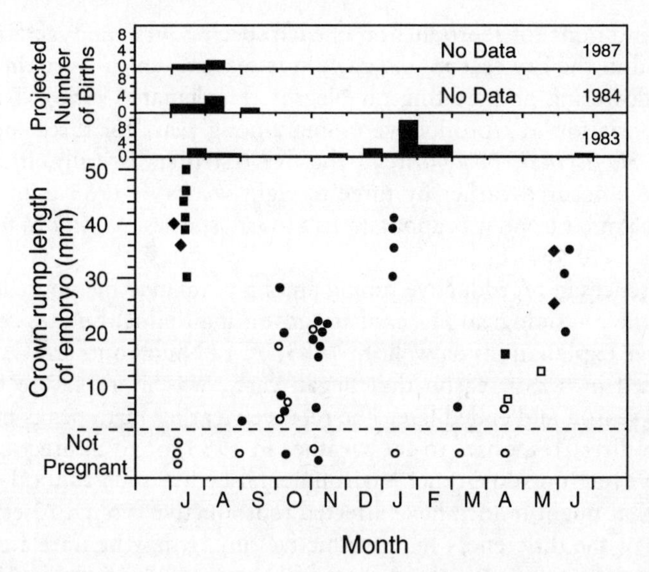

Fig. 3. Reproductive records and estimates of births from pregnant adult *Harpyionycteris whiteheadi* on Southern Negros Island in 1982/83 (circles), 1984 (squares) and 1987 (diamonds). Open symbols indicate lactation. At the top of the figure are frequency distributions of births for 1982/83 (lower panel), 1984 (middle panel) and 1987 (upper panel).

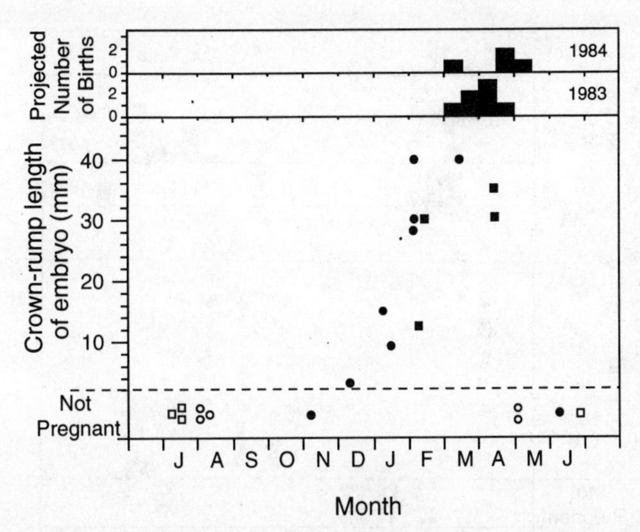

Fig. 4. Reproductive records and estimates of births from pregnant adult *Nyctimene rabori* on Southern Negros Island 1982/83 (circles) and 1984 (squares). Open symbols indicate lactation. At the top of the figure are frequency distributions of births for 1983 (lower panel) and 1984 (upper panel).

optimum conditions for reproduction in each species. Birth and lactation were concentrated in the dry season and early wet season, which were the periods of highest flowering and fruiting on Negros (Heideman 1989b). There was significant variation in reproductive timing among years for three species (*C. brachyotis, H. fischeri, P. jagorii*) of the five tested statistically. In all three species, births occured earlier by three to eight weeks in 1983 than in other years, and a similar trend was apparent in a fourth species, *N. rabori*, for which sample sizes were low (Fig. 4).

The differences in reproductive timing among years may merely reflect inaccuracies in the physiological mechanisms governing timing, but are consistent with adaptive explanations as well. In 1983, an El Niño Southern Oscillation event resulted in a dry season that began early, was more severe than the average dry season and ended late. The relatively earlier birth peaks may have been due to direct responses to the weather in 1983, or to indirect effects of the El Niño event on some other environmental factor, such as food quantity or quality, that might in turn have affected reproductive timing. Alternatively, some or all of the differences in reproductive timing may be unrelated to the El Niño event. In particular, the August birth peak of *P. jagorii* in 1982 was significantly earlier than that in 1984, but the 1982 peak occurred before the effects of the El Niño event.

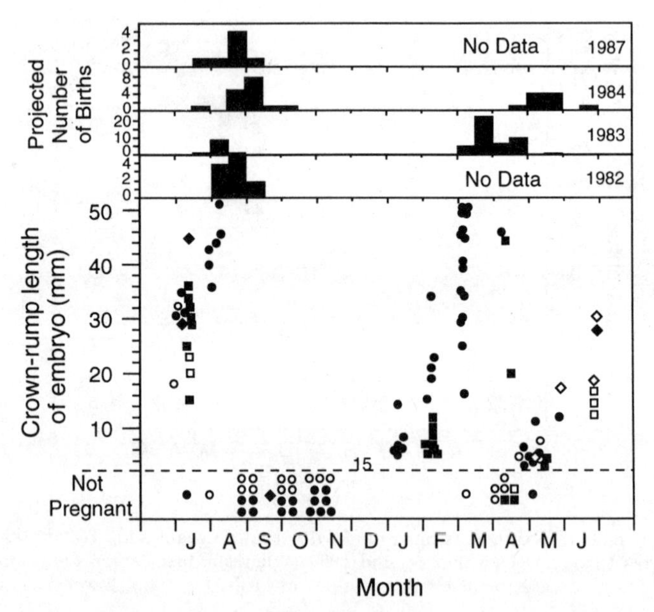

Fig. 5. Reproductive records and estimates of births from pregnant adult *Ptenochirus jagorii* on Southern Negros Island in 1982/83 (circles), 1984 (squares), and 1987 (diamonds). Open symbols indicate lactation. At the top of the figure are frequency distributions of births for 1982, 1983, 1984 and 1987.

The many field studies on the timing of reproduction in tropical bats have provided a broad picture of the patterns of their reproductive timing. It is clear that most tropical species have synchronous and, probably, seasonal peaks in births (reviewed by Racey 1982; and see Thomas & Marshall 1984; Willig 1985; Graham 1987; Heideman, Deoraj & Bronson 1992), as was true of all species in this study for which sample sizes were sufficient for analysis (Tables 2 and 3). There is abundant correlational evidence suggesting that rainfall is one of the primary causes of seasonal birth peaks in tropical bats, probably in large part through its effects on food production (reviewed by Racey 1982; and see Dinerstein 1986; McWilliam 1987; Happold & Happold 1990; Wilson, Handley & Gardner 1991). However, rainfall patterns alone are insufficient to account for all of the variation in reproductive timing in tropical bats (e.g. Heideman 1988), and it seems likely that other causes of seasonal reproduction play important roles in many populations.

In the tropics, seasonal periods assessed as optimal in terms of food abundance (or other environmental factors) are generally synchronized with the three reproductive stages, late pregnancy, lactation, or weaning, that demand the most energy or are most stressful (Bradbury & Vehrencamp 1977; Racey 1982; Dinerstein 1986; McWilliam 1987). On southern Negros, periods of birth and lactation during the dry season and early wet season coincided with peaks in flower and fruit abundance in the Balinsasayao forest (Heideman 1989b). Thus, reproduction may be timed in part to match resource peaks with late pregnancy, lactation or weaning. Bradbury & Vehrencamp (1977) presented evidence that bat species may vary in the stages that are timed to match optimal conditions. This latter point is generally difficult to assess because of the possibility of subtle differences among species in the ultimate causes shaping seasonality, as well as in the physiological mechanisms bats use to achieve seasonality.

Little is known about how tropical bats time seasonal reproduction. It seems likely that bats routinely use predictive cues to time reproduction (e.g. Beasley & Zucker 1984; McGuckin & Blackshaw 1992; Heideman & Bronson 1994), rather than following an opportunist strategy of mating only after the onset of optimal conditions. The lag between mating induced by the onset of optimal conditions and parturition would be sufficiently long in bats (gestation periods from about 2–11 months) for opportunistic births often to occur after the end of optimal conditions. In contrast, the use of an appropriate predictive cue can accurately synchronize reproduction with optimal conditions. The predictability of wet and dry seasons in much of the tropics should favour the use of predictive cues that initiate reproduction in advance of optimal conditions, if bats can detect such cues. Recent studies show that male *Anoura geoffroyi* reproduce seasonally on Trinidad (Heideman *et al.* 1992), and must use some predictive environmental cue to time their reproduction (Heideman & Bronson 1994).

Little is known about which environmental cues tropical bats use to enforce seasonal reproduction. The annual cycle of change in photoperiod is used by many temperate zone mammals to time their reproduction, and there is some evidence that temperate and subtropical populations of bats may also rely upon photoperiod as a seasonal cue (Racey 1978; Beasley & Zucker 1984; McGuckin & Blackshaw 1992). In the deep tropics, bats may rely upon non-photoperiodic environmental cues for seasonal timing. On the equator, photoperiod is constant and therefore could not be used as a seasonal cue. It is unknown how far from the equator photoperiod can be used as a seasonal cue (Heideman & Bronson 1993; Bronson & Heideman 1994), but a tropical population of bats from 10° latitude (Heideman *et al.* 1992; Heideman & Bronson 1994) and a subtropical/tropical population of *Pteropus* from about 25° latitude (O'Brien, Curlewis & Martin 1993) apparently do not use photoperiod to regulate their reproduction seasonally. Heideman & Bronson (1994) discussed potential non-photoperiodic environmental cues that might regulate reproductive timing in tropical bats.

We know little about whether tropical bats use environmental cues to trigger particular reproductive events directly, or whether environmental cues entrain endogenous reproductive rhythms that in turn regulate reproduction. Endogenous reproductive rhythms are common in long-lived mammals that reproduce seasonally (reviewed by Bronson & Heideman 1994), and have been reported in bats from the temperate zone (Beasley & Zucker 1984). Heideman & Bronson (1994) described an endogenous circannual reproductive rhythm in a tropical phyllostomid, *Anoura geoffroyi*, and data from zoo or laboratory colonies suggest that similar rhythms may occur in other tropical bat species (Häussler, Möller & Schmidt 1981; Taft & Handley 1991).

It is not known how precisely tropical bats detect the seasons, nor whether tropical bats can modify, adaptively, the timing of reproduction in a particular year. The fact that many birth peaks in this study were similar among years, with the midpoint of peaks varying only by 1–3 weeks, implies that considerable precision in reproductive timing is possible for tropical bats. More studies on the synchrony of reproductive timing within years and on differences in reproductive timing among years, carried out in different parts of the tropics, are necessary in order to delimit the variation we need to explain. Multi-year studies that correlate or manipulate resources or other environmental variables with reproductive timing would allow us to test whether variation is adaptive. The recent availability of computers which can run randomization statistical tests on moderate to large samples will enhance our ability to analyse these data.

There are conspicuous gaps in our understanding of the evolution and physiology of seasonality in tropical bats (and in tropical mammals generally). There are abundant field data describing peaks, most of which tell us little or nothing about their regulation or annual variation in timing. Detailed field work is necessary in order to clarify the relationship between reproductive timing and environmental variation. It will be particularly valuable to obtain detailed information from two or more years, perhaps by returning to populations that have been studied previously. Similarly, laboratory studies to discover the proximate cues that regulate reproduction in tropical bats, together with studies of the physiological pathways that carry this information to the reproductive system, will be vital to understanding the evolution and ecology of synchrony and seasonality.

Acknowledgements

Extensive logistical support was provided by Dr Angel C. Alcala and the late Dr Luz Ausejo of Silliman University, as well as The Institute of Philippine Culture of Ateneo de Manila University. T. Batal, O. J. I. Delalamon, K. R. Erickson, L. R. Heaney, S. M. G. Hoffman, S. M. J. Hoffman, J. H. S. Klompen, V. LaRoche, C. Lumhod, E. A. Rickart, J. A. Schneider, L. Tagat, R. L. Thomas and R. C. B. Utzurrum were of great help in the field. I am grateful to R.C.B. Utzurrum for collecting the February 1984 sample. F. H. Bronson and P. A. Racey provided helpful comments on an earlier draft of the manuscript. This research was supported in part by National Institutes of Health grant HD 24177 and by NSF grant BSR–8514223.

References

Beasley, L. J. & Zucker, I. (1984). Photoperiod influences the annual reproductive cycle of the male pallid bat (*Antrozous pallidus*). *J. Reprod. Fert.* 70: 567–573.

Batschelet, E. (1981). *Circular statistics in biology*. Academic Press, New York.

Bradbury, J. W. & Vehrencamp, S. L. (1977). Social organization and foraging in emballonurid bats. IV. Parental investment patterns. *Behav. Ecol. Sociobiol.* **2**: 19–29.

Bronson, F. H. & Heideman, P. D. (1994). Seasonal regulation of reproduction in mammals. In *The physiology of reproduction*: 541–583. (Eds Knobil, E. & Neill, J. D.). (2nd edn). Raven Press, New York.

Dinerstein, E. (1986). Reproductive ecology of fruit bats and the seasonality of fruit production in a Costa Rican cloud forest. *Biotropica* **18**: 307–318.

Graham, G. L. (1987). Seasonality of reproduction in Peruvian bats. *Fieldiana, Zool.* (N. S.) **39**: 173–186.

Happold, D. C. D. & Happold, M. (1990). Reproductive strategies of bats in Africa. *J. Zool., Lond.* **222**: 557–583.

Häussler, U., Möller, E. & Schmidt, U. (1981). Zur Haltung und Jugendentwicklung von *Molossus molossus* (Chiroptera). *Z. Säugetierk.* **46**: 337–351.

Heaney, L. R. & Heideman, P. D. (1987). Philippine fruit bats: endangered and extinct. *Bats* 5: 3–5.

Heaney, L. R., Heideman, P. D., Rickart, E. A., Utzurrum, R. B. & Klompen, J. S. H. (1989). Elevational zonation of mammals in the central Philippines. *J. trop. Ecol.* **5**: 259–280.

Heideman, P. D. (1987). *The reproductive ecology of a community of Philippine fruit bats (Pteropodidae. Megachiroptera)*. Unpubl. PhD thesis: University of Michigan.

Heideman, P. D. (1988). The timing of reproduction in the fruit bat *Haplonycteris fischeri* (Pteropodidae): geographic variation and delayed development. *J. Zool., Lond.* **215**: 577–595.

Heideman, P. D. (1989a). Delayed development in Fischer's pygmy fruit bat, *Haplonycteris fischeri*, in the Philippines. *J. Reprod. Fert.* **85**: 363–382.

Heideman, P. D. (1989b). Temporal and spatial variation in the phenology of flowering and fruiting in a tropical rainforest. *J. Ecol.* 77: 1059–1079.

Heideman, P. D. & Bronson, F. H. (1993). Responses of Syrian hamsters, *Mesocricetus auratus*, to amplitudes and rates of change in photoperiod typical of the tropics. *J. Biol. Rhythms* **8**: 325–327.

Heideman, P. D. & Bronson, F. H. (1994). An endogenous circannual rhythm of reproduction in a tropical bat, *Anoura geoffroyi*, is not entrained by photoperiod. *Biol. Reprod.* **50**: 607–614.

Heideman, P. D., Deoraj, P. & Bronson, F. H. (1992). Seasonal reproduction of a tropical bat, *Anoura geoffroyi*, in relation to photoperiod. *J. Reprod. Fert.* **96**: 765–773.

Heideman, P. D. & Erickson, K. R. (1987). The climate and hydrology of the Lake Balinsasayao watershed, Negros Oriental, Philippines. *Silliman J.* **34**: 82–107.

Heideman, P. D. & Heaney, L. R. (1989). Population biology and estimates of abundance of fruit bats (Pteropodidae) in Philippine submontane rainforest. *J. Zool., Lond.* **218**: 565–586.

Higgins, L. V. (1993). The nonannual, nonseasonal breeding cycle of the Australian sea lion, *Neophoca cinerea*. *J. Mammal.* **74**: 270–274.

Manly, B. F. J. (1991). *Randomization and Monte Carlo methods in biology.* Chapman and Hall, London.

McGuckin, M. A. & Blackshaw, A. W. (1992). Effects of photoperiod on the reproductive physiology of male flying foxes, *Pteropus poliocephalus. Reprod. Fert. Devel.* **4**: 43–53.

McWilliam, A. N. (1987). The reproductive and social biology of *Coleura afra* in a seasonal environment. In *Recent advances in the study of bats*: 324–350. (Eds Fenton, M. B., Racey, P. A. & Rayner, J. M. V.). Cambridge University Press, Cambridge.

Mudar, K. M. & Allen, M. S. (1986). A list of bats from northeastern Luzon, Philippines. *Mammalia* **50**: 219–225.

O'Brien, G. M., Curlewis, J. D. & Martin, L. (1993). Effect of photoperiod on the annual cycle of testis growth in a tropical mammal, the little red flying fox, *Pteropus scapulatus*. *J. Reprod. Fert.* **98**: 121–127.

Rabor, D. S. (1977). *Philippine birds and mammals.* Science Education Center, University of the Philippines Press, Quezon City.

Racey, P. A. (1978). The effect of photoperiod on the initiation of spermatogenesis in pipistrelle bats, *Pipistrellus pipistrellus*. In *Proceedings of the fourth international bat research conference*: 255–258. (Eds Olembo, R. J., Castelino, J. B. & Mutere, F. A.). Kenya Literature Bureau, Nairobi.

Racey, P. A. (1982). Ecology of bat reproduction. In *Ecology of bats*: 57–104. (Ed. Kunz, T. H.). Plenum Press, New York & London.

Rickart, E. A., Heaney, L. R., Heideman, P. D. & Utzurrum, R. C. B. (1993). The distribution and ecology of mammals on Leyte, Biliran, and Maripipi Islands, Philippines. *Fieldiana, Zool.* (N. S.) **72**: 1–62.

Taft, L. T. & Handley, C. O. (1991). Reproduction in a captive colony. In *Demography and natural history of the common fruit bat.* Artibeus jamaicensis, *on Barro*

Colorado Island. Panama: 19–42. (Eds Handley, C. O., Wilson, D. E. & Gardner, A. L.). Smithsonian Institution Press, Washington, D. C.

Thomas, D. W. & Marshall, A. G. (1984). Reproduction and growth in three species of West African fruit bats. *J. Zool., Lond.* **202**: 265–281.

Willig, M. R. (1985). Reproductive patterns of bats from Caatingas and Cerrado biomes in northeast Brazil. *J. Mammal.* **66**: 668–681.

Wilson, D. E., Handley, C. O. & Gardner, A. L. (1991). Reproduction on Barro Colorado Island. In *Demography and natural history of the common fruit bat*, Artibeus jamaicensis, *on Barro Colorado Island. Panama*: 43–52. (Eds Handley, C. O., Wilson, D. E. & Gardner, A. L.). Smithsonian Institution Press, Washington, D. C.

Symp. zool. Soc. Lond. (1995) No. 67: 167–184

The reproductive biology of Australian flying-foxes (genus *Pteropus*)

L. MARTIN, J. H. KENNEDY, L. LITTLE, H. C. LUCKHOFF, G. M. O'BRIEN, C. S. T. POW, P. A. TOWERS, A. K. WALDON and D. Y. WANG

Department of Physiology & Pharmacology
The University of Queensland
Queensland 4072, Australia

Synopsis

Three Australian species of *Pteropus* occupying tropical to temperate habitats have similar distributions of birth-times. A fourth, *P. scapulatus*, overlaps the others' ranges yet breeds six months out of phase. Captive males show annual testicular cycles which shift little between years and are unresponsive (*P. scapulatus*), or only sluggishly responsive (*P. poliocephalus*), to altered photoperiod. In the breeding-season, males increase their weight, odour and marking behaviour, copulating repeatedly with intact and ovariectomized females; out of season, copulations occur well into pregnancy. Females are not simple reflex ovulators. Lactation does not delay conception. Corpora lutea do not persist to the next pregnancy and there is no strong evidence for ovulation alternating between ovaries. Female *P. poliocephalus* exposed prematurely to short photoperiod do not conceive early, but females isolated until after the breeding season conceive and give birth late. With no evidence of delayed embryo implantation or development, natural late births probably reflect late conceptions. Endocrine regulation of ovulation remains enigmatic. There are no useful markers of ovarian function, peripheral progesterone levels are anomalous and endometrial growth associated with the preovulatory follicle is confined to the ipsilateral horn. Unilateral growth probably involves preferential local transfer of steroids from ovary to uterus via the unusual ovarian vasculature. However, it is difficult to explain or to reconcile with ovarian steroid feedback on the hypothalamus, without postulating novel ovarian steroids, weak receptor-binding, active metabolic inactivation of steroids by the uterus, or high-affinity steroid-binding plasma proteins. Nonetheless, *Pteropus* plasma binds progesterone and corticoids with high affinity, but not oestradiol–17β.

Introduction

Marshall's Physiology of reproduction (Rowlands & Weir 1984) states that

ZOOLOGICAL SYMPOSIUM No. 67
ISBN 0 19 854945 8

Megachiroptera have

> 'a reproductive pattern . . . characterised by a monoestrous or polyoestrous cycle in which seasonality . . . is more closely allied to rainfall pattern than photoperiodicity, the changes of which are very slight in their tropical environment. All pteropodids are monovular and the . . . young is born at the onset of the rainy season and weaned at the height of the fruiting season. In an environment of prolonged fruiting a second or even third . . . cycle is made possible by . . . post-partum oestrus . . . both ovaries have equal potential; the first ovulation is random but ovulation thereafter alternates . . . the CL remaining functional to the end of gestation'.

Views of ovarian function and reproductive asymmetry in Megachiroptera derive largely from Marshall's (1947, 1949) work with *Pteropus giganteus*—which provided the only description of an endocrine enigma, pre-gestational unilateral endometrial growth. Views of the regulation of *Pteropus* seasonality derive largely from tropical island populations (Baker & Baker 1936; Marshall 1947). Australian *Pteropus* differ greatly from these paradigms. We update a previous review (Martin, Towers, McGuckin, Little, Luckhoff & Blackshaw 1987) in relation to reproductive seasonality.

The distribution and movements of Australian flying-foxes

Pteropus conspicillatus is confined to eastern rainforests above 19° S (Richards 1990). *P. alecto*, found also in Sulawesi (0–5° S) and New Guinea, extends in Australia from 25° S on the west coast, across the north and down the east coast, recently extending its range from 25 to 29° S. Its camps empty and fill seasonally but there are no data on long-distance movements, nor evidence of seasonal north–south migrations. The endemic *P. poliocephalus* extends down the east coast from about 23–39° S (Hall 1987). In captivity it hybridizes with *P. alecto* (e.g. a [female *P. poliocephalus* × male *alecto*] female hybrid mated with a *P. poliocephalus*, producing a male offspring). DNA studies confirm this and indicate that hybridization occurs in the wild (N. J. Webb & C. R. Tidemann pers. comm.). From seasonal emptying of camps, Ratcliffe (1931) thought that *P. poliocephalus* underwent regular south–north spring–autumn migrations. Nelson (1965a) found no evidence of this and suggested autumn post-conception dispersal to the hinterland, with individual nomadism in search of food. Our observations support Nelson. Long-distance radio-tracking of *P. poliocephalus* (Spencer, Palmer & Parry-Jones 1991; Eby 1991) show individuals moving several hundred kilometres over relatively short periods, but the limited data provide no evidence for mass long-distance spring–autumn migrations.

Pteropus scapulatus, also endemic, extends from 26° S in the west, across the north in a broad band and through eastern Australia, reaching 39° S in summer (Hall 1987). It is the only species to extend far inland, and undergoes extensive seasonal migrations which, according to Ratcliffe (1931), showed 'great irregularity'. In recent years *P. scapulatus* have moved into camps around Brisbane (28–30° S) from December onwards and leave by April, with camps then forming in central Queensland. Nelson (1965a) suggested that *P. scapulatus* became non-gregarious in winter. We have seen winter camps in central Queensland, as also has Prociv (1983); we believe that *P. scapulatus* remain gregarious and that this is essential for survival of young born in May or June. *P. scapulatus* roost close together and, when cold, cluster in compact cones of many bats.

Husbandry of captive *Pteropus*

Colonies are housed at 28° S in adjoining open galvanized wire mesh cages (2 × 2 × 12 or 6 m), each with several feeding stations and an insulated shelter area at one end with curtains for privacy and protection from weather. In winter, only the *P. scapulatus* shelter is warmed at night. A mix of diced fruit is provided daily *ad libitum*, plus milk substitute, with access to vitamin and mineral-enriched molasses blocks (O'Brien, Martin & Curlewis 1993) and water. Most mature *P. poliocephalus* and *P. alecto* females conceive, give birth and rear young to independence: subsets have been routinely bled via leg vein through the breeding season and into pregnancy without obvious effect on reproduction, as have subsets of male *P. poliocephalus* and *P. scapulatus* (McGuckin & Blackshaw 1991).

Distribution of births

There are no major differences in the distributions of births of captive *P. poliocephalus* and *P. alecto*, nor in their bled and unbled subsets (Fig. 1): each has a steep leading edge and shallow trailing edge, with some post-peak late births. Data from conservation groups raising orphan *Pteropus* indicate that similar distributions occur in wild populations, with little difference in dates of onset and peak between years or between *P. conspicillatus* at 16° S,

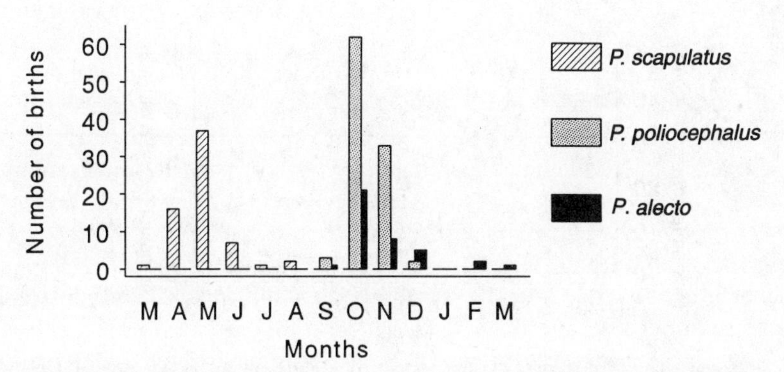

Fig. 1. The distribution of births in captive *P. poliocephalus, P. alecto* and *P. scapulatus*.

P. alecto at 19 and 28° S and *P. poliocephalus* at 28 and 33° S. *P. scapulatus* births are similarly distributed but peak in April–May. This is consistent with observations of wild populations (Ratcliffe 1931; Nelson 1965a; Prociv 1983), although there are no precise data. In Kakadu (12° S) C.R. Tidemann (pers. comm.) observed new-born *P. scapulatus* and *P. alecto* with older young in April and M. McCoy (pers. comm.) states that *P. alecto* births begin in October, peak in November and spread to April, while *P. scapulatus* births peak in May, spreading over succeeding months.

The apparent consistency in breeding times of each species over such a range of habitats is intriguing, in relation both to mechanisms for maintaining constancy and to the varying season quoted for *P. giganteus* (7° N, May–June: Marshall 1947; 13° N, March: Neuweiler 1969; 21° N, January–February: Moghe 1951).

Seasonal changes in males

Captive *P. poliocephalus*, *P. alecto* and *P. scapulatus* show annual testicular cycles which shift little in successive seasons (O'Brien 1993) and correspond to those in the wild (Nelson 1965a; McGuckin & Blackshaw 1987). Our most complete data are for *P. poliocephalus*. Sexual behaviour matches that in the wild (Nelson 1965b) and is nearly identical in *P. alecto*. Male–female associations begin in December while females are suckling young. By January, males establish territories. The frequency of mating increases from February, peaking in April. Over this period there is repeated copulation by individual pairs and considerable promiscuity; ovariectomized females are mated as frequently as are intact females. Plasma testosterone (T) peaks in March–April, as in the wild (Fig. 2). From February, seminal vesicles (SVs) secrete globular bodies that later constitute much of the semen and may contribute to a vaginal

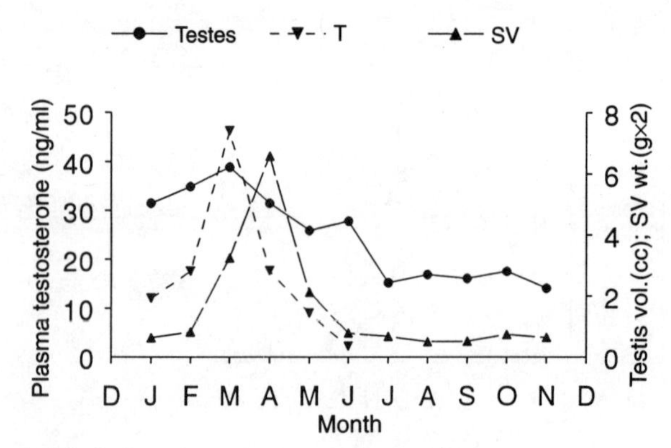

Fig. 2. Seasonal changes in testes, plasma testosterone and seminal vesicles of wild male *P. poliocephalus*; redrawn from McGuckin (1988) and McGuckin & Blackshaw (1987).

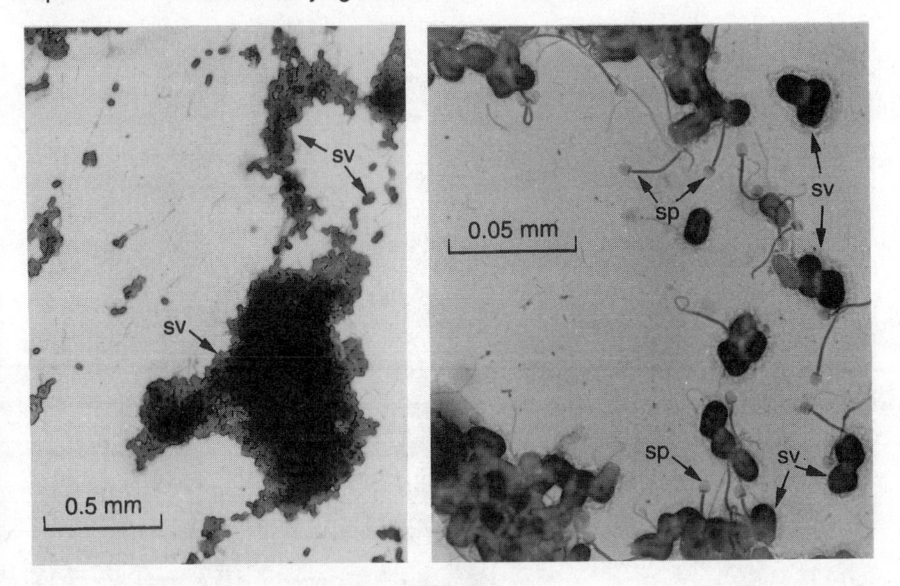

Fig. 3. Vaginal smear from female *P. poliocephalus*, stained with methylene blue and showing spermatozoa (sp) and SV secretion (sv).

plug (Fig. 3). Spermatogenesis and epididymal spermatozoa peak before SV secretion and early in the season spermatozoa alone appear in vaginal smears. This may account for absence of conception from February matings. Aside from maintaining the viability of spermatozoa, SV secretion may affect female reproduction in other ways.

Intromission and ejaculation are relatively brief, but are preceded by long periods in which the male vocalizes at the female or grooms her genitalia deeply with his tongue. Males attempt copulation throughout the year and successful repeated intromissions occur until August, when most females are pregnant. From McGuckin's (1988) data, one would then expect spermatogenesis and epididymal spermatozoa to be minimal and SV secretion to be non-existent. Nonetheless we often find spermatozoa and SV secretion in females mated in July or August.

Breeding-season males become highly odorous. Androgen-sensitive sebaceous neck-glands (Fig. 4; Nelson's (1965b) scapular glands) enlarge (Fig. 5) and are used to mark self, territory and females—so much so that cage galvanizing is worn away and glands become ulcerated. Breeding-season males repeatedly anoint themselves with urine. As they hang with wings slightly cupped and penis three quarters erect, urine is dribbled onto chest and neck and spread by vigorous sideways rubbing with the chin onto the fur and wings. These male tactile, aural, pheromonal factors and secretions may determine mating success, induce or synchronize follicle maturation and ovulation, or establish luteal function.

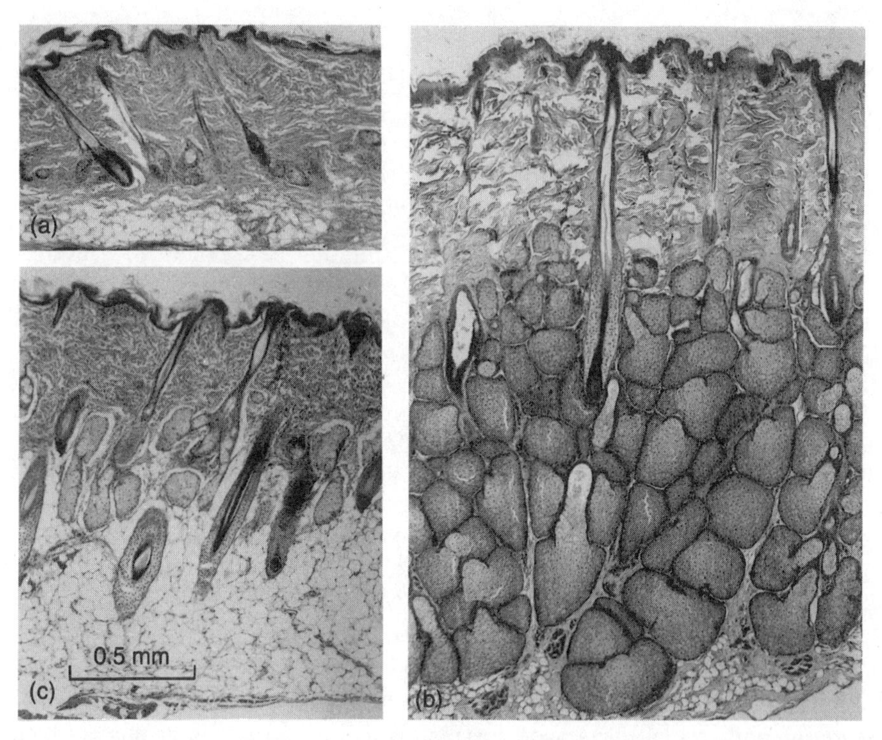

Fig. 4. Sections (7 μm, H & E) of skin of male *P. scapulatus* (a) and neck glands of *P. poliocephalus*; (b) male; (c) female.

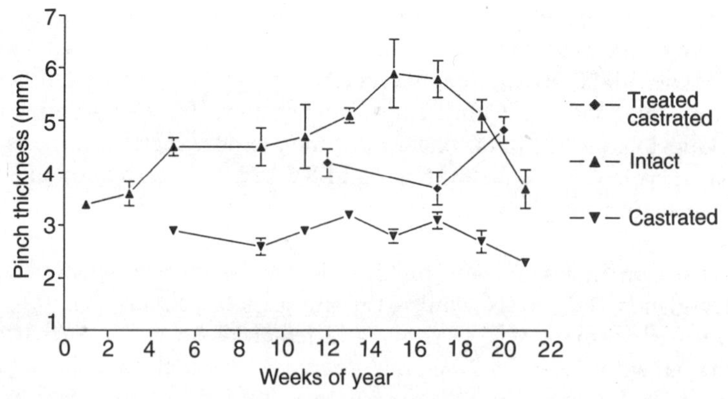

Fig. 5. Seasonal changes in neck gland thickness of adult intact and castrated male *P. poliocephalus* and males castrated at C and given 2.5mg/kg testosterone propionate 3 times a week from T until 3 days before final measurement. Results are means ± SEMs with n=4. Each gland was pinched between thumb and forefinger and the double thickness measured by vernier calipers.

Regulation of ovulation

Both ovaries function, and pregnancy occurs with equal frequency on right and left sides (Baker & Baker 1936; Marshall 1947). Marshall (1947) believed that *P. giganteus* 'resembles . . . mammals in which ovulation is conditioned by mating or erotic play between the sexes'. Nonetheless he noted abundant spermatozoa in females without an 'egg . . . differentiated to a stage where . . . it will ovulate'. Towers & Martin (1985a) found uterine spermatozoa, but no ovulations or corpora lutea (CLs), in mature female *P. poliocephalus* examined in March. Thus ovulation is not induced by copulatory stimuli *per se*.

In December 1987, five female *P. poliocephalus* were isolated from males, within sight, sound and smell of them. In July 1988, they were returned to three males whose season had been delayed by alternating two-month periods of long (16L) and short (8L) photo-period from November 1987. Within minutes, males were grooming females and numerous intromissions were observed. Mating, confirmed by spermatozoa and SV secretion in smears, continued until September. One female, ovariectomized 48h after return to males, had a fresh CL. The others gave birth in January–February 1989, three months after the main colony, plasma progesterone increasing at the appropriate time (Fig. 6). Elsewhere, an injured female, returned to a community cage in October, mated, then gave birth in May (L. Collins pers. comm.).

Our four 'late' mothers, suckling young during the 1989 mating season, conceived, and gave birth in October 1989. In our colony, spontaneous late births and subsequent lactation do not produce late births the following season. Thus lactation *per se* does not delay conception, though in the wild associated nutritional stress might inhibit the pituitary–hypothalamic axis.

Three adult female *P. poliocephalus*, isolated from males in March 1989,

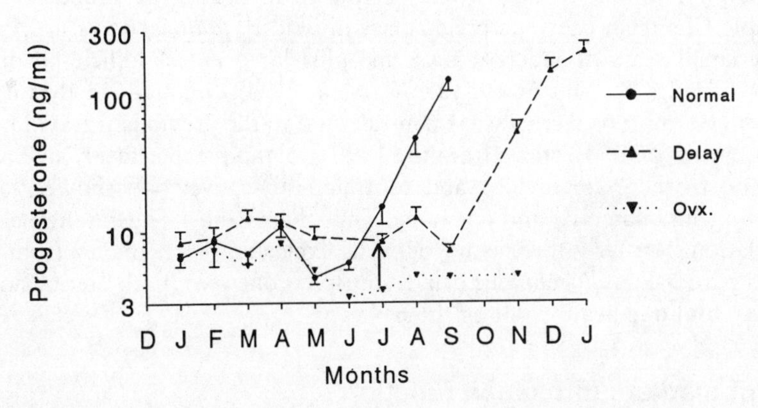

Fig. 6. Plasma progesterone (means & SEMs; RIA of hexane extracts) in nine intact and three ovariectomized (ovx.) adult captive female *P. poliocephalus* housed as normal in the breeding colony and four females whose contact with males was delayed until July (arrow).

were ovariectomized in May. One had a fresh CL, one had a luteinized follicle contralateral to an old CL, and one had large antral follicles. In March 1990, eight adult *P. alecto* and eight *P. poliocephalus* females were isolated 2km from the breeding colony. Four of each species were returned to males in April and ovariectomized seven days later, together with isolated females. Four isolated females contained unilateral CLs (one multiple), one a preovulatory follicle, one a fresh ovulation, one bilateral small antrals and one a cystic follicle contralateral to a luteinized follicle. Six 'returned' females contained unilateral CLs (one multiple), one a cystic follicle and one bilateral large antral follicles: histology and plasma hormone profiles indicated that some CLs had formed before return to the males. Thus, isolated females ovulate spontaneously, though they conceive rapidly when placed with males. We have no evidence of delayed embryo implantation or development in captive or wild bats (L. Martin & P. A. Towers unpublished): spontaneous late births probably reflect late conceptions.

Alternation of ovulation

Alternation of ovulation between ovaries is regarded as the typical pteropodid reproductive asymmetry (Wimsatt 1979). It undoubtedly occurs in *Cynopterus sphinx* (Krishna & Dominic 1983; Sandhu 1984) and *Rousettus leschenaulti* (Gopalakrishna & Choudhari 1977), where the CL of one pregnancy persists into the next. In four late pregnant *P. scapulatus* (plasma oestradiol, 159±23 pg/ml; plasma progesterone, 93±9 ng/ml), four days' treatment with 4iu pFSH plus three days with 250 iu hCG induced follicle development in the non-gravid-side ovaries but not the gravid. This suggests that a CL locally inhibits follicle growth, although foetal or placental factors transported ipsilaterally via the ovarian vascular complex (see below) might be responsible.

The *P. poliocephalus* CL persists to the end of pregnancy, but appears increasingly non-functional (Towers & Martin in press). We found no recognizable CLs from previous pregnancies in wild *P. poliocephalus* in March —only small scars of indeterminate age plus large antral follicles in both ovaries (Towers & Martin 1985a). Marshall (1949) maintained that early *P. giganteus* embryos were always contralateral to the previously gravid horn (PGH). However, in 12 tracts from the 1990 'isolation' experiment, a CL was ipsilateral to the PGH in four and contralateral in five; two showed large antral follicles bilaterally and one none. Thus there is no stringent alternation of ovulation. Similarly there is no rigorous exclusion of pregnancy from the PGH: a parous *P. poliocephalus* delivered triplets one year (both uterine horns involved) and one healthy young the next.

Lack of markers of ovarian function

The vagina, apparently unresponsive to changing hormone levels, remains cornified in the breeding season, late pregnancy and after ovariectomy:

vaginal smears have limited value in diagnosing mating because males frequently tongue-groom the vagina. There is no restricted period of oestrus; ovariectomized and pregnant females mate readily. The ovary is encapsulated; follicles and CL remain internal and cannot be staged visually, only by histology.

Anomalous steroid levels

Radioimmunoassays (RIAs) of plasma oestradiol (E2) and progesterone (P) in wild and captive *P. poliocephalus* showed no meaningful changes in the breeding season (when large antral follicles develop) or in early pregnancy, when the CL forms (Figs 6 & 7). Both hormones increase later in pregnancy with placental secretion (Towers & Martin in press). While sensitivity of E2 RIAs is a limiting factor, endometria in the breeding season show little growth before a preovulatory follicle forms. Thus a preovulatory oestrogen surge may be brief and easily missed. Plasma androstenedione (A) and T RIAs did not indicate ovarian changes any better than did P (Kennedy 1993).

The problem with P is not RIA sensitivity but the high levels found in bats without CLs or ovaries. Three separate studies gave consistent results. Towers & Martin (in press; Fig. 7) confirmed the identity of *P. poliocephalus* plasma P by HPLC and mass spectroscopy. Kennedy (1993; Fig. 6) demonstrated similar levels in female *P. poliocephalus* and *P. alecto* and suggested the adrenal as a major source. Wang, Martin, Kennedy, O'Brien & Bathgate (1992; Fig. 8) confirmed Kennedy's results and found high plasma levels of P (confirmed by HPLC) in intact and castrated males.

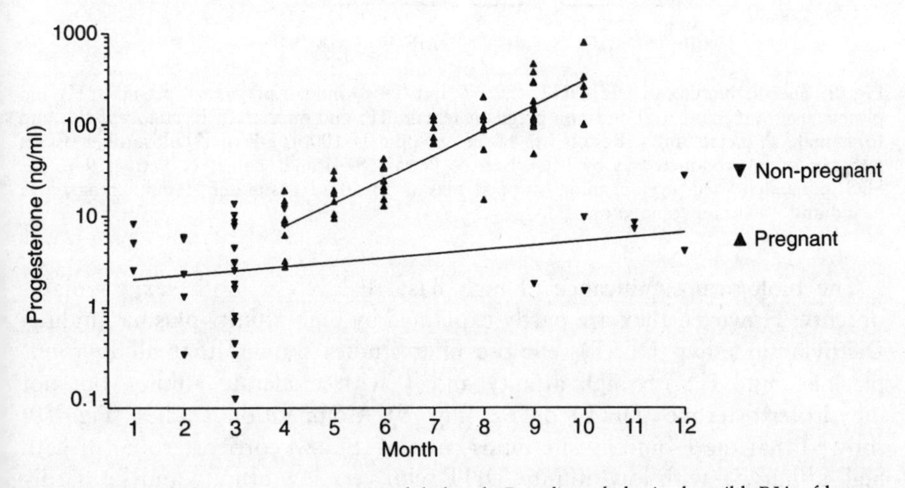

Fig. 7. Peripheral plasma progesterone in adult female *P. poliocephalus* in the wild; RIA of hexane extracts; fitted linear regression (non-pregnant; slope, $b = 0.60 \pm 0.26$, $r = 0.39$, $P<0.05$) and power curve (pregnant; $Y = aX^b$, $b = 3.8 \pm 0.3$, $r = 0.89$, $P<0.0001$); from Towers & Martin (in press).

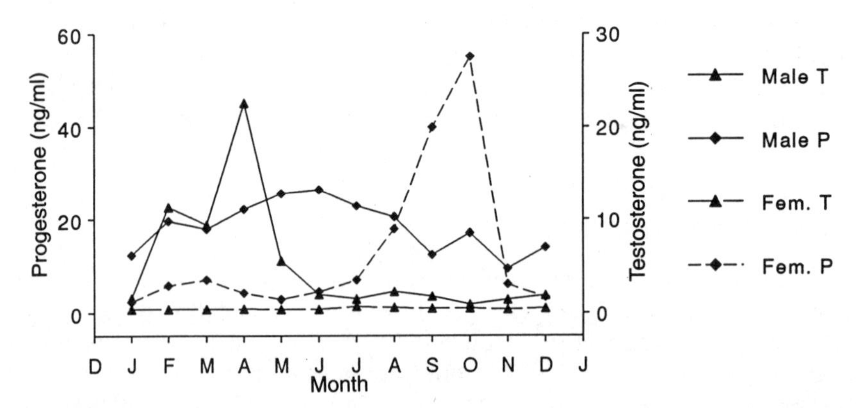

Fig. 8. Plasma progesterone and testosterone in captive male and female *P. poliocephalus*; RIA of plasma ether extracts. Each point is the mean of two separate pools of plasma from 3–5 bats.

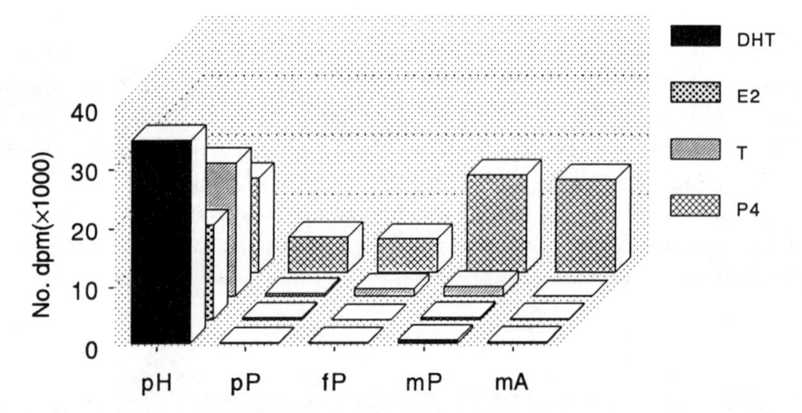

Fig. 9. Specific binding of [^{3}H]-DHT, -E2, -T and -P4 to human pregnancy plasma (pH) and plasma from pregnant (pP) and non-pregnant female (fP) and male (mP) *P. poliocephalus* and from male *P. alecto* (mA). Results are mean no. dpm (×1000). DEAE cellulose filter assays (×5 replicates), conducted as by Mickelson & Petra (1974) and Schiller & Petra (1976) for SBG and CBG, used 1 μl charcoal-stripped plasma in 50 μl tris-buffer, 4 nmol radioactive ligand and 400 nmol competitor.

The biological significance of high basal P levels in both sexes remains obscure. However, they are partly explained by high-affinity plasma binding. Diethylaminoethyl (DEAE)-cellulose filter studies showed that all *Pteropus* plasmas bind P with high affinity and T with moderate affinity, but not dihydrotestosterone (DHT) or E2 (Fig. 9). Competition studies (Fig. 10) showed that the P-binding site binds cortisol (F) and corticosterone (B) with high affinity, T with low affinity, DHT with very low affinity and E2 hardly at all. Thus the protein resembles corticosteroid binding globulin (CBG) rather than steroid binding globulin. This CBG-like moiety, in higher concentration

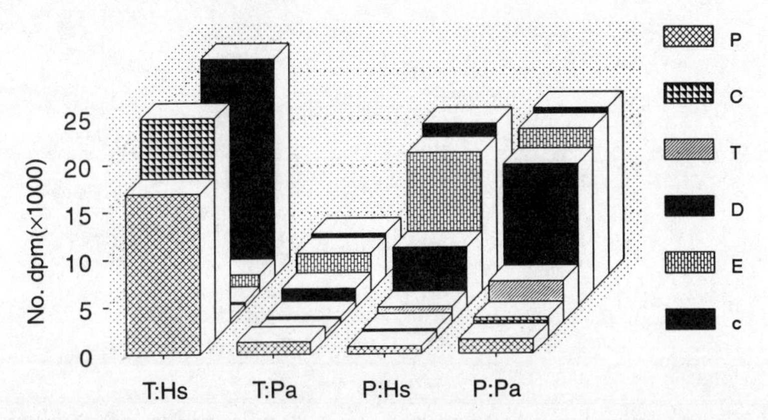

Fig. 10. Competition for binding of [^{3}H]-T and [^{3}H]-P4 to human-pregnancy- and male *P. alecto*-plasma (respectively, T: Hs, T: Pa, P:Hs and P:Pa) by P4, corticosterone, T, DHT and E2 (respectively, P, C, T, D, E): DEAE filter assays (×5/steroid) conducted as in Fig. 9 with 4 nmol radioactive ligand alone (c) or radioactive ligand plus 400 nmol unlabelled ligand.

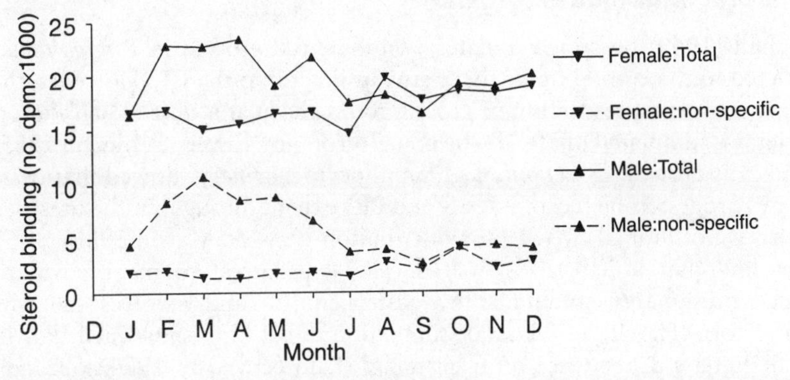

Fig. 11. Total and non-specific binding of [^{3}H]-cortisol + [^{3}H]-progesterone; plasma pools of male and female *P. poliocephalus* as in Fig. 8; cortisol and progesterone binding to DEAE filters was averaged; assays (×5/steroid) conducted as in Fig. 9 with 4 nmol radioactive ligand alone (total binding) or radioactive ligand plus 400 nmol unlabelled ligand (non-specific binding).

in males, appears to increase in breeding-season males but not in pregnant females (Fig. 11).

Our findings confirm and extend those of Kwiecinski, Damassa, Gustafson & Armao (1987). RIAs and HPLC showed that all *Pteropus* have very high levels of circulating F and B in a ratio of approximately 3:1 (*P. scapulatus* plasma pools: 978±100:256±114 ng/ml, n = 5; *P. alecto* plasma pools: 1090±191:440±124 ng/ml, n =5). In *P. poliocephalus*, F and B levels are higher in males than in females and increase in breeding-season males but show no circannual change in females (Fig. 12).

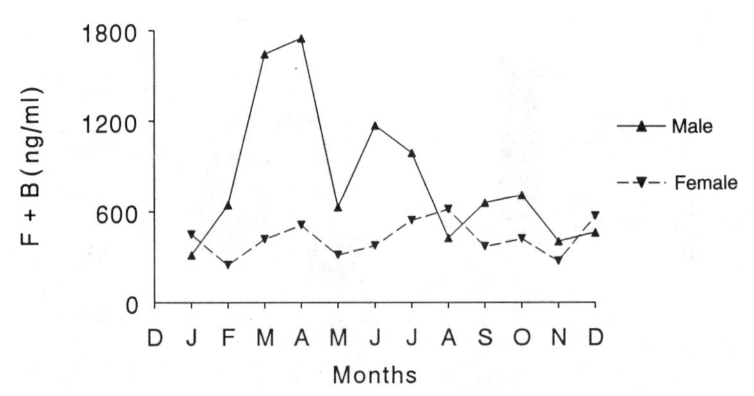

Fig. 12. Plasma cortisol (F) + corticosterone (B) in captive male and female *P. poliocephalus*; RIA of ether extracts of separate pools of plasma from 3–5 bats as in Fig. 8.

Unilateral endometrial growth

Marshall (1949) described unilateral endometrial growth in *P. giganteus* as restricted to the cranial tip of the uterus ipsilateral to the CL. However, there is no fundamental restriction of growth to one horn at a time: when *Pteropus* ovaries are stimulated bilaterally by gonadotrophin (Towers & Martin 1985b), both horns grow. In *P. scapulatus*, immunocytochemistry showed that oestrogen and progesterone receptors (ER and PR) extend throughout the uterus, i.e. growth is not limited by receptor distribution (Pow & Martin 1988).

We find that unilateral growth develops ipsilateral to the preovulatory follicle, presumably stimulated by oestrogen (E), and involves the whole uterine horn (Fig. 13). Unilateral stimulation might limit subsequent P action via the latter's dependence on E 'priming' (Finn & Martin 1974). Secondary unilateral growth modulation might subsequently arise from endometrial refractoriness, like that induced in mice by E and P (Martin, Hallowes, Finn & West 1973).

Ovarian structure and vasculature

Hypotheses to explain unilateral growth include transfer of ovarian hormones from ovarian veins or lymphatics to arteries supplying the uterus (Wimsatt 1979). Using injection casts and histology (Fig. 14), Pow & Martin (1994) showed that the ovarian artery coils just cranial to the ovary, provides a branch to the ovary and continues caudally as the major supply to the cranial region of the uterus, where it anastomoses with the smaller uterine artery. The ovarian arterial coil is enclosed by a venous sinus draining the ovary. The ovary is heavily encapsulated, with primordial follicles confined to a caudal zone and follicles or CLs placed cranially, close to the ovarian arterial coil. This would

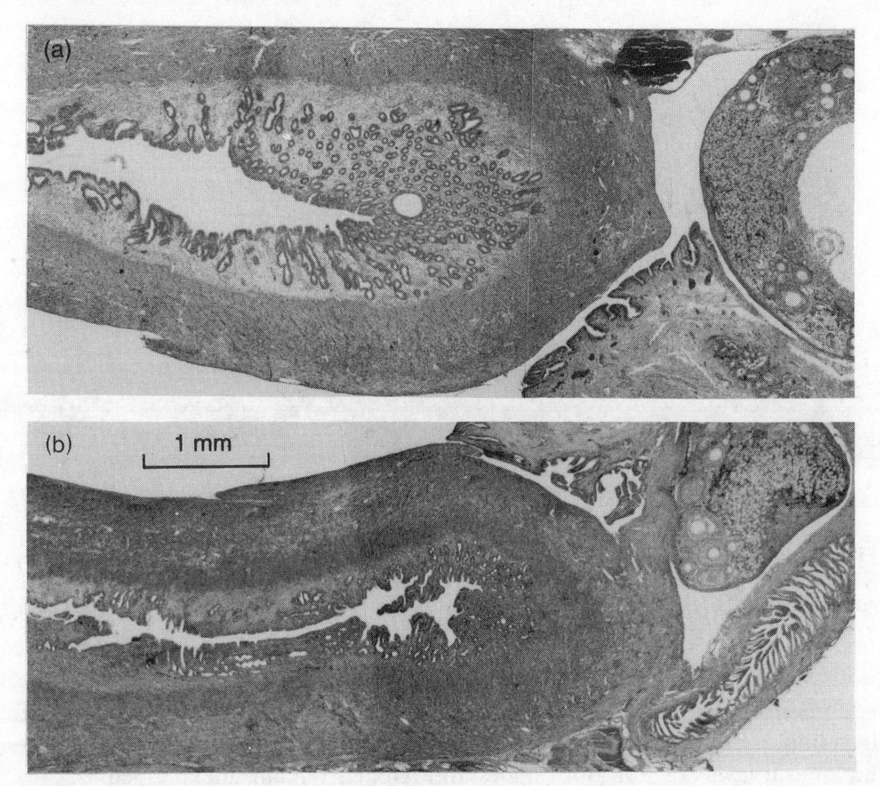

Fig. 13. Sections (7μm, H & E) of uterine horns and ovaries of *P. poliocephalus*: (a) hypertrophy of whole horn ipsilateral to the preovulatory follicle; (b) atrophic contralateral horn; bar = 1 mm.

facilitate counter-current transfer of ovarian steroids via ovarian vein and artery to ipsilateral uterus, the locally high levels stimulating endometrial growth. Cranial to the coil, the ovarian artery is enclosed by a venous sinus derived from both uterine and ovarian veins. This would allow transfer of uterine factors to ipsilateral ovary.

An encapsulated ovary, with segregated primordial follicles and ovarian venous sinus enclosing a coiled ovarian artery, is present in nine species from nine megachiropteran genera (C.S. Pow & L. Martin unpublished). Such structures are not evident in six microchiropteran species from six families studied by us, nor in any published pictures of microchiropteran and mammalian ovaries. We postulate that this unusual ovarian anatomy and vasculature is unique to Megachiroptera.

Unilateral endometrial growth and peripheral steroid levels

Preferential transfer from CL to ipsilateral endometrium may explain why peripheral P does not increase in early pregnant *Pteropus*: if P underwent

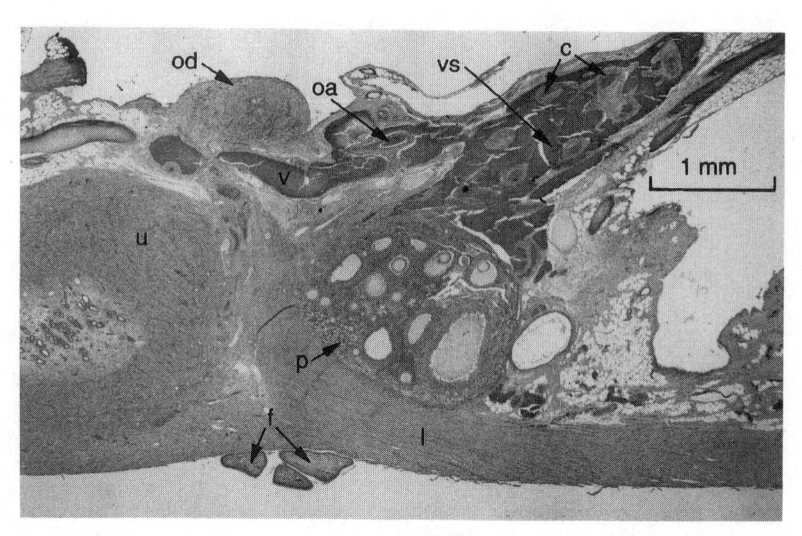

Fig. 14. Section (7μm, H & E) of ovary, oviduct (od), uterus (u) and coiled ovarian artery (c) of *P. poliocephalus*: shows junction of uterine vein (v) with ovarian venous sinus (vs), their enclosure of the ovarian artery (oa) and primordial follicles (p) confined to caudal zone close to fimbria (f) and ligament (1); bar = 1 mm.

endometrial metabolism, as in rodents (Clark 1975), luteal secretion would have minimal effect on peripheral levels and contralateral endometrium. The high basal levels of P in both males and females remain unexplained.

Unilateral endometrial growth associated with the preovulatory follicle presumably involves E, which usually does not undergo significant endometrial metabolism. In Jensen & Jacobson's (1962) experiments, injected [^{3}H]-E2 was taken up preferentially by rat uterine ER in amounts sufficient to stimulate growth. In *Pteropus*, we find hypertrophied endometria ipsilateral to long-standing cystic follicles, with contralateral endometria remaining atrophic. If *Pteropus* ER bound E2 as strongly as does rat ER, one might expect that, with prolonged ovarian secretion, sufficient E2 would leak into the general circulation eventually to saturate uterine ER bilaterally. Preferential local transport is probably not sufficient to explain unilateral growth. Our hypothesis that saturation of contralateral ER is prevented by high-affinity E2-binding in plasma was disproved (Wang *et al.* 1992). Metabolism of E2 by *Pteropus* endometrium needs to be explored.

To examine ovaro-uterine transport of steroids in *Pteropus*, we infused [^{3}H]-E2 into one ovary and measured uterine levels 1 h later (Pow & Martin 1991). We based the experiment on studies in mice (Martin 1964), which showed that physiological doses of [^{3}H]-E2 instilled into target organs were retained unchanged for long periods bound to ER. Our experiment demonstrated [^{3}H]-E2 transfer from ovary to ipsilateral uterus (though not the actual route), but [^{3}H]-E2 recovery was very low, suggesting that *Pteropus* ER might have low affinity for E2. A uterine low-affinity ER would be saturated by locally high E2 concentration from the ipsilateral ovary, but not by low systemic levels. Operationally the same would obtain if *Pteropus* ER resembled rat ER in affinity for E2, but the ovary secreted a novel oestrogen which was bound with low affinity by uterine ER.

Pteropus are spontaneous ovulators and such animals usually require ovarian E2 to induce the LH surge essential for ovulation. Thus we must question how ovary communicates with hypothalamus if it does not communicate with contralateral uterus. Again, one is led to postulate a novel ovarian steroid as messenger.

Environmental factors regulating breeding season

Photoperiod manipulation does not disrupt the testicular cycle of *P. scapulatus* (McGuckin & Blackshaw 1988; O'Brien *et al.* 1993). This is consistent with the small pineal and low-amplitude nocturnal changes in melatonin in this species (McGuckin 1988). *P. poliocephalus* has a relatively large pineal; nocturnal peaks in melatonin accurately reflect duration of the scotophase and males do respond to altered photoperiod (McGuckin & Blackshaw 1992). Nevertheless, their response is so sluggish compared with other short-day breeders (O'Brien, Martin & Curlewis 1991) that photoperiod is probably not a major proximate factor entraining their cycle in the wild (O'Brien 1993). Similar experiments indicate that *P. alecto* responds to photoperiod much like *P. poliocephalus*.

In January 1989, six female *P. poliocephalus* were placed in open cages with males whose season had been advanced by photoperiod manipulation. Births were not advanced, confirming that females are not induced ovulators capable of conceiving at any time that males are fertile. In November 1991, six female *P. poliocephalus* were placed in short daylength (11L:13D ≡ mid-April at 28°S) until 1 March 1992, when they were placed in the outdoor breeding colony until 1 July, then returned to short daylength indoors. The maximal possible advance in conception time was 4-6 weeks. Two females gave birth in September and one on October 2, suggesting an advance in conception time. Accordingly, 13 females and two males were placed in 11L:13D in November 1992 and maintained there. To date (November 1993) there have been no births.

Discussion

Australian *Pteropus* present many endocrine and reproductive puzzles, including retention of reproductive synchrony from year to year over large and changing ranges. Although photoperiod appears to have no regulatory role in the nomadic *P. scapulatus*, it may have in *P. poliocephalus* and *P. alecto*—at least in males. Its role in females remains unknown. Our experiments and the distribution of births indicate that female *P. poliocephalus* cannot ovulate before a certain date, but thereafter may do so repeatedly. Thus hypothalamic gonadotrophin secretion may be 'gated' by day-length, perhaps mediated by pineal melatonin, with responsiveness requiring a sequence of slowly increasing and decreasing photoperiod. Nonetheless, our experiments do not exclude male factors having a major role in regulating female hypothalamic sensitivity and ovarian function. Females may also affect male function and we need additional photoperiod experiments with male–female groups rather than isolated males. In interpreting experiments with such species, there are always questions of stress, adequacy of housing, food etc. In this respect, *P. poliocephalus* is remarkably resilient—most adult females housed outdoors in various group-mixes and cage-sizes continue to conceive and give birth seasonally.

Acknowledgements

We thank the University of Queensland for scholarships (CP & JK), the ARC for generous long-term support plus GO'B's Fellowship, and Queensland Primary Industries inspectors who helped to provide bat food for many years.

References

Baker, J. R. & Baker, Z. (1936). The seasons in a tropical rain forest (New Hebrides). Part 3. Fruit bats (Pteropodidae). *J. Linn. Soc. Lond.* **40**: 123–141.

Clark, B. F. (1975). Effect of oestrogen priming on the pattern of progesterone metabolism in the mouse uterus. *J. Endocr.* **66**: 293–294.

Eby, P. (1991). Seasonal movements of grey-headed flying-foxes, *Pteropus poliocephalus* (Chiroptera: Pteropodidae), from two maternity camps in Northern New South Wales. *Wildl. Res.* **18**: 547–559.

Finn, C. A. & Martin, L. (1974). The control of implantation. *J. Reprod. Fert.* **39**: 195–206.

Gopalakrishna, A. & Choudhari, P. N. (1977). Breeding habits and associated phenomena in some Indian bats Part I—*Rousettus leschenaulti* (Desmarest)—Megachiroptera. *J. Bombay nat. Hist. Soc.* **74**: 1–16.

Hall, L. S. (1987). Identification, distribution and taxonomy of Australian flying-foxes (Chiroptera: Pteropodidae). *Aust. Mammal.* **10**: 75–79.

Jensen, E. V. & Jacobson, H. I. (1962). Basic guides to the mechanism of estrogen action. *Rec. Prog. Horm. Res.* **18**: 387–414.

Kennedy, J. H. (1993). *Regulation of ovarian function in Australian flying-foxes.* PhD thesis: University of Queensland.

Krishna, A. & Dominic, C. J. (1983). Reproduction in the female short-nosed fruit bat *Cynopterus sphinx* Vahl. *Period. biol.* **85**: 23–30.

Kwiecinski, G. G., Damassa, D. A., Gustafson, A. W. & Armao, M. E. (1987). Plasma sex steroid binding in Chiroptera. *Biol. Reprod.* **36**: 628–635.

Marshall, A. J. (1947). The breeding cycle of an equatorial bat *(Pteropus giganteus* of Ceylon). *Proc. Linn. Soc. Lond.* **159**: 103–111.

Marshall, A. J. (1949). Pre-gestational changes in the giant fruit bat *(Pteropus giganteus)*, with special reference to an asymmetric endometrial reaction. *Proc. Linn. Soc. Lond.* **161**: 26–36.

Martin, L. (1964). The uptake of locally applied [6, 7–^{3}H]-oestradiol by the vagina of the ovariectomized mouse. *J. Endocr.* **30**: 337–346.

Martin, L., Hallowes, R. C., Finn, C. A. & West, D. G. (1973). Involvement of the uterine blood vessels in the refractory state of the uterine stroma which follows oestrogen stimulation in progesterone-treated mice. *J. Endocr.* **56**: 309–314.

Martin, L., Towers, P. A., McGuckin, M. A., Little, L., Luckhoff, H. & Blackshaw, A. W. (1987). Reproductive biology of flying foxes (Chiroptera: Pteropodidae). *Aust. Mammal.* **10**: 115–118.

McGuckin, M. A. (1988). *Seasonal changes in the reproductive physiology of male flying-foxes (*Pteropus *spp.) in south east Queensland.* PhD thesis: University of Queensland.

McGuckin, M. A. & Blackshaw, A. W. (1987). Seasonal changes in spermatogenesis (including germ cell degeneration) and plasma testosterone concentration in the grey-headed fruit bat, *Pteropus poliocephalus. Aust. J. biol. Sci.* **40**: 211–220.

McGuckin, M. A. & Blackshaw, A. W. (1988). Different responses to the same environmental cues regulate long and short day breeding seasons in two species of *Pteropus* (Megachiroptera) in south-east Queensland. *Proc. int. Congr. Anim. Reprod. artif. Insem.* **11**: 412.

McGuckin, M. A. & Blackshaw, A. W. (1991). Seasonal changes in testicular size, plasma testosterone concentration and body weight in captive flying foxes (*Pteropus poliocephalus and P. scapulatus. J. Reprod. Fert.* **92**: 339–346.

McGuckin, M. A. & Blackshaw, A. W. (1992). Effects of photoperiod on the reproductive physiology of male flying foxes, *Pteropus poliocephalus. Reprod. Fert. Dev.* **4**: 43–53.

Mickelson, K. E. & Petra, P. H. (1974). A filter assay for the sex steroid binding protein of human serum. *FEBS Lett.* **44**: 34–38.

Moghe, M. A. (1951). Development and placentation of the Indian fruit bat, *Pteropus giganteus giganteus* (Brünnich). *Proc. zool. Soc. Lond.* **121**: 703–721.

Nelson, J. E. (1965a). Movements of Australian flying foxes (Pteropodidae: Megachiroptera). *Aust. J. Zool.* **13**: 53–73.

Nelson, J. E. (1965b). Behaviour of Australian Pteropodidae (Megachiroptera). *Anim. Behav.* **13**: 544–557.

Neuweiler, G. (1969). Verhaltensbeobachtungen an einer indischen Flughundkolonie (*Pteropus giganteus* Brunn.). *Z. Tierpsychol.* **26**: 166–199.

O'Brien, G. M. (1993). Seasonal reproduction in flying foxes, reviewed in the context of other tropical mammals. *Reprod. Fert. Dev.* **5**: 499–521.

O'Brien, G. M., Martin, L. & Curlewis, J. (1991). Unusual reproductive photoresponsiveness of male greyheaded flying foxes (*Pteropus poliocephalus*). *Proc. Aust. Soc. Reprod. Biol.* **23**: 160.

O'Brien, G. M., Martin, L. & Curlewis, J. (1993). Effect of photoperiod on the annual cycle of testis growth in a tropical mammal, the little-red flying-fox, *Pteropus scapulatus. J. Reprod. Fert.* **98**: 121–127.

Pow, C. S. & Martin, L. (1988). Distribution of oestrogen receptors in the female reproductive tract of the flying fox *Pteropus scapulatus. Proc. Aust. Soc. Reprod. Biol.* **20**: 55.

Pow, C. S. & Martin, L. (1991). Preferential transfer of [^{3}H]-oestradiol from ovary to ipsilateral uterus in the grey headed flying fox (*Pteropus poliocephalus*). *Proc. Aust. Soc. Reprod. Biol.* **23**: 17.

Pow, C. S. T. & Martin, L. (1994). The ovarian-uterine vasculature in relation to unilateral endometrial growth in flying foxes (genus *Pteropus*, suborder Megachiroptera, order Chiroptera). *J. Reprod. Fert.* **101**: 247–255.

Prociv, P. (1983). Seasonal behaviour of *Pteropus scapulatus* (Chiroptera: Pteropodidae). *Aust. Mammal.* **6**: 45–46.

Ratcliffe, F. N. (1931). The flying fox in Australia. *Bull. Coun. scient. ind. Res. Melb.* No. 53: 1–81.

Richards, G. C. (1990). The spectacled flying-fox, *Pteropus conspicillatus*, (Chiroptera: Pteropodidae) in North Queensland. 1. Roost sites and distribution patterns. *Aust. Mammal.* **13**: 17–24.

Rowlands, I. W. & Weir, B. J. (1984). Reproductive cycles. Mammals: non-primate eutherians. In *Marshall's Physiology of reproduction* (4th edn): 455–658. (Ed. Lamming, G. E.). Churchill Livingstone, Edinburgh.

Sandhu, S. (1984). Breeding biology of the Indian fruit bat, *Cynopterus sphinx* (Vahl) in central India. *J. Bombay nat. Hist. Soc.* **81**: 600–612.

Schiller, H. S. & Petra, P. H. (1976). A filter assay for the corticosteroid binding globulin of human serum. *J. Steroid Biochem.* **7**: 55–59.

Spencer, H. J., Palmer, C. & Parry-Jones, K. (1991). Movements of fruit-bats in eastern Australia, determined by using radio-tracking. *Wildl. Res.* **18**: 463–468.

Towers, P. A. & Martin, L. (1985a). Some aspects of female reproduction in the grey-headed flying-fox, *Pteropus poliocephalus* (Megachiroptera: Pteropodidae). *Aust. Mammal.* **8**: 257–263.

Towers, P. A. & Martin, L. (1985b). PMSG induced ovulation in the flying fox *Pteropus scapulatus. Proc. Aust. Soc. Reprod. Biol.* **17**: 10.

Towers, P. A. & Martin, L. (In press). Peripheral plasma progesterone levels in pregnant and nonpregnant greyheaded flying-foxes (*Pteropus poliocephalus*) and little red flying-foxes *(P. scapulatus). Reprod. Fert. Dev.*

Wang, D., Martin, L., Kennedy, J. H., O'Brien, G. M. & Bathgate, R. (1992). Steroid binding in the plasma of flying foxes. *Proc. Aust. Soc. Reprod. Biol.* **24**: 108.

Wimsatt, W. A. (1979). Reproductive asymmetry and unilateral pregnancy in Chiroptera. *J. Reprod. Fert.* **56**: 345–357.

Part IV

Ecology and ecophysiology of Microchiroptera

Symp. zool. Soc. Lond. (1995) No. 67: 187–201

Chiropteran nocturnality

John R. SPEAKMAN

Department of Zoology
University of Aberdeen
Aberdeen AB9 2TN, Scotland, UK

Synopsis

Despite their diverse feeding habits almost all bats are exclusively nocturnal. Explanations of chiropteran nocturnality have focused on three potentially negative consequences of feeding in daylight: competition with insectivorous (and frugivorous) birds, risk of avian predation and risk of hyperthermia. A survey of daylight flying by bats in the UK suggested that the risk of avian predation might be the most significant factor, and this was supported by observed predation rates on bats deliberately released to fly during daylight in south-eastern Australia. There are many island groups where bats are found but where there are no avian predators likely to take bats. If avian predation is the most significant factor influencing nocturnality then I predict that these bats should fly in daylight. One group of islands where this behaviour would be predicted is the Azores. Early observations of the indigenous Azorean bat (*Nyctalus azoreum*) suggested that this bat flies in daylight. However, more recent detailed studies of temporal patterning of activity in the Azorean bat indicate that it is primarily nocturnal, thus contradicting the predation hypothesis. Anecdotal records of several other island populations of insectivorous bats also indicate that they have retained nocturnality in the absence of diurnal predators. In contrast, several studies of activity patterns in pteropodid bats suggest that they fly during the day when predators are absent. A heat balance model was constructed to evaluate the significance of the hyperthermia hypothesis. Although there is a high endogenous heat production during flight, incoming solar radiation may place a heat burden on a flying bat an order of magnitude greater. Flying bats can dilate blood vessels in their wings to dissipate the heat burden convectively. However, there is a critical air temperature above which this physiological mechanism fails to dissipate all the incoming heat, and fatal hyperthermia results. A multitude of factors influences the critical air temperature. Global variation in these factors leads to a prediction that hyperthermia could constrain diurnal activity of small (9 g) bats between 20 and 40° N and S, and of large (900 g) bats at all latitudes lower than 50° N and S. None of the proposed hypotheses can explain all the observed features of chiropteran nocturnality.

Introduction

There are almost 950 species of bats (Hill & Smith 1984), making them the second most diverse order within the class Mammalia (Corbet 1978).

ZOOLOGICAL SYMPOSIUM No. 67
ISBN 0 19 854945 8

Furthermore, the diversity of bat feeding habits is the greatest for any order of mammals (Hill & Smith 1984). In spite of this wealth of phylogenetic and ecological diversity, it is remarkable that almost all bat species are exclusively nocturnal. By restricting their foraging activity periods to darkness, bats fail to exploit abundant resources which are available during the day. This is particularly so for small temperate-zone insectivorous bats during summer, when the night is relatively short and the peak aerial insect availability occurs during the late afternoon or early evening, before it gets dark (Fig. 1; and Williams 1961; Rydell 1992). A direct consequence of nocturnality for these bats is that they frequently fail to meet their energy requirements and they must use torpor to balance their energy budgets (Kunz 1980; Kurta, Johnson & Kunz 1987; Speakman & Racey 1987; Audet & Fenton 1988; Kurta, Bell, Nagy & Kunz 1989). Being on a knife edge of energy balance is probably the most important factor restricting chiropteran litter sizes (Kurta & Kunz 1987; but see Barclay this volume pp. 245–258). The greater aerial insect resources available by day in the temperate zone are reflected in the large clutch sizes which are raised by many diurnal insectivorous birds. Hirundines, for example, typically raise two clutches of four to seven eggs (Bruun & Singer 1970) over the same summer period as that in which most bats raise only a single young or, more rarely, twins. Nocturnal insectivorous birds, like Caprimulgiformes,

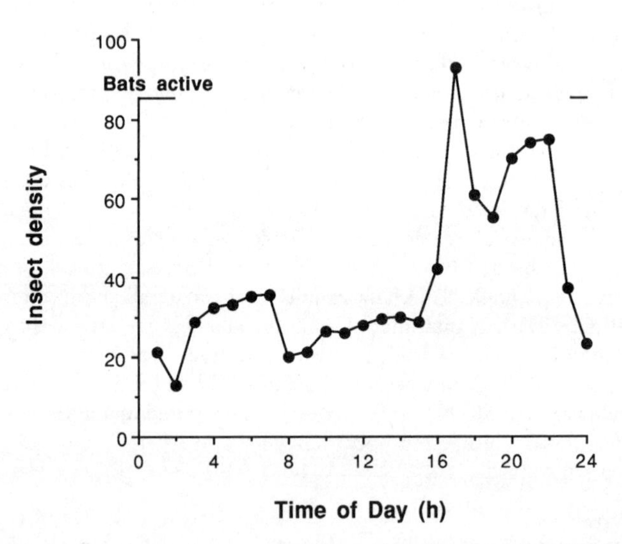

Fig. 1. The diel pattern of aerial insect availability at Aberdeen, UK (57° N). Data collected by J. Rydell (unpublished and used with permission) using a Johnson Taylor suction trap. The data represent the mean numbers of insects trapped in each hour averaged across three consecutive days in midsummer (June 1993). The peak insect abundance occurred between 16:00 and 22:00 during daylight. The bats emerged to feed around 22:45 and had returned to roost by 14:00. This activity period is denoted by a bar at the top of the figure. The activity of bats did not coincide with the period when the maximum numbers of insects were available.

on the other hand, also have small clutches of usually two eggs (Bruun & Singer 1970). The only insectivorous birds which do not conform to this pattern are the Apodiformes, which are diurnal but also raise small clutches of two to three eggs. This probably reflects the high costs of their more continuously aerial lifestyle, since time in flight is an important factor influencing avian daily energy demands (Bryant & Tatner 1991).

Given the energetic and nutritional disadvantages of restricting foraging activity to the night, chiropteran nocturnality is a key behavioural feature which requires an evolutionary explanation. Attempts to explain it have focused on direct fitness disadvantages which might outweigh the potential energetic or nutritional benefits of diurnality. First, bats may be outcompeted by diurnal birds exploiting the same food sources (Tugendhat 1966; Moore 1975). Second, bats may be exposed to increased risks of predation by diurnal predatory birds (e.g. Gillette & Kimbourgh 1970). Finally, bats may be unable to dissipate the radiative heat burden from the sun at the relatively high diurnal temperatures, and consequently may fatally overheat (S.P. Thomas & Suthers 1972; S.P. Thomas, Follette & Farabaugh 1991).

Caraco, Martindale & Whitman (1980) suggested that foraging animals might be sensitive in their foraging behaviour to both the mean and variance in energy returns, and coined the terms risk-averse (feeding at low variance sites and times) and risk-prone (feeding at high variance sites and times) foraging strategies. This concept was extended to include non-energetic risks, for example predation, that might be associated with foraging at given places and times (Milinski & Heller 1978). The hypotheses that bats forage at night to avoid competition, predation or hyperthermia are all equivalent to interpreting nocturnality as a risk-averse foraging strategy. Risk-averse explanations of nocturnality predict that bats should occasionally fly in daylight (become risk-prone in their foraging behaviour). This is because there are situations where bats would inevitably die if they rigidly adopted nocturnality. Consider, for example, a prolonged cold spell during the summer. Nocturnal and diurnal insect availabilities will be depressed and bats flying exclusively at night would fail to meet an energy balance and would thus fall torpid during the day. If the cold spell continued then the bats would run down their energy reserves to a point where they would die during the next daytime period, even if they fell torpid. This would happen rapidly early in the year when the bats have few fat reserves (Ransome 1990), but more slowly later in the summer. Despite its risks, flying in daylight would be a profitable strategy in this situation because the probability of survival by flying in daylight ($P > 0$) would exceed that of remaining nocturnal ($P = 0$).

Another situation favouring daylight flying occurs in winter when insectivorous bats in the temperate zone are hibernating. A bat rousing from torpor early in the day (Twente & Twente 1987; D. W. Thomas 1993 and this volume pp. 233–244) would have to wait several hours until the next dark period. Waiting until it was dark before emerging to feed (Avery 1985;

Hays, Speakman & Webb 1992) or drink (Speakman & Racey 1989, 1990; D. W. Thomas this volume pp. 233–244) would be extremely costly, and the resulting use of energy might compromise survival during the subsequent period of prolonged torpor. The bat might do better to risk flying in the day, and this behaviour might be particularly expected if it was warm and the bat could cover the costs of its flight by some feeding when it emerged.

Bats *are* occasionally observed flying during daylight. This infrequent behaviour, however, could represent sporadic acts of disturbance to day-roosting bats, forcing them to emerge and fly. Does the pattern of these occasional daylight flights correspond with that expected from the risk-prone or risk-averse theory, or is it random? To gather data on this subject I organized a survey of daylight flying behaviour of bats throughout the UK between 1986 and 1989 (Speakman 1990, 1991a). I sent circulars to bat conservation groups and local ornithological societies soliciting information on sightings of bats flying in daylight. A total of 420 records had been received by the end of 1989, when formal recording ended. Bats observed flying in daylight in winter were significantly more likely to be seen on warm days than on cold days (Speakman 1990). During summer, however, the daytime temperature was unimportant but, at least early in the summer, days when bats were seen flying in daylight were preceded by colder nights than were days when they were not seen (Speakman 1990). These survey data therefore indicated that diurnal activity of insectivorous bats in the UK was consistent with a risk-prone strategy, and thus by implication that nocturnality was its risk-averse equivalent. What these data do not indicate, however, is the nature of the risk that bats were routinely avoiding by being nocturnal. In the remainder of this paper I will consider the evidence for and against each of the alternative disadvantages suggested for chiropteran diurnality.

The avian competition hypothesis

The principal data in favour of the avian competition hypothesis are direct observations that bats which are occasionally observed flying in daylight are attacked by diurnal flying competitive birds (e.g. Tugendhat 1966). These attacks, however, are generally not fatal, and insectivorous bats easily outmanoeuvre any attacking insectivorous bird. Although the apparent direct consequences of attacks may be relatively trivial, the actual consequences may be more serious because a bat continually avoiding attack by an avian competitor may be unable to feed effectively. Unfortunately there are currently no measurements of the feeding rates of diurnally feeding bats either in the absence of, or during, attack by a competitor.

Attacks by competitive birds on daylight-flying bats are rarely seen, when compared with all the records of day-flying bats. For example, among the 420 reports submitted in the UK survey (Speakman 1990) there were only 19 records of hirundines or apodids using the same airspace, and in only six of these were antagonistic interactions observed (Speakman 1991a). Although most observed bats used airspace generally not occupied by hirundines or apodids, this does not indicate that competition is unimportant, since the bats may have been forced into such airspace by competition and this airspace may have had lower densities of aerial insects.

Vernier (1990) observed that in the early evening above the town of Padua in Northern Italy (35° N), both Kuhl's pipistrelles (*Pipistrellus kuhlii*) and swifts (*Apus apus*) overlapped in their activity for about 30 min before nightfall. The bats generally fed above the swifts until the birds departed, at which time the bats descended to feed at the lower level. These observations are consistent with competitive displacement of the bats by the swifts. However, the same pattern of bat movement was observed when the swifts were absent, suggesting that competition was not the most important factor influencing the spatial exploitation patterns of the bats. Indeed, if the bats were exploiting the most abundant food source independently of the presence or absence of swifts, then it could be reasoned that the bats were competitively displacing the swifts.

If we consider non-insectivorous species of bats, then there are clearly some situations where it is not possible to suggest that competition with other diurnal animals is the cause of nocturnality. For example, the sanguinivorous vampire bats (*Desmodus* and *Diadema* sp.) have no known sympatric diurnal counterparts with which to compete. For these species there may be other risks that are more important, or they may be unable to feed on their diurnally active prey until these become quiescent at night. There are, however, many diurnal frugivorous, nectarivorous and carnivorous birds and mammals which could compete with frugivorous, nectarivorous and carnivorous bats. I am not aware of any reports of interactions between such animals and the occasionally observed diurnally active bats in these trophic groups.

The avian predation hypothesis

There are many records of bats falling prey to diurnal predatory birds (see reviews in Macy & Macy 1969; Gillette & Kimbourgh 1970; Speakman 1991b). Taken alone, the volume of these records by far exceeds the numbers of published records of interactions between daylight flying bats and competitive birds. There are, however, two problems with these records. First, the number of records alone may bear no relation to the ecological and evolutionary importance of the phenomenon. Attacks by predatory birds on daylight flying bats are dramatic events which are, perhaps, more likely to be noticed, written up and published than less dramatic interactions between bats and avian competitors. This may give a biased indication of the importance of predation as a factor restricting chiropteran diurnality. Second, it is not the number of reported attacks which is important, but rather the rate at which these attacks occur relative to the rate of nocturnal predation. There are far more records of bats being preyed upon by owls (*Strigiformes*), for example (e.g. Bauer 1956; Ruprecht 1979), than by diurnal avian predators. Probably about 200 000 bats are killed by predators annually in the UK (*c.* 10–12% of total mortality) and approximately 93% of these fall prey to owls, particularly tawny owls (*Strix aluco*), with only 5% falling prey to diurnal predatory birds (accipiters and falcons; Speakman 1991b).

I estimated the rate of predation on diurnal flying bats using the 420 records in the British survey (Speakman 1990, 1991a). These 420 records included 13 attempted acts of predation, at least four of which were successful. Using the reported times for which bats were observed, this leads to an estimated probability of being killed of 0.01 per hour. This translates to an average life expectancy for a daylight-flying bat of only about 14.3 h. We can compare this with a maximum estimate for the probability of being preyed on when flying around at night (after Speakman 1991a). This estimate is based on the pipistrelle bat (*Pipistrellus pipistrellus*) which comprised about 65% of the daylight flying records in the UK survey. The annual survival of the pipistrelle bat is about 0.33 (Thompson 1987). Pipistrelle bats fly on average for 225 min per night (Swift 1980), for a summer which lasts around 150 days. The total annual flight time at night (ignoring winter activity; Avery 1985) is thus about 450 h. Assuming that all the annual mortality is a consequence of predation, then the maximum hourly risk of mortality is around 0.0001, or one hundredth the risk of flying in the day. A more realistic estimate of the contribution of predation to total annual mortality is about 10% (see above; Speakman 1991b). Thus the risk of nocturnal predation may be as low as one thousandth the risk of flying in daylight. This is strong evidence favouring the hypothesis that predation is the principal constraint on diurnal activity.

This calculation, however, may also be compromised by the problems of over-reporting of dramatic predation events in the survey. To test this, I analysed the recorded fates of 1023 bats deliberately released during daylight in south-eastern Australia by Lindy Lumsden (Speakman, Lumsden & Hays 1994). Of these released bats, four were killed by predators. The interpretation of these data depends on how they are calculated and on what assumptions are made about the actual risks of predation at both locations. If it is assumed that the actual risks of predation in south-eastern Australia and in the UK were equal and if predation events are expressed as a proportion of total releases, then the calculated predation rates indicate that there was a bias towards reporting predation events in the UK survey, where 1% of bats seen were preyed upon, compared with 0.4% in south-eastern Australia. On the other hand the rate per unit time for which bats were in view was about 10 times greater in Australia than in the UK. The most likely scenario explaining these data is that reporters in the UK were biased in their reporting of predation events but that they also over-estimated the times for which the bats were in sight. The two biases in the British survey probably do not cancel exactly, and this may mean that the estimated risks of diurnal predation, at 100 to 1000 times the risks of nocturnal predation, were underestimated. This assessment of risks, however, takes no account of the potentially different risks of predation while roosting at night and in the day, which are thus assumed to be trivial relative to the predation risks whilst flying.

For non-insectivorous bats there are also numerous data suggesting that predation by diurnal predatory birds is potentially significant. For example, records of kills brought to nests suggest that peregrine falcons (*Falco peregrinus*) in Fiji prey regularly on pteropodid fruit bats (Clunie 1972). Again, however, such records alone are only circumstantial evidence that predation might be important, and a full assessment of the balance of diurnal and nocturnal predation risks is required for these bats. Constructing such a risk assessment for pteropodids would be more complex than for most insectivorous bats because there is not only the risk of predation whilst flying, but also significant risks of predation in trees whilst roosting and feeding, mostly from arboreal snakes. These snakes can have significant population impacts on bats (e.g. in Guam : Wiles 1987), and different species of these predators are potentially active both diurnally and nocturnally. Activity patterns in fruit bats might then reflect a complex trade-off in the predation risks associated with roosting, feeding and flying in daylight and at night. Such a risk assessment is not currently available.

A potential test of the predation hypothesis would be to examine the activity patterns of bats inhabiting remote islands where there are no diurnal birds likely to prey upon them. Several island populations of bats potentially meet these requirements. However, in the majority of these islands there have been no detailed studies of the activity patterns of the resident bats, although for many islands there are anecdotal records. The exceptions are detailed studies of the Samoan flying fox (*Pteropus samoensis*) on Western Samoa (Cox 1983; Wilson & Engbring 1992), *Pteropus melanotus* on Christmas Island in the Indian Ocean (Tidemann 1987) and *Pteropus livingstonii* and *P. seychellensis* in the Comores (Trewhella & Reason 1992). In all of these situations where fruit bats are released from predation, except *P. seychellensis* in the Comores, the bats are mostly diurnally active, supporting the predation hypothesis. In 1988 Peter Webb and I visited the Azores archipelago to assess the extent of diurnal and nocturnal flight in the indigenous insectivorous Azorean bat (*Nyctalus azoreum*; Speakman & Webb 1993). Previous records of this bat, which date back to the turn of the century, indicated that it was diurnal (Ulfstrand 1961; Bannerman & Bannerman 1966; Moore 1975). We found that this bat flies frequently in daylight (Speakman & Webb 1993). However, this diurnal activity, at least on the main island of São Miguel, was restricted to the area around the upland caldera lakes (> 500 m elevation), and at lowland sites, despite extensive searches and continuous tape-recording linked to a bat detector, we found no diurnally active bats (Speakman & Webb 1993). In contrast, bats were more active in these lowland sites, and to a much lesser extent at the upland sites, during the night. Nocturnal flying in this bat was found over an area 1.5 times greater than diurnal flying and the maximum number of bats seen was almost 10 times greater at night than during the day. Overall, therefore, although this bat is more diurnally active than many species, the diurnal activity was spatially restricted, and the routine activity of this bat was clearly nocturnal. Detailed studies of the activity patterns of insectivorous bats inhabiting other islands where there are anecdotal records of diurnality (for example São Tome: P. Jones pers. comm.) are clearly required. For the only detailed study of activity cycles of insectivorous bats inhabiting islands where the predation hypothesis would predict they should fly in daylight (Speakman & Webb 1993), the overwhelming evidence points to nocturnality being retained as the dominant activity pattern. There are also several anecdotal records of insectivorous bats on predator-free islands that are nocturnal despite the release from predation : notably the hoary bat (*Lasiurus cinereus*) on Hawaii (R.M. Barclay, J. Fullard, D. Jacobs pers. comm. and pers. obs.) and the hoary and red bats (*Lasiurus brachyotis*) on the Galapagos (G. MacCraken pers. comm.).

Pteropodid fruit bats seem to fit the expectations of the predation hypothesis better than do insectivorous bats in that three of the four well-studied species living on predator-free islands *do* fly in daylight in accord with the expectation. The only notable discrepancy is *P. seychellensis* living on the Comores. However, there is considerable anecdotal evidence that many islands harbour pteropodids which are nocturnal despite the absence of avian predators (e.g. on Guam : G. Wiles pers. comm.), so perhaps here also the predation hypothesis does not stand up to close scrutiny.

The hyperthermia hypothesis

Daylight-flying bats may experience hyperthermia because of the difficulty in dissipating the high endogenous heat production which occurs during bat flight (S. P. Thomas 1975; Carpenter 1986; Speakman & Racey 1991) when it is combined with a high external short-wave radiative heat load from sunlight. Short-wave radiation is more readily absorbed across the naked wing membranes of bats because they have a low albedo (Speakman & Hays 1992). This contrasts with the highly insulated wings of birds (S. P. Thomas & Suthers 1972; S.P. Thomas *et al.* 1991) that play a minor role in heat balance of avian flight (Martineau & Larochelle 1988).

Several behavioural observations are consistent with the hyperthermia hypothesis. First, the incidence of daylight flying activity relative to the local populations of bats within the UK increases with increases in latitude and is greatest in areas where it is cooler (Speakman 1990). Second, the winter activity of bats in daylight, when ambient temperatures are lower than in summer, is almost as frequent as that observed in midsummer, even though the numbers of bats active at night in winter (Avery 1985) are very much reduced. Third, in the Azores, where diurnality is more frequent than

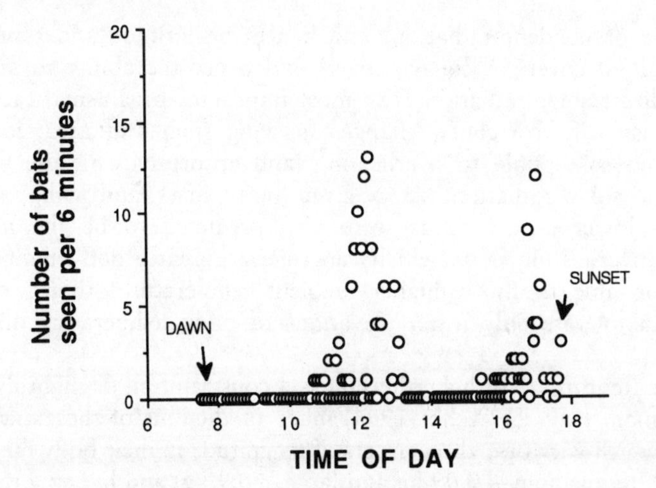

Fig. 2. The diurnal activity of the Azorean bat (*Nyctalus azoreum*) at high altitude (500 m) on the island of São Miguel, Azores. Observations were made continuously and the maximum number of bats that had been simultaneously in the air were recorded at the end of each minute. These data were then summed over 6-min periods (J. R. Speakman and P. I. Webb unpubl. obs.). There was a lull in activity in mid-afternoon when ambient temperatures were at their greatest.

in other areas (Speakman & Webb 1993), we recorded a lull in activity of diurnally flying bats in the mid-afternoon, when temperatures were at their highest (J.R. Speakman & P.I. Webb pers. obs. ; Fig. 2). Fourth, bats flown in wind tunnels, where they do not even experience solar radiation, appear unable to regulate their body temperatures when ambient temperatures exceed 28–35 °C (Carpenter 1985; S. P. Thomas *et al.* 1991).

Speakman, Hays & Webb (1994) constructed a bio-physical model which evaluated the potential of a bat to dissipate heat convectively to the environment under different conditions of heat load from solar radiation. The model generated values for a critical ambient air temperature (T_{acrit}) above which bats would be unable to dissipate the combined net radiative heat load and endogenous heat production of flight, and would consequently fatally overheat. This model suggests that the exogenous heat load from radiation may amount to up to 10 times the endogenous heat load from flight metabolism. This balance is surprising because we generally consider flight to be a very costly activity generating a great deal of endogenous heat. However, it is consistent with the evaluated heat balance of flying butterflies (*Colias* sp. ; Tsuji, Kingsolver & Watt 1986). The important question is, at what air temperature would dissipating this heat load by convection become untenable?

Speakman, Hays & Webb (1994) identified the important environmental and organismal variables which largely determined what ambient temperature would restrict daytime bat flight. The most important environmental factors were the angle of incident radiation, and hence the latitude and time of day, and the cloud cover (or foliage cover) and hence the ability to avoid the input of direct solar radiation. The most important organismal factors were body mass with correlated changes in wing span and area, larger bats being more susceptible to overheating, and absorptivity of the wing membranes for solar radiation. At a given body mass and wing area, the bats with lower aspect ratios were also predicted to be the more susceptible to overheating. Bats which can tolerate greater body temperatures would be able to fly at higher ambient temperatures than would bats which can tolerate only lower elevations of body temperature above resting levels.

To assess the importance of hyperthermia as a constraint on daylight flying activity Speakman, Hays & Webb (1994) made predictions of the expected T_{acrit} for bats which vary over three orders of magnitude in their body masses (small—0.009 kg, medium—0.09 kg and large—0.9 kg) and live in a range of different latitudes (0 to 60° N or S). Because of the interaction of the variables, the estimated T_{acrit} was remarkably constant over a wide range of latitudes; hence between the equator and 50° N or S the T_{acrit} for small bats (9 g) varied between 307.0 and 307.8 K, for medium sized bats (90 g) between 304.7 and 305.5 K and for the largest bats (900 g) between 300.4 and 301.8 K. At latitudes above 50° N or S the critical temperatures for all size classes increased by between 2 K (9 g bat) and 3.5 K (900 g bat).

We collated data on the mean maximum daily shade air temperatures in summer (July in the northern and January in the southern hemisphere) at 554 sites between the equator and 80° latitude. Sites around the equator were hot (*c.* 306 K) but very consistent in their temperatures (standard deviation across sites approx. 1 K). Temperatures increased to a maximum around 20 to 30° N or S, corresponding to the hot arid desert regions. In this zone the temperatures were also more variable, with the standard deviation across sites at 40 to 50° N averaging 7 K. We combined the spatial variations in daylight temperature with the predicted T_{acrit} values for bats of 9, 90 and 900 g to predict at each latitude the proportion of sites at which the ambient temperature would exceed T_{acrit} and hence bats would be constrained from flying in daylight by hyperthermia (Fig. 3).

These estimates suggest that hyperthermia is likely to be a constraint on the daylight flying behaviour of large bats at about 85% of sites between the equator and approximately 40° N or S. For intermediate-sized bats, hyperthermia is also likely to be a significant constraint at about 60% of sites between 20 and 30° N, but at only 10% of sites at the equator. For the smallest bats, hyperthermia is likely only to constrain daylight flying at, on average, about 40% of sites between 20 and 30° N or S. At the equator, hyperthermia would almost never be a constraint. For small and medium-sized bats (9 and 90 g), hyperthermia would be a constraint on diurnal activity at less than 0.5% of sites greater than 50° N or S and for large bats at less than 0.5% of

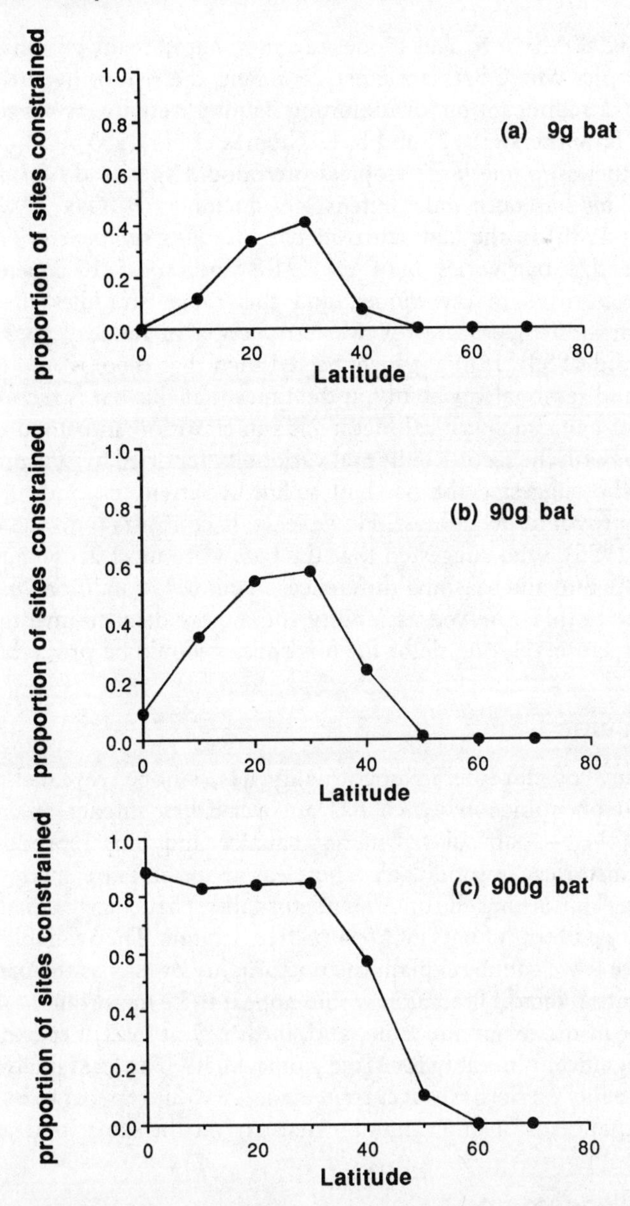

Fig. 3. The probability that flying in daylight will be constrained at a given site as a function of latitude for bats of three body masses representative of typical bats (9 g, 90 g and 900 g). Calculations were made from the probability density function of a normal distribution using a *z* score (calculated as the predicted critical ambient air temperature calculated for that bat at that latitude minus the mean ambient air temperature across a range of sites at the appropriate latitude, and this difference divided by the standard deviation of air temperature estimates across the sites). (After calculations in Speakman, Hays & Webb 1994.)

sites above 60° N or S. This model suggests that in many regions, particularly in the tropics where bats are most abundant, the risk of hyperthermia might represent a significant factor inhibiting daytime activity, as suggested by S. P. Thomas & Suthers (1972) and S. P. Thomas *et al.* (1991).

Nevertheless, some large tropical pteropodid species do fly diurnally (see above). This has been most extensively documented (Cox 1983; Wilson & Engbring 1992) in the Samoan fruit bat *Pteropus samoensis*. The calculated T_{acrit} for this bat varies between 301.8 and 304.5 K depending on the assumptions used in the model, and this range straddles the mean daily maximum air temperature for Western Samoa in January (303.8 K: Oliver & Fairchild 1984). It might be expected then that there would be significant diurnal and seasonal variability in the times that this bat is recorded in flight, dependent upon such critical factors as sun elevation and cloud cover. This is consistent with the recorded diurnal variability reported by Wilson & Engbring (1992) who suggested the bat had a 'lull in activity around mid-day' when solar input would be greatest. However, this contrasts with the observations of Cox (1983), who suggested that the bats were most active between 10:00 and 14:00. Site and seasonal differences in climatic conditions may, however, contribute to this observed variability and a more detailed investigation of the thermal relations during flight for this species would be profitable.

Conclusions

This review of chiropteran nocturnality has, I hope, revealed that it is an important phenomenon which has an over-riding impact on many aspects of bat biology—particularly energy balance and thus reproductive output and life histories. It should also be clear that there are many gaps in our knowledge concerning chiropteran nocturnality, particularly detailed studies of the activity patterns of bats in predator-free habitats. The available information suggests that we cannot explain the nocturnality of bats as the consequence of a single causal factor. Predation would appear to be important at some sites, in particular in the temperate zone, and, in theory at least, hyperthermia might also be significant, notably for large tropical bats. The least likely explanation of nocturnality on the basis of current evidence would appear to be competition with diurnal birds or other animals that exploit the same food sources.

Acknowledgements

I am grateful to Paul Racey for inviting me to speak at this symposium. My thoughts on chiropteran nocturnality have benefited greatly from stimulating discussions with many colleagues, in particular Peter Webb, Graeme Hays, Jack Hayes, Jens Rydell, Gareth Jones, Bill Rainey, Dixie Pierson, Robert Barclay, James Fullard, Susan Thomson, Richard Stone and Regina McDevitt. I thank Jens Rydell for allowing me to use his data on 24h patterns of aerial insect availability in Aberdeen (Fig. 1). Peter Webb and I visited the Azores in 1988 with financial assistance from the BES, Carnegie Trust, TAP (Air Portugal), Waterstones and Bissets bookshops and Zonal tapes Ltd.

References

Audet, D. & Fenton, M. B. (1988). Heterothermy and the use of torpor by the bat *Eptesicus fuscus* (Chiroptera: Vespertilionidae): a field study. *Physiol. Zool.* **61**: 197–204.

Avery, M. I. (1985). Winter activity of pipistrelle bats. *J. Anim. Ecol.* **54**: 721–738.

Bannerman, D. A. & Bannerman, W. M. (1966). *Birds of the Atlantic islands* 3. A *history of the birds of the Azores*. Oliver and Boyd, Edinburgh and London.

Bauer, K. (1956). Scleiereule (*Tyto alba* Scop.) als Fledermausjäger. *J. Orn.* **97**: 335–340.

Bruun, B. & Singer, A. (1970). *The Hamlyn guide to the birds of Britain and Europe*. Hamlyn, London.

Bryant, D. M. & Tatner, P. (1991). Interspecies variation in avian energy expenditure: correlates and constraints. *Ibis* **133**: 236–245.

Caraco, T., Martindale, S. & Whitman, T. S. (1980). An empirical demonstration of risk sensitive foraging preferences. *Anim. Behav.* **28**: 820–830.

Carpenter, R. E. (1985). Flight physiology of flying foxes, *Pteropus poliocephalus*. *J. exp. Biol.* **114**: 619–647.

Carpenter, R. E. (1986). Flight physiology of intermediate-sized fruit bats (Pteropodidae). *J. exp. Biol.* **120**: 79–103.

Clunie, F. (1972). Fijian birds of prey. *Fiji Mus. educ. Ser.* 3.

Corbet, G. B. (1978). The mammals of the Palaearctic region : a taxonomic review. *Publs Br. Mus. nat. Hist.* No. 788 : 1–314.

Cox, P. A. (1983). Observations on the natural history of Samoan bats. *Mammalia* **47**: 519–523.

Gillette, D. D. & Kimbourgh, J. D. (1970). Chiropteran mortality. In *About bats*: 262–281. (Eds Slaughter, B. H. & Walton, D. W.). Dallas Southern Methodist University Press, Dallas.

Hays, G. C., Speakman, J. R. & Webb, P. I. (1992). Why do brown long-eared bats (*Plecotus auritus*) fly in winter? *Physiol. Zool.* **65**: 554–567.

Hill, J. E. & Smith, J. D. (1984). Bats: a natural history. *Publs Br. Mus. nat. Hist.* No. 877: 1–243.

Kunz, T. H. (1980). Daily energy budgets of free-living bats. In *Proceedings fifth international bat research conference* : 369–392. (Eds Wilson, D. E. & Gardner, A. L.). Texas Tech Press, Lubbock.

Kurta, A., Bell, G. P., Nagy, K. A. & Kunz, T. H. (1989). Energetics of pregnancy and lactation in free-ranging little brown bats (*Myotis lucifugus*). *Physiol. Zool.* **62**: 804–818.

Kurta, A., Johnson, K. A. & Kunz, T. H. (1987). Oxygen consumption and body temperature of female little brown bats (*Myotis lucifugus*) under simulated roost conditions. *Physiol. Zool.* **60**: 386–397.

Kurta, A. & Kunz, T. H. (1987). Size of bats at birth and maternal investment during pregnancy. *Symp. zool. Soc. Lond.* No. 57: 79–106.

Macy, R. N. & Macy, R. W. (1969). Hawks as enemies of bats. *J. Mammal.* **20**: 252.

Martineau, L. & Larochelle, J. (1988). The cooling power of pigeon legs. *J. exp. Biol.* **136**: 193–208.

Milinski, M. & Heller, R. (1978). Influence of a predator on the optimal foraging behaviour of sticklebacks (*Gasterosteus aculeatus* L.). *Nature, Lond.* **275**: 642–644.

Moore, N. W. (1975). The diurnal flight of the Azorean bat (*Nyctalus azoreum*) and the avifauna of the Azores. *J. Zool., Lond* **177**: 483–506.

Oliver, J. E. & Fairchild, R. W. (1984). *The encyclopedia of climatology.* Van Nostrand Reinhold, New York.

Ransome, R. (1990). *The natural history of hibernating bats.* Croom Helm, London.

Ruprecht, A. L. (1979). Bats (Chiroptera) as constituents of the food of barn owls *Tyto alba* in Poland. *Ibis* **121**: 489–494.

Rydell, J. (1992). Occurrence of bats in northernmost Sweden (65°N) and their feeding ecology in summer. *J. Zool., Lond.* **227**: 517–529.

Speakman, J. R. (1990). The function of daylight flying in British bats. *J. Zool., Lond.* **220**: 101–113.

Speakman, J. R. (1991a). Why do insectivorous bats in Britain not fly in daylight more frequently? *Funct. Ecol.* **5**: 518–524.

Speakman, J. R. (1991b). The impact of predation by birds on bat populations in the British Isles. *Mammal Rev.* **21**: 123–142.

Speakman, J. R. & Hays, G. C. (1992). Albedo and transmittance of short wave radiation for bat wings. *J. thermal Biol.* **17**: 317–321.

Speakman, J. R., Hays, G. C. & Webb, P. I. (1994) Hyperthermia, a constraint on the diurnal activity of bats: a bio-physical model. *J. theor. Biol.* **171**: 325–341.

Speakman, J. R., Lumsden, L. F. & Hays, G. C. (1994). Predation rates on bats released to fly during daylight in south-eastern Australia. *J. Zool., Lond.* **233**: 318–321.

Speakman, J. R. & Racey, P. A. (1987). The energetics of pregnancy and lactation in the brown long-eared bat, *Plecotus auritus.* In *Recent advances in the study of bats*: 367–393. (Eds Fenton, M. B., Racey, P. A. & Rayner, J. M. V.). Cambridge University Press, Cambridge.

Speakman, J. R. & Racey, P. A. (1989). Hibernal ecology of the pipistrelle bat: energy expenditure, water requirements and mass loss, implications for survival and the function of winter emergence flights. *J. Anim. Ecol.* **58**: 797–813.

Speakman, J. R. & Racey, P. A. (1990). Why do bats emerge from hibernation? In *European bat research 1987*: 625–626. (Eds Hanák, V., Horáćek, I. & Gaisler, J.). Charles University Press, Prague.

Speakman, J. R. & Racey, P. A. (1991). No cost of echolocation for bats in flight. *Nature, Lond.* **350**: 421–423.

Speakman, J. R. & Webb, P. I. (1993). Taxonomy, status and distribution of the Azorean bat (*Nyctalus azoreum*). *J. Zool., Lond.* **231**: 27–38.

Swift, S. M. (1980). Activity patterns of pipistrelle bats (*Pipistrellus pipistrellus*) in north-east Scotland. *J. Zool., Lond.* **190**: 285–295.

Thomas, D. W. (1993). Lack of evidence for a biological alarm clock in bats hibernating under natural conditions. *Can. J. Zool.* **71**: 1–3.

Thomas, S. P. (1975). Metabolism during flight in two species of bats, *Phyllostomus hastatus* and *Pteropus gouldii. J. exp. Biol.* **63**: 273–293.

Thomas, S. P., Follette, D. B. & Farabaugh, A. T. (1991). Influence of air temperature on ventilation rates and thermoregulation of a flying bat. *Am. J. Physiol.* **260**: R960–R968.

Thomas, S. P., & Suthers, R. A. (1972). The physiology and energetics of bat flight. *J. exp. Biol.* **57**: 317–335.

Thompson, M. J. A. (1987). Longevity and survival of female pipistrelle bats (*Pipistrellus pipistrellus*) on the Vale of York, England. *J. Zool, Lond.* **211**: 209–214.

Tidemann, C. R. (1987). Notes on the flying-fox, *Pteropus melanotus* (Chiroptera : Pteropodidae) on Christmas Island, Indian Ocean. *Aust. Mammal.* **10**: 89–91.

Trewhella, W. J. & Reason, P. F. (Eds). (1992). *The final report of the University of Bristol Comoros '92 Expedition*. Unpublished report. University of Bristol.

Tsuji, J. S., Kingsolver, J. G. & Watt, W. B. (1986). Thermal physiological ecology of *Colias* butterflies in flight. *Oecologia* **69**: 161–170.

Tugendhat, M. (1966). Swallows mobbing pipistrelle bat. *Br. Birds* **59**: 435.

Twente, J. W. & Twente, J. (1987). Biological alarm clock arouses hibernating big brown bats, *Eptesicus fuscus*. *Can. J. Zool.* **65**: 1668–1674.

Ulfstrand, S. (1961). On the vertebrate fauna of the Azores. *Bolm Mus. munic. Funchal* **14**: 75–86.

Vernier, M. (1990). Ecological observations on the evening flight of *Pipistrellus kuhlii* in the town of Padova, Italy. In *European bat research 1987*: 537–541. (Eds Hanák, V., Horáček, I. & Gaisler, J.). Charles University Press, Prague.

Wiles, G. J. (1987). Current research and future management of Marianas fruit bats (Chiroptera: Pteropodidae) on Guam. *Aust. Mammal.* **10**: 93–95.

Williams, C. B. (1961). Studies on the effect of weather conditions on the activity and abundance of insect populations. *Phil. Trans. R. Soc. (B)* **244**: 331–378.

Wilson, D. E. & Engbring, J. (1992). History of fruit bat use, research and protection in the northern Mariana islands. *U. S. Fish Wildl Serv. biol. Rep.* **90**: 74–101.

Symp. zool. Soc. Lond. (1995) No. 67: 203–218

The comparative ecophysiology of water balance in microchiropteran bats

Peter I. WEBB[1]
Department of Zoology
University of Aberdeen
Aberdeen AB9 2TN, UK

Synopsis

Aspects of water balance were studied in the three species of bat commonly found in north-east Scotland: *Plecotus auritus*, *Myotis daubentonii* and *Pipistrellus pipistrellus*. One of these species (*M. daubentonii*) has a close association with open water in the wild, while one (*P. pipistrellus*) has a much lower mean body mass (*c.* 6 g) than the other two (*c.* 10 g). Faecal water content and maximum urine-concentrating ability were not significantly different between *P. auritus* and *M. daubentonii*. However, urine loss was higher (by *c.* 140 μl/day) and resting evaporative water loss lower (also by *c.* 140 μl/day) in *M. daubentonii* than in *P. auritus*. In free-flying captive colonies of bats, total water intake was highest in *M. daubentonii*, intermediate in *P. auritus* and lowest in *P. pipistrellus*. It seems possible that *M. daubentonii* demonstrate high water flux outside the day roost and compensate through a low water flux within the day roost. A low rate of water intake in captive *P. pipistrellus* may simply be a function of small body size.

A brief review of ecophysiological aspects of water balance in bats in general is presented. Faecal water content of insectivorous bats is approximately 70% of wet weight and falls at the upper end of the range found in vegetarian rodents. Maximum urine-concentrating ability in bats is dependent on both diet and habitat aridity. How maximum urine-concentrating ability in bats compares with that in non-volant mammals is still unclear. Resting evaporative water loss in bats appears higher than that in similar-sized non-volant mammals. Evaporative water loss from flying bats varies between about 1.1% and 5.7% of body mass per hour, depending on species. Water flux in captive bats is not significantly different to that in captive rodents of equivalent body mass.

Introduction

There are several reasons why differences in water balance between bats and non-volant mammals might be predicted. Most prominent of these is that bats not only can fly, but can and do use sustained flight. Flight in mammals requires the possession of large areas of naked or sparsely

[1]Current address: Mammal Research Institute, University of Pretoria, Pretoria, South Africa

ZOOLOGICAL SYMPOSIUM No. 67
ISBN 0 19 854945 8

furred skin in the form of the flight membranes; we might therefore expect cutaneous evaporation to be comparatively high in bats. Flight is also energetically expensive (e.g. Speakman & Racey 1991) and we might therefore expect pulmonary evaporation to be comparatively high in bats. Many bats are also insectivorous, with a high dietary protein content and a concomitant requirement for the excretion of large quantities of nitrogenous waste in the urine. All these factors may lead to a suspicion that bats will have comparatively high rates of water flux when compared to non-volant mammals. However, there are compensatory factors which may help to reduce water flux in bats. Examples are the presence of capillary sphincters (Nicoll & Webb 1946) that can reduce blood flow to the flight membranes, a comparatively high efficiency for the extraction of oxygen from the lungs (Chappell & Roverud 1991) and, in many species, the use of short or prolonged periods of torpor in which evaporation is considerably reduced (e.g. Herreid & Schmidt-Nielsen 1966; Webb, Speakman & Racey 1995).

The three species of bat common in north-east Scotland are the brown long-eared bat, *Plecotus auritus*, Daubenton's bat, *Myotis daubentonii*, and the pipistrelle bat, *Pipistrellus pipistrellus* (Speakman, Racey, Catto, Webb, Swift & Burnett 1991). All three species are insectivorous vespertilionids. *P. auritus* is a species generally associated with woodland habitats, roosting in tree holes, in nest boxes on trees and in buildings (Swift & Racey 1983; Boyd & Stebbings 1989; Fuhrmann & Seitz 1992), foraging with a slow fluttering flight in and around woodland edges and gleaning much of its prey from leaf and tree surfaces (Anderson & Racey 1991). *M. daubentonii* is a species closely associated with open water, roosting in cracks and crevices in walls, under bridges, in water-filled tunnels (Park 1988; Richardson 1989; Speakman *et al.* 1991) and also in buildings, sometimes sharing roost sites with *P. auritus* (Swift & Racey 1983). *M. daubentonii* forage low over open water (Miller & Degn 1981; Swift & Racey 1983) and gaff much of their prey from the water surface (Jones & Rayner 1988). The roosts of *M. daubentonii* are often characterized by heavy urine staining, implying that bats of this species produce large quantities of urine within the roost site. With a body mass of 4 to 7 g in the wild, *P. pipistrellus*, the smallest, commonest and most ubiquitous British bat (Racey 1991), is much smaller than the other two species, which generally have a body mass of between 7 and 13 g.

The intention of the present study was at least partly to answer two questions. First, are there interspecific differences in water flux between captive individuals of the three species of bat commonly found in north-east Scotland and, if so, can these differences be related to differences in their ecology? In answering this question I shall concentrate primarily on a comparison of *P. auritus* with *M. daubentonii*. Second, does water balance in bats differ from that in other mammal groups?

Water loss in non-reproductive bats

As in all non-reproductive mammals, water loss in non-reproductive bats occurs along three major routes: free water loss in the faeces, water loss in the urine, and water loss by evaporation.

Water loss in the faeces

The water content of faeces from *P. auritus* and *M. daubentonii* fed on mealworm larvae (*Tenebrio* spp.) is 73.3% and 72.3% of wet weight respectively and is not significantly different between the two species (Webb, Speakman & Racey 1993). This compares with a faecal water content of approximately 70% in some other insectivorous bats (Vogel & Vogel 1972), 73% in the mainly marine-invertebrate-eating bat *Pizonyx vivesi* (Carpenter 1968), 64% in the sanguinivorous bat *Desmodus rotundus* (McFarland & Wimsatt 1969), and 44 to 74% in 12 species of rodent fed on a dry vegetarian diet (see references in Fyhn 1979).

As mealworm larvae are, on average, 61.1% wet weight water and the apparent dry mass assimilation efficiency of mealworms by both *P. auritus* and *M. daubentonii* is 85.3% (Webb *et al.* 1993), this implies that approximately 25% of free water in the food of these bats is lost as free water in the faeces. Also, when food consumption is the same in both species, total loss of water in the faeces will not be significantly different between them.

Water loss in the urine

Urine flow rate

In bats fed similar quantities of mealworms and denied access to drinking water, urine loss during the first 12 h after feeding was found by Webb, Speakman & Racey (1994) to be significantly greater in *M. daubentonii* than in *P. auritus* (Fig. 1). The difference between the two species in the volume of urine produced was apparently independent of food consumption (Fig. 2) and was due to an exceptionally high rate of urine loss by *M. daubentonii* during the first 3–4 h after the meal (Webb *et al.* 1994; Fig. 1). Urine flow rate during these first few hours was so high that urine flow in *M. daubentonii* had virtually ceased 7 h after the meal (Fig. 1) and led to severe dehydration unless individuals were provided with drinking water (Webb *et al.* 1994).

Maximum urine flow rate after feeding in the temperate-zone vespertilionid *Myotis lucifugus* (approx. body mass 7 g), when euthermic, is approximately 0.03 to 0.05 μl/g.min, while minimum or 'basal' urine flow rate in the same species when euthermic and (apparently) not dehydrated is 0.008 to 0.010 μl/g.min (Kallen & Kanthor 1967; Bassett & Wiebers 1980). These values are similar to those noted in *P. auritus* by Webb *et al.* (1994). Maximum urine flow rate after feeding in the vampire bat *Desmodus rotundus* is 4 μl/g.min (approx. 110 μl/min; McFarland & Wimsatt 1969), the highest mass-specific

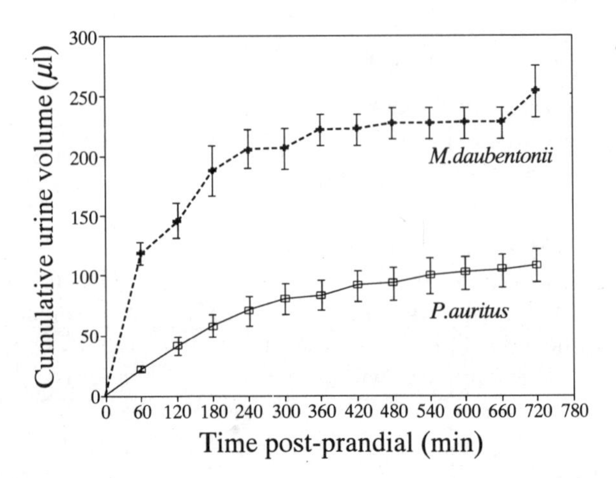

Fig. 1. Cumulative urine loss after feeding in *Plecotus auritus* (□) (n=12) and *Myotis daubentonii* (+) (n = 4). Bats were deprived of food for 12 h prior to the experiment and then fed between time −15 min and time zero, after which they were denied access to both food and water. Food consumption was not significantly different between the two species. Taken with kind permission from Webb *et al.* (1994).

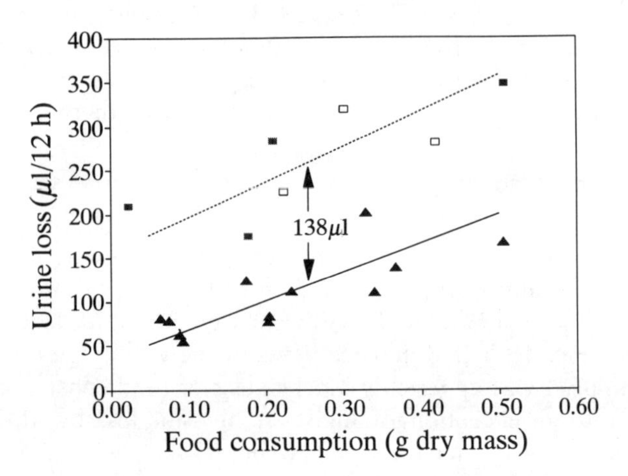

Fig. 2. Total urine loss during the first 12 h after feeding in *Plecotus auritus* denied water (▲) and *Myotis daubentonii* denied water (■) or provided with water (□), shown as a function of food consumption. There was no significant difference in urine loss between the two groups of *M. daubentonii* nor in the slope of the relationship between the two species; however, urine loss in *M. daubentonii* was on average 138 μl greater than it was in *P. auritus*. The lines represent the reduced major axis regression equations for *M. daubentonii* (upper, dotted line: $y = 156+402x$, $r^2 = 0.61$, $F_{1,5} = 7.7$, $P = 0.039$) and *P. auritus* (lower, solid line: $y = 35+330x$, $r^2 = 0.63$, $F_{1,10} = 17.1$, $P = 0.002$). Taken with kind permission from Webb *et al.* (1994).

urine flow rate recorded in any mammal and approximately 18 times that in *M. daubentonii*.

Urine-concentrating ability

Maximum urine-concentrating ability has been empirically determined in 21 species of microchiropteran bat (Beuchat 1990; Webb 1992), of which 16 are insectivorous. It appears that, in general, bats are not able to produce urine any more concentrated than that of non-volant mammals of similar body mass (Fig. 3). However, insectivorous species of bat are able to produce more concentrated urine than are plant-eating species (Studier, Wisniewski, Feldman, Dapson, Boyd & Wilson 1983; Fig. 3), and maximum urine-concentrating ability is greater in species or populations of bats inhabiting arid environments than in species or populations of bats inhabiting more mesic environments (Geluso 1978; Bassett 1982; Studier *et al.* 1983). When the kidney morphology of insectivorous bats is compared with that of non-volant insectivorous mammals from habitats of similar aridity, the differences between the two groups of mammals suggest that the bats will have maximum urine-concentrating abilities higher than those of the non-volant insectivores (Geluso 1980). Further research clearly needs to be undertaken before a valid comparison can be made between the abilities of bats and non-volant mammals to produce concentrated urine.

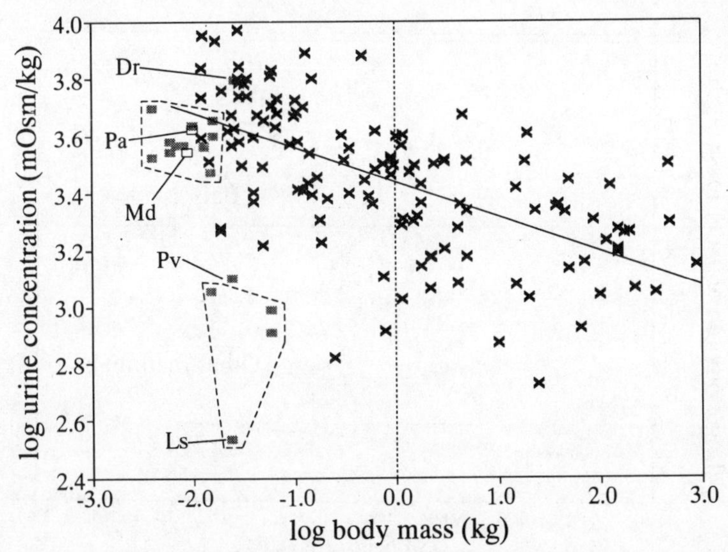

Fig. 3. A log–log plot of maximum urine-concentrating ability against body mass in bats (■) and non-volant mammals (×). Data taken from Beuchat (1990) and the present study. The upper boxed group of bats are insectivorous species, the lower group are plant-eating species. Ls–*Leptonycteris sanborni*, Pv – *Pizonyx vivesi*, Md – *Myotis daubentonii*, Pa – *Plecotus auritus*, Dr – *Desmodus rotundus*.

One possible reason for a higher rate of urine loss after feeding in *M. daubentonii* than in *P. auritus* could be a difference in urine-concentrating ability between the two species. However, at 3500 and 4200 mOsm/kg respectively, maximum urine-concentrating ability was not significantly different between them (Webb *et al.* 1994). Alternatively, the difference between the two species in the rate of urine loss after feeding may represent a mechanism for rapid dumping of excess water by *M. daubentonii* during and after foraging on the assumption that drinking water will always be readily available outside the roost, a situation that cannot be guaranteed for *P. auritus*.

Water loss by evaporation

Hattingh (1972) found that area-specific transepidermal water loss was higher in the bat *Miniopterus schreibersii* than in any of 10 other conscious mammalian species. A log–log plot of resting evaporative water loss in bats (Studier 1970; present study) and in non-volant mammals (Crawford & Lasiewski 1968) clearly demonstrates an elevated rate of evaporation in the bats (Fig. 4). Even though the conditions under which the measurements in Fig. 4 were made varied, the difference in resting evaporative water loss between the two groups would appear large enough to be a real one.

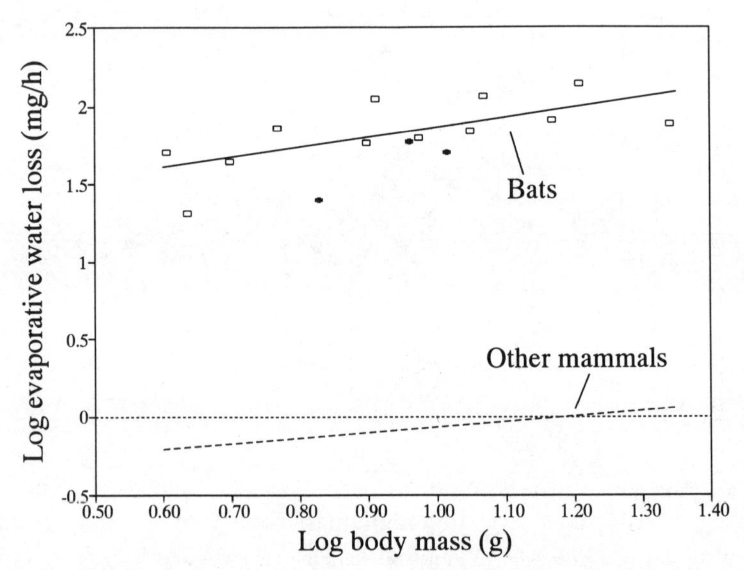

Fig. 4. A comparison of resting evaporative water loss in single bats (open rectangles and solid line) (data from Studier 1970 and present study) and other mammals (dotted line) (data from Crawford & Lasiewski 1968). The crosses represent data for the three species of bat in the present study.

Resting evaporative water loss in non-reproductive adult *P. auritus* and *M. daubentonii* increases linearly with local vapour pressure deficit and oxygen consumption and decreases linearly with the product of ambient temperature and local vapour pressure deficit (Webb *et al.* 1995; Table 1). Despite there being no significant differences in the slopes of these three relationships between the two species, resting evaporative water loss was significantly higher in *P. auritus* than it was in *M. daubentonii* under equivalent environmental conditions (Webb *et al.* 1995). The mean difference between the two species was 0.1 μl/min or 144 μl/day regardless of the environmental conditions. Thus, although urinary water loss in *M. daubentonii* is approximately 140 μl/day greater than it is in *P. auritus* when food consumption is the same in both species, the effect of this difference on water balance appears to be counteracted by a depression in evaporative water loss in *M. daubentonii* of approximately 140 μl/day when compared to *P. auritus* under equivalent environmental conditions. As total faecal water loss under conditions of equivalent food intake is also not significantly different between the two species, this implies that minimum total water loss is not significantly different between the two species when behavioural and environmental conditions are the same.

Resting evaporative water loss from *P. pipistrellus* increases linearly with oxygen consumption and air flow rate, and decreases linearly with the product of ambient (i.e. not local) relative humidity and air flow rate (P. I. Webb unpubl. data; Table 1). Under equivalent conditions of environment and oxygen consumption there was no significant difference in resting evaporative water loss between pregnant, lactating and post-lactating females or between adult female and juvenile (1–21 days old) *P. pipistrellus*. Previous studies have suggested in other bat species that resting evaporative water loss in pregnant individuals may be depressed (Procter & Studier 1970; Studier 1970).

Correcting resting evaporative water loss in *P. auritus, M. daubentonii* and *P. pipistrellus* to standardized conditions of oxygen consumption (2.5 times basal metabolic rate as predicted allometrically from body mass: Lechner 1978), ambient temperature (25 °C), air flow rate (14.9 cm/min), ambient (in-flowing) relative humidity (45.0%), and local vapour pressure deficit (2673 N/m^2), using the regression coefficients given in Table 1, shows that under these standardized conditions, resting evaporative water loss was highest in *P. auritus*, intermediate in *P. pipistrellus* and lowest in *M. daubentonii* (Fig. 5). The body mass difference between *P. pipistrellus* (mean = 6.20 g) and *P. auritus* (mean = 9.12 g) suggests that the difference in resting evaporative water loss between these two species may simply be a function of body size. However, it seems likely that the comparatively low rate of evaporation in *M. daubentonii* has some real physiological basis.

Evaporative water loss in flight has been determined from mass loss in several species of bats (Table 2). Since the conditions under which these measurements were made are relatively consistent between species, it can

Table 1. The influence of environmental variables, oxygen consumption and body mass on evaporative water loss (in mg/min) in resting *Plecotus auritus*, *Myotis daubentonii* and *Pipistrellus pipistrellus* as determined by using open-flow respirometry techniques. The table shows the regression coefficients for multiple regression equations generated through full-factorial analysis of covariance. Data for *P. auritus* and *M. daubentonii* were analysed together with species as a factor; hence where there was no significant difference between the two species the regression coefficient quoted represents the mean across both species. Data for *P. pipistrellus* were analysed separately. A zero implies that the variable did not have a significant effect in a particular species, a dash implies that the influence of the variable was not tested in a particular species. VP— Local vapour pressure deficit. T_a—Ambient temperature. Flow—Air flow rate. RH—Ambient relative humidity.

Variable	*P. auritus*	*M. daubentonii*	*P. pipistrellus*
Constant	−0.28	−0.38	−0.04
Body mass (g)	0	0	0
Oxygen consumption (ml/min)	0.67	0.67	1.33
Ambient temperature (°C)	$-0.000018 \times VP$	$-0.000018 \times VP$	—
Local vapour pressure deficit (N/m^2)	$0.0011-0.000018 \times T_a$	$0.0011-0.000018 \times T_a$	—
Ambient relative humidity (%)	—	—	$-0.002 \times Flow$
Air flow rate (cm/s)	—	—	$0.14-0.002 \times RH$

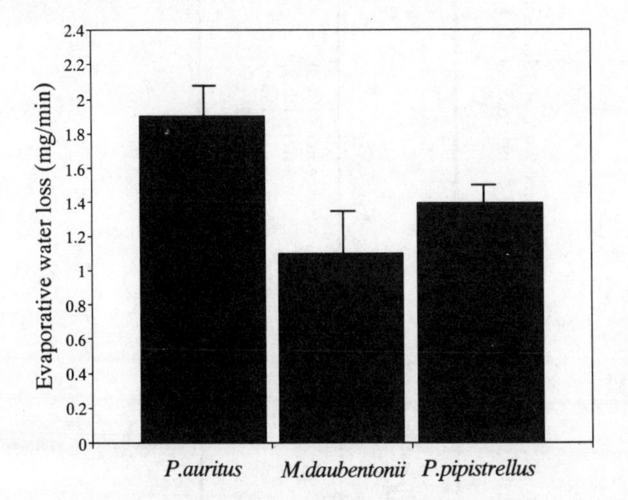

Fig. 5. Resting evaporative water loss under standardized conditions (see text) in *Plecotus auritus, Myotis daubentonii* and *Pipistrellus pipistrellus*. Resting evaporative water loss was significantly higher in *P. auritus* than in *P. pipistrellus* and significantly higher in *P. pipistrellus* than in *M. daubentonii*.

only be assumed that interspecific differences in evaporative water loss in flight are due to differences in physiology or flight mode.

Water intake

Water intake in bats, as in other mammals, takes place in three major ways: by production of metabolic water, by consumption of free water in the food and by drinking water. In the present study a comparison is made of total water influx in captive *P. auritus*, *M. daubentonii* and *P. pipistrellus*. All bats were fed on mealworms that had a mean free water content of 61.1% wet weight and the potential to produce 0.530 ml of metabolic water per gram dry mass (Webb *et al.* 1993). Under standardized conditions of energy intake and environment, differences in total water loss between the three species should therefore be reflected by differences in total water intake expressed as differences in drinking water consumption.

Table 3 shows a summary of water flux parameters in bats of all three study species maintained as small single-species colonies in a room (3m×3m×3m) in which they could fly freely whenever they wanted. The room contained a wooden roost box and was maintained under a 12L:12D photoperiod. Flight activity within the room was monitored by a system of infra-red light beams. Food and water were always available *ad lib.* and food and water consumption were determined on a daily basis (Hays, Speakman & Webb 1992). Determinations of food consumption were converted to an estimate of net daily energy intake assuming a calorific content of the food

Table 2. A summary of evaporative water loss in flying bats and the conditions under which the measurements were made. All measurements were determined as mass loss during flight once mass changes due to the respiratory uptake of O_2 and loss of CO_2 had been accounted for assuming an RQ of 0.78 (as for flying birds: Tucker 1968).

Species	Mean body mass (g)	Ambient temperature (°C)	Ambient relative humidity (%)	Evaporative water loss (mg/min)	Evaporative water loss (% body mass/h)	Reference
Pizonyx vivesi	25	?	?	23.80	5.7	Carpenter (1968)
Eptesicus fuscus	13.3	23	45–55	6.85	3.1	Carpenter (1969)
Antrozous pallidus	19.5	23	45–55	10.41	3.2	Carpenter (1969)
Leptonycteris sanborni	24.4	23	45–55	15.90	3.9	Carpenter (1969)
Plecotus auritus	10.2	21	50	1.83	1.1	Present study

Table 3. A summary of water flux in captive colonies of free-flying *Plecotus auritus* (n=9 bats), *Myotis daubentonii* (n=4 bats) and *Pipistrellus pipistrellus* (n=22 bats) and the conditions under which the measurements were made. Daily energy expenditure was calculated from food consumption and daily body mass change, assuming an energy content of food of 28.6 kJ/g dry mass, an assimilation efficiency for energy of 90% (Webb *et al.* 1993), that 79.3% of variation in body mass was attributable to variation in body water and the remainder to variation in body fat (from data in Studier & Ewing 1971), and an energy content for body fat of 39.4 kJ/g (Ewing, Studier & O'Farrell 1970). Free-water intake with the food was calculated assuming a water content for food of 61.1% of wet weight (Webb *et al.* 1993). Metabolic water production was estimated from food consumption assuming a metabolic water production of 0.530 g/g dry weight (Webb *et al.* 1993). All data are given ± 1 standard deviation.

	P. auritus	*M. daubentonii*	*P. pipistrellus*
Days (n)	28	29	11
Body mass (g)	9.56±0.49	9.20±0.36	6.0±0.5
Time in flight (min/day)	48.4±24.0	9.0±3.0	48.5±15.7
Ambient temperature (°C)	16.3±2.0	14.9±4.1	22.2[a]
Ambient relative humidity (%)	63.9±8.0	33.8±11.2	59.9[a]
Food consumption (g dry/day)	1.50±0.46	0.96±0.20	0.59±0.16
Energy expenditure (kJ/day)	34.8±10.5[b]	23.6±10.4	16.9±6.8
Drinking water consumption (g/day)	0.86±0.27	1.76±0.42	0.77±0.19
Free water intake with the food (g/day)	2.22±0.64	1.43±0.29	0.89±0.43
Metabolic water production (g/day)	0.80± 0.25	0.51±0.11	0.31±0.09
Total water influx (g/day)	3.88±0.95	3.70±0.74	1.97±0.43

[a]Errors not available.[b]n=25.

of 28.6 kJ/g dry mass and a dry mass apparent absorption efficiency of 90% (Webb *et al.* 1993). Daily energy expenditure was estimated from net daily energy intake and daily mean body mass change, assuming 79.3% of variation in body mass to be attributable to variation in body fat levels, the remainder to be attributable to water (from data in Studier & Ewing 1971), and the energy content of body fat to be 39.4 kJ/g (Ewing, Studier & O'Farrell 1970).

The apparent differences in total water influx between the three species (Table 3) are difficult to interpret because of the interspecific differences in the conditions under which the measurements were made (Table 3). A clearer comparison is obtained by correcting the data so that mean time in flight and mean vapour pressure deficit were the same in all three species, by assuming evaporative water loss in flight to be 1.83 μl/min (as for *P. auritus*; P.I. Webb unpubl. data) and using the regression coefficients for variables affecting resting evaporative water loss given in Table 1. On plotting the corrected values of total water flux against daily energy expenditure, it is apparent that, although the slope of the relationship between water influx and energy expenditure was not significantly different between the three species, at any given value of energy expenditure, water influx was highest in *M. daubentonii*, intermediate in *P. auritus* and lowest in *P. pipistrellus* (Fig. 6).

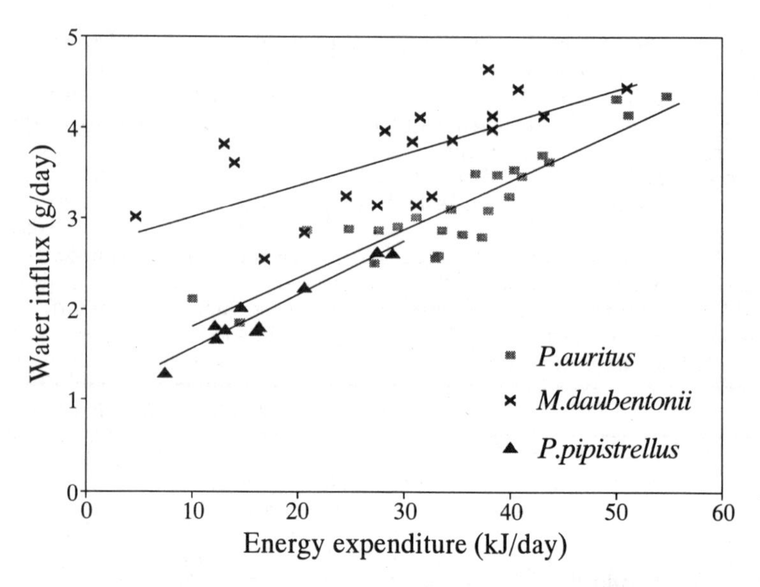

Fig. 6. The influence of daily energy expenditure on total water influx corrected for interspecific variation in behavioural and environmental conditions (see text) in captive colonies of *Plecotus auritus, Myotis daubentonii* and *Pipistrellus pipistrellus*. There was no significant difference in the slope of the relationship between the three species but at any given value of energy expenditure water influx was on average 0.81 g/day greater in *M.daubentonii* than in *P. auritus* and 0.35 g/day greater in *P. auritus* than in *P. pipistrellus*.

As there are no published data available on water flux in free-living *P. auritus* or *M. daubentonii*, it is impossible to confirm whether the interspecific differences in total water influx observed in the laboratory reflect interspecific differences in the wild. Nevertheless such differences are what we might expect from the general ecology and morphology of these three species. *M. daubentonii* has a strong association with open water in terms of both roost sites (Park 1988; Richardson 1989; Speakman *et al.* 1991) and foraging activity (Miller & Degn 1981; Jones & Rayner 1988). The foraging mode of *M. daubentonii* involves either catching insects low above the water surface (Miller & Degn 1981; Jones & Rayner 1988) (and hence possibly of high body water content) or gaffing prey from the water surface itself (Jones & Rayner 1988), possibly with an incidental intake of water. Neither *P. auritus* nor *P. pipistrellus* is reported to have any such association with open water. *M. daubentonii* may therefore be expected to have easy access to drinking water while foraging and on leaving or before re-entering the day roost, and also to have a high free-water influx in comparison to the other two species. The low water flux in captive *P. pipistrellus* in comparison to the other two species may simply be a function of body size differences.

A comparison of total water flux in captive *P. auritus*, *M. daubentonii* and *P. pipistrellus* with that in captive rodents shows that water flux was not

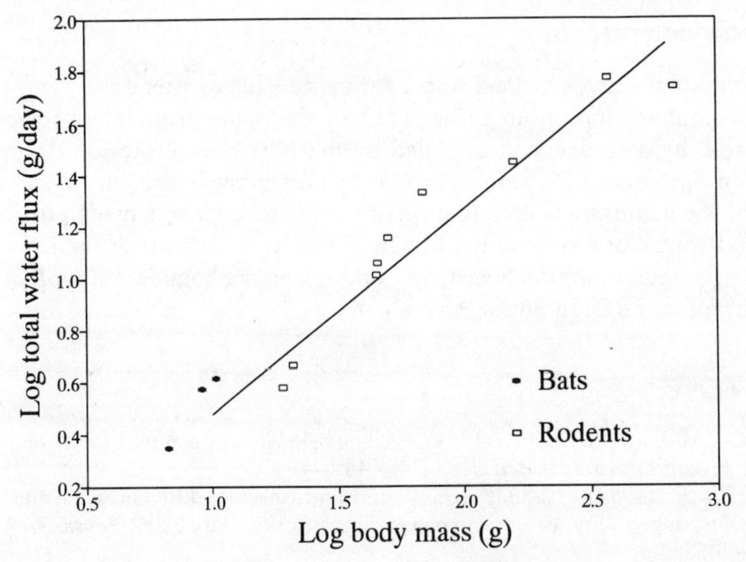

Fig. 7. Total water flux in captive rodents (□) and bats (+). The solid line represents the model I regression for rodents: $y = -0.31 + 0.79x$, $r^2 = 0.93$, $F_{1,8} = 100.8$, $P < 0.001$. Once body mass effects had been removed, water flux in the bats was not significantly different from that in the rodents (ANCOVA: $F_{1,10} = 0.30$, $P = 0.60$). The data for the bats are taken from Fig. 6, assuming an energy expenditure equivalent to free-living daily energy expenditure as predicted allometrically from body mass using the equation of Nagy (1987). Daily water flux in the bats was not significantly different from that in the rodents once differences in body mass had been accounted for (ANCOVA: $F_{1,10} = 0.30$, $P = 0.60$).

significantly different between the two groups once the effect of body mass had been accounted for (Fig. 7).

Summary

In summary, data on water balance in bats are patchy, and controlled comparisons between bats and non-volant mammals of aspects of physiology associated with water flux are few. More data are needed before the particular adaptations of faecal and urinary water loss in bats can be accurately determined. Bats do show comparatively high rates of resting evaporative water loss, but total water flux in captive individuals does not differ significantly from that in captive rodents of similar body mass. Physiological aspects of water balance such as maximum urine-concentrating ability, urine loss after feeding, evaporative water loss at rest and in flight, and total water flux in captive individuals vary tremendously between different bat species. This variability is associated with diet, the availability of free water in the environment and also with other ecological parameters such as foraging mode and, possibly, with environmental conditions within roosting sites.

Acknowledgements

I am grateful to Professor Paul Racey for inviting me to attend this symposium and for commenting on an earlier draft of the manuscript. The project was supported by a studentship awarded by the NERC to Professor Racey and Dr John Speakman. Dr Sara Frears was constructively critical of an earlier draft of the manuscript. Presentation of the manuscript was made possible by financial contributions from the Mammal Research Institute at the University of Pretoria, the Zoological Society of London and the Foundation for Research Development (FRD) in South Africa.

References

Anderson, M. E. & Racey, P. A. (1991). Feeding behaviour of captive brown long-eared bats, *Plecotus auritus*. *Anim. Behav.* **42**: 489–493.

Bassett, J. E. (1982). Habitat aridity and intraspecific differences in the urine concentrating ability of insectivorous bats. *Comp. Biochem. Physiol. (A)* **72**: 703–708.

Bassett, J. E. & Wiebers, J. E. (1980). Urine concentration dynamics in the postprandial and the fasting *Myotis lucifugus lucifugus*. *Comp. Biochem. Physiol. (A)* **64**: 373–379.

Beuchat, C. A. (1990). Body size, medullary thickness, and urine concentrating ability in mammals. *Am. J. Physiol.* **258**: R298–R308.

Boyd, I.L. & Stebbings, R. E. (1989). Population changes of brown long-eared bats (*Plecotus auritus*) in bat boxes at Thetford Forest. *J. appl. Ecol.* **26**: 101–112.

Carpenter, R. E. (1968). Salt and water metabolism in the marine fish-eating bat, *Pizonyx vivesi*. *Comp. Biochem. Physiol.* **24**: 951–964.

Carpenter, R. E. (1969). Structure and function of the kidney and the water balance of desert bats. *Physiol. Zool.* **42**: 288–302.

Chappell, M. A. & Roverud, R. C. (1991). Temperature effects on metabolism, ventilation, and oxygen extraction in a Neotropical bat. *Respir. Physiol.* **81**: 401–412.

Crawford, E. C. & Lasiewski, R. C. (1968). Oxygen consumption and respiratory evaporation of the emu and rhea. *Condor* **70**: 333–339.

Ewing, W. G., Studier, E. H. & O'Farrell, J. (1970). Autumn fat deposition and gross body composition in three species of *Myotis*. *Comp. Biochem. Physiol.* **36**: 119–129.

Fyhn, H. J. (1979). Rodents. In *Comparative physiology of osmoregulation in animals* **2**: 95–144. (Ed. Maloiy, G. O.). Academic Press, London.

Fuhrmann, & Seitz, A. (1992). Nocturnal activity of the brown long-eared bat (*Plecotus auritus* L., 1758): data from radio-tracking in the Lenneberg Forest near Mainz (Germany). In *Widlife telemetry: remote monitoring and tracking of animals*: 538–548. (Eds Priede, I. G. & Swift, S. M.). Ellis Horwood, New York, London etc.

Geluso, K. N. (1978). Urine concentrating ability and renal structure of insectivorous bats. *J. Mammal.* **59**: 312–323.

Geluso, K. N. (1980). Renal form and function in bats: an ecophysiological appraisal. In *Proceedings, fifth international bat research conference*: 403–414. (Eds Wilson, D. E. & Gardner, A. L.). Texas Tech Press, Lubbock.

Hattingh, J. (1972). A comparative study of transepidermal water loss through the skin of various animals. *Comp. Biochem. Physiol. (A)* **43**: 715–718.

Hays, G. C., Speakman, J. R., & Webb, P. I. (1992). Why do brown long-eared bats (*Plecotus auritus*) fly in winter? *Physiol. Zool.* **65**: 554–567.

Herreid, C. F. & Schmidt-Nielsen, K. (1966). Oxygen consumption, temperature, and water loss in bats from different environments. *Am. J. Physiol.* **211**: 1108–1112.

Jones, G. & Rayner, J. V. (1988). Flight performance, foraging tactics and echolocation in free-living Daubenton's bats *Myotis daubentoni* (Chiroptera: Vespertilionidae). *J. Zool., Lond.* **215**: 113–132.

Kallen, F. C. & Kanthor, H. A. (1967). Urine production in the hibernating bat. In *Mammalian hibernation III:* 280–294. (Eds Fisher, K. C., Dawe, A. R., Lyman, C. P., Schonbaum, E. & South, F. E.). Oliver & Boyd, Edinburgh & London.

Lechner, A. J. (1978). The scaling of maximal oxygen consumption and pulmonary dimensions in small mammals. *Respir. Physiol.* **34**: 29–44.

McFarland, W. N. & Wimsatt, W. A. (1969). Renal function and its relation to the ecology of the vampire bat, *Desmodus rotundus. Comp. Biochem. Physiol.* **28**: 985–1006.

Miller, L. A. & Degn, H. -J. (1981). The acoustic behavior of four species of vespertilionid bats studied in the field. *J. comp. Physiol. (A)* **142**: 67–74.

Nagy, K. A. (1987). Field metabolic rate and food requirement scaling in mammals and birds. *Ecol. Monogr.* **57**: 111–128.

Nicoll, P. A. & Webb, R. L. (1946) Blood circulation in the subcutaneous tissue of the living bat's wing. *Ann. N. Y. Acad. Sci.* **46**: 697–711.

Park, E. (1988). *Distribution of bat roosts along river systems in north-east Scotland.* Honours thesis: University of Aberdeen.

Procter, J. W. & Studier, E. H. (1970). Effects of ambient temperature and water vapor pressure on evaporative water loss in *Myotis lucifugus. J. Mammal.* **51**: 799–804.

Racey, P. A. (1991). Pipistrelle. In *The handbook of British mammals* (3rd edn): 124–128. (Eds Corbet, G. B. & Harris, S.). Blackwell Scientific Publications, Oxford.

Richardson, P. (1989). Activity at a summer roost site of Daubenton's bat (*Myotis daubentoni*). In *European bat research 1987*: 623–624. (Eds Hanak, V., Horacek, I. & Gaisler, J.). Charles University Press, Prague.

Speakman, J. R. & Racey, P. A. (1991). No cost of echolocation for bats in flight. *Nature, Lond.* **350**: 421–423.

Speakman, J. R., Racey, P. A., Catto, C. C., Webb, P. I., Swift, S. & Burnett, A. (1991). Minimum summer populations and densities of bats in N. E. Scotland, near the northern borders of their distributions. *J. Zool., Lond.* **225**: 327–345.

Studier, E. H. (1970). Evaporative water loss in bats. *Comp. Biochem. Physiol.* **35**: 935–943.

Studier, E. H. & Ewing, W. G. (1971). Diurnal fluctuation in weight and blood composition in *Myotis nigricans* and *Myotis lucifugus. Comp. Biochem. Physiol. (A)* **38**: 129–139.

Studier, E. H., Wisniewski, S. J., Feldman, A. T., Dapson, R. W., Boyd, B. C. & Wilson, D. E. (1983). Kidney structure in neotropical bats. *J. Mammal.* **64**: 445–452.

Swift, S. & Racey, P. A. (1983). Resource partitioning in two species of vespertilionid bats (Chiroptera) occupying the same roost. *J. Zool., Lond.* **200**: 249–259.

Tucker, V. A. (1968). Respiratory exchange and evaporative water loss in the flying budgerigar. *J. exp. Biol.* **48**: 67–87.

Vogel, V. B. & Vogel, W. (1972). Über das Konzentrationsvermögen der Nieren zweier

Fledermausarten (*Rhinopoma hardwickei* und *Rhinolophus ferrum-equinum*) mit unterschiedlich langer Nierenpapille. *Z. vergl. Physiol.* **76**: 358–371.

Webb, P. I. (1992). *Aspects of the ecophysiology of some vespertilionid bats at the northern borders of their distribution*. PhD thesis: University of Aberdeen.

Webb, P. I., Speakman, J. R. & Racey, P. A. (1993). Defecation, apparent absorption efficiency, and the importance of water obtained in the food for water balance in captive brown long-eared (*Plecotus auritus*) and Daubenton's (*Myotis daubentoni*) bats. *J. Zool., Lond.* **230**: 619–628.

Webb, P. I., Speakman, J. R. & Racey, P. A. (1994). Post-prandial urine loss and its relation to ecology in brown long-eared (*Plecotus auritus*) and Daubenton's (*Myotis daubentoni*) bats (Chiroptera : Vespertilionidae). *J. Zool., Lond.* **233**: 165–173.

Webb, P. I., Speakman, J. R. & Racey, P. A. (1995). Evaporative water loss in two sympatric species of vespertilionid bat, *Plecotus auritus* and *Myotis daubentoni*: relation to foraging mode and implications for roost site selection. *J. Zool., Lond.* **235**: 269–278.

Symp. zool. Soc. Lond. (1995) No. 67: 219–231

Dispersal and philopatry in colonial animals: the case of *Miniopterus schreibersii*

Jorge M. PALMEIRIM — *Dept. Zoologia, Faculdade de Ciências Universidade de Lisboa P–1700 Lisboa, Portugal*

and Luisa RODRIGUES — *Instituto da Conservação da Natureza Rua Filipe Folque, 46–2 P–1000 Lisboa, Portugal*

Synopsis

The dispersal behaviour of *Miniopterus schreibersii*, a colonial cave-dwelling bat, was studied with data obtained by recovery of animals ringed in the eight maternity colonies known in Portugal. The results showed that this species has a strictly philopatric behaviour during the nursing season; females return to give birth in the colony where they were born and very seldom change maternity colony during their lifetime. Males also showed a high level of attachment to their birth site, although during the nursing season most of them roosted in other locations with nonbreeding females. This strictly philopatric behaviour is due to strong imprinting, rather than to lack of knowledge of alternative roosts. Imprinting to maternity colonies takes place during the first weeks of life, and not during the first breeding season.

Several factors that may have played a role in the evolution of dispersal behaviour are discussed. Low levels of dispersal (philopatry) should be favoured by large colonies (with low risk of inbreeding and of colony extinction), stable roosts, low availability of vacant roosts, widely spaced colonies, intense competition, strong social segregation, high risk of settling in roosts unsuitable for a colony, and intense social interdependence. Most of these factors are likely to be important selective forces in both colonial and non-colonial species, but the two last are particular to colonial animals.

The persistence of colonial species is analysed by using metapopulation models. These models predict that species with large stable colonies, like *M. schreibersii*, can persist even in the presence of minimal dispersal rates.

Introduction

The level of attachment of an animal to its birth or breeding site is a character that varies from species to species and is thought to be the result of natural

ZOOLOGICAL SYMPOSIUM No. 67
ISBN 0 19 854945 8

selection (Hamilton & May 1977). Some species have strongly philopatric behaviour, returning to breed in the same sites year after year, whereas others often disperse to breed in new areas. Although there is an extensive literature on this subject, very few works address the particular situation of colonial species.

Miniopterus schreibersii is a migratory bat that congregates in large colonies during the nursing season. In this paper we (1) describe the results of a study of the dispersal behaviour of this species, (2) examine factors that may determine the evolution of dispersal behaviour in colonial animals and (3) discuss the amount of dispersal necessary for the survival of colonial species.

Dispersal is here defined as any movement that results in an animal first breeding in a colony different from the one where it was born (pre-breeding dispersal), or in an adult animal changing breeding colony during its lifetime (post-breeding dispersal). This definition, and the discussion in the paper, only concern inter-colony dispersal.

In many animal species the males are present in the colonies where the young are born and raised, or are closely associated with them. In this case the above definition of dispersal applies to both males and females. But the situation of *M. schreibersii* and of many other bat species is quite different (e.g. Humphrey & Cope 1976; Tuttle 1976; Fenton & Thomas 1985). Mating takes place during the autumn and winter and not in the spring and summer, when the maternity colonies assemble. During the mating season the roosts occupied by the males can be used by females from several different maternity colonies and therefore an individual male is not reproductively associated with any particular female maternity colony. In the roost, or group of roosts, used by a male during the mating season the universe of females with which the animal may mate is likely to include females from various colonies. In this context, male dispersal can be defined as any shift in the centre of activity of the male that results in a change of the universe of females with which he can mate. Male dispersal cannot, therefore, be evaluated by studying the exchange of individuals among maternity colonies. For this reason the study of *M. schreibersii* described in this paper only concerns the dispersal of females.

An animal can disperse to a site occupied by another colony or to a vacant site. In the discussion that follows we assume that selection favours either philopatry or dispersal, without distinguishing between the two types of destination of the dispersal movement, an assumption that may not always be true. In the study of *M. schreibersii* we only measured the potential dispersal to sites used by other colonies.

The fact that some artificial roosts, such as abandoned mine galleries, have been colonized by this species shows that there is a certain amount of dispersal to unused roosts. However, since new colonies are very seldom formed, we know that either the number of animals dispersing in this way is very small or their colonizing success is extremely low.

Dispersal is very important in the evolutionary process because it is a major vector of gene flow and therefore can affect the rate of evolution, which in turn influences geographic variation, ecotypic adaptation, speciation and long-term evolution (Mayr 1963). In the case of *M. schreibersii*, and of many other bat species, the mixing of animals from different maternity colonies during the mating season may be more important than dispersal in the control of gene flow (J.M. Palmeirim & L. Rodrigues unpubl.). Dispersal is also essential for the persistence of species with patchy distributions, such as those of most colonial animals; a flux of immigrants can avoid local extinctions (J.H. Brown & Kodric-Brown 1977), and dispersal is essential in the colonization (or recolonization) of habitat patches (Andrewartha & Birch 1954).

Dispersal behaviour of *Miniopterus schreibersii*

Miniopterus schreibersii is the most widespread bat species in the world. Its range includes large areas in tropical and temperate Europe, Asia, Africa, Australia and Oceania. Maeda (1982) suggested that it should be split into several species. It exists throughout the Portuguese mainland territory, where it is the most abundant cave-dwelling bat.

Miniopterus schreibersii is a very social species, forming large colonies throughout the year, although individuals can also be found isolated or in small groups. In Portugal it gives birth in June, in eight maternity colonies (Fig. 1) that range in number from about 1000 to over 20 000 individuals (Palmeirim & Rodrigues 1992). Most of the maternity colonies are abandoned in August, after all the young are flying, and the animals disperse until the winter, when most of them assemble again in a reduced number of roosts. In the spring they leave the hibernation shelters and, after a period of frequent movements, the adult females settle in the nursing roosts, often just before giving birth. Some males and non-breeding females can also be found with them, but most remain in other shelters. The migratory behaviour of this species has also been studied in north-east Spain and south-west France (Serra-Cobo & Balcells 1985), South Africa (Van der Merwe 1975) and Australia (Dwyer 1966, 1969).

Methods

Bats were caught from 1987 to 1993 with harp traps placed at the entrances to roosts. To minimize disturbance to the maternity colonies, bats were only trapped there after all the young were flying. The bats were ringed with rings made by Lambournes, using the model recommended by the British Mammal Society. In this paper only two age classes are considered: young of the year and adults. The two classes are easy to distinguish by the state of ossification of the hand bones (e.g. Baagøe 1977).

All females found in a maternity colony during a nursing season were assumed to have given birth there. An animal was considered a disperser if it was found in two or more different maternity colonies during any

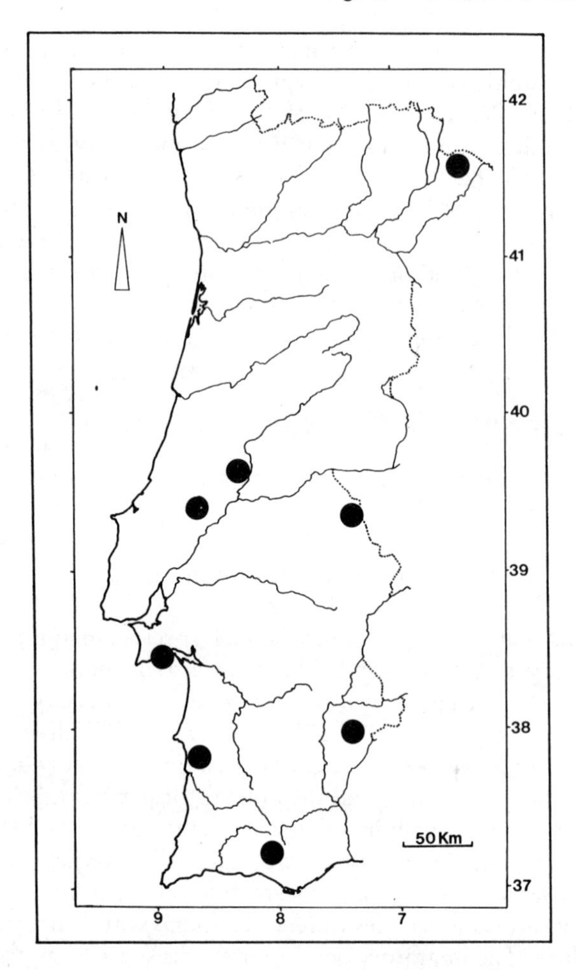

Fig. 1. Location of the eight maternity colonies of *Miniopterus schreibersii* known in Portugal and included in this study.

of the seven nursing seasons (1 June to 31 July) included in the study. This is, however, not always correct. We observed that the females of one maternity colony that lost most of its young during one nursing season (owing to pesticide poisoning) abandoned the colony earlier than usual. A few of those females were recaptured, during the same nursing season, in a nearby maternity colony. Had we not observed that their young had died during nursing elsewhere, we would have assumed that they were breeding in the colony where we found them. Such mistakes may lead to overestimates of the rates of dispersal. This may be partly avoided by checking if the recaptured females are producing milk, which we did not always do.

Table 1. Bats recaptured in maternity roosts during the nursing seasons.

Adult males	Young males	Adult females	Young females
Returned to same maternity colony where ringed			
43	25	324	41
Found in another maternity colony			
3	1	1	0

In southern Portugal there are two caves used by *M. schreibersii* during the nursing season, which are just 15 km apart. Because thc bats were caught when flying at the entrance of the roost in the evening, and because they often fly between the two roosts, our data for those roosts are unclear. The two caves may actually be alternative roosts of the same maternity colony. Therefore, in this analysis, the animals of the two roosts were pooled.

Results

Table 1 shows the results of the recaptures of bats ringed in the maternity colonies. No cases of natal dispersal of females were observed, since they all returned to give birth in the colony where they were born. In the case of the adult females, one may have switched maternity colonies. But this case may also correspond to a female that lost its nursing young and subsequently moved to another maternity colony, as explained above. In a large number of animals a few such situations are likely to occur.

A few of the males were found in different maternity colonies during the nursing season. But this fact does not really imply dispersal, as explained before. In any case, the fact that during the nursing season the males are far more likely to be present in the colony where they were born than in any other maternity colony shows a strong attachment to that colony or, at least, to the area where they were born.

Bats know the location of roosts of foreign maternity colonies and regularly visit them. Of the 499 animals (sexes and ages combined) recaptured in roosts of maternity colonies outside the nursing season, 201 were in the roosts of foreign colonies.

Factors determining the evolution of dispersal behaviour in colonial animals

A central assumption in the discussion of the evolution of the rates of dispersal is that there are potential costs and benefits associated with both philopatry and dispersal, which are listed and discussed in various papers (e.g. Shields

1987; Greenwood 1987). The benefits of dispersal most commonly referred to are reduced risk of inbreeding and an escape from competition in the area of origin. The obvious costs of dispersal are an increased risk during the dispersal and adaptation periods, and related energy costs.

Inbreeding

Colonial animals have a clumped distribution and therefore the colonies are, to a varying extent, isolated from each other. When the level of isolation is high, inbreeding may become deleterious and this may cause selection for dispersal (Pusey 1987).

If inbreeding avoidance is a factor determining dispersal frequency, then species with smaller colonies, where the risks of inbreeding are higher, should show higher rates of dispersal. Bats that aggregate in large colonies, such as *M. schreibersii*, should have low risk of inbreeding and therefore low dispersal rates.

In the case of *M. schreibersii* and of other migratory bat species the risk of inbreeding is further reduced by the fact that individuals from different colonies meet during the mating season, as described above.

Distance between colonies

If the colonies are very far apart they may be located in regions with substantially different ecological conditions. It is then possible that the individuals of each colony show adaptations to the local environment. In this case there may be selection against dispersal, since a dispersing individual that tries to settle in a different colony is likely to be outcompeted there by the locals. Under this model the amount of dispersal should be inversely proportional to the distance between the colonies.

In these situations high rates of dispersal may also be detrimental to the colonies; genetic isolation may allow the retention of adaptations to the local environment, which are lost if there is enough gene flow (Mayr 1963).

This factor is not likely to be important for bats, which usually have several colonies in the same ecological region, but may, for example, be of great relevance for pelagic seabirds. Randi, Spina & Massa (1989) suggested that morphological differences among colonies of the Cory's shearwater (*Calonectris diomedea*) in the Mediterranean Sea are linked to local variations in food availability. This may act as a reinforcement to philopatry, since dispersing individuals would be less adapted to the conditions of the new colony than would the locals.

Roost stability

If a roost of a colony is very stable the probability that it will be usable during the entire life of an individual and by many generations of its descendants is high. This should lead to the selection of philopatric behaviour, since dispersing is likely to be less adaptive than remaining in the same roost. In contrast, if the roosts are ephemeral dispersal becomes advantageous.

This factor may have been important in the evolution of the strong philopatry observed in *M. schreibersii*, which roosts in large caves, and some pelagic seabirds that nest in oceanic islands. Both types of roosts are very stable.

Roost availability

In general, roosts for colonies have to fulfil more requirements than do those for individuals. To shelter a colony a roost needs to have, for example, enough physical 'locations' for a large number of individuals and there must be enough food resources within an efficient foraging distance for them. Furthermore, since colonies are much more obvious to predators than are individuals or small groups, some type of passive barrier to predators is often required.

Because of these requirements, it is particularly difficult for dispersing colonial animals to find a suitable roost. Lower roost availability decreases the success of dispersal of individuals and, therefore, should result in a higher degree of philopatry.

Roost selection (chances of settling in unsuitable roosts)

The success of a dispersing animal is dependent not only on finding an appropriate site to breed, but also on the capability to evaluate the actual suitability of each potential site. Settling in an inappropriate site will lower the reproductive success of the animal, or even result in its death. This evaluation is a comparatively difficult task for colonial animals; dispersing colonial animals may settle and successfully reproduce in a poor site, but colonies founded in such sites are likely to become extinct in a short time. Dispersing bats may, for example, settle in a cave surrounded by an amount of foraging habitat insufficient to support enough animals to form a stable colony. Likewise, seabirds may, unknowingly, colonize a site accessible to terrestrial predators, but the colony is likely to be wiped out as soon as the population is large enough to be obvious to predators. (Large colonies of seabirds are usually located in sites free of terrestrial predators, while those nesting in sites accessible to these predators usually nest solitarily or in small colonies: Clode 1993.) If the chances of founding these abnormally short-lived colonies are high then this factor should reinforce philopatric behaviour.

Dynamics of the populations (colony extinction rate)

Because of stochastic environmental and demographic factors, all animal colonies have a finite lifetime. However, the frequency of colony extinctions certainly varies dramatically from species to species. These variations are likely to influence the frequency of dispersal behaviour in two ways:

1. If colony extinctions are rare, then the relative fitness of the individuals that do not abandon the colony (thus avoiding the risks associated with dispersal) is increased. In contrast, frequent colony extinctions should reduce the relative fitness of the individuals that do not disperse.
2. When an animal colony becomes extinct, a roost suitable for the establishment of a new colony becomes available (unless the roost was destroyed). Higher colony extinction rates result in a larger number of available roosts for new colonies and should, therefore, increase the fitness of dispersal (E. S. Brown 1951; Southwood 1962). The rate of dispersal is expected to evolve to a level that compensates for the extinctions (Hanski 1991).

In both cases lower colony extinction rates should increase the level of philopatry. Large colonies of cave-dwelling bats, like those of *M. schreibersii*, are probably quite stable, and therefore low dispersal rates are likely to evolve in these species.

Social interdependence

Animal colonies are more than simple aggregates of individuals. There are usually mutual advantages that increase the fitness of each of the individuals in the colony. In the case of seabirds, social facilitation in the location of patchy food resources over large foraging areas may actually be the cause of colonial behaviour (Clode 1993). In bats, advantages of coloniality may also include social facilitation in the location of foraging sites (Howell 1979) and changes in the microclimate of the roosts (e.g. Kunz 1982; Racey 1982). Dwyer & Harris (1972) demonstrated the presence of thermal social facilitation in parturition colonies of *M. schreibersii*. If the reproductive success of the animal is very dependent on the advantages provided by a colony, then getting established in a vacant site becomes extremely difficult. Social interdependence should, therefore, decrease the fitness of dispersal. This, of course, does not apply to dispersals to existing colonies, nor if large numbers of individuals disperse together and can form fully functional colonies upon arrival at the new sites, a strategy that may be common in bats.

Social segregation

If the individuals of each maternity colony recognize each other and behave as a social group they may obstruct the assimilation of new animals into these colonies. This should reduce the chances of dispersing individuals and, therefore, select against dispersal. In the case of *M. schreibersii* this behaviour, if present, is only active during the nursing season, since until the onset of parturition females are often found in foreign colonies. Even during the nursing season we found five females in the roost of a foreign colony, although they were not breeding there (the young in their own maternity colony had died earlier). But since the animals were caught when flying out of the cave they may have been roosting at a considerable distance from the maternity colony. Exclusion of strange males during this season may also help to explain the small number of males that were found with maternity colonies other than the one in which they were born.

Competition

In the case of species whose colonies grow to the point of depleting essential limiting resources, there may be a strong selection for dispersal. However, to establish the actual importance of this factor, it is necessary to evaluate how common intraspecific competition is. Although this is particularly difficult to study in the case of bats, a few studies have shown strong evidence for competition for resources in this group (Husar 1976; Thomas 1985; Fleming 1988; Palmeirim, Gorchov & Stoleson 1989). Regrettably there are too few case studies to evaluate how common this situation is. There are no data on this subject for *M. schreibersii*.

Dispersal rates required for the survival of colonial species

Levins (1969) developed the concept of metapopulation, referring to a set of partially isolated populations, interconnected by dispersing individuals. According to metapopulation models, colonial species will only survive if the extinction rate of the colonies is compensated for by the colonization rate.

Levins' model assumes that the rate of colonization is proportional to the fraction of potential roost sites in use (p), which are the sources of dispersing individuals, and the fraction of unoccupied roost sites ($1-p$), the targets for colonization by the dispersing individuals. Changes of p in time are given by:

$$\mathrm{d}p/\mathrm{d}t = mp\,(1-p)-ep,$$

where e and m represent extinction and colonization, respectively.

If m is too small then p will have a negative growth and the metapopulation will become extinct. Dispersal is the factor that controls m. Therefore colonial species with colonies that often become extinct (large e) will only survive if there are high rates of dispersal. In contrast, species with long-lasting colonies can survive in the presence of strongly philopatric behaviour.

A group of maternity colonies of *M. schreibersii* cannot strictly be considered a metapopulation, because genetically they are interconnected not only by dispersing individuals but also by mating between individuals of different colonies. But the dynamics of the extinction/colonization processes should follow metapopulation models.

Under natural conditions the colonies of *M. schreibersii* are likely to last for very long periods; the large caves in which they roost are very stable, and the large numbers of individuals that compose each colony protect it from extinction due to stochastic population events. The survival of the species can, therefore, be guaranteed by a minimal dispersal rate.

The situation of pelagic seabirds is quite similar to that of cave-dwelling bats that form large colonies. Their colonies are also often large and their roosts, usually small islands without terrestrial predators, are very stable. Like cave bats these species can, therefore, survive in the presence of very low dispersal rates.

Dispersal can influence the dynamics of colonial animals, not only by allowing the colonization of empty sites, but also by decreasing the extinction rate of the occupied sites (J.H. Brown & Kodric-Brown 1977). A colony that regularly receives immigrants that dispersed from other colonies is less likely to become extinct than one that is completely isolated. The original model proposed by Levins (1969) does not take this effect into consideration, but various models that incorporate it have now been proposed (Hanski 1982; Gotelli 1991; Gotelli & Kelley 1993).

Discussion and conclusions

Both pre-breeding and post-breeding dispersal of females of *M. schreibersii* to other maternity colonies are very low (in our data 0% and 0.3% respectively). Dispersal to vacant sites is also likely to be minimal or unsuccessful, as discussed above. This is a very low dispersal rate, especially if we take into consideration that the only dispersal movement of a female registered in this study may not have been a real case of dispersal. Such strongly philopatric behaviour suggests that this species is subjected to strong selection against dispersal. This is in agreement with the results of the above discussion on the factors that control the evolution of dispersal behaviour. In particular, low risk of inbreeding, low colony extinction rates and high roost stability should favour philopatry in *M. schreibersii*. This is also likely to be the case for most of the remaining factors discussed, although there are fewer data to evaluate them.

We found very few quantitative data on inter-colony dispersal of colonial animals. However, the few available studies with bats show that, like *M. schreibersii*, these animals seldom change maternity colonies. This is the case in *Myotis lucifugus* (Humphrey & Cope 1976), *Myotis grisescens* (Tuttle 1976), *Eptesicus nilssonii* (Rydell 1989), and *Pipistrellus pipistrellus* (Thompson 1992).

Few colonial animal species have dispersal rates as low as those observed in *M. schreibersii* and in other bats, but we found similar levels of philopatry in pelagic seabirds (e.g. Harris 1972; Harper 1976).

In some bat species each maternity colony may occupy various alternative roosts during the same nursing season. This behaviour has been observed in species nursing in both caves (*Myotis grisescens*: Tuttle 1976), and buildings (*Eptesicus nilssonii*: Rydell 1989; *Pipistrellus pipistrellus:* Thompson 1992).

The colonies that we studied remained in the same roost year after year, although the location of the bats within the same cave or mine system was not always the same. (One of our colonies may also have an alternative roost, but we have not been able to confirm this.)

Outside the nursing season individuals of *M. schreibersii* of both sexes often visit the roosts of foreign maternity colonies. This shows that the very high level of philopatry observed is due to strong imprinting, rather than to lack of knowledge of alternative roosts.

Imprinting to the breeding roost in some colonial seabirds takes place during the development of the nestling (Kress & Nettleship 1988; Serventy, Gunn, Skira, Bradley & Wooler 1989). In *M. schreibersii*, the fact that there is no pre-breeding dispersal shows that imprinting also takes place during nursing, rather than during the first breeding season.

Table 2 summarizes the main conclusions of the discussion on the potential factors controlling the frequency of dispersal behaviour in colonial species. These factors were discussed separately for the sake of clarity, but they are not likely to act independently in the selection processes.

Most of the factors discussed are likely to be important selective forces in both colonial and non-colonial animals, although their relative importance may be different in the two types of species. However, some are particular to colonial animals and therefore deserve to be mentioned here: social interdependence, and the risk of settling in roosts that are unsuitable for the establishment of a colony. This does not mean, however, that they are the most important factors acting upon colonial species.

Natural selection genetically fixed very low levels of dispersal in various colonial species, including many bats. But the persistence of these species was probably only possible in the presence of low rates of colony extinction. According to the metapopulation models discussed above, *M. schreibersii* metapopulations can persist with minimal levels of dispersal, but the low capacity of these species to recolonize sites in which the colonies became

Table 2. Factors that may influence the selection of dispersal or philopatric behaviour.

Dispersal rare (Philopatry)	Dispersal frequent
Low risk of inbreeding (large colonies)	High risk of inbreeding (small colonies)
Colonies far apart, in ecologically different areas	Colonies close to each other, in ecologically similar areas
Stable roosts	Unstable roosts
Low vacant roost availability	High vacant roost availability
High chances of settling in unsuitable roosts	Low chances of settling in unsuitable roosts
Low colony extinction rate	High colony extinction rate
High social interdependence	Low social interdependence
Strong social segregation	Weak social segregation
Low competition	High competition

extinct may prove to be fatal for many species with populations threatened by the recent destructive activities of man.

Acknowledgements

We are grateful to the many people that helped us in the field. Much of the cost of this project was supported by the 'Instituto da Conservação da Natureza'.

References

Andrewartha, H. G. & Birch, L. C. (1954). *The distribution and abundance of animals.* University of Chicago Press, Chicago.

Baagøe, H. J. (1977). Age determination in bats (Chiroptera). *Vidensk. Meddr dansk naturh. Foren.* **140**: 53–92.

Brown, E. S. (1951). The relation between migration-rate and type of habitat in aquatic insects, with special reference to certain species of Corixidae. *Proc. zool. Soc. Lond.* **121**: 539–545.

Brown, J. H & Kodric-Brown, A. (1977). Turnover rate in insular biogeography: effects of immigration on extinction. *Ecology* **58**: 445–449.

Clode, D. (1993). Colonially breeding seabirds: predators or prey? *Trends Ecol. Evol.* **8**: 336–338.

Dwyer, P. D. (1966). The population pattern of *Miniopterus schreibersii* (Chiroptera) in north-eastern New South Wales. *Aust. J. Zool.* **14**: 1073–1137.

Dwyer, P. D. (1969). Population ranges of *Miniopterus schreibersii* (Chiroptera) in south-eastern Australia. *Aust. J. Zool.* **17**: 665–686.

Dwyer, P. D. & Harris, J. A. (1972). Behavioral acclimatization to temperature by pregnant *Miniopterus* (Chiroptera). *Physiol. Zool.* **45**: 14–21.

Fenton, M. B. & Thomas, D. W. (1985). Migrations and dispersal of bats (Chiroptera). *Contr. mar. Sci.* **68** (Suppl.): 409–424.

Fleming, T. H. (1988). *The short-tailed fruit bat. A study in plant–animal interactions.* The University of Chicago Press, Chicago.

Gotelli, N. J. (1991). Metapopulation models: the rescue effect, the propagule rain, and the core-satellite hypothesis. *Am. Nat.* **138**: 768–776.

Gotelli, N. J. & Kelley, W. G. (1993). A general model of population dynamics. *Oikos* **68**: 36–44.

Greenwood, P. J. (1987). Inbreeding, philopatry and optimal outbreeding in birds. In *Avian genetics—a population and ecological approach*: 207–222. (Eds Cooke, F. & Buckley, P. A.). Academic Press, London.

Hamilton, W. D. & May, R. M. (1977). Dispersal in stable habitats. *Nature, Lond.* **269**: 578–581.

Hanski, I. (1982). Dynamics of regional distribution: the core and satellite species hypothesis. *Oikos* **38**: 210–221.

Hanski, I. (1991). Single-species metapopulation dynamics: concepts, models and observations. *Biol. J. Linn. Soc.* **42**: 17–38.

Harper, P. C. (1976). Breeding biology of the fairy prion (*Pachyptila turtur*) at the Poor Knights Islands, New Zealand. *New Zealand J. Zool.* **3**: 351–371.

Harris, M. P. (1972). Inter-island movements of Manx shearwaters. *Bird Study* **19**: 167–171.

Howell, D. J. (1979). Flock foraging in nectar-feeding bats: advantages to the bats and to the host plants. *Am. Nat.* **114**: 23–49.

Humphrey, S. R. & Cope, J. B. (1976). Population ecology of the little brown bat, *Myotis lucifugus*, in Indiana and north-central Kentucky. *Spec. Publs Am. Soc. Mammal.* No. **4**: 1–88.
Husar, S. L. (1976). Behavioral character displacement: evidence of food partitioning in insectivorous bats. *J. Mammal.* 57: 331–338.
Kress, S. W. & Nettleship, D. N. (1988). Re-establishment of Atlantic puffins (*Fratercula arctica*) at a former breeding site in the Gulf of Maine. *J. Fld Orn.* **59**: 161–170.
Kunz, T. H. (1982). Roosting ecology of bats. In *Ecology of bats*: 1–55. (Ed. Kunz, T. H.). Plenum Press, New York.
Levins, R. (1969). Some demographic and genetic consequences of environmental heterogeneity for biological control. *Bull. ent. Soc. Am.* **15**: 237–240.
Maeda, K. (1982). Studies on the classification of *Miniopterus* in Eurasia, Australia and Melanesia. *Honyurui Kagaku Suppl.* No. **1**: 1–176.
Mayr, E. (1963). *Animal species and evolution.* Harvard University Press, Cambridge, Mass.
Palmeirim, J. M. & Rodrigues, L. (1992). Plano nacional de conservação dos morcegos cavernícolas. *Estudos Biol. Conserv. Natur.* No. **8**: 1–165.
Palmeirim, J. M., Gorchov, D. L. & Stoleson, S. (1989). Trophic structure of a Neotropical frugivore community: is there competition between birds and bats? *Oecologia* **79**: 403–411.
Pusey, A. E. (1987). Sex-biased dispersal and inbreeding avoidance in birds and mammals. *Trends Ecol. Evol.* **2**: 295–299.
Racey, P. A. (1982). Ecology of bat reproduction. In *Ecology of bats*: 57–104. (Ed. Kunz, T. H.). Plenum Press, New York.
Randi, E., Spina, F. & Massa, B. (1989). Genetic variability in Cory's shearwater (*Calonectris diomedea*). *Auk* **106**: 411–417.
Rydell, J. (1989). Site fidelity in the northern bat (*Eptesicus nilssoni*) during pregnancy and lactation. *J. Mammal.* **70**: 614–617.
Serra-Cobo, J. & Balcells, E.R. (1985). Mise à jour des résultats des campagnes de baguage de *Miniopterus schreibersii* dans le N. E. espagnol et le S. E. français. *Colloque francophone de Mammalogie, Rouen* **9**: 85–98.
Serventy, D. L., Gunn, B. M., Skira, I. J., Bradley, J.S. & Wooler, R.D. (1989). Fledgling translocation and philopatry in a seabird. *Oecologia* **81**: 428–429.
Shields, W. M. (1987). Dispersal and mating systems: investigating their causal connections. In *Mammalian dispersal patterns: the effects of social structure on population genetics*: 3–24. (Eds Chepko-Sade, B. D. & Halpin, Z. T.). University of Chicago Press, Chicago.
Southwood, T. R. E. (1962). Migration of terrestrial arthropods in relation to habitat. *Biol. Rev.* **37**: 171–214.
Thomas, D. W. (1985). Apparent competition between two species of West African bats. *Bat Res. News* **26**: 73.
Thompson, M. J. A. (1992). Roost philopatry in female pipistrelle bats *Pipistrellus pipistrellus*. *J. Zool., Lond.* **228**: 673–679.
Tuttle, M. D. (1976). Population ecology of the gray bat (*Myotis grisescens*): philopatry, timing and patterns of movement, weight loss during migration, and seasonal adaptive strategies. *Occ. Pap. Mus. nat. Hist. Univ. Kans.* No. **54**: 1–38.
Van der Merwe, M. (1975). Preliminary study on the annual movements of the Natal clinging bat. *S. Afr. J. Sci.* **71**: 237–241.

Symp. zool. Soc. Lond. (1995) No. 67: 233–244

The physiological ecology of hibernation in vespertilionid bats

Donald W. THOMAS

Groupe de Recherche en Écologie, Nutrition et Énergétique
Département de Biologie
Université de Sherbrooke
Sherbrooke, Québec, Canada J1K 2R1
and
Mušee du Séminaire de Sherbrooke
Sherbrooke, Québec, Canada J1H 1J9

Synopsis

Chiropteran hibernation is characterized by prolonged bouts of torpor punctuated by periodic returns to endothermy. Laboratory studies indicate that these arousals are energetically very costly and, in the absence of feeding, they account for over 75% of winter fat depletion. Thus, knowledge of the organization and factors causing arousals is central to any analysis of winter energy balance.

To examine the organization of torpor and arousals, I placed temperature-sensitive radio tags on 18 hibernating *Myotis lucifugus* and monitored their superficial body temperatures over a 30–40-day period. Bats roused during all times of the day and there was no evidence for any significant temporal clustering of arousals. The first arousal provoked by handling lasted significantly longer (13.9 h) than subsequent natural arousals (5.0 h). During most first torpor bouts (10 of 11), but few subsequent bouts (one of 17), body temperature showed a slow positive drift of 0.82 °C/day up to a mean of 3.0 °C above ambient temperature. Torpor bouts accompanied by a drift in body temperature were significantly shorter (4.9 days) than those where body temperature fluctuated around 0.5 °C above ambient temperature (15.1 days). This study indicates that hibernating bats do not exhibit an endogenous circadian rhythm regulating the timing of arousals. Handling results in an increase in energy expenditure of 11.8 kJ above that typical of a 15.3 day torpor–arousal cycle. I examine the regulation of arousals and the energetic constraints that bats face during hibernation.

Introduction

Winter presents a severe metabolic challenge to warm-blooded animals. Just as food availability drops to its annual minimum, low ambient temperatures increase thermoregulatory costs and hence food requirements. Many mammals

ZOOLOGICAL SYMPOSIUM No. 67
ISBN 0 19 854945 8

escape this energy bottleneck by hibernating, whereby they depress body temperature to near-ambient levels. Reduction in the body to ambient temperature gradient substantially reduces the cost of thermoregulation and so reduces metabolic rate by as much as 99% (Wang & Hudson 1971; Cranford 1983; Thomas, Cloutier & Gagné 1990).

The suppression of metabolic rate through torpor has obvious adaptive value when feeding opportunities are sporadic or non-existent and fat reserves are limited. Hibernators, and particularly young of the year, appear to have difficulties meeting the energy demands of up to eight months of hibernation. Davis & Hitchcock (1965), for example, noted that juvenile *Myotis lucifugus* suffered disproportionately high mortality late in the hibernation season, which suggests that they depleted their fat reserves prematurely.

Hibernators, however, do not exploit the energy saving associated with torpor to the maximum; all hibernating mammals studied to date rouse periodically (Lyman, Willis, Malan & Wang 1982). These arousals are energetically very costly and represent 80–90% of the total energy requirement for hibernation (Kayser 1953; Wang 1978; Thomas, Dorais & Bergeron 1990). Arousals are thus the single most important factor determining hibernation energy budgets and possibly causing over-winter mortality due to energy shortfalls. However, arousals are also among the least understood of hibernation phenomena.

Understanding the regulation of arousals and their impact on winter energy budgets is hampered by the fact that data from laboratory studies do not always accord well with those from field studies. In the field, the timing and frequency of arousals often differ from those observed in the laboratory. Twente & Twente (1987) found that arousals of captive hibernating *Eptesicus fuscus* were strongly clustered around 18:00. They thus proposed that an endogenous alarm clock regulated the timing of arousals to coincide with dusk feeding opportunities. However, Thomas (1993) could find no clustering of activity in a natural hibernating population of *M. lucifugus* and Speakman (1990) noted that most winter flight activity of bats occurred during the day rather than at dusk. Torpor bouts for hibernating *Spermophilus columbianus* and for *M. lucifugus* are 56% and 55% longer, respectively, in the field than in the laboratory (Twente & Twente 1965; Brack & Twente 1985; Twente, Twente & Brack 1985; Young 1990). In the field, *S. columbianus* do not exhibit the progressive shortening of torpor bouts at the end of winter that is typically seen in the laboratory (Pengelley & Fisher 1961; Geiser, Hiebert & Kenagy 1990; Young 1990).

The discrepancy between laboratory and field studies suggests that stress and subtle stimuli associated with captivity modify the natural rhythm of torpor and arousal, and underlines the necessity for reliable field observations. Although easy to perform, direct observations of torpor and arousals or movements in bats may not be reliable because presence of an observer may provoke premature arousals and so over-estimates of their frequency.

The purpose of this study was to examine the temporal organization of torpor and arousals in a naturally hibernating population of *M. lucifugus* not subjected to disturbance by an observer. To achieve this, I relied on radio-telemetry of body temperatures.

Methods and materials

This study was conducted between 5 January and 15 March 1993 in a 150 m-long abandoned mine tunnel near Windsor, Québec. Eighteen adult male *M. lucifugus* were fitted with precalibrated, temperature-sensitive radio transmitters (Holohil Systems Inc., Woodlawn, Ontario, Canada; Model BD-2T; 0.9 g mass) to allow me to monitor changes in body temperature during hibernation and arousal. I collected bats at random from hibernating clusters, trimmed the hair of the interscapular region and secured the transmitter with Skin-Bond surgical adhesive. After a 5–10-min handling period, individuals were either released (three bats) or held as groups of two or three bats (15 bats total) in small wire cylinders (25 cm diameter × 20 cm height) and placed within 20 cm of a large hibernating cluster. A single temperature-sensitive transmitter was placed beside the cages to track ambient temperature.

Skin temperature (hereafter referred to as body temperature) was encoded as the inter-pulse interval and transmitted on individual frequencies. Transmitters were read at 12-min intervals over a 30–40-day period by a computerized radio receiver (LOTEK Engineering Ltd, Aurora, Ontario, Canada; model W-18). The receiver sequentially scanned the frequencies, measured the inter-pulse interval (±1 ms), and calculated body temperature from pre-programmed regressions. Body temperature records were accumulated in memory and downloaded to a computer during brief visits to the mine at two-week intervals. This system allowed me to monitor body temperature without the risk of provoking arousals artificially.

In the laboratory, I compiled body temperature records for each bat. I obtained complete records for 11 bats over 28–39-day periods. Two of the uncaged bats moved beyond reception distance after 1–3 days, transmitters failed on two caged bats and the receiver detected an additional two bats only sporadically. Thus there were partial records for these six additional bats. Temperature records, up to 4680 individual measurements per bat, were manipulated by means of Systat. All time measurements are accurate to 12 min. When transmitter batteries became weak, body temperatures appeared to decline progressively below ambient temperatures (Fig. 1). When this occurred, I discarded records after the decline was noted. For three bats, measurements were terminated before the bat roused. In these cases, I recorded the torpor bout duration as the time between arousal and the end of the record to provide a conservative estimate of minimum torpor duration. For the calculation of the timing of arousals, I used all increases in body temperature to above 10 °C to indicate arousal attempts (n=25). For measurements of the duration

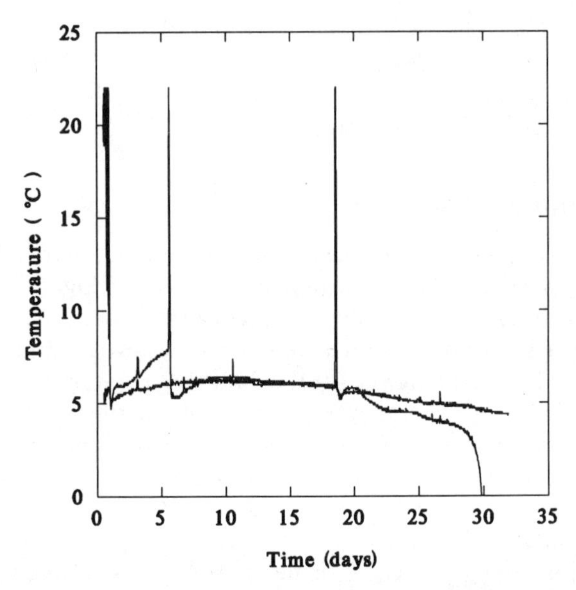

Fig. 1. A typical body temperature record for a *M. lucifugus* hibernating over a 35-d period. The lower, nearly stable trace indicates ambient temperature. Note that the initial provoked arousal was followed by re-entry into torpor and subsequently by a drift in body temperature leading to a second arousal after 6 d. The second and third arousals are separated by a natural torpor bout lasting 13 d. Body temperature declined below ambient temperature as the transmitter batteries weakened after 20 d.

of arousals after the first provoked arousal, I only used cases where body temperature increased to above 20 °C (n=17). All values are mean ± SE unless otherwise stated.

Preliminary tests indicated that transmitters measured body temperature to ±0.5 °C accuracy when body temperature was within 10 °C of ambient temperature. Measurement error increased progressively to approximately 10 °C as body temperature rose to about 30 °C above ambient temperature.

Results

Figure 1 shows a typical 35-day body temperature record for a caged bat overlaid on ambient temperature. The single uncaged bat for which I successfully collected data had body temperatures and durations of torpor bouts that fell within the range of those of caged bats, which suggests that confinement did not substantially alter normal behaviour. Body temperature records allowed me to identify (1) the timing of arousals, (2) the duration of the initial arousal provoked by handling and subsequent natural arousals, (3) variation in body temperature during torpor bouts and (4) the duration of the first and subsequent torpor bouts.

Table 1. Durations of the warming, endothermic and re-entry components of arousals. The warming phase of provoked arousals was not measured because bats were held in the hand for part of the period. Probability indicates significance levels and non-significance (NS) for Anova tests comparing provoked and natural arousals. Sample sizes are indicated in parentheses.

Arousal phase	Provoked arousals (n=11) (h)	Natural arousals (n=17) (h)	Probability
Warming	—	1.5±0.2	—
Endothermy	10.1±0.5	1.8±0.4	<0.001
Re-entry	2.3±1.2	1.6±0.2	NS
Total	13.9±0.8[a]	5.0±0.5	<0.001

[a]Total was calculated assuming 1.5 h warming.

The first arousals were all provoked by handling at approximately 11:45; however, the timing of all subsequent arousals (n = 25) represents the natural or endogenous cycle. Natural arousals were distributed throughout the 24 hours, with 48% occurring between 24:00 and 12:00 and 52% between 12:00 and 24:00. Two independent tests of temporal clustering (Watson's U^2 and Rayleigh's Z test; Zar 1984) indicate that arousals were randomly distributed throughout the day ($U^2 = -1.0$, $P > 0.05$; $Z = 1.9$, $P > 0.05$).

Handling had a profound effect on the total duration of arousals (Table 1). Initial, provoked arousals lasted significantly longer than did subsequent natural arousals ($F = 97.6$, $P < 0.001$). This difference was due to bats remaining endothermic longer when arousals were provoked than when they occurred naturally ($F = 183.6$, $P < 0.001$). I cannot compare the duration of the warming phase of provoked and natural arousals because bats were hand-held while transmitters were attached; however, re-entry into torpor did not differ significantly between the two ($F = 2.9$, $P > 0.05$; Table 1).

Re-entry into torpor was preceded by a series of declines (test drops) in body temperature ranging from 1–10 °C in amplitude (Fig. 2). Because of their extreme variability and the progression from small to large declines approaching re-entry, it was difficult to set a single criterion to identify and quantify test drops. However, provoked arousals typically terminated in five to 10 test drops while natural arousals commonly had only one to three.

With re-entry into torpor, body temperature fell to within ±0.5 °C of ambient temperature (*c.* 5.5 °C). Following re-entry, there existed two fundamental patterns of body temperature. Either body temperature remained relatively stable, fluctuating around ambient temperature and never exceeding it by more than 1.5 °C (mean difference: +0.5 °C), or body temperature underwent a progressive, near-linear drift to ≥2 °C above ambient. This drift in body temperature appeared to be a prolonged effect of the initial handling

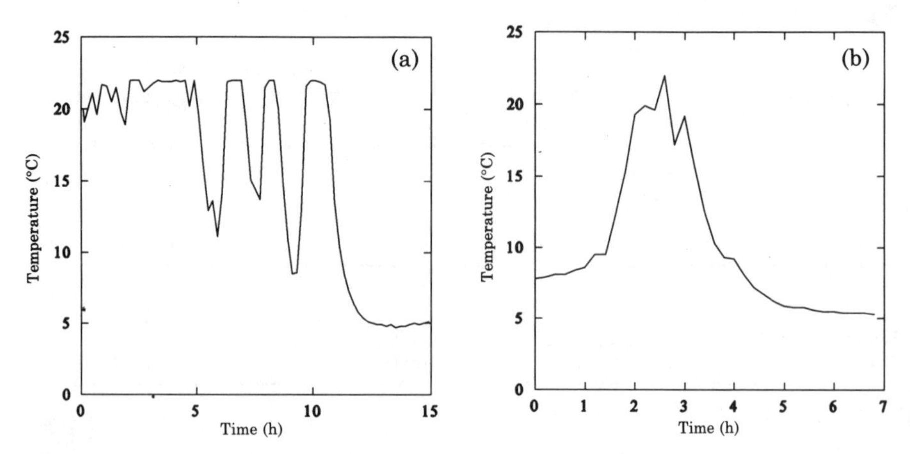

Fig. 2. Body temperature of a *M. lucifugus* during a provoked arousal (a) and a subsequent natural arousal (b). Note the prolonged endothermic period and more numerous test-drops leading to re-entry in the provoked arousal compared with the natural arousal. The warming phase of the provoked arousal was not recorded.

disturbance because 10 of 11 first torpor bouts (91%), but only one of 17 subsequent bouts (6%), exhibited this pattern.

When drift occurred, body temperature increased by 0.82 ± 0.12 °C/day, leading to a total increase of 3.0 ± 0.4 °C (range 2.0–5.2 °C) above ambient temperature prior to arousal. Body-temperature drift had a marked effect on torpor bout duration. When body temperature drifted by ⩾2 °C, torpor bout duration (4.9 ± 0.4 days) was significantly shorter than when drift was ⩽1.5 °C (15.1 ± 1.4 days; $F = 91.8$, $P < 0.001$). Although body temperature drift and short torpor-bout duration both typically occurred immediately after the provoked arousal, there is circumstantial evidence that drift rather than handling was the proximate cause of short torpor-bout duration. The single bat that showed a body-temperature drift in a subsequent torpor bout also had a short torpor bout duration (2.7 days).

Discussion

Studies of hibernation may be subject to the Heisenberg uncertainty principle, whereby the simple act of observing modifies the phenomenon being observed. By being present to observe hibernating bats, observers may provoke arousals and so document an entirely unnatural phenomenon. The only means to detect whether this problem exists is to develop alternative means of monitoring torpor and arousals. The use of radio-telemetry and automated data collection minimizes observer disturbance and presumably allows us to detect the effect of disturbance on the natural pattern of torpor and arousal.

Although based on a relatively small sample size, this study sheds new light on (1) the timing of arousals under natural conditions, (2) the impact of handling on arousals and torpor and (3) the duration of natural torpor bouts.

Timing of arousals and effect of disturbance

On the basis of a laboratory study of *E. fuscus*, Twente & Twente (1987) proposed that arousals in bats were regulated by an endogenous alarm clock. They argued that an endogenous control of arousal timing would have adaptive value if it permitted bats to synchronize activity with dusk and potential foraging opportunities. Although some bats may leave hibernation sites and forage when weather conditions are favourable (Avery 1985; Speakman & Racey 1989), the absence of any clear environmental cues makes it difficult to understand how such an endogenous rhythm could remain synchronized with the environment. Endogenous circadian rhythms are typically free-running and become progressively desynchronized in the absence of external environmental cues (Erkert 1982). Bats hibernating deep in mines or caves should thus become progressively desynchronized and not show any consistent timing of arousal.

This study and that of Thomas (1993) show that no such precise regulation of the timing of arousals is expressed in naturally hibernating populations. Whether data are collected on individuals (radio-telemetry of individual body temperatures) or populations (daily patterns of activity in a hibernating population; Thomas 1993), there is no significant clustering of arousals around dusk (18:00). These studies show that an endogenous biological alarm clock does not exist and that the temporal clustering of arousals was a laboratory artifact.

The temporal clustering of arousals in captivity, but not in the field, suggests that bats responded to some subtle, non-tactile stimulus present in the laboratory setting. Just how sensitive hibernating bats are to non-tactile stimuli is unclear. Harrison (1965) found that nervous activity in the posterior colliculus ceased in *M. lucifugus* as body temperature dropped below 12 °C, indicating that they were deaf. However, Speakman, Webb & Racey (1991) found a 2.6–5.6% response rate to sounds and a 6.3–11.1% response to prolonged exposure to light in hibernating bats and Thomas (1993) found that entry into a mine without touching the bats provoked a large number of arousals and increased activity over the following 24 h.

These latter two studies suggest that at least some bats are sensitive to sound or light stimuli and that some of these will rouse. If even a small proportion of a hibernating population (frequently numbering in the thousands) rouses and moves in response to non-tactile stimuli, then this would create a cascade effect. It is well known that tactile stimuli arouse bats (Speakman *et al.* 1991); active bats trying to reintegrate into clusters would disturb adjacent bats, arousing them in turn.

This study shows that the stress associated with handling has a profound effect on the following arousal–torpor cycle. After being handled, not only did bats remain endothermic 8.3 h longer than normal, but the following torpor bout was shortened from 15.1 to 4.9 days. This appeared to be mediated by an alteration of the normal regulation of temperature, as was indicated by the drift in body temperature.

Taking the natural torpor–arousal cycle to be 15.3 days (15.1 days torpor; 5.0 h arousal), we can estimate the energetic cost of disturbance for a 6.5 g *M. lucifugus*. A normal 15.3-day cycle includes 362.4 h of torpor with a $\dot{V}_{O2}$ of 0.02 ml O_2/g/h (Thomas, Cloutier *et al.* 1990). Since fat is the oxidized substrate (Dodgen & Blood 1956), liberating 20.1 J/ml O_2 (Nagy 1983), 15.3 days of torpor cost 947 J. Warming and re-entry into torpor cost 563 J and 378 J respectively (Thomas, Dorais *et al.* 1990), while 1.8 h of endothermy ($\dot{V}_{O2}$ =8.29 ml O_2/g/h; Thomas, Dorais *et al.* 1990) costs 1733 J. Thus, a complete torpor–arousal cycle costs 3.6 kJ.

Following handling, the same 15.3-day period includes a provoked arousal with 10.1 h of endothermy, a first torpor bout lasting 4.9 days and a natural arousal followed by 9.6 days (230.9 h) of torpor. The total cost of 15.3 days following disturbance is thus 15.4 kJ. This represents an increase of 11.8 kJ in energy expenditure due to handling, primarily due to the long endothermic period during the provoked arousal. The increase of 11.8 kJ is the equivalent of 0.3 g fat, 3.3 natural torpor–arousal cycles, or 50 days hibernation. This analysis accords qualitatively with that of Speakman *et al.* (1991), who showed that tactile stimuli elicited arousals lasting up to 15 h and costing up to 10.1 kJ.

These data indicate that handling bats during the hibernation period can have a marked effect on their use of winter fat stores and possibly on over-winter survival. Currently, nothing is known about the residual fat reserves that *M. lucifugus* have at the end of hibernation, so it is impossible to assess the impact of a 0.3 g increase in fat metabolism on over-winter survival. However, the fact that juveniles suffer disproportionately high mortality over winter suggests that they have little flexibility in the supply end of their energy budget and that they would be most affected by handling disturbance.

These data also raise questions about the accuracy of using winter mass change to estimate the natural rate of depletion of fat reserves. Because weighing involves handling, it will artificially increase mass loss and result in an over-estimate of the natural rate of fat metabolism.

Natural torpor bout duration and the energetic cost of hibernation

In this study, second and subsequent torpor bouts provide an estimate of the minimum duration of natural, undisturbed torpor. Because three bouts (19.5 days, 17.1 days and 23.8 days) were incomplete, the mean duration of 15.1 days (SD = 4.3 days; range = 9.4–23.8 days) must be viewed as a conservative estimate.

Twente *et al.* (1985) and Brack & Twente (1985) reported torpor bout durations of 12.7 ±12.1 days and 19.7 days for *M. lucifugus* hibernating in captivity (5 °C) and in the wild (6 °C), respectively. Although these lengths are not statistically different owing to the extreme variation in bout durations, the tendency for torpor bouts to be longer in the field suggests that many bouts are terminated prematurely under laboratory conditions. Indeed, in the laboratory 30% of 169 bouts measured at 5 °C were less than 4 days (Twente *et al.* 1985), whereas in the current study no natural bouts shorter than 9.4 days were observed. The high coefficient of variation in the laboratory data (CV = 95%) compared with that from this study (CV = 28.4%) also suggests that many torpor bouts were truncated, increasing the variation in the data and reducing the estimate of torpor bout duration.

My results and those of Brack & Twente (1985) indicate that *M. lucifugus* hibernating at an ambient temperature of 5 °C to 6 °C have torpor bouts lasting between 15 and 20 days. This means that during a hibernating season of about 200 days, bats will rouse between 10 and 13 times. This confirms my original calculation (Thomas, Dorais *et al.* 1990) that *M. lucifugus* require a minimum of 1.9–2.0 g fat to survive winter and that arousals account for approximately 85% of fat depletion.

One of the most intriguing problems that remains to be addressed is the effect of latitude and length of winter on the total fat reserves available at the start of hibernation and their depletion through winter. Near the northern limit of their distribution, one would expect parturition to be later than in more southerly locations. This should decrease not only the time that juveniles have available to complete growth but also the time that reproductive females and juveniles have to accumulate fat reserves prior to winter.

Nothing is known about what sets the upper limit on the rate of fat deposition and the amount of fat accumulated prior to winter. The fact that juveniles arrive at hibernation sites on average later than adult females (D. W. Thomas unpubl. data) suggests that they require longer to complete growth and deposit fat. This would suggest that later parturition and an earlier winter would reduce fat reserves, at least in juveniles. There is, however, some evidence that juveniles can compensate by initiating fat deposition prior to completion of muscle growth (T. H. Kunz pers. comm.). Juveniles may reduce their metabolic costs by limiting lean or metabolic mass and so sustain the energy requirement for hibernation on smaller fat reserves. Whether it was an adaptive developmental strategy or simply the result of a time constraint, one would expect a negative correlation between latitude and juvenile size at the onset of hibernation, unless a compensatory increase in pre-or post-weaning growth rate occurred (Audet & Fenton 1988).

A longer winter, however, should require a proportionally greater fat index (g fat/g lean mass) in both adults and juveniles, independent of any adjustments in lean mass. This would entail greater wing loading and increased flight costs concurrent with a reduction in manoeuvrability and hunting efficiency (Aldridge 1987). We do not sufficiently understand bat flight to know whether individuals could carry more than 30% of body mass in fat and still hunt profitably at a time when insect densities decline, but this seems unlikely. The fact that maximum fat masses in southern Quebec (about 2.5 to 3 g in adult males; unpubl. data) correspond closely with peak fetal mass suggests that 2.5 to 3 g may be the functional ceiling for load-carrying in *M. lucifugus*.

With an upper limit on fat reserves set by either the time available for fattening or constraints imposed by flight and manoeuvrability, the only component of the hibernation energy budget that could be adjusted would be torpor bout duration (and hence arousal frequency). This raises the important issue of what regulates arousals. Are torpor duration and arousal frequency under individual control or are arousals provoked by an uncontrollable degradation of homeostasis during torpor? A detailed analysis of the regulation of arousals is beyond the scope of this paper; however, three lines of evidence indicate that arousals result from cumulative pulmonary and cutaneous water loss leading to dehydration. Firstly, in a detailed analysis of food and water budgets of hibernating *Pipistrellus pipistrellus*, Speakman & Racey (1989) concluded that water rather than food was the limiting factor under almost all conditions. Secondly, using rates of evaporative water loss measured for *M. lucifugus* (Thomas & Cloutier 1992), I calculate that the total water loss during average short and long torpor bouts (4.9 and 15.1 days respectively) fall within 7% of each other. Because of the positive drift in body temperature during short torpor bouts, bats experience increased evaporative water loss and are forced to rouse. Finally, we recently modelled the effects of metabolic rate ($\dot{V}_{O2}$), body temperature and evaporative water loss on documented torpor bout duration in *S. lateralis* over ambient temperatures ranging from −2 °C to 8 °C. Only evaporative water loss accurately predicted torpor bout duration over the entire range of temperatures. This analysis suggests that arousal frequency is fixed by environmental factors beyond individual control. Although bats may cluster to reduce cutaneous evaporative water loss, they do so already so we cannot expect a large additional effect of clustering. This in turn suggests that latitude should have an important and additive effect on winter fat requirements.

Studies of hibernating bats, such as *M. lucifugus*, whose ranges cover a large latitudinal span, will tell us much about how adaptable species are in the face of environmental change and how important morphological and energetic constraints are in determining species distributions.

Acknowledgements

This research was supported by grants from the Natural Sciences and Engineering Research Council of Canada and the Fonds pour la Formation de Chercheurs et à l'Aide à la Recherche de Québec. This is contribution 91 of the Groupe de Recherche en Écologie, Nutrition et Énergétique.

References

Aldridge, H. D. J. N. (1987). Turning flight of bats. *J. exp. Biol.* **128**: 419-425.

Audet, D. & Fenton, M. B. (1988). Heterothermy and the use of torpor by the bat *Eptesicus fuscus* (Chiroptera: Vespertilionidae): a field study. *Physiol. Zool.* **61**: 197-204.

Avery, M. I. (1985). Winter activity of pipistrelle bats. *J. Anim. Ecol.* **54**: 721-738.

Brack, V., Jr., & Twente, J. W. (1985). The duration of the period of hibernation of three species of vespertilionid bats. I. Field studies. *Can. J. Zool.* **63**: 2952–2954.

Cranford, J. A. (1983). Body temperature, heart rate and oxygen consumption of normothermic and heterothermic western jumping mice (*Zapus princeps*). *Comp. Biochem. Physiol.* **74A**: 595-599.

Davis, W. H. & Hitchcock, H. B. (1965). Biology and migration of the bat, *Myotis lucifugus*, in New England. *J. Mammal.* **46**: 296-313.

Dodgen, C. L. & Blood, F. R. (1956). Energy sources in the bat. *Am. J. Physiol.* **87**: 151-154.

Erkert, H. G. (1982). Ecological aspects of bat activity rhythms. In *Ecology of bats*: 201–242. (Ed. Kunz, T. H.). Plenum Press, New York.

Geiser, F., Hiebert, S., & Kenagy, G. J. (1990). Torpor bout duration during the hibernation season of two sciurid rodents: interrelations with temperature and metabolism. *Physiol. Zool.* **63**: 489-503.

Harrison, J. B. (1965). Temperature effects on responses in the auditory system of the little brown bat, *Myotis l. lucifugus*. *Physiol. Zool.* **38**: 34-48.

Kayser, C. (1953). L'hibernation des mammifères. *Année biol.* **29**: 109-150.

Lyman, C. P., Willis, J. S., Malan, A. & Wang, L. C. H. (1982). *Hibernation and torpor in mammals and birds*. Academic Press, New York.

Nagy, K. A. (1983). The doubly-labelled water ($^{3}HH^{18}O$) method: a guide to its use. *UCLA Publ.*. No. 12-1417: 1-45.

Pengelley, E. T. & Fisher, K. C. (1961). Rhythmical arousal from hibernation in the golden-mantled ground squirrel, *Citellus lateralis tescorum*. *Can. J. Zool.* **39**: 105-120.

Speakman, J. R. (1990). The function of daylight flying in British bats. *J. Zool., Lond.* **220**: 101-113.

Speakman, J. R. & Racey, P. A. (1989). Hibernal ecology of the pipistrelle bat: energy expenditure, water requirements and mass loss, implications for survival and the function of winter emergence flights. *J. Anim. Ecol.* **58**: 797-813.

Speakman, J. R., Webb, P. I. & Racey, P. A. (1991). Effects of disturbance on the energy expenditure of hibernating bats. *J. appl. Ecol.* **28**: 1087-1104.

Thomas, D. W. (1993). Lack of evidence for a biological alarm clock in bats (*Myotis* spp.) hibernating under natural conditions. *Can. J. Zool.* **71**: 1-3.

Thomas, D. W. & Cloutier, D. (1992). Evaporative water loss by hibernating little brown bats, *Myotis lucifugus*. *Physiol. Zool.* **65**: 443-456.

Thomas, D. W., Cloutier, D. & Gagné, D. (1990). Arrhythmic breathing, apnea and non-steady-state oxygen uptake in hibernating little brown bats (*Myotis lucifugus*). *J. exp. Biol.* **149**: 395-406.

Thomas, D. W., Dorais, M. & Bergeron, J.-M. (1990). Winter energy budgets and cost of arousals for hibernating little brown bats, *Myotis lucifugus*. *J. Mammal.* **71**: 475–479.

Twente, J. W. & Twente, J. A. (1965). Regulation of hibernating periods by temperature. *Proc. natn. Acad. Sci. USA* **54**: 1058-1061.

Twente, J. W. & Twente, J. (1987). Biological alarm clock arouses hibernating big brown bats, *Eptesicus fuscus*. *Can. J. Zool.* **65**: 1668-1674.
Twente, J. W., Twente, J. & Brack, V., Jr. (1985). The duration of the period of hibernation of three species of vespertilionid bats. II. Laboratory studies. *Can. J. Zool.* **63**: 2955–2961.
Wang, L. C. H. (1978). Energetic and field aspects of mammalian torpor: the Richardson's ground squirrel. In *Strategies in cold: natural torpidity and thermogenesis*: 109–145. (Eds Wang, L. C. H. & Hudson, J. W.) Academic Press, New York.
Wang, L. C. H. & Hudson, J. W. (1971). Temperature regulation in normothermic and hibernating eastern chipmunks, *Tamias striatus*. *Comp. Biochem. Physiol.* **38A**: 59-90.
Young, P. J. (1990). Hibernating patterns of free-ranging Columbian ground squirrels. *Oecologia* **83**: 504-511.
Zar, J. H. (1984). *Biostatistical analysis*. Prentice-Hall Inc., Englewood Cliffs, N. J.

Symp. zool. Soc. Lond. (1995) No. 67: 245–258

Does energy or calcium availability constrain reproduction by bats?

Robert M. R. BARCLAY

Behavioural Ecology Group
Ecology Division
Department of Biological Sciences
University of Calgary
Calgary, Alberta, Canada T2N 1N4

Synopsis

Bats are unusual mammals in being small but having long lives and small litters (typically only one or two young). I hypothesize that litter size is constrained by the need to raise young to near adult size before they can be independent. Our studies, and those of others, on a variety of species of bats indicate that juveniles typically start to fly at over 70% of adult mass and over 95% of adult skeletal size. This constraint appears to be associated with flight in vertebrates, since young birds also do not fly until fully grown. This means that each young is very costly and restricts the number that can be raised. Although energetic demands may be the proximate constraint, I argue that calcium is more important. For bats, calcium demand on reproductive females is high and calcium availability in most diets (insects, fruit, pollen) is low. Birds can at least partially overcome this by supplementing their diet with calcium-rich inanimate objects that are unavailable to bats because of their inability to forage on the ground and detect such items. This may help to explain why the reproductive output of birds exceeds that of bats. If the hypothesis is correct, bat foraging strategies may be based on the calcium content of prey in addition to energy content, and female and male foraging strategies may be based on different currencies. Vertebrate-pollinated and seed-dispersed plants may attract bats by offering high calcium rewards. In addition, however, it would mean that flight could only have evolved in bats in association with long lifespans, thereby constraining the possible life histories available to these mammals.

Introduction

Amongst mammals, many life-history traits correlate with body size (Millar 1977, 1981; Harvey & Read 1988; Read & Harvey 1989). In general, large mammals live long lives and produce litters of few, large, slow-growing, late-maturing offspring. Small mammals live short lives and have litters of many, small, rapidly growing, early-maturing offspring. These correlations may simply result from common selective forces acting independently on

ZOOLOGICAL SYMPOSIUM No. 67
ISBN 0 19 854945 8

body size and life-history characteristics (Read & Harvey 1989; Promislow & Harvey 1990). Nonetheless, even when body size is factored out, life-history traits still correlate with one another (Read & Harvey 1989); some mammals produce litters of few, large, slow-growing offspring and live long lives for their body size, whereas others produce large litters of fast-growing young and die at an early age.

In the debate over the evolution of mammalian life-history variation, the second largest group of mammals, the Chiroptera, has been either completely ignored (Sacher & Staffeldt 1974; Western 1979; Millar 1981; Western & Ssemakula 1982), or severely under-represented (Millar 1977; Blueweiss, Fox, Kudzma, Nakashima, Peters & Sams 1978; Promislow & Harvey 1990). Bats can be used to argue against a simple allometric constraint on life-history variation. Despite the small size of bats (most have body masses under 100 g: Barclay & Brigham 1991), they are long-lived (Tuttle & Stevenson 1982) and have small litters. Most species produce only a single young and only eight are known regularly to produce more than two young (Tuttle & Stevenson 1982).

Recent laboratory and field studies on the growth, development and nutritional requirements of bats, and the availability of energy and nutrients in their diets, suggest that there are unique constraints on bat reproduction that may explain why this group of mammals has evolved the life-history pattern it has (Barclay 1994).

Pre- versus post-natal constraints

Since bats are the only mammals to have evolved true flight, a reasonable assumption might be that small litters are somehow linked to the ability to fly. For example, litter mass may be constrained, thereby limiting the number of viable neonates that can be produced (Millar 1977). The mass of a near-term litter influences the ability of a female to fly and forage, since increased mass and wingloading increase flight costs and reduce the ability to fly slowly, manoeuvrably and with agility (Norberg & Rayner 1987). This could impair a female's foraging efficiency, especially for aerial insectivorous bats which rely on manoeuvrability to capture their prey. However, bats and similar-sized terrestrial mammals both produce litters with a mass averaging 25% that of the female (Kurta & Kunz 1987). Female bats thus carry as large a load as do other female mammals. The difference is that female bats put this mass into very few large offspring while terrestrial mammals divide it amongst many small neonates.

Flight may constrain litter size of bats, but for post-natal, not pre-natal, reasons. Young terrestrial mammals are weaned when they are on average 37% of adult body mass and some, such as lagomorphs, are weaned at 16% of adult mass (Millar 1977). In contrast, young bats begin to fly and become independent of the female only when they are at a mean of 70.9 ± (SD) 15.7%

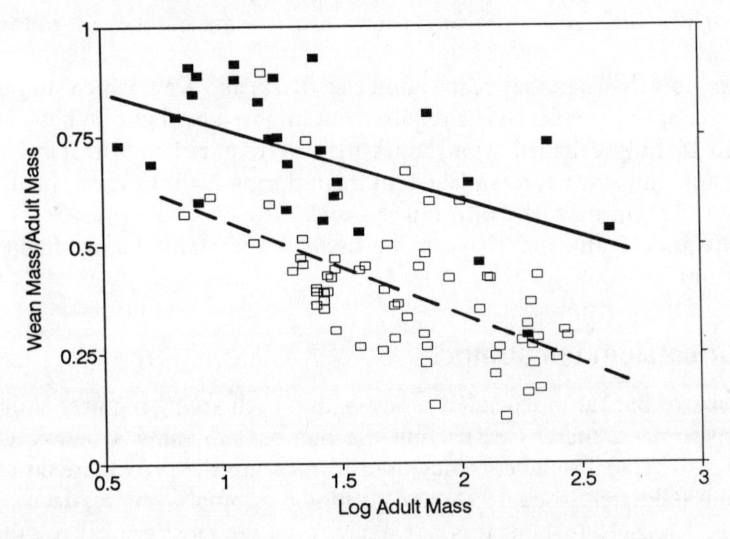

Fig. 1. Ratio of mass at first flight or weaning to adult mass versus the log of adult body mass, for terrestrial mammals (open squares) and bats (closed squares). Lines were calculated using least squares (bats: ratio = 0.94 − 0.17 × log adult mass; terrestrial mammals: ratio = 0.78 − 0.22 × log adult mass).

(n = 27; Appendix) of adult mass. In an analysis of covariance (ANCOVA) comparing the relative size at 'independence' for bats, and terrestrial mammals of the same size (<300 g: Millar 1977), with the log of adult mass as the covariate, both taxon ($F = 62.4$, $d.f. = 1, 86$, $P < 0.001$) and log of adult mass ($F = 43.5$, $d.f. = 1, 86$, $P < 0.001$) significantly influenced relative size at independence (the interaction term was not significant and was removed from the model). The relative size of young at first flight or weaning declines with adult size for both bats and terrestrial mammals, but bats start to fly at a significantly larger relative mass ($\bar{x} = 70.9 \pm 15.7\%$) than that at which similarly sized terrestrial mammals are weaned ($\bar{x} = 39.1 \pm 14.9\%$, $n = 62$; Fig. 1). Even compared to other mammals with small litters (≤2 young), bats raise their young to a significantly larger size (other mammals, $\bar{x} = 38.3 \pm 17.5\%$, $n = 13$ (data from Millar 1977); $t = 5.66$, $d.f. = 38$, $P < 0.001$; arcsine squareroot transformed proportions). The difference in weaning mass between bats and other mammals is even larger, since female bats continue to nurse their young after they first fly while the young learn to fly, forage and echolocate (Jones 1967; Kunz 1973; Tuttle & Stevenson 1982; Brown, Brown & Grinnell 1983; Koehler 1991). In addition, maternal milk is the sole energy and nutrient source for non-volant bats, while nursing young of many terrestrial mammals obtain some of their nutrition by foraging for themselves. Each young is thus more expensive (in terms of energy and nutrients) to a female bat than is each young to an equivalently sized terrestrial mammal. I suggest that this high cost of raising young restricts the number of young a bat can rear.

Birds also only fly when they reach adult size (Ricklefs 1979), which suggests that large size at independence is a requirement imposed by flight. In bats, and possibly birds, fully calcified wing bones may be required to withstand the unique torsion and shear forces placed on them during flight (Swartz, Bennett & Carrier 1992). In bats, the fifth finger provides wing camber necessary to generate lift and, again, may have to be ossified and stable before flight is possible (Kunz 1973).

Energy or calcium constraint?

The large size of bats at independence means that each young requires a large parental investment, thereby restricting the number of young a female can raise. Energy is typically the currency used to measure the proximate cost of reproduction (Clutton-Brock 1991) and to model optimal foraging decisions (Stephens & Krebs 1986), but it is not likely to be the most critical resource restricting litter size in bats. At weaning, young bats are large not only in terms of mass, but also in terms of their skeletal system (Medway 1972; Kunz 1973). Forearm length, a standard measure of growth and size in bats, averages 91.2 ± 5.9% (n = 30) of adult size at first flight (Fig. 2; Appendix), and other skeletal elements are similarly large (Maeda 1972; O'Farrell & Studier 1973; Pagels & Jones 1974). Indeed, while relative mass at first flight varies (coefficient of variation (CV) = 22.1%), and females of

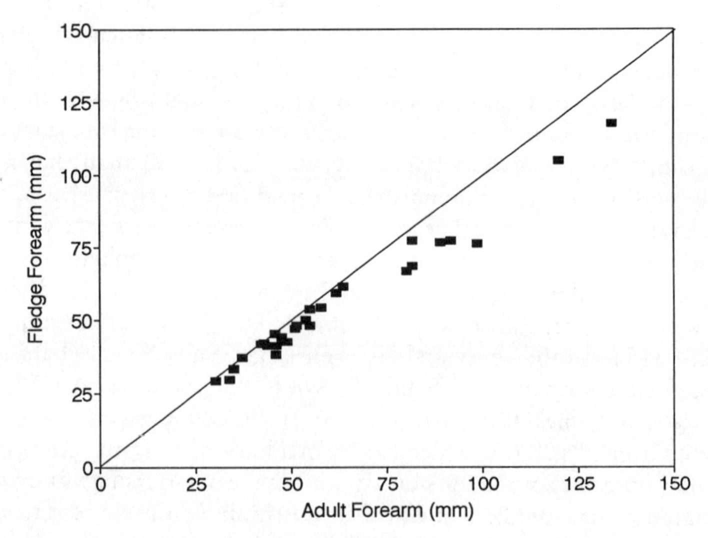

Fig. 2. Length of forearms of bats at first flight versus length of adult forearm. The line indicates a 1:1 relationship.

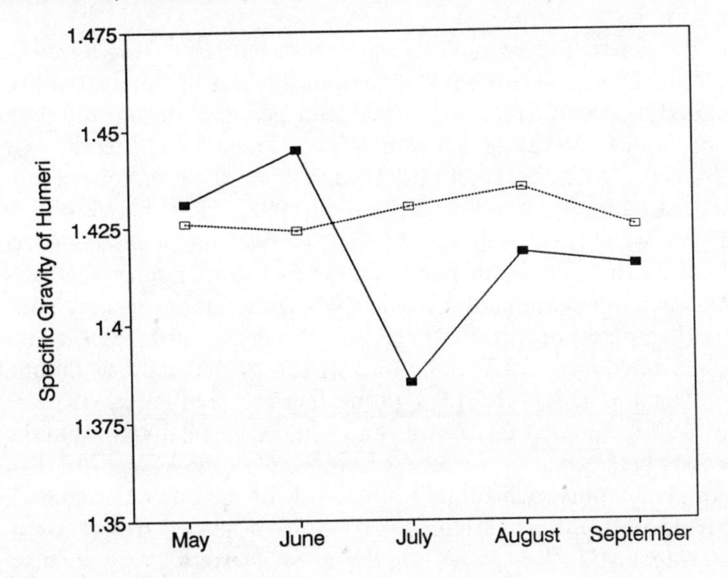

Fig. 3. Change in mean specific gravity of the humeri of adult male (open squares) and adult female (closed squares) little brown bats, *Myotis lucifugus*. Peak lactation occurs in July. Data from Kwiecinski *et al.* (1987).

some species start transferring the energetic costs of their young to the young themselves at 50–60% of adult mass, forearm length at first flight varies much less (CV = 6.4%). Females thus bear almost the entire nutrient cost of producing the skeleton of their offspring. Calcium demand is particularly stressful and causes significant structural changes in the bones of lactating bats (Kwiecinski, Krook & Wimsatt 1987). In the small (7–8 g) insectivorous little brown bat, *Myotis lucifugus*, for example, the specific gravity of the humeri of lactating females declines significantly, while no such decline is seen in males (Fig. 3; Kwiecinski *et al.* 1987). Females deplete their own calcium stores to meet the demand of their growing offspring. This osteoporosis is particularly evident in the mandible and the long bones of the wings and could reduce a female's fitness, owing to tooth loss or increased risk of wing-bone fractures (G. Kwiecinski pers. comm.). *M. lucifugus* bears only one young per year. The problems inherent in low calcium availability will be even greater for females of species that produce larger litters.

Calcium availability

The problem that *M. lucifugus* and other insectivorous bats face is that insects have a low calcium content (Maxson & Oring 1980; Turner 1982 and pers. comm.; Ormerod, Bull, Cummins, Tyler & Vickery 1988; Studier & Sevick 1992). A reproductive female relying on insects for calcium cannot obtain as much as she needs. For example, a big brown bat (*Eptesicus fuscus*), which is a small (15-20 g) insectivorous species with litters of one or two, requires an estimated 11.56 to 23.12 mg of calcium per day during mid-pregnancy (Keeler & Studier 1992). If calcium is assimilated as efficiently as energy (75%; Barclay, Dolan & Dyck 1991), and a random sample of insects is consumed (1.62 mg/g; Studier & Sevick 1992), then between 9.48 and 18.96 g, dry weight, or 31.6 and 63.2 g, live weight, of insects is required per night to satisfy calcium demand. This is two to four times the bat's body mass! At an energy assimilation efficiency of 75% and an energy content of 7.25 kJ/g (live mass) of insect (Kurta, Bell, Nagy & Kunz 1989), the female would obtain between 171.7 and 343.2 kJ energy per night while obtaining sufficient calcium. This is 3.5 to 7.0 times her required energy (Kurta, Kunz & Nagy 1990). Compared to calcium, energy is relatively abundant for insectivorous bats.

Calcium availability is also low in the foods of the other two main groups of bats, frugivores and nectarivores. Vertebrate-dispersed fruits have a mean calcium content of 2.91 ± 2.74 mg/g dry mass (Herrera 1987; see also Duke & Atchley 1986). Pollen is ingested by nectar-eating bats as a protein source, and has a mean calcium content of 3.15 ± 2.69 mg/g dry mass (Stanley & Linskens 1974).

Nor can calcium be obtained in sufficient quantity by drinking fresh or salt water. Even fresh-water lakes classified as 'hard' contain only 25 mg Ca/l, or somewhat higher, and the surface water typically contains even less calcium (Reid & Wood 1976). At 25 mg Ca/l (= 0.025 mg/g), water contains 20 times less calcium than the average live insect does and a pregnant female *E. fuscus* would need to drink a minimum of 0.46 l/day to meet her requirements! Although seawater is richer in calcium, it still only averages 0.41 mg/g, or less than the average content of insects (Moran, Morgan & Wiersma 1986).

It could be argued that if bats have lighter skeletons than do terrestrial mammals, as an adaptation for flight, calcium demand on a reproductive female would be less than expected. I obtained mass data from museum-prepared skeletons of bats from 13 species (six families), ranging in adult body mass from 8 to 122 g (Barclay 1994). I compared these data with those from similarly sized terrestrial mammals (Prange, Anderson & Rahn 1979) which I supplemented with measurements from other rodents and insectivores housed in the University of Calgary Museum of Zoology (Barclay 1994). In an ANCOVA, with taxon as the main effect and the log of adult body mass as the covariate, only log of adult mass ($F = 334.8$, $d.f. = 1, 23$, $P < 0.001$) significantly influenced skeletal mass (Fig. 4). Skeletal mass was not different between bats and terrestrial mammals ($F = 0.93$, $d.f. = 1, 23$, $P > 0.3$) and skeletal mass changed relative to log of adult mass in a similar manner in the two groups ($F = 0.07$, $d.f. = 1, 23$, $P > 0.7$). Thus, assuming that bats and other mammals have similar calcium content in their bones, the calcium demand on female bats should not be lower than expected. A recent study by E. H. Studier and T. H. Kunz (pers. comm.) on mineral accretion in juvenile Mexican free-tailed bats (*Tadarida brasiliensis*) and cave bats (*Myotis velifer*), suggests that calcium requirements for growth are somewhat less than those for rodents. However, they conclude that, even in these species which produce only one young, calcium is the nutritional factor limiting growth and females are in calcium debt during lactation.

Although the difference in offspring size at weaning, and the associated calcium costs, may explain the difference in litter size between bats and small terrestrial mammals, it cannot explain the difference in reproductive output between bats and birds, since birds also fledge at adult size (Ricklefs 1979). Differences in calcium availability are the likely explanation. Most birds are relatively adept at terrestrial locomotion and are primarily visual predators. This allows them to locate and consume inanimate, calcium-rich items such as calcareous grit, eggshells, snail shells, bone fragments and ash, as well as calcium-rich invertebrates such as amphipods (Maclean 1974; Beasom & Pattee 1978; Mayoh & Zach 1986; Ficken 1989; Repasky, Blue & Doerr 1991; St. Louis & Breebaart 1991). Birds (and terrestrial mammals) can thus supplement their calcium intake.

Most bats are not adept at terrestrial locomotion. In addition, micro-chiropteran bats rely primarily on echolocation to detect prey. These features preclude bats from locating and/or obtaining the widely dispersed, inanimate

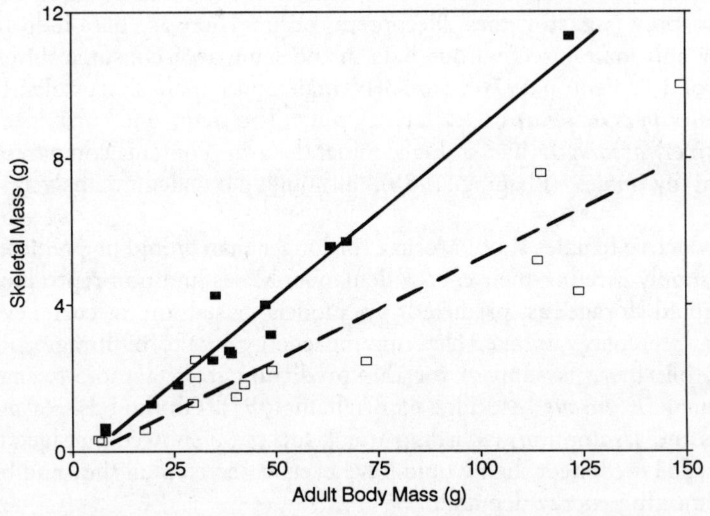

Fig. 4. Mass of adult skeletons in relation to adult body mass for terrestrial mammals (open squares) and bats (closed squares). Lines were calculated using least squares (terrestrial mammals: skeleton = −0.29 + 0.06 × adult body mass; bats: skeleton = −0.46 + 0.09 × adult body mass).

calcium-rich items consumed by birds. Even gleaning insectivorous bats rely on prey movement and/or sounds to detect insects on surfaces (Faure & Barclay 1992); they cannot distinguish inanimate objects from the substrate. Bats must thus rely on prey to provide calcium for reproduction and the low concentration in prey constrains reproductive output.

Predictions

A number of testable predictions arise from the calcium-constraint hypothesis and positive results of such tests would lend support to it.

1. Female bats in species or populations to which more calcium is available should have larger litters, and/or more litters per year, and/or faster juvenile growth rates. Increased calcium intake could arise from consuming vertebrate prey, or by ingesting inanimate sources of calcium, as might be possible for bats roosting in limestone caves.

2. Reproductive females should base their foraging decisions on optimizing calcium intake while males and non-reproductive females would be expected to forage in a manner consistent with maximizing net energy intake. Since the calcium content of insects (Studier & Sevick 1992), fruit (Herrera 1987), and pollen (Stanley & Linskens 1974) varies, reproductive females could select prey on the basis of its calcium content rather than its energy content. Recent field experiments indicate that echolocating insectivorous bats do not make such fine-detailed discriminations between prey (Barclay & Brigham 1994). An alternative is that females select foraging areas where calcium-rich prey (e.g. stoneflies, Plecoptera) occur. There are data indicating that female and male insectivorous bats in the same area consume different prey (Belwood & Fenton 1976). Similarly, male and female short-tailed fruit bats (*Carollia perspicillata*) eat different fruits. The main one consumed by females, *Piper amalago*, has a high mineral (ash) content compared to those eaten by males (Fleming 1988), although no calcium analysis has been done.

3. Reproductive females should forage for longer than would be predicted if they were simply meeting their energy demands. Males and non-reproductive females should forage as predicted by models based on a currency of maximizing net energy intake. Determining energy intake of foraging bats is difficult, but there is support for this prediction from a radio-telemetry study of female *E. fuscus* (Aldridge & Brigham 1991). On the basis of mean attack rates and assumptions regarding attack success, it showed that lactating females foraged for longer than would have been expected if all they had been doing was meeting energy demand.

4. Females should have higher rates of tooth loss and bone fractures than males have. It is possible that the higher mortality rate amongst females of several species (e.g. Keen & Hitchcock 1980) reflects such differences between

the sexes, but less severe differences (finger bone fractures, tooth loss) might be evident in populations of captured bats.

5. Plants that compete for bats as pollinators or seed-dispersers should have higher calcium content in their pollen and/or nectar, and fruit, than invertebrate-pollinated plants do. Although protein content is often proposed as an important reward for pollinators and seed dispersers (Thomas 1984; Fleming 1988), for bats calcium may be even more important.

6. Bats that hibernate should have greater calcium constraints on reproduction than do bats that migrate. This prediction stems from the fact that hibernating bats resorb calcium from their bones and emerge from hibernation with depleted reserves (Kwiecinski *et al.* 1987). It is thus crucial that reproductive females replace the calcium lost during lactation prior to hibernation. Migrating bats, which are active all year, do not face this problem and may be able to cope with more severe or more prolonged calcium depletion, given that they can replenish their supply over the winter in preparation for the next litter. Among the Vespertilionidae, the only family of bats with a range of litter sizes, litters larger than one are more common amongst migrating species and those that are active all year round in tropical areas, than among hibernators (Koehler 1991).

There thus appears to be at least some support for the calcium-constraint hypothesis. Further field and experimental studies will be necessary in order to substantiate it more fully or refute it. If it is correct, however, it has broad implications for the biology of bats. Not the least of these is that if flight requires large size at fledging, the evolution of flight in mammals may only have been possible in conjunction with long life spans. The more typical life-history pattern of small mammals, involving short lives and large litters, is not one open to bats.

Acknowledgements

H. C. Smith (Alberta Provincial Museum) and M. D. Engstrom and W. E. Hlywka (Royal Ontario Museum) kindly provided skeleton data for specimens in their collections. A. Hickey allowed me access to specimens in the University of Calgary Museum of Zoology. S. L. Holroyd, C. E. Koehler, E. H. Studier, J. O. Keeler, A. K. Turner, S. Steele, A. Allgaier and T. H. Kunz provided unpublished data and pre-prints which helped in piecing together information. Discussions with M. B. Fenton, G. G. Kwiecinski, R. M. Brigham, D. W. Thomas, T. H. Fleming, B. R. McMahon, C. E. Koehler, S. L. Holroyd, L. S. Johnson, T. H. Kunz and E. H. Studier clarified my thoughts. The above and the University of Calgary's Ecology Division BELCH group made comments that improved the manuscript. My own research has been funded by the Natural Sciences and Engineering Research Council of Canada and the University of Calgary.

References

Aldridge, H. D. J. N. & Brigham, R. M. (1991). Factors influencing foraging time in two aerial insectivores: the bird *Chordeiles minor* and the bat *Eptesicus fuscus*. *Can. J. Zool* **69**: 62–69.

Barclay, R. M. R. (1994). Constraints on reproduction by flying vertebrates: energy and calcium. *Am. Nat.* **144**: 1021–1031.

Barclay, R. M. R. & Brigham, R. M. (1991). Prey detection, dietary niche breadth, and body size in bats: why are aerial insectivorous bats so small? *Am. Nat.* **137**: 693–703.

Barclay, R. M. R. & Brigham, R. M. (1994). Constraints on optimal foraging: a field test of prey discrimination by echolocating insectivorous bats. *Anim. Behav.* **48**: 1013–1021.

Barclay, R. M. R., Dolan, M.-A. & Dyck, A. (1991). The digestive efficiency of insectivorous bats. *Can. J. Zool.* **69**: 1853–1856.

Beasom, S. L. & Pattee, O. H. (1978). Utilization of snails by Rio Grande turkey hens. *J. Wildl. Mgmt* **42**: 916–919.

Belwood, J. J. & Fenton, M. B. (1976). Variation in the diet of *Myotis lucifugus* (Chiroptera: Vespertilionidae). *Can. J. Zool.* **54**: 1674–1678.

Bergmans, W. (1979). Taxonomy and zoogeography of the fruit bats of the People's Republic of Congo, with notes on their reproductive biology (Mammalia, Megachiroptera). *Bijdr. Dierk.* **48**: 161–186.

Blueweiss, L., Fox, H., Kudzma, V., Nakashima, D., Peters, R. & Sams, S. (1978). Relationships between body size and some life history parameters. *Oecologia* **37**: 257–272.

Bradbury, J. W. (1977). Lek mating behavior in the hammer-headed bat. *Z. Tierpsychol.* **45**: 225–255.

Brown, P. E., Brown, T. W. & Grinnell, A. D. (1983). Echolocation, development, and vocal communication in the lesser bulldog bat, *Noctilio albiventris*. *Behav. Ecol. Sociobiol.* **13**: 287–298.

Burnett, C. D. & Kunz, T. H. (1982). Growth rates and age estimation in *Eptesicus fuscus* and comparison with *Myotis lucifugus*. *J. Mammal.* **63**: 33–41.

Clutton-Brock, T. H. (1991). *The evolution of parental care*. Princeton University Press, Princeton.

Davis, R. (1969). Growth and development of young pallid bats, *Antrozous pallidus*. *J. Mammal.* **50**: 729–736.

de Paz, O. (1986). Age estimation and postnatal growth of the Greater Mouse bat *Myotis myotis* (Borkhausen, 1797) in Guadalajara, Spain. *Mammalia* **50**: 243–251.

Duke, J. A. & Atchley, A. A. (1986). *CRC handbook of proximate analysis tables of higher plants*. CRC Press, Boca Raton.

Faure, P. A. & Barclay, R. M. R. (1992). The sensory basis of prey detection by the long-eared bat, *Myotis evotis*, and the consequences for prey selection. *Anim. Behav.* **44**: 31–39.

Ficken, M. S. (1989). Boreal chickadees eat ash high in calcium. *Wilson Bull.* **101**: 349–351.

Fleming, T. H. (1988). *The short-tailed fruit bat. A study in plant–animal interactions*. University of Chicago Press, Chicago.

Funakoshi, K. & Uchida, T. A. (1981). Feeding activity during the breeding season and postnatal growth in the Namie's frosted bat, *Vespertilio superans superans*. *Jap. J. Ecol.* **31**: 67–77.

Harvey, P. H. & Read, A. F. (1988). How and why do mammalian life histories vary? In *Evolution of life histories of mammals: theory and pattern*: 213–232. (Ed. Boyce, M. S.). Yale University Press, New Haven.

Herrera, C. M. (1987). Vertebrate-dispersed plants of the Iberian peninsula: a study of fruit characteristics. *Ecol. Monogr.* **57**: 305–331.

Holroyd, S. L. (1993). *Influences of some extrinsic and intrinsic factors on reproduction by big brown bats* (Eptesicus fuscus) *in southeastern Alberta.* MSc thesis: University of Calgary.

Hoying, K. M. (1983). *Growth and development of the eastern pipistrelle bat,* Pipistrellus subflavus. MA thesis: Boston University.

Hughes, P. M., Ransome, R. D. & Jones, G. (1989). Aerodynamic constraints on flight ontogeny in free-living greater horseshoe bats, *Rhinolophus ferrumequinum.* In *European bat research 1987*: 255–262. (Eds Hanak, V., Horacek, I. & Gaisler, J.). Charles University Press, Praha.

Jones, C. (1967). Growth, development, and wing loading in the evening bat, *Nycticeius humeralis* (Rafinesque). *J. Mammal.* **48**: 1–19.

Keeler, J. O. & Studier, E. H. (1992). Nutrition in pregnant big brown bats (*Eptesicus fuscus*) feeding on June beetles. *J. Mammal.* **73**: 426–430.

Keen, R. & Hitchcock, H. B. (1980). Survival and longevity of the little brown bat (*Myotis lucifugus*) in southeastern Ontario. *J. Mammal.* **61**: 1–7.

Kleiman, D. G. (1969). Maternal care, growth rate and development in the Noctule (*Nyctalus noctula*), Pipistrelle (*Pipistrellus pipistrellus*), and Serotine (*Eptesicus serotinus*) bats. *J. Zool., Lond.* **157**: 187–211.

Kleiman, D. G. & Davis, T. M. (1979). Ontogeny and maternal care. In *Biology of bats of the New World family Phyllostomatidae. Part III*: 387–402. (Eds Baker, R. J., Jones, J. K. Jr. & Carter, D. C.). Texas Tech Press, Lubbock. (*Spec. Publs Mus. Texas tech. Univ.* No.16.)

Koehler, C. E. (1991). *The reproductive ecology of the hoary bat* (Lasiurus cinereus) *and its relation to litter size variation in vespertilionid bats.* MSc thesis: University of Calgary.

Kunz, T. H. (1973). Population studies of the cave bat (*Myotis velifer*): reproduction, growth, and development. *Occ. Pap. Mus. nat. Hist. Univ. Kans.* No. **15**: 1–43.

Kunz, T. H. & Anthony, E. L. P. (1982). Age estimation and post-natal growth in the bat *Myotis lucifugus. J. Mammal.* **63**: 23–32.

Kurta, A., Bell, G. P., Nagy, K. A. & Kunz, T. H. (1989). Energetics of pregnancy and lactation in free-ranging little brown bats (*Myotis lucifugus*). *Physiol. Zool.* **62**: 804–818.

Kurta, A. & Kunz, T. H. (1987). Size of bats at birth and maternal investment during pregnancy. *Symp. zool. Soc. Lond.* No. 57: 79–106.

Kurta, A., Kunz, T. H. & Nagy, K. A. (1990). Energetics and water flux of free-ranging big brown bats (*Eptesicus fuscus*) during pregnancy and lactation. *J. Mammal.* **71**: 59–65.

Kwiecinski, G. G., Krook, L. & Wimsatt, W. A. (1987). Annual skeletal changes in the little brown bat, *Myotis lucifugus lucifugus*, with particular reference to pregnancy and lactation. *Am. J. Anat.* **178**: 410–420.

Maclean, S. F. (1974). Lemming bones as a source of calcium for arctic sandpipers (*Calidris* spp.). *Ibis* **116**: 552–557.

Maeda, K. (1972). Growth and development of large noctule, *Nyctalus lasiopterus* Schreber. *Mammalia* **36**: 269–278.

Maxson, S. J. & Oring, L. W. (1980). Breeding season time and energy budgets of the polyandrous spotted sandpiper. *Behaviour* **74**: 200–263.

Mayoh, K. R. & Zach, R. (1986). Grit ingestion by nestling tree swallows and house wrens. *Can. J. Zool.* **64**: 2090–2093.

Medway (Lord) (1972). Reproductive cycles of the flat-headed bats *Tylonycteris pachypus* and *T. robustula* (Chiroptera: Vespertilioninae) in a humid equatorial environment. *Zool. J. Linn. Soc.* **51**: 33–61.

Millar, J. S. (1977). Adaptive features of mammalian reproduction. *Evolution, Lawrence, Kans.* **31**: 370–386.

Millar, J. S. (1981). Pre-partum reproductive characteristics of eutherian mammals. *Evolution, Lawrence, Kans.* **35**: 1149–1163.

Moran, J. M., Morgan, M. D. & Wiersma, J. H. (1986). *Introduction to environmental science.* W. H. Freeman and Co., New York.

Norberg, U. M. & Rayner, J. M. V. (1987). Ecological morphology and flight in bats (Mammalia: Chiroptera): wing adaptations, flight performance, foraging strategy and echolocation. *Phil. Trans. R. Soc. (B)* **316**: 335–427.

O'Farrell, M. J. & Studier, E. H. (1973). Reproduction, growth, and development in *Myotis thysanodes* and *Myotis lucifugus* (Chiroptera: Vespertilionidae). *Ecology* **54**: 18–30.

Okia, N. O. (1974). Breeding in Franquet's bat, *Epomops franqueti* (Tomes), in Uganda. *J. Mammal.* **55**: 462–465.

Ormerod, S. J., Bull, K. R., Cummins, C. P., Tyler, S. J. & Vickery, J. A. (1988). Egg mass and shell thickness in dippers *Cinclus cinclus* in relation to stream acidity in Wales and Scotland. *Envir. Poll.* **55**: 107–121.

Pagels, J. F. & Jones, C. (1974). Growth and development of the free-tailed bat, *Tadarida brasiliensis cynocephala* (LeConte). *SWest.Nat.* **19**: 267–276.

Prange, H. D., Anderson, J. F. & Rahn, H. (1979). Scaling of skeletal mass to body mass in birds and mammals. *Am. Nat.* **113**: 103–122.

Promislow, D. E. L. & Harvey, P. H. (1990). Living fast and dying young: a comparative analysis of life-history variation among mammals. *J. Zool., Lond.* **220**: 417–437.

Rakhmatulina, I. K. (1972). The breeding, growth, and development of pipistrelles in Azerbaidzhan. *Soviet J. Ecol.* **2**: 131–136.

Read, A. F. & Harvey, P. H. (1989). Life history differences among the eutherian radiations. *J. Zool., Lond.* **219**: 329–353.

Reid, G. K. & Wood, R. D. (1976). *Ecology of inland waters and estuaries.* (2nd edn). D. Van Nostrand Co., New York.

Repasky, R. R., Blue, R. J. & Doerr, P. D. (1991). Laying red-cockaded woodpeckers cache bone fragments. *Condor* **93**: 458–461.

Ricklefs, R. E. (1979). Adaptation, constraint, and compromise in avian postnatal development. *Biol. Rev.* **54**: 269–290.

Sacher, G. A. & Staffeldt, E. F. (1974). Relation of gestation time to brain weight for placental mammals: implications for the theory of vertebrate growth. *Am. Nat.* **108**: 593–615.

St. Louis, V. & Breebaart, L. (1991). Calcium supplements in the diet of nestling tree swallows near acid sensitive lakes. *Condor* **93**: 286–294.

Stanley, R. G. & Linskens, H. F. (1974). *Pollen biology, biochemistry, management.* Springer-Verlag, Berlin.

Stephens, D. W. & Krebs, J. R. (1986). *Foraging theory.* Princeton University Press, Princeton.

Studier, E. H. & Sevick, S. H. (1992). Live mass, water content, nitrogen and mineral levels in some insects from south-central lower Michigan. *Comp. Biochem. Physiol. (A)* **103**: 579–595.

Swartz, S. M., Bennett, M. B. & Carrier, D. R. (1992). Wing bone stresses in free flying bats and the evolution of skeletal design for flight. *Nature, Lond.* **359**: 726–729.
Thomas, D. W. (1984). Fruit intake and energy budgets of frugivorous bats. *Physiol. Zool.* **57**: 457–467.
Thomas, D. W. & Marshall, A. G. (1984). Reproduction and growth in three species of West African fruit bats. *J. Zool., Lond.* **202**: 265–281.
Turner, A. K. (1982). Timing of laying by swallows (*Hirundo rustica*) and sand martins (*Riparia riparia*). *J. Anim. Ecol.* **51**: 29–46.
Tuttle, M. D. (1976). Population ecology of the gray bat (*Myotis grisescens*): factors influencing growth and survival of newly volant young. *Ecology* **57**: 587–595.
Tuttle, M. D. & Stevenson, D. (1982). Growth and survival of bats. In *Ecology of bats*: 105–150. (Ed. Kunz, T. H.). Plenum Press, New York.
Western, D. (1979). Size, life history and ecology in mammals. *Afr. J. Ecol.* **17**: 185–204.
Western, D. & Ssemakula, J. (1982). Life history patterns in birds and mammals and their evolutionary interpretation. *Oecologia* **54**: 281–290.
Yokoyama, K., Ohtsu, R. & Uchida, T. A. (1979). Growth and LDH isozyme patterns in the pectoral and cardiac muscles of the Japanese Lesser horseshoe bat, *Rhinolophus cornutus cornutus* from the standpoint of adaptation for flight. *J. Zool., Lond.* **187**: 85–96.

Appendix

Mean mass and forearm measurements of adult and juvenile (at first flight) bats from field (*) and laboratory studies.

Species	Mass (g)			Forearm (mm)			Reference[e]
	Adult	Juvenile	%	Adult	Juvenile	%	
Vespertilionidae							
*Myotis lucifugus**	7.7	6.5	84.4	37.4	37.1	99.2	1
*M. grisescens**				43.0	42.1	97.9	2
*M. thysanodes**				43.8	41.2	94.1	3
*M. velifer**	11.6	10.6	91.4	46.1	41.3	89.6	4
*M. myotis**	24.9	23.1	92.8	63.7	61.2	96.1	5
Eptesicus fuscus[a]*	16.1	12.0	74.5	45.8	45.2	98.7	6
E. fuscus[b]*	19.2	13.2	68.8	47.6	44.1	92.6	7
E. serotinus	27.0	15.0	55.6	51.0	47.0	92.2	8
*Lasiurus cinereus**	30.0	18.3	61.0	55.1	53.6	97.3	9
*Lasionycteris noctivagans**	11.5	10.1	87.8	42.2	41.8	99.1	9
*Antrozous pallidus**	17.4	13.0	74.7	53.9	50.0	92.8	10
Nyctalus noctula	26.5	19.0	71.7	51.5	48.0	93.2	8
N. lasiopterus				62.0	59.0	95.2	11
Vespertilio superans	18.9	11.0	58.2	48.9	42.5	86.9	12
*Pipistrellus subflavus**	6.5	5.2	80.0	35.1	33.2	94.6	13
*P. pipistrellus**	5.1	3.5	68.6	30.5	29.1	95.4	14
Nycticeius humeralis	8.0	4.8	60.0	34.0	29.5	86.8	15,16
*Tylonycteris robustus**	7.4	6.7	90.5				17
*T. pachypus**	3.7	2.7	73.0				17
Rhinolophidae							
*Rhinolophus cornutus**	8.0	7.1	88.8				18
R. ferrumequinum[c]*						98.0	19

Appendix (continued)

Species	Mass (g)			Forearm (mm)			Reference[e]
	Adult	Juvenile	%	Adult	Juvenile	%	
Molossidae							
*Tadarida brasiliensis**	14.5	12.0	82.8	46.0	38.0	82.6	20
Phyllostomidae							
Carollia perspicillata[d]	17.0	15.0	88.2			93.4	21
*Phyllostomus hastatus**	74.1	59.1	79.8	81.9	77.0	94.0	27
Noctilionidae							
*Noctilio albiventris**	37.8	20.0	52.9	58.0	54.0	93.1	22
Pteropodidae							
*Micropteropus pusillus**	31.0	20.0	64.5	55.0	48.0	87.3	23
*Epomops franqueti**	109.0	70.0	64.2	89.3	76.4	85.6	24,25
*E. buettikoferi** (female)	120.0	55.0	45.8	92.0	77.0	83.7	23
(male)	190.0	55.0	28.9	99.0	76.0	76.8	23
Epomophorus							
*wahlbergi** (female)				80.4	66.4	82.6	25
(male)				81.9	68.2	83.3	25
Hypsignathus							
*monstrosus** (female)	234.0	172.0	73.5	120.0	105.0	87.5	26
(male)	420.0	225.0	53.6	134.0	117.5	87.7	26
	n=27 $\bar{x}$= 70.9 ± 15.7%			n = 30 $\bar{x}$ = 91.2 ± 5.88%			

[a]From an eastern North America population in which twins are produced.
[b]From a western North America population in which single young are produced. [c] Based on wing span. [d] Absolute values not given.
[e]References: 1. Kunz & Anthony (1982); 2. Tuttle (1976); 3. O'Farrell & Studier (1973); 4. Kunz (1973); 5. de Paz (1986); 6. Burnett & Kunz (1982); 7. Holroyd (1993); 8. Kleiman (1969); 9. Koehler (1991); 10. Davis (1969); 11. Maeda (1972); 12. Funakoshi & Uchida (1981); 13. Hoying (1983); 14. Rakhmatulina (1972); 15. S. Steele (pers. comm.); 16. Jones (1967); 17. Medway (1972); 18. Yokoyama, Ohtsu & Uchida (1979); 19. Hughes, Ransome & Jones (1989); 20. Pagels & Jones (1974); 21. Kleiman & Davis (1979); 22. Brown *et al.* (1983); 23. Thomas & Marshall (1984); 24. Okia (1974); 25. Bergmans (1979); 26. Bradbury (1977); 27. A. Allgaier (pers. comm.).

Symp. zool. Soc. Lond. (1995) No. 67: 259–273

Echolocation signal design, foraging habitats and guild structure in six Neotropical sheath-tailed bats (Emballonuridae)

Elisabeth K. V. KALKO

University of Tübingen
Animal Physiology
Auf der Morgenstelle 28
D–72076 Tübingen, Germany[1];
Smithsonian Tropical Research Institute
PO Box 2072, Balboa
Republic of Panama

Synopsis

The three major families of aerial insectivorous bats that occur in both Old and New World tropics (Emballonuridae, Molossidae, Vespertilionidae) show great variety in echolocation signal design. To understand better the function of this variation, I used an ultrasound detector and night-vision goggles to study echolocation and foraging behaviour of six species of emballonurids that occur sympatrically at each of five field sites spanning Costa Rica, Panama and Venezuela. At all five localities I found a consistent association of specific signal types with three distinct habitat types in which these species foraged, suggesting that the guild of insectivorous bats is highly structured. First, the ghost bat (*Diclidurus albus*) and a bat tentatively identified as *Peropteryx* sp. both hunted in open, uncluttered space and emitted low-frequency, shallow-modulated echolocation signals. Second, two white-lined bats (*Saccopteryx bilineata, S. leptura*) and Wagner's sac-winged bat (*Cormura brevirostris*) foraged in cluttered habitats close to vegetation and used medium-frequency echolocation signals with a shallow-modulated middle part starting and ending with a steep frequency-modulated component. Third, the proboscis bat (*Rhynchonycteris naso*) restricted its foraging to within a metre or two of the surface of small streams. Its high-frequency echolocation signals were similar in structure to those of noctilionid bats, which use similar habitats.

Signal design of echolocation calls in emballonurids gives more information on foraging strategies and habitat use than do morphology and phylogenetic relationships alone. Association of call structure with habitat, and phylogenetic flexibility of signal

[1]Address for offprint requests

ZOOLOGICAL SYMPOSIUM No. 67
ISBN 0 19 854945 8

type, appear to extend to vespertilionids and molossids as well. This contrasts sharply with leaf-nosed bats (Phyllostomidae) which form the largest group of bats in the Neotropics yet show rather uniform signal design. Unlike the aerial insectivores that depend almost entirely on echolocation for foraging, phyllostomids supplement sonar with other sensory modalities (e.g., passive hearing, vision and olfaction).

Introduction

Bats are ecologically the most diverse mammals of tropical ecosystems. This is particularly true in the New World tropics, where some localities may be inhabited by more than 100 species of microchiropterans which forage on a variety of food, ranging from insects, small vertebrates and blood to fruits, nectar and pollen. These bats occur in large numbers and their foraging affects a wide range of other organisms (e.g. through pollination, seed dispersal and control of insect populations). Therefore, characterizing the patterns of community structure and understanding the factors underlying bat diversity pose important practical problems for conservation (e.g. habitat requirements), as well as for theoretical biology.

At present, studies of bat community structure have not been able to refute the idea that bat communities represent random assemblages of species (Fleming 1986; Willig & Moulton 1989; Willig, Camilo & Noble 1993). The lack of support for the idea of structured communities is largely based on comparisons of species lists across different geographical areas, comparisons of ecomorphological characteristics of sympatric bat species with null distributions of these characters and comparisons of diet overlap in sympatric species (e.g. Fleming 1986; Findley 1993; Willig *et al.* 1993). However, as the authors of these studies state, the basic life cycles and ecology of most bat species are very poorly known and the techniques that have been employed thus far suffer from severe limitations. To take one example, aerial insectivorous bats are notoriously difficult to catch in mist nets and are therefore commonly under-represented on species lists. I propose that the available data base is still insufficient properly to address the question of whether bat communities represent random assemblages or not. We do not possess enough information to rule out the possibility that guilds, and probably also communities, are more structured than is commonly believed (e.g. Crome & Richards 1988). Here I show how studies of echolocation signal design can be used as a tool to gather further information with which to address these important questions.

I will describe studies in which I used an ultrasound detector to characterize the echolocation signals of six species of sheath-tailed bats (Emballonuridae) at several locations in Latin America. In all of these sites I also used night-vision goggles to observe foraging behaviour. I was then able to relate signal design to the foraging behaviour of these bats. The few echolocation signals previously described for the emballonurids indicate clear differences in signal design between species (e.g. Griffin & Novick 1955; Novick 1962; Bradbury &

Vehrencamp 1976; Fenton, Bell & Thomas 1980; Pye 1980; Barclay 1983; Heller 1989). It is generally accepted that design of echolocation signals determines to a large degree where a bat is feeding and what it is feeding on (e.g. Neuweiler 1989; Fenton 1990). I will use the results of those studies to argue that the design of those signals allows testable predictions to be made about habitat preferences and foraging strategies. Further, echolocation signals can be used to reveal the presence or absence of certain species that are difficult to sample otherwise. Moreover, my results suggest a previously unappreciated degree of structure to Neotropical insectivorous bat communities.

Methods and materials

Study sites

I studied Neotropical emballonurids during several field trips totalling 10 months from 1991 to 1993 at five field sites: Tortuguero, Costa Rica; the field station, Barro Colorado Island (BCI), Panamá; Nuri, province of Bocas del Toro, Panama; Guatopo National Park, state of Miranda, Venezuela; and Imataca National Forest Reserve, province of Bolivar, Venezuela. Most of the recordings and observations were made in Panama.

Identification of species

I positively identified four species (*Saccopteryx bilineata, S. leptura, Cormura brevirostris* and *Rhynchonycteris naso*). Initially, echolocation signals were linked to species by recording during release from mist nets and during take-offs from day roosts. *Diclidurus albus* was identified by high, fast flight, by large size and white coloration seen in the beam of a headlight, and by the audible 'twittering' during foraging mentioned by most collectors of the species (e.g. Starrett & Casebeer 1968). *D. albus* is the only high-flying, white aerial insectivorous bat known to occur in Panama, where my best observations were made.

I tentatively identify the sixth species of my study as *Peropteryx* sp. I observed this medium-sized, rather fast-flying bat at each of my study sites except Guatopo National Park in Venezuela. The pattern of its echolocation signals (see results) suggests an emballonurid. The shape of its wings (C. O. Handley pers. comm.) seems to adapt *Peropteryx* sp. to rather swift flight, which is in accordance with my observations. *Peropteryx* sp. is known to forage around trees in open areas (Starrett & Casebeer 1968; C.O. Handley pers. comm.). *Cyttarops alecto*, an emballonurid similar in size and wingshape to *Peropteryx*, is supposedly rare. Further, observations by Fiona A. Reid (pers. comm. to C.O. Handley) state that *Cyttarops alecto* usually starts foraging after dusk and emits high-pitched echolocation signals. Both observations contrast sharply with my observations of the unknown bat. The bat that I tentatively identify as *Peropteryx* sp. starts to forage at dusk and emits rather low-frequency signals. The only other emballonurid known to occur in my study area is *Centronycteris maximiliani*. It forages low, with slow flight, in clutter-rich situations (Starrett & Casebeer 1968; C. O. Handley pers. comm.).

My taxonomic identifications and signal descriptions are based on recordings which I made in Panama. At my other study sites I found signals similar to the ones I analysed and I observed similar behaviour in the bats. Further analysis of the signals will determine the possibility and extent of intraspecific and geographical variation.

Field observations of foraging bats

After detection and identification of foraging emballonurid bats, I observed them with 3D night-vision goggles (Typ WILD) and simultaneously recorded their echolocation signals. I defined foraging habitats by making transects along established trails through forests and adjoining open areas. At some sites I looked over the forest canopy from hill tops. On BCI (Panama) I used a 42-m canopy tower to record bats hunting at various heights in the forest and above the canopy. All of my field sites were in, or close to, primary or secondary forest with almost unbroken canopy, interspersed by treefall gaps and occasional small to large clearings.

Sound recording and analysis

I recorded echolocation signals of foraging bats with a custom-made bat detector. The signals were amplified, fed into a transient recorder and read out at 1/15 reduced speed on a Sony Walkman professional. I analysed sound sequences on a laptop (Amstrad) with a custom-made sound-analysing program (Sona-PC by B. Waldmann; University of Tübingen, Germany). I measured signal parameters with cursors on the screen. For analysis I used a Hanning Window 256 and displayed the signals in 35 ms steps with a resolution of 400 lines vertically and 500 lines horizontally and a frequency range of 112 kHz. This procedure gave a frequency resolution of 281 Hz, and a time resolution of 0.069 ms (interpolated). I restricted the dynamic range of the frequency-analyser (−72 dB) to −60 dB in order to eliminate background noise and set the measurement points 40 dB below maximum. To characterize the echolocation behaviour, I described signals used by the bats in the search phase as defined by Griffin, Webster & Michael (1960). I named signal components with frequency sweeps < 1 kHz/ms and bandwidths < 4 kHz 'quasi-constant-frequency' (QCF) and signal components with bandwidths < 400 Hz 'constant-frequency' (CF).

Results

Foraging behaviour

I defined two situations in which the six species of emballonurids foraged: open space away from obstacles (uncluttered) and space with obstacles in close proximity to the bats (cluttered). I used these definitions to classify three distinct habitat types: open space (uncluttered), space near or within vegetation (cluttered) and space close to water surfaces (cluttered).

First, I found two species, *Diclidurus albus* and *Peropteryx* sp., foraging exclusively in open, uncluttered space (Fig. 1). *Diclidurus albus* appeared irregularly at the observation sites. Characteristically, it flew very high (> 15–20 m) above the ground, often above the canopy, and very fast. It occasionally made long dives in pursuit of insects into forest clearings or large open spaces.

Typically, *Peropteryx* sp. appeared at dusk, foraging rather high (> 5–10 m), in open space, in circuits up to 20–30 m long. These circuits were rather straight and were only briefly varied by short dives or upward swoops during insect pursuits. *Peropteryx* sp. flew fast, usually parallel to forest edges and occasionally in larger forest gaps. It kept a minimum distance of 5–10 m from the vegetation. After an initial feeding interval of about 10–15 min the bats began foraging above the canopy and disappeared. During the night I encountered *Peropteryx* sp. only rarely. Frequently, I noticed a second peak in hunting activity of *Peropteryx* sp. just before dawn.

Second, I found three species, *S. bilineata, S. leptura* and *C. brevirostris*, in cluttered space close to vegetation (Fig. 1). I often encountered both *Saccopteryx* species at dusk, sometimes in late afternoon, foraging low—no

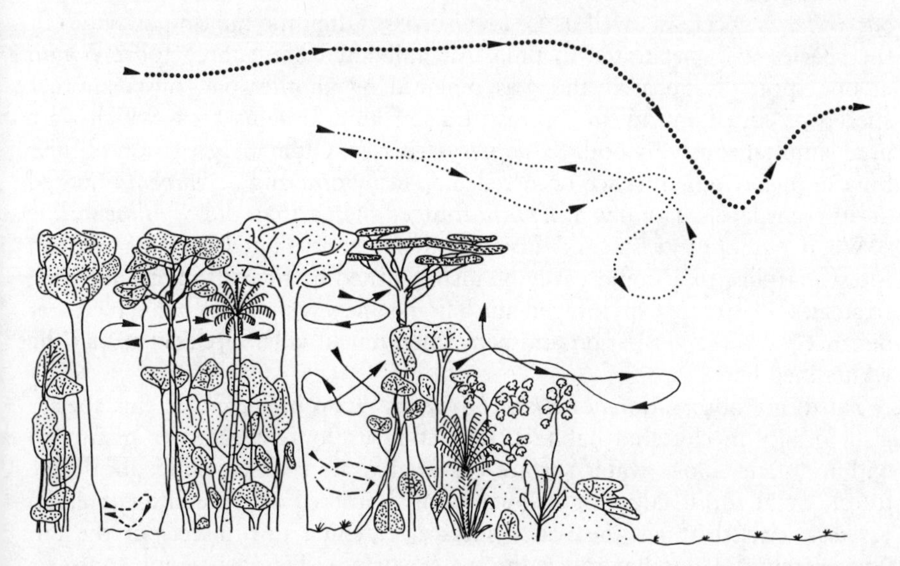

Fig. 1. Schematic drawing of characteristic foraging patterns of six Neotropical emballonurids.
••••••• *Diclidurus albus*
.... *Peropteryx* sp.
— · — · — *Saccopteryx bilineata* & *S. leptura*
———— *S. bilineata, S. leptura* & *Cormura brevirostris*
– – – – – – – *Rhynchonycteris naso.*

more than 0.5–3 m above ground level—inside the forest. Both species preferred small openings in the forest, especially trails and treefall gaps, and avoided dense vegetation. Typically they hunted in a stereotyped way, patrolling in slow and manoeuvrable flight, parallel to dense vegetation. They often flew in somewhat elliptical circles which were interrupted by sudden dives or ascents during insect pursuits. After the first hunting period, which usually ended with nightfall, both white-lined bats increased their foraging height. During the night I observed *S. bilineata* and *S. leptura* hunting in openings between subcanopy and canopy strata inside the forest, above streams bordered by forest, around forest edges and occasionally around trees and shrubs in the open (Fig. 1). Within the forest they mostly flew within 0.5–2 m of vegetation. Outside the forest they rarely foraged more than 5 m away from trees or shrubs. I did not find any *Saccopteryx* hunting above canopy level.

I also observed and recorded *C. brevirostris* foraging inside the forest, usually in openings between canopy and subcanopy (Fig. 1). In contrast to *S. bilineata* and *S. leptura, C. brevirostris* was not found hunting at low levels (< 3m), either early in the evening or later. Also, unlike *S. bilineata* and *S. leptura, C. brevirostris* seemed to be more restricted to the forest interior and small forest gaps. Only rarely did I find this species on forest edges bordering larger open areas.

I found several foraging sites within the forest which were used by both *Saccopteryx* species as well as *C. brevirostris*. Often the hunting activities of the species were separated in time. An individual bat hunted for 2–5 min at one spot, disappeared and was replaced by another bat, often another species of emballonurid. In contrast, I also found foraging areas which were used simultaneously by both *Saccopteryx* species. Often a segregation of their hunting behaviour in space occurred; i.e., *S. leptura* and *S. bilineata* hunted at different levels. Usually, *S. leptura* foraged higher than did *S. bilineata*.

When *Saccopteryx* sp. and *Peropteryx* sp. hunted in the same areas on forest margins, they could easily be distinguished by their disparate foraging strategies. *Peropteryx* sp. foraged much higher, flew faster, kept a much greater distance from the vegetation and performed much wider circles than did the white-lined bats.

Third, my observations of *Rhynchonycteris naso* indicate that this species also forages in cluttered habitats, restricting its foraging activity mainly to within 1–2 m above water surfaces of small watercourses, either inside the forest, or at forest edges or in open areas bordered by vegetation (Fig. 1). *R. naso* circled above the water surface, catching tiny insects in the air. Only rarely did this bat touch the water surface. This behaviour contrasts with the behaviour observed in other bats feeding close to the water surface. For example, the vespertilionid *Myotis daubentonii* (G. Jones & Rayner 1988; Kalko & Schnitzler 1989) regularly takes insects directly from the water surface.

The descriptions of foraging habitats and foraging strategies for *S. leptura, S. bilineata, R. naso, D. albus* and presumably *Peropteryx* sp., are largely in accordance with observations made by J. K. Jones (1966), Bradbury & Emmons (1974) and Bradbury & Vehrencamp (1976).

Echolocation behaviour

In all five localities I found a consistent association of distinct habitat types with specific types of search signals. First, the two species hunting in the open, *Diclidurus albus* and *Peropteryx* sp., emitted rather long (9–14 ms), low-frequency (24–32 kHz) echolocation signals which consisted of a quasi-constant-frequency (QCF) component (Fig. 2A–B; Table 1). The main energy was concentrated in the second harmonic. However, as a weak first harmonic was present in both species, the echolocation calls were often audible to humans. In both species, the QCF signals were upwardly modulated. In *Peropteryx* sp., the beginning and the end of the QCF signal were slightly more steeply modulated than the main QCF part. *Peropteryx* sp. emitted search signals in triplets, with each subsequent call being 2–3 kHz higher than the preceding one (Table 1). The data base for *D. albus* was not sufficient to decide whether the succeeding pulses differed in frequency.

Second, the three species (*Saccopteryx bilineata, S. leptura*, and *Cormura*

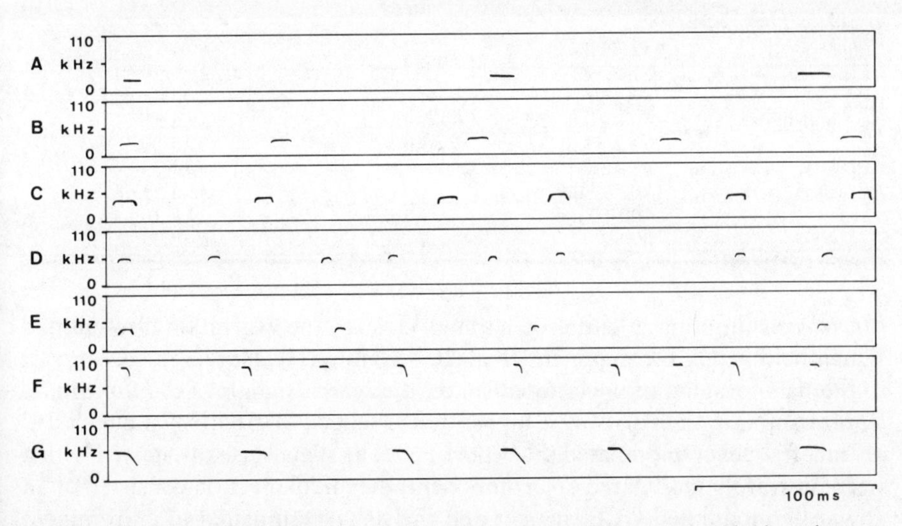

Fig. 2. Sonagrams (frequency versus time) of echolocation sequences in search phase from six Neotropical emballonurids and a noctilionid. (A) *Diclidurus albus* + (B) *Peropteryx* sp. forage in open space. Notice the dominance of the QCF component. (C) *Saccopteryx bilineata* + (D) *Saccopteryx leptura* + (E) *Cormura brevirostris* forage close to vegetation. Notice the distinct FM components in the signals of the three bats. (F) *Rhynchonycteris naso* + (G) *Noctilio leporinus* forage over water surfaces. Notice the similarity in distinctive CF and terminal FM component.

Table 1. Parameters of search phase echolocation calls in six Neotropical emballonurids given as mean ± standard deviation, minimum and maximum value, and sample size *n*.

	Best frequency of QCF or CF component (kHz)	Sound duration (ms)	Pulse interval (ms)
Diclidurus albus	24.3±1.2 22.5;26.7 *n*=20	11.3±2.4 7.2;16.2 *n*=17	183.3±96.1 69.9;365.7 *n*=17
Peropteryx sp.			
First signal	25.4±0.5 24.4;26.1 *n*=28	9.3±0.9 7.0;11.3 *n*=17	84±13.1 48.8;105.1 *n*=17
Second signal	28.7±0.6 27.3;29.8 *n*=30	10.1±1.1 7.7;11.9 *n*=21	82.5±14.5 53.3;120.8 *n*=20
Third signal	32.1±0.5 30.6;33.4 *n*=27	10.0±1.2 8.1;12.8 *n*=19	85.8±10.8 62.1;100.5 *n*=18
Cormura brevirostris	42±0.3 41.3;42.7 *n*=60	5.2±0.9 3.7;6.9 *n*=54	102.3±11.6 79.9;136.0 *n*=48
Saccopteryx bilineata			
First signal	45.1±0.5 43.8;45.8 *n*=60	9.4±1.3 7.0;12.4 *n*=41	85.8±20.1 68.2;114.8 *n*=41
Second signal	47.1±0.6 46.1;48.1 *n*=57	9.0±1.7 5.3;11.6 *n*=38	52.0±9.1 27.7;65.3 *n*=38
Saccopteryx leptura			
First signal	52.7±0.6 51.1;53.7 *n*=41	5.3±1.0 3.5;6.5 *n*=31	60.1±20.5 25.4;107.1 *n*=25
Second signal	56.0±0.8 54.2;57.6 *n*=42	5.4±0.8 3.9;8.3 *n*=29	49.6±17.1 31.9;90.4 *n*=27
Rhynchonycteris naso	102.5±0.9 100;103.6 *n*=70	5.0±0.8 3.2;6.7 *n*=90	53.5±19.8 27.5;107.4 *n*=82

brevirostris) hunting in cluttered habitats close to the vegetation also emitted signals with a QCF component (Fig. 2C–E; Table 1). However, in contrast to the bats foraging in uncluttered space, the search signals of *S. bilineata, S. leptura* and *C. brevirostris* were higher in frequency (42–56 kHz) and included distinct frequency-modulated (FM) portions. The signals usually started with a steep, upwardly modulated FM component, the middle section consisted of an upwardly modulated QCF element and the signal terminated in a prominent steep downwardly modulated FM component. All three species concentrated the main energy of their signals on the second harmonic. Occasionally, the third harmonic was visible. Often, the first harmonic was also present. Typically, both *Saccopteryx* emitted signals in doublets during search phase (Fig. 2C,D). The first signal was thus shorter and lower in frequency than the second one. Also, the pulse interval between the doublets was longer than among the constituent signals of a group. In comparison, *S. bilineata* emitted signals which were lower in frequency and had longer sound durations and pulse intervals than those of *S. leptura* (Table 1).

In contrast to both *Saccopteryx, C. brevirostris* did not emit the signals in groups during search phase (Fig. 2E). Also, the signals did not differ in frequency.

Third, *Rhynchonycteris naso*, which also foraged in cluttered habitats over water surfaces, emitted very high-frequency (100–105 kHz) signals (Fig. 2F; Table 1). In contrast to the other emballonurids of this study, the search signals of *R. naso* consisted of CF instead of QCF components. In the search phase the proboscis bat emitted CF signals with a prominent downward modulated terminal FM component (CF–FM) interspersed with pure CF signals. No initial upward modulated FM component was present in *R. naso*. The main signal energy of *R. naso* was concentrated in the fourth harmonic. The signal design of *R. naso* resembled very closely the echolocation behaviour of bulldog bats (Noctilionidae; Fig. 2G), which forage in similar habitats. However, it should be noted that similar habitats do not necessarily predict similar signal types, as some vespertilionids such as *Myotis daubentonii* and the leaf-nosed bat *Macrophyllum macrophyllum* (Phyllostomidae) also use this kind of habitat but employ different signal types (e.g. G. Jones & Rayner 1988; Kalko & Schnitzler 1989; E. K. V. Kalko unpubl. results).

Signal duration of emballonurids foraging in cluttered habitats (with the exception of *S. bilineata*) was shorter (4–6 ms) than signal duration of emballonurids such as *D. albus* and *Pteropteryx* sp. foraging in open space. The rather long signals of *Saccopteryx bilineata* (8–11 ms) given in the paper originate from recordings which were made from bats flying at a forest edge into less cluttered, open space, whereas the recordings from the other three emballonurids (*S. leptura*, *C. brevirostris* and *R. naso*) were made of bats foraging within cluttered spaces inside the forest or low over water surfaces.

The basic characteristics of echolocation behaviour that I observed in *S. bilineata, S. leptura, C. brevirostris* and *Rhynchonycteris naso* agree with the descriptions of Barclay (1983), Bradbury & Vehrencamp (1976) and Pye (1980). Differences in sound duration, pulse interval and ranges of call frequencies among the studies are presumably related to different recording situations and to intraspecific and geographic variability. The echolocation signals of the species that I tentatively identify as *Peropteryx* sp. are similar to the signals of an unknown emballonurid described by Barclay (1983) who also assumed that this species is likely to be *Peropteryx* sp.

Discussion

In all five localities, the major result of my study confirms the prediction that species-specific signal types of emballonurids are associated with characteristic habitat types, foraging strategies and, possibly, food types. This result also suggests that the guild of aerial insectivorous emballonurids is highly structured, with clear differences separating species that forage in open space, at forest edges and in gaps, and over water surfaces. Emballonurids foraging in uncluttered habitats (open space) use signals which are low in frequency and rather long and have no prominent FM-components, whereas emballonurids hunting in cluttered habitats (e.g. forest edge or gap, confined space, or over water) emit signals which are shorter and higher in frequency and have prominent FM-components.

It is reasonable to assume that echolocation signal design in bats has been largely shaped by the need to detect, localize and classify potential prey in different environments (e.g. Schnitzler & Henson 1980; Neuweiler 1989; Fenton 1990). This task differs dramatically for aerial insectivorous bats foraging in the open with no obstacles in close proximity to their flight paths (uncluttered), and for bats foraging close to and within vegetation or above water surfaces (cluttered). In the latter case, bats must separate echoes of interest (e.g. prey) from echoes produced by the surrounding background clutter. I will discuss how the physical properties of the different signal types, observed in the emballonurids, correspond to the challenge posed by foraging in different habitats.

First, the search signals of the emballonurids all share narrowband components (QCF, CF). This common feature is true regardless of the specific foraging habitats of the taxa, even though signal design among the genera of emballonurids that I studied is quite variable. In QCF and CF signals energy is concentrated in a small frequency band which facilitates the detection of weak insect echoes (Schnitzler 1987). Detection is further improved in the presence of acoustic glints. Acoustic glints are created when an echolocation signal hits a flying insect at the instant its wings are perpendicular to the impinging sound wave (Kober & Schnitzler 1990). Acoustic glints can enhance the sound pressure level (SPL) of echoes from flying insects by up to 20 dB in comparison with echoes of non-flying insects (Kober & Schnitzler 1990). QCF and CF signals are optimally suited to carry this type of information (Schnitzler 1987). Chances to perceive glints depend on the duty cycle (percentage of time occupied by sound) of the bat's echolocation sequence and the wing beat rate of the insect. All emballonurids of this study except the ghost bat are small and hunt mainly for small insects (Bradbury & Vehrencamp 1976; pers. obs.) which typically have rapid wingbeats. With an average duty cycle of 9–15% and estimated wingbeat rates of the prey at about 100 Hz, sheath-tailed bats could perceive 9–15 glints (duty cycle × wing beat rate) per second. The emballonurids of this study have average repetition rates of 9–20 signals per second, which would enable them to perceive up to one glint in every signal. Hence, the QCF and CF components are of adaptive value to emballonurids to improve the detection of flying insects.

However, QCF and CF signals are not well suited for accurate location of objects, an essential component of information for bats flying in obstacle-rich environments. In contrast to the emballonurids foraging in open space, which emit pure QCF signals, prominent FM components are present in the search signals of the four species foraging close to vegetation or over water surfaces (*S. bilineata, S. leptura, C. brevirostris* and *R. naso*). It is generally accepted that broad-band, steep FM components are best suited for determining range and angle of objects (e.g. Simmons & Stein 1980; Schnitzler & Henson 1980). Hence the search signals of *S. bilineata, S. leptura, C. brevirostris* and *R. naso* can be interpreted as an adaptation to cluttered habitats, where the bats employ a mixed strategy, using QCF and CF components to improve the detection of flying insects and FM components to monitor the bat's position relative to the clutter-producing background in order to avoid it.

Second, another parameter associated with habitat type and foraging strategy is sound duration. Characteristically, emballonurids foraging in open space emit longer signals than do those foraging in confined space. It has been shown for several vespertilionid bats shifting from uncluttered to cluttered habitat that they avoid an overlap of outgoing signal and returning echoes from obstacles and prey by reducing sound duration (Schnitzler, Kalko, Miller & Surlykke 1987; Kalko & Schnitzler 1989, 1993). Such an overlap presumably masks important information from the bats. Thus, bats foraging close to clutter-producing objects should have shorter sound durations in order to avoid this overlap. This might explain why *Cormura brevirostris*, which forages almost exclusively inside the forest close to obstacles, has much shorter sound durations than the similar-sized *Saccopteryx bilineata* recorded while foraging at forest edges bordering a large gap.

Third, another important parameter in echolocation signal design is call frequency. Call frequency is often correlated with size, i.e., smaller bats often emit higher frequencies than larger bats do (e.g. Heller & Helversen 1989). For emballonurids, this is valid within genera (e.g. *Saccopteryx*), where the smaller species, *S. leptura*, emits signals which are similar in structure but higher in frequency than those of the larger species, *S. bilineata* (Table 1). However, the size-to-frequency rule is generally not applicable within the emballonurids. For example, *Peropteryx* sp. is in about the same size class as *S. bilineata* and *C. brevirostris* (7–11 g), but emits signals which are much lower in frequency (Table 1). Also, *R. naso* is about the same size as *S. leptura* (3–5 g), but its call frequency is much higher (Table 1).

Therefore, ecological constraints imposed by the environment and foraging strategy appear best to explain the observed pattern in echolocation signals. The frequency of the calls determines the detection range of objects. Higher frequencies are attenuated in air much more rapidly than are lower frequencies. Hence, the higher call frequencies in the four emballonurids of this study foraging in cluttered habitats indicate a short-range detection strategy. This is typical for bats foraging in confined spaces which have to monitor obstacles and to detect potential prey at close range. Conversely, the lower call frequencies in *D. albus* and *Peropteryx* sp. indicate a long-range detection strategy. This is characteristic of bats foraging in uncluttered space which detect prey at long range (e.g. Barclay 1985, 1986) and do not have to monitor obstacles at close range.

Call frequency might also be expected to reflect the size of potential prey. In order to get distinct echoes from small targets, high frequencies are necessary. High frequency should thus allow the exploitation of smaller prey. The few data available for emballonurids corroborate this hypothesis. *R. naso*, which has the highest frequency in search signals of all emballonurids, feeds on insects which are significantly smaller than the prey eaten by *Saccopteryx* sp. (Bradbury & Vehrencamp 1976). Emballonurids with lower frequencies should concentrate on larger prey and should have longer detection ranges. Preliminary observations in Panama, where I observed *D. albus* catching mainly medium-sized moths, support this prediction.

In this diversity of echolocation signals and their apparent association with foraging habitats, aerial insectivorous, Neotropical emballonurid bats contrast sharply with the leaf-nosed bats (Phyllostomidae) which form the largest group of bats in the Neotropics. Signal design in phyllostomids is rather uniform, consisting mainly of short, multiharmonic, frequency-modulated signals emitted at low intensities (e.g. Bradbury 1970; Barclay, Fenton, Tuttle & Ryan 1981; Belwood 1988). However, no other families of bats (or other mammals) have the variety of diets found in leaf-nosed bats, which range from insects, small vertebrates and blood to fruit, nectar and pollen (e.g. Gardner 1977). The low plasticity in signal design within this family may reflect constraints imposed by the foraging behaviour and habitats of its members. With few exceptions, phyllostomids forage in highly cluttered habitats, in and around vegetation. As they mostly glean food from surfaces (e.g. large insects, fruit), they face the problem of having to separate echoes from stationary targets from those from background clutter. Recent studies show that leaf-nosed bats often supplement their sonar with other sensory cues (e.g. passive hearing, olfaction) to detect, localize and classify potential food (e.g. Tuttle & Ryan 1981; Laska 1990; E. K. V. Kalko, unpubl. results). Therefore, in leaf-nosed bats echolocation signal design does not reflect the diversity of food but it does reflect the characteristic way in which these bats forage, i.e. taking food items from surfaces in highly cluttered habitats.

Nonetheless, in the case of aerial insectivorous bats such as emballonurids, echolocation signal design proves to be a good indicator of habitat diversity and foraging strategies. Similar associations of call structure with habitat and phylogenetic flexibility of signal types appear also to extend to vespertilionids and molossids (e.g. Simmons *et al.* 1978; Neuweiler 1989; Rydell 1990; Fenton 1990; Kalko & Schnitzler 1993). Thus, signal design of echolocation calls of aerial insectivorous bats gives more information on foraging strategies and habitat use than do morphology and phylogenetic relationships alone, and appears to be a powerful tool for gathering more ecological information on species which are otherwise difficult to monitor in the field.

Attempts have already been made at other localities to identify bats by acoustic monitoring (e.g. Fenton & Bell 1981; Weid & Helversen 1987), to assess bat activity in various habitats (e.g. Fenton 1982; Fenton, Merriam & Holroyd 1983; Fenton, Tennant & Wyszecki 1987) and to integrate morphology (e.g. wing shape) and echolocation behaviour (Aldridge & Rautenbach 1987; Norberg & Rayner 1987; Crome & Richards 1988). Acoustic monitoring of emballonurids with standardized recording and analysing techniques will allow a refined analysis of microhabitat selection by sympatric species. This will be especially rewarding for the study of closely related species foraging in similar habitats. Comparisons of microhabitat selection in species-rich sites in Amazonia, where up to 12 emballonurids including four species of *Saccopteryx* might occur (Handley 1967 and pers. comm.), with that in relatively species-poor sites such as Panama should give valuable data on possible differences in the degree of niche overlap and niche separation among sympatric species. So far, such data are not available for aerial insectivorous bats in the Neotropics. We can look forward to the insights that such data will bring.

Acknowledgements

I wish to thank the many people and institutions who helped to accomplish this work, especially Charles Handley, Allen Herre, Egbert Leigh and Hans-Ulrich Schnitzler for many valuable suggestions, discussions and language assistance, José Ochoa for the introduction to Venezuelan bats, and the staff of the Smithsonian Tropical Research Institute (STRI, Panama), INRENARE (Panama) and INPARQUES (Venezuela) for logistic support and permits. The research was supported by a NATO postdoctoral fellowship to Kalko and by a grant of the Deutsche Forschungsgemeinschaft (SPP 'Mechanismen der Aufrechterhaltung tropischer Diversität') to Schnitzler and Kalko.

References

Aldridge, H. D. J. N. & Rautenbach, I. L. (1987). Morphology, echolocation and resource partitioning in insectivorous bats. *J. Anim. Ecol.* **56**: 763–778.

Barclay, R. M. R. (1983). Echolocation calls of emballonurid bats from Panama. *J. comp. Physiol.* **151**: 515–520.

Barclay, R. M. R. (1985). Long- versus short-range foraging strategies of hoary (*Lasiurus cinereus*) and silver-haired (*Lasionycteris noctivagans*) bats and the consequences for prey selection. *Can. J. Zool.* **63**: 2507–2515.

Barclay, R. M. R. (1986). The echolocation calls of hoary (*Lasiurus cinereus*) and silver-haired (*Lasionycteris noctivagans*) bats as adaptations for long- versus short-range foraging strategies and the consequences for prey selection. *Can. J. Zool.* **64**: 2700–2705.

Barclay, R. M. R., Fenton, M. B., Tuttle, M. D. & Ryan, M. J. (1981). Echolocation calls produced by *Trachops cirrhosus* (Chiroptera: Phyllostomatidae) while hunting for frogs. *Can. J. Zool.* **59**: 750–753.

Belwood, J. J. (1988). Foraging behavior, prey selection, and echolocation in phyllostomine bats (Phyllostomidae). In *Animal sonar*: 601–605. (Eds Nachtigall, P. E. & Moore, P. W. B.). Plenum Press, New York. (*NATO adv. Stud. Ser. Ser. A Life Sci.* No. 156.)

Bradbury, J. W. & Emmons, L. H. (1974). Social organisation of some Trinidad bats. I. Emballonuridae. *Z. Tierpsychol.* **36**: 137–183.

Bradbury, J. W. & Vehrencamp, S. L. (1976). Social organization and foraging in emballonurid bats. I. Field studies. *Behav. Ecol. Sociobiol.* **1**: 337–381.

Bradbury, J. W. (1970). Target discrimination by the echolocating bat *Vampyrum spectrum*. *J. exp. Zool.* **173**: 23–46.

Crome, F. H. J. & Richards, G. C. (1988). Bats and gaps: microchiropteran community structure in a Queensland rain forest. *Ecology* **69**: 1960–1969.

Fenton, M. B. (1982). Echolocation calls and patterns of hunting and habitat use of bats (Microchiroptera) from Chillagoe, north Queensland. *Aust. J. Zool.* **30**: 417–425.

Fenton, M. B. (1990). The foraging behavior and ecology of animal-eating bats. *Can. J. Zool.* **68**: 411–422.

Fenton, M. B. & Bell, G. P. (1981). Recognition of species of insectivorous bats by their echolocation calls. *J. Mammal.* **62**: 233–243.

Fenton, M. B., Bell, G. P. & Thomas, D. W. (1980). Echolocation and feeding behaviour of *Taphozous mauritianus* (Chiroptera: Emballonuridae). *Can. J. Zool.* **58**: 1774–1777.

Fenton, M. B., Merriam, H. G. & Holroyd, G. L. (1983). Bats of Kootenay, Glacier, and Mount Revelstoke National Parks in Canada: identification by echolocation calls, distribution and biology. *Can. J. Zool.* **61**: 2503–2508.

Fenton, M. B., Tennant, D. C. & Wyszecki, J. (1987). Using echolocation calls to measure the distribution of bats: the case of *Euderma maculatum*. *J. Mammal.* **68**: 142–144.

Findley, J. S. (1993). *Bats: a community perspective.* Cambridge University Press, Cambridge, New York & Oakleigh.

Fleming, T. H. (1986). The structure of Neotropical bat communities: a preliminary analysis. *Revta chil. Hist. nat.* **59**: 135–150.

Gardner, A. L. (1977). Biology of bats of the New World family Phyllostomatidae. Part 2. Feeding habits. *Spec. Publs Mus. Texas tech. Univ.* No. 13: 293–350.

Griffin, D. R. & Novick, A. (1955). Acoustic orientation of neotropical bats. *J. exp. Zool.* **130**: 251–299.

Griffin, D. R, Webster, F. A. & Michael, C. R. (1960). The echolocation of flying insects by bats. *Anim. Behav.* **8**: 141–154.

Handley, C. O. Jr. (1967). Bats of the canopy of an Amazonian forest. In *Atas do simpósio sôbre a biota Amazônica* **5**: 211–221. (Ed. Lent, H.). Conselho Nacional de Pesquisas, Rio de Janeiro.

Heller, K.-G. (1989). Echolocation calls of Malaysian bats. *Z. Säugetierk.* **54**: 1–8.

Heller, K.-G. & Helversen, O. V. (1989). Resource partitioning of sonar frequency bands in rhinolophid bats. *Oecologia* **80**: 178–186.

Jones, G. & Rayner J. M. V. (1988). Flight performance, foraging tactics and echolocation in free-living Daubenton's bats, *Myotis daubentoni* (Chiroptera: Vespertilionidae). *J. Zool., Lond.* **215**: 113–132.

Jones, J. K. Jr. (1966). Bats from Guatemala. *Univ. Kans. Publs Mus. nat. Hist.* **16**: 439–472.

Kalko, E. K. V. & Schnitzler, H.-U. (1989). The echolocation and hunting behaviour of Daubenton's bat, *Myotis daubentoni*. *Behav. Ecol. Sociobiol.* **24**: 225–238.

Kalko, E. K. V. & Schnitzler, H.-U. (1993). Plasticity in echolocation signals of European pipistrelle bats in search flight: implications for prey detection and habitat use. *Behav. Ecol. Sociobiol.* **33**: 415–428.

Kober, R. & Schnitzler, H.-U. (1990). Information in sonar echoes of fluttering insects available for echolocating bats. *J. acoust. Soc. Am.* **87**: 882–896.

Laska, M. (1990). Olfactory sensitivity to food odor components in the short-tailed fruit bat, *Carollia perspicillata* (Phyllostomatidae, Chiroptera). *J. comp. Physiol. (A)* **166**: 395–399.

Neuweiler, G. (1989). Foraging ecology and audition in echolocating bats. *Trends Ecol. Evol.* **4** (6): 160–166.

Norberg, U. M. & Rayner, J. M. V. (1987). Ecological morphology and flight in bats (Mammalia: Chiroptera): wing adaptations, flight performance, foraging strategy and echolocation. *Phil. Trans. R. Soc. (B)* **316**: 335–427.

Novick, A. (1962). Orientation in Neotropical bats. I. Natalidae and Emballonuridae. *J. Mammal.* **43**: 449–455.

Pye, J. D. (1980). Adaptiveness of echolocation signals in bats. *Trends Neurosci.* **3**: 232–235.

Rydell, J. (1990). Behavioural variation in echolocation pulses of the northern bat, *Eptesicus nilssoni*. *Ethology* **85**: 103–113.

Schnitzler, H.-U. (1987). Echoes of fluttering insects: information for echolocating bats. In *Recent advances in the study of bats*: 226–243. (Eds Fenton, M. B., Racey, P. & Rayner, J. M. V.). Cambridge University Press, Cambridge.

Schnitzler, H.-U. & Henson, O. W., Jr. (1980). Performance of airborne animal sonar systems. I. Microchiroptera. In *Animal sonar systems*: 109–181. (Eds Busnel, R.-G. & Fish, J. F.). Plenum Press, New York.

Schnitzler, H.-U., Kalko, E., Miller, L. & Surlykke, A. (1987). The echolocation and hunting behavior of the bat, *Pipistrellus kuhli*. *J. comp. Physiol. (A)* **161**: 267–274.

Simmons, J. A. & Stein, R. A. (1980). Acoustic imaging in bat sonar: echolocation signals and the evolution of echolocation. *J. comp. Physiol.* **135**: 61–84.

Simmons, J. A., Lavender, W. A., Lavender, B. A., Childs, J. E., Hulebak, K., Ridgen, M. R., Sherman, J., Woolman, B. & O'Farrell, M. J. (1978). Echolocation by free-tailed bats (*Tadarida*). *J. comp. Physiol.* **125**: 291–299.

Starrett, A. & Casebeer, R. S. (1968). Records of bats from Costa Rica. *Contr. Sci.* No. 148: 1–21.

Tuttle, M. D. & Ryan, M. J. (1981). Bat predation and the evolution of frog vocalizations in the Neotropics. *Science* **214**: 677–678.

Weid, R. & Helversen, O. vd. (1987). Ortungsrufe europäischer Fledermäuse beim Jagdflug im Freiland. *Myotis* **25**: 5–27.

Willig, M. R. & Moulton, M. P. (1989). The role of stochastic and deterministic processes in structuring Neotropical bat communities. *J. Mammal.* 70: 323–329.

Willig, M. R., Camilo, G. R. & Noble, S. J. (1993). Dietary overlap in frugivorous and insectivorous bats from edaphic cerrado habitats of Brazil. *J. Mammal.* **74**: 117–128.

Part V

Microchiropteran behaviour and ecology

Symp. zool. Soc. Lond. (1995) No. 67: 277–289

Constraint and flexibility—bats as predators, bats as prey

M. Brock FENTON

Department of Biology
York University
North York, Ontario
Canada M3J 1P3

Synopsis

Viewed as predators or as prey, bats offer excellent examples of how constraints on performance are ameliorated by behavioural flexibility. Bats hunting airborne targets often eat different things to those taking prey from surfaces. These behavioural differences determine prey species and hence energetic input per prey item. Limitations translate less directly to dietary differences when similar species are compared. Flexibility often means exploiting concentrations of prey, such as the insects around lights, but can include different hunting behaviours and patterns of habitat use. Heterothermy, the ability to reduce body temperatures, is another facet of flexibility only available to some bats. The echolocation calls of bats function as releasers for defensive behaviour by some insects, affecting the bats' hunting success. Some arctiid moths further affect bat hunting by producing clicking sounds as they are attacked. Some bats circumvent hearing-based insect defences by producing quiet, short echolocation calls or using other cues to find their prey. The same blend of constraints and flexibility is shown by some predators of bats. For diurnal raptors, light levels and feeding behaviour limit intake of bats. Bats emerging from roosts show behaviour patterns predicted by the selfish herd principle.

Introduction

Individual flexibility is a common feature of the behaviour of many different kinds of predators. The literature abounds with examples of predators adjusting their behaviour according to their prey at any given time (e.g. spotted hyaenas, *Crocuta crocuta*: Kruuk 1972; red foxes, *Vulpes vulpes*: Henry 1986). There are also surprising examples of opportunism in predators, for example sea otters (*Enhydra lutris*) taking octopods from discarded beverage cans (McCleneghan & Ames 1976), or roadrunners (*Geococcyx californicus*) grabbing hummingbirds visiting feeders (Spofford 1976).

The range of flexibility exhibited by a predator, however, is constrained by factors such as morphology. Werdelin (1989), for example, showed how a

ZOOLOGICAL SYMPOSIUM No. 67
ISBN 0 19 854945 8

basic canid structure limited the bone-cracking and meat-cutting abilities of *Osteoborus*, a fossil canid. In spite of constraints, there are many instances of body size limiting the prey accessible to predators. Behavioural mechanisms, however, can permit predators to operate under a wide range of circumstances (e.g. Kruuk 1972; Henry 1986).

Bats provide interesting examples of the complexity of factors that limit or constrain performance and interact with flexibility. The purpose of this review is to illustrate constraints and flexibility, using data drawn from studies of bats as predators and as prey. The examples I present have been published elsewhere and the details of materials, methods and study sites are provided in the cited literature.

Bats as predators

The diversity of microchiropteran bats provides strong circumstantial evidence that species use a variety of approaches to foraging. Whether the focus of attention is wing morphology or echolocation (e.g. Aldridge & Rautenbach 1987; Norberg & Rayner 1987; Fenton 1990), it is clear that different microchiropterans forage in different ways. Different foraging strategies can influence the prey available to, and taken by, different bats.

An excellent demonstration of this diversity comes from the work of Swift & Racey (1983) who compared the diets of *Myotis daubentonii* and *Plecotus auritus* that roosted in the same building. While *M. daubentonii* hunted over water, taking airborne prey, *P. auritus* foraged in woodland, taking mainly moths and beetles and occasionally flightless prey. I suspect that while the *Plecotus* could exploit some of the prey taken by the *Myotis*, the reverse was less true (see also Fenton 1990), indicating more flexible behaviour on the part of *Plecotus*.

Barclay (1991) used field data from three sympatric *Myotis* to provide another example of how foraging behaviour influences prey selection and, in turn, reproduction. He found that while *Myotis evotis* bore young in his study area in the mountains west of Calgary (Alberta, Canada), *Myotis lucifugus* and *Myotis volans* lived, but did not reproduce, there. The *M. evotis* regularly ate non-airborne prey, while the other two species foraged mainly, if not exclusively, on airborne insects. In cool, damp areas, the ability to take non-airborne prey translated into more opportunities for energetic input and reproductive output.

Fenton (in press) used data on the moths taken by two 30-g bats, *Lasiurus cinereus* and *Nycteris grandis*, to illustrate how different foraging strategies affected diet (Fig. 1). While *L. cinereus* took only airborne moths (Hickey 1993) which it invariably consumed on the wing, *N. grandis* fed on airborne and non-airborne prey (Fenton, Rautenbach, Chipese, Cumming, Musgrave, Taylor & Volpers 1993), usually (always?) at a feeding perch. It is clear that *N. grandis* eat a much larger size range of moths than *L. cinereus* do (Fig. 1).

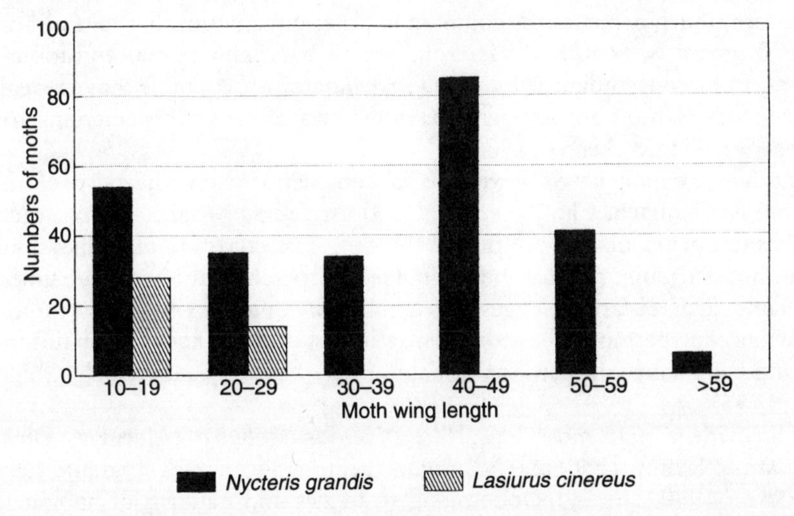

Fig. 1. Two 30-g bat species take different-sized (by wing length) sphingid and noctuid moths. While *Nycteris grandis* often hunts from a perch, taking airborne and non-airborne prey which it eats at a feeding perch, *Lasiurus cinereus* takes only airborne prey which it eats on the wing. Data from Fenton (in press).

These examples demonstrate how differences in foraging behaviour can translate into differences in diet composition and, in one case, differences in reproductive output in a given study area. The importance of differences, however, becomes blurred when more similar species are compared. For example, Herd & Fenton (1983) reported how *Myotis lucifugus* and *Myotis yumanensis* tended to feed in different areas, which coincided with differences in diet composition although both species fed exclusively on airborne insects. Saunders & Barclay (1992) found considerable overlap between two other species of *Myotis (Myotis lucifugus* and *Myotis volans*) that hunt airborne prey. Even though there are various suggestions that bats specialize on one kind of insect or another (e.g. hard-bodied versus soft-bodied: Strait 1993), Findley (1993) reviewed much of the literature on community structure in bats and concluded that there was little evidence from bats that their communities were organized by competition for resources. Behavioural flexibility may be a key factor in accounting for overlap between species.

One component of flexibility is the tendency to exploit concentrations of prey. There are many published reports of bats hunting in rich patches of prey such as insects at lights (e.g. Hickey & Fenton 1990; Acharya & Fenton 1992; Rydell 1992; Rydell & Racey this volume pp. 291–307) or in mating or emerging swarms of insects (e.g. Gould 1978). Bell (1980) used bat responses to insects at lights to investigate structure in a desert bat community in southern Arizona. He found that most species there quickly exploited rich patches of insect prey. Such behaviour is common in other insectivores that take airborne prey (Brigham & Fenton 1991). Still, not all bats hunt in concentrations of insects. Extensive studies of *Euderma maculatum*, for example, have revealed no tendency to hunt around lights or in the swarms of insects emerging from water (Wai-Ping & Fenton 1989).

Another component of flexibility is demonstrated by diversity of diet. Fenton, Rautenbach, Chipese *et al.* (1993) used remains taken below feeding perches to document the variation in the diet of *Nycteris grandis*. They found significant variation in diet between feeding perches at the same time in the same general area, suggesting great individual flexibility in foraging behaviour. Furthermore, these differences in diet often coincided with different hunting behaviour and patterns of habitat use (Fenton, Rautenbach, Chipese *et al.* 1993).

It is evident that most bats, like *Nycteris thebaica* (Aldridge, Obrist, Merriam & Fenton 1990) and *N. grandis* (Fenton, Swanepoel, Brigham, Cebek & Hickey 1990) adjust their foraging strategies and patterns of habitat use according to prevailing conditions (such as insect availability). Vaughan (1976) observed *Cardioderma cor* switching from pursuit of airborne to non-airborne prey. Similar observations have been made of *Myotis emarginatus* which also changed its echolocation behaviour dramatically according to whether it was hunting airborne or stationary insects (Schumm, Krull & Neuweiler 1991), as did *Antrozous pallidus* (Krull 1992).

A longer-term (five-year) study of two species of *Lasiurus* in south-western Ontario (Canada) has demonstrated how the diets of *L. borealis* (12–15 g) and *L. cinereus* (30 g) varied according to prey availability (L. Acharya & M. B. C. Hickey pers. comm.). The two species ate moths in the same places at the same time and the degree of niche overlap between the two species varied from 0.26 to 0.46, with niche breadths of 4.95 to 6.52 (*Lasiurus borealis*) and 7.75 to 10.10 (*Lasiurus cinereus*). Values for niche breadth and niche overlap varied according to local weather conditions and moth availability.

Heterothermy, the ability to lower body temperatures, is yet another component of flexibility for some insectivorous bats, notably species in the families Rhinolophidae and Vespertilionidae. Using telemetry to monitor body temperature, field studies of vespertilionids (*Eptesicus fuscus*: Audet & Fenton 1988; *Myotis myotis*: Audet 1992; *Lasiurus cinereus*: Hickey 1993) have demonstrated how each of these species uses reduced body temperatures to cut energetic costs of thermoregulation during cool weather.

Heterothermy is the reason that vespertilionid and rhinolophid bats dominate the temperate-zone bat faunas in the Old World (both families) and New World (vespertilionids). In Africa, for example, the diversity of two other animal-eating bats declines rapidly away from the equator compared to that of vespertilionids and rhinolophids (Fig. 2). In many situations, increased flexibility is balanced by reduced reproductive output, which is one important

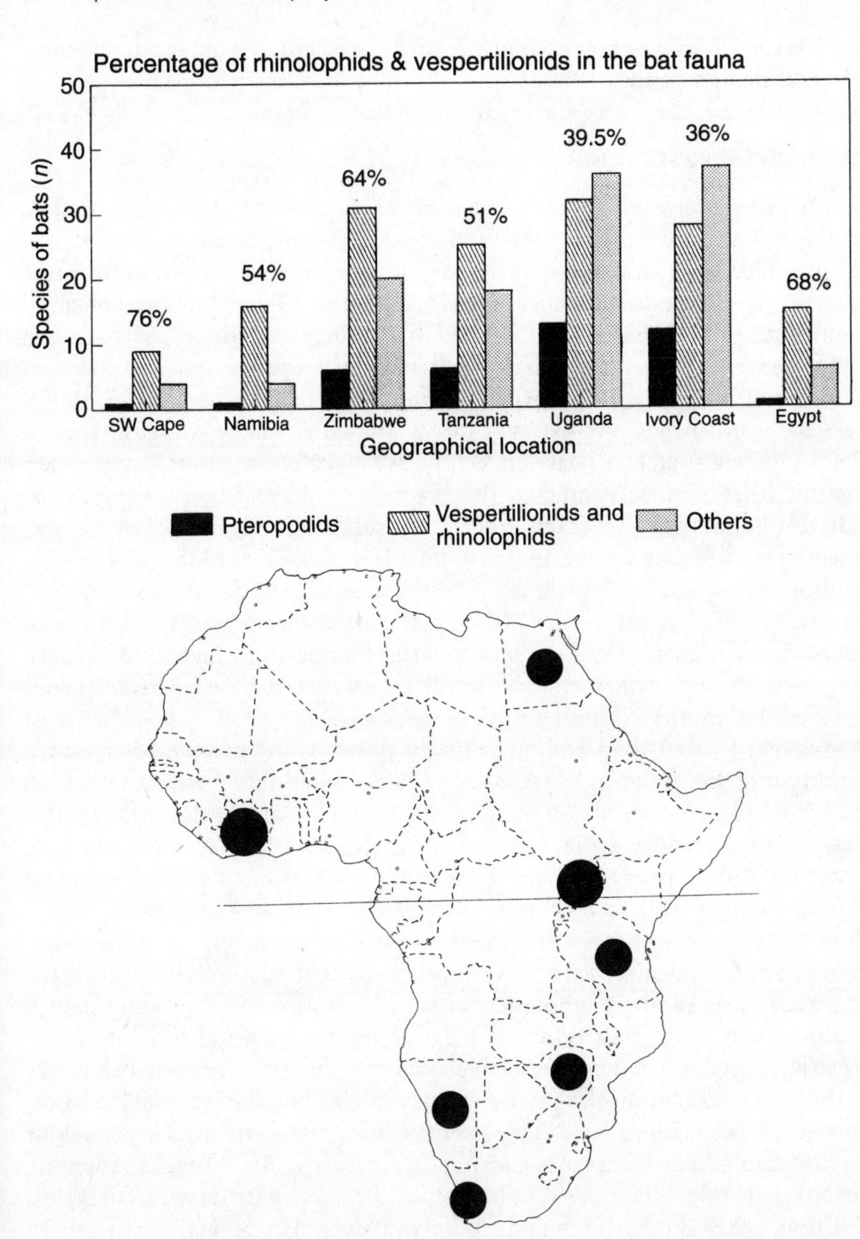

Fig. 2. A comparison of the bat faunas from some parts of Africa (South-Western Cape, Namibia, Zimbabwe, Tanzania, Uganda, Ivory Coast and Egypt) showing the numbers of species in the families Vespertilionidae and Rhinolophidae and the percentage of the total bat fauna represented by there two families. The numbers of species in the Pteropodidae are also shown, as are those in 'other' families (Rhinopomatidae, Emballonuridae, Nycteridae, Hipposideridae and Molossidae). The data are from Kingdon (1974), Qumsiyeh (1985), Rosevear (1965) and Smithers (1983). The map below shows the location of the areas compared relative to the equator.

cost of heterothermy as practised by pregnant or lactating females, for example in *Myotis myotis* (Audet 1992).

The insects' responses

The prey's response to an attack can affect a predator's foraging success, and the interactions between bats and insects provide excellent examples of this dynamic interaction. Many insectivorous bats use echolocation to locate and assess targets, especially airborne ones (Fenton 1990). A disadvantage of echolocation is that the calls are dependable and conspicuous signals (Fenton 1994) used by many insects as a means to detect and avoid hunting bats (Belwood & Morris 1987; Surlykke 1988; Yager, May & Fenton 1990). Insect defensive behaviour exerts a direct influence on the foraging success of bats. Insects that can hear bat echolocation calls appear to adjust their defensive behaviour according to the strength of the bats' signals, showing negative phonotaxis to distant (faint) bats, and flight cessation or evasive maneouvres in reaction to close (loud) bats (e.g. Surlykke 1988).

Although few studies have examined the question in detail, insectivorous bats appear to succeed about 40% of the time when attacking airborne targets (e.g. Vaughan (1977)—*Hipposideros commersoni*; Hickey & Fenton (1990)—*Lasiurus borealis*; Hickey (1992)—*Lasiurus cinereus*). Roeder (1967) suggested that moths exhibiting evasive manoeuvring had 40% less chance of being caught by bats than had those moths not showing evasive behaviour.

From his observations at lights, Rydell (1992) found that *Eptesicus nilssonii* succeeds in all of its attacks on dung beetles and 35% of its attacks on moths. Acharya (1992) reported that moths with ears successfully evaded attacking *L. borealis* and *L. cinereus* about 40% of the time, while deafened moths of the same families were caught more often on the bats' first attacks.

Acharya's (1992) study also revealed that bats hunting at lights made repeated attacks on moths there, so that moths that successfully evaded the first attack rarely survived subsequent attacks. This situation may be one factor explaining why many bats forage among the insects attracted to lights.

Hearing-based insect defences against bats can affect bat foraging behaviour in other ways. Some moths (many species in the family Arctiidae) click in response to close (loud) bats. These moth clicks affect the attack behaviour of echolocating bats, sometimes causing them to abort their attacks (Dunning & Roeder 1965; Miller 1991). In the field, some bats take arctiids less often than other moths (Dunning, Acharya, Merriman & Dal Ferro 1992), which presumably reflects the role of sound in the defensive behaviour of the arctiids.

Work with captive *Eptesicus fuscus* has demonstrated that these bats adjust their reactions to arctiid clicks according to their prior experience (Bates & Fenton 1990). Naive bats are startled by arctiid clicks, while bats that have learned to associate the clicks with unpalatable food treat the clicks as aposematic signals. In the field it is obvious that *Lasiurus borealis* use the clicks of the arctiid *Hypoprepia fucosa* as warnings of unpalatability (Acharya & Fenton 1992).

By using short, low-intensity echolocation calls (Faure, Fullard & Barclay 1990; Faure, Fullard & Dawson 1993) or by switching to vision (Bell & Fenton 1986) or other forms of passive orientation (Fiedler 1979; Tuttle, Ryan & Belwood 1985; Ryan & Tuttle 1987; Anderson & Racey 1991; Faure & Barclay 1992), bats can avoid alerting some insects to their presence. Use of passive orientation may be more feasible when the prey is moving on surfaces than when it is airborne because movement on a surface is noisier than movement in air.

Bats as prey

Although predation is often invoked to explain different aspects of bat behaviour, beyond anecdotal reports there are few studies of predation on bats, although Speakman (1991) provides a refreshing exception. It should be no surprise that predators of bats show the same combination of flexibility and limitations as the bats do themselves.

Predation on bats often occurs in or near their roosts. One striking example of flexibility by predators feeding on bats is the tendency for some New World monkeys to hunt tent-roosting bats. The monkeys (*Saimiri oerstedi*) scout the tent from above and, by leaping on to it, grab some bats and knock others to the ground and then catch them there (Boinski & Timm 1985). Other, less spectacular, examples are often reported in the literature.

The pelage patterns and behaviour of bats that roost in foliage may combine to provide some protection against predators (Fenton 1992). Bats that roost in large numbers, often in hollow structures (the roofs of buildings, caves, or the trunks of trees) are more often the targets of such predators as small carnivores and raptors, usually as the bats are leaving or returning to the sites (e.g., Fenton, Rautenbach, Smith, Swanepoel, Grossell & Van Jaarsveld 1994).

The tendency for groups of bats to emerge together from roosts could represent anti-predator behaviour, simply reflect the physical 'bottleneck' of the roost entrance (Speakman, Bullock, Eales & Racey 1992) or indicate information-sharing (Wilkinson 1992). In Kruger National Park in January 1993, I found a group of molossid bats (*Tadarida pumila*) roosting in a crevice above the entrance to a cave. Just after noon, when I reached into the crevice and tried to grab a bat, 15 to 20 of them emerged together, in less than a second, and flew off in different directions. This burst emergence was accompanied by the sounds of fluttering wings and shrill cries and, because it was disorienting to me, appeared to be defensive.

Fenton, Rautenbach, Smith *et al.* (1994) showed how burst-emergences conferred an important advantage on individual bats hunted by raptors by

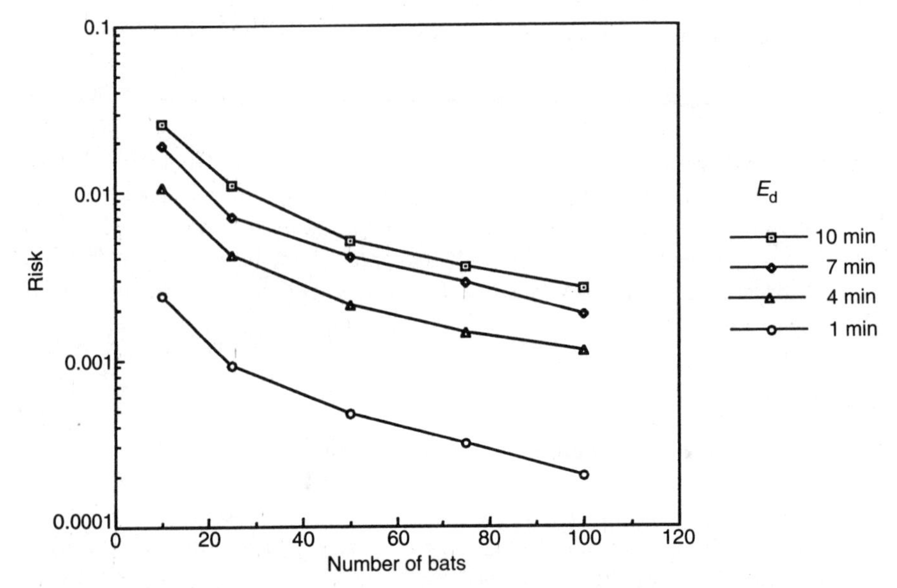

Fig. 3. The risk posed to individual bats by a raptor waiting outside the roost is reduced both by the duration of the emergence (E_d in minutes) and by the numbers of bats emerging (after Fenton, Rautenbach, Smith *et al.* 1994).

limiting the chances of attack on any individual, both through the dilution of risk to any individual and through reduction of the time taken for the bats to emerge (Fig. 3). In larger colonies, increased group size also confers some protection as predicted by the 'selfish herd effect' (Hamilton 1971).

By roosting in smaller groups, bats could be less conspicuous to predators, and by postponing their emergence until later in the evening they could minimize the risk of attack by diurnal or crepuscular raptors. Their occurrence in large colonies suggests that the bats are exploiting a beneficial roost situation (Fig. 4), while early emergence may be related to prey availability and the numbers of feeding bats (Fenton, Rautenbach, Smith *et al.* 1994).

Light levels affect the visual acuity of many diurnal raptors (Fox, Lehmkuhle & Westendorf 1976), in turn limiting the time available for them to hunt bats. This physical constraint is amplified by the birds' hunting behaviour. Many raptors occasionally take flying bats and, after catching them in their talons, eat them at a roost. Fenton, Rautenbach, Smith *et al.* (1994) illustrated how the time a raptor spent processing its catch significantly affected the rate at which it could harvest bats (Fig. 5). Larger raptors (*Aquila wahlbergi*) achieved more rapid handling times than smaller ones (*Accipiter tachiro*), reflecting the basic mechanics of size. Bat hawks, *Macheirhamphus alcinus*, have minimized their handling time by feeding on the wing and swallowing bats whole (Black, Howard & Stjernstedt 1979). The hunting behaviour of bat hawks coincides

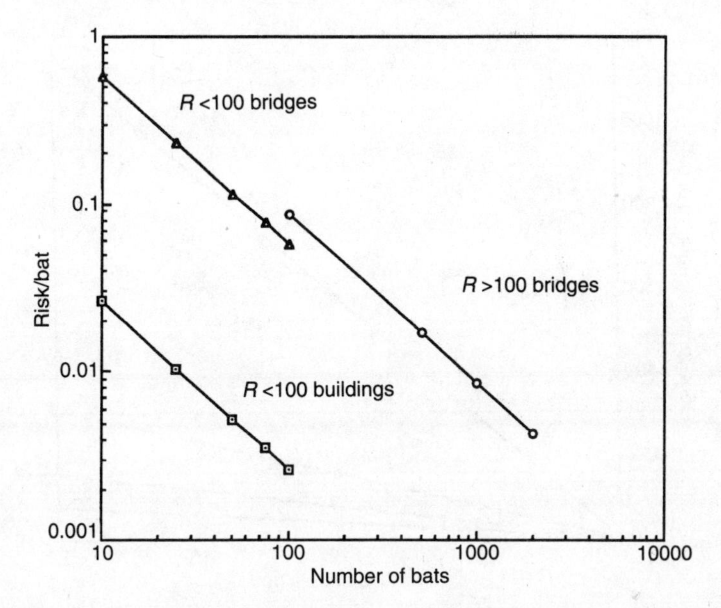

Fig. 4. The risk posed to individual bats by raptors hunting outside roosts is reduced by colony size (numbers of bats). Differences in risks (R) between different situations (buildings/bridges) reflects the tendency of raptors to hunt outside larger colonies (after Fenton, Rautenbach, Smith *et al.* 1994).

with large eyes and a proportionally large gape, apparently specializations for feeding on bats.

Constraints ameliorated by flexibility

Animals are limited in their range of activities by various morphological, neurological and physiological factors. These constraints are balanced, to some extent, by the organisms' abilities to adjust their behaviour according to the situation. Bats as predators and bats as prey illustrate some of the complexities of interactions between constraint and flexibility.

Acknowledgements

I thank L. Acharya, H. de la Cueva and M. B. C. Hickey for sharing unpublished data with me, and L. Acharya, R. M. R. Barclay, R. Csada, J. Dunlop, C. Merriman, P. A. Racey, J. Rydell and D. M. Syme for reading the manuscript and making helpful suggestions about it. My work on bats has been supported by grants from the Natural Sciences and Engineering Research Council of Canada. I am particularly grateful to the York University Ad Hoc

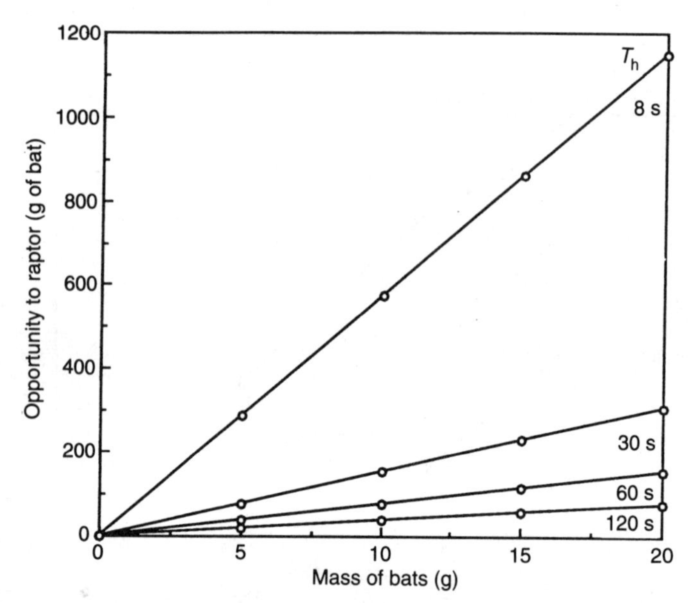

Fig. 5. The opportunity bats present to raptors, in terms of numbers of grams of bats harvested, is strongly influenced by the time it takes the raptor to handle its catch (T_h) as well as by the size (in g) of the bat (after Fenton, Rautenbach, Smith *et al.* 1994).

Fund for supporting my travel to the meetings in London and to the Zoological Society of London for their support and invitation.

References

Acharya, L. (1992). Are ears valuable to moths flying around lights? *Bat Res. News* **33**: 47.

Acharya, L. & Fenton, M. B. (1992). Echolocation behaviour of vespertilionid bats (*Lasiurus cinereus* and *Lasiurus borealis*) attacking airborne targets, including arctiid moths. *Can. J. Zool.* **70**: 1292–1298.

Aldridge, H. D. J. N. & Rautenbach, I. L. (1987). Morphology, echolocation and resource partitioning in insectivorous bats. *J. Anim. Ecol.* **56**: 763–778.

Aldridge, H. D. J. N., Obrist, M., Merriam, H. G. & Fenton, M. B. (1990). Roosting, vocalizations, and foraging by the African bat, *Nycteris thebaica*. *J. Mammal.* **71**: 242–246.

Anderson, M. E. & Racey, P. A. (1991). Feeding behaviour of captive brown long-eared bats, *Plecotus auritus*. *Anim. Behav.* **42**: 489–493.

Audet, D. (1992). *Roost quality, foraging and young production in the mouse-eared bat,* Myotis myotis*: a test of the ESS model of group size selection.* PhD diss.: York University, Ontario.

Audet, D. & Fenton, M. B. (1988). Heterothermy and the use of torpor by the bat *Eptesicus fuscus* (Chiroptera: Vespertilionidae): a field study. *Physiol. Zool.* **61**: 197–204.

Barclay, R. M. R. (1991). Population structure of temperate zone insectivorous bats in relation to foraging behaviour and energy demand. *J. Anim. Ecol.* **60**: 165–178.

Bates, D. L. & Fenton, M. B. (1990). Aposematism or startle? Predators learn their responses to the defenses of prey. *Can. J. Zool.* **68**: 49–52.

Bell, G. P. (1980). Habitat use and response to patches of prey by desert insectivorous bats. *Can. J. Zool.* **58**: 1876–1883.

Bell, G. P. & Fenton, M. B. (1986). Visual acuity, sensitivity and binocularity in a gleaning insectivorous bat, *Macrotus californicus* (Chiroptera: Phyllostomidae). *Anim. Behav.* **34**: 409–414.

Belwood, J. J. & Morris, G. K. (1987). Bat predation and its influence on calling behavior in neotropical katydids. *Science, Wash.* **238**: 64–67.

Black, H. L., Howard, G. & Stjernstedt, R. (1979). Observations on the feeding behavior of the bat hawk (*Macheirhamphus alcinus*). *Biotropica* **11**: 18–21.

Boinski, S. & Timm, R. M. (1985). Predation by squirrel monkeys and double-toothed kites on tent-making bats. *Am. J. Primatol.* **9**: 121–127.

Brigham, R. M. & Fenton, M. B. (1991). Convergence in foraging strategies by two morphologically and phylogenetically distinct nocturnal aerial insectivores. *J. Zool., Lond.* **223**: 475–489.

Dunning, D. C., Acharya, L., Merriman, C. & Dal Ferro, L. (1992). Interactions between bats and arctiid moths. *Can. J. Zool.* **70**: 2218–2223.

Dunning, D. C. & Roeder, K. D. (1965). Moth sounds and the insect-catching behavior of bats. *Science, N. Y.* **147**: 173–174.

Faure, P. A. & Barclay, R. M. R. (1992). The sensory basis of prey detection by the long-eared bat, *Myotis evotis*, and the consequences for prey selection. *Anim. Behav.* **44**: 31–39.

Faure, P. A., Fullard, J. H. & Barclay, R. M. R. (1990). The response of tympanate moths to the echolocation calls of a substrate gleaning bat, *Myotis evotis*. *J. comp. Physiol. (A)* **166**: 843–849.

Faure, P. A., Fullard, J. H. & Dawson, J. W. (1993). The gleaning attacks of the northern long-eared bat, *Myotis septentrionalis*, are relatively inaudible to moths. *J. exp. Biol.* **178**: 173–189.

Fenton, M. B. (1990). The foraging behaviour and ecology of animal-eating bats. *Can. J. Zool.* **68**: 411–422.

Fenton, M. B. (1992). Pelage patterns and crypsis in roosting bats: *Taphozous mauritianus* and *Epomophorus* species. *Koedoe* **35**: 49–55.

Fenton, M. B. (1994). Assessing signal variability and reliability: 'to thine own self be true'. *Anim. Behav.* **47**: 756–764.

Fenton, M. B. (In press). Natural history and biosonar signals. In *Hearing by bats*. (Eds Popper, A. N. & Fay, R. R.). Springer Verlag, New York. (*Springer Handb. audit. Res.* **11**.)

Fenton, M. B., Rautenbach, I. L., Chipese, D., Cumming, M. B., Musgrave, M. K., Taylor, J. S. & Volpers, T. (1993). Variation in foraging behaviour, habitat use and diet of large slit-faced bats (*Nycteris grandis*). *Z. Säugetierk.* **58**: 65–74.

Fenton, M. B., Rautenbach, I. L., Smith, S. M., Swanepoel, C. M., Grossell, J. & van Jaarsveld, J. (1994). Bats and raptors: threats and opportunities. *Anim. Behav.* **48**: 9–18.

Fenton, M. B., Swanepoel, C. M., Brigham, R. M., Cebek, J. E. & Hickey, M. B. C. (1990). Foraging behavior and prey selection by large slit-faced bats (*Nycteris grandis*; Chiroptera: Nycteridae). *Biotropica* **22**: 2–8.

Fiedler, J. (1979). Prey catching with and without echolocation in the Indian false vampire (*Megaderma lyra*). *Behav. Ecol. Sociobiol.* **6**: 155–160.

Findley, J. S. (1993). *Bats: a community perspective.* Cambridge University Press, Cambridge.

Fox, R., Lehmkuhle, S. W. & Westendorf, B. H. (1976). Falcon visual acuity. *Science, N. Y.* **192**: 263–265.

Gould, E. (1978). Opportunistic feeding by tropical bats. *Biotropica* **10**: 75–76.

Hamilton, W. J. III. (1971). Geometry of the selfish herd. *J. theor. Biol.* **31**: 295–311.

Henry, J. D. (1986). *Red fox, the catlike canine.* Smithsonian Institution Press, Washington.

Herd, R. M. & Fenton, M. B. (1983). An electrophoretic, morphological, and ecological investigation of a putative hybrid zone between *Myotis lucifugus* and *Myotis yumanensis* (Chiroptera: Vespertilionidae). *Can. J. Zool.* **61**: 2029–2050.

Hickey, M. B. C. (1992). Effect of radiotransmitters on the attack success of hoary bats, *Lasiurus cinereus*. *J. Mammal.* **73**: 344–346.

Hickey, M. B. C. (1993). *Thermoregulatory and foraging behaviour of hoary bats,* Lasiurus cinereus. PhD diss.: York University, Ontario.

Hickey, M. B. C. & Fenton, M. B. (1990). Foraging by red bats (*Lasiurus borealis*): do intraspecific chases mean territoriality? *Can. J. Zool.* **68**: 2477–2482.

Kingdon, J. (1974). *East African mammals. An atlas of evolution in Africa* **2A**. Academic Press, London & New York.

Krull, D. (1992). *Jagdverhalten und Echoortung bei* Antrozous pallidus (*Chiroptera: Vespertilionidae*). PhD diss.: Ludwig-Maximilians Universität München.

Kruuk, H. (1972). *The spotted hyaena: a study of predation and social behaviour.* University of Chicago Press, Chicago & London.

McCleneghan, K. & Ames, J. A. (1976). A unique method of prey capture by a sea otter, *Enhydra lutris*. *J. Mammal.* **57**: 410–412.

Miller, L. A. (1991). Arctiid moth clicks can degrade the accuracy of range difference discrimination in echolocating big brown bats, *Eptesicus fuscus*. *J. comp. Physiol. (A)* **168**: 571–579.

Norberg, U. M. & Rayner, J. M. V. (1987). Ecological morphology and flight in bats (Mammalia: Chiroptera): wing adaptations, flight performance, foraging strategy and echolocation. *Phil. Trans. R. Soc. (B)* **316**: 335–427.

Qumsiyeh, M. B. (1985). The bats of Egypt. *Spec. Publs Mus. Texas tech. Univ.* No. **23**: 1–102.

Roeder, K. D. (1967). *Nerve cells and insect behavior.* (Revised edn). Harvard University Press, Cambridge.

Rosevear, D. R. (1965). *The bats of West Africa.* British Museum (Natural History), London.

Ryan, M. J. & Tuttle, M. D. (1987). The role of prey-generated sounds, vision and echolocation in prey localization by the African bat, *Cardioderma cor* (Megadermatidae). *J. comp. Physiol. (A)* **161**: 59–66.

Rydell, J. (1992). Exploitation of insects around streetlamps by bats in Sweden. *Funct. Ecol.* **6**: 744–750.

Saunders, M. B. & Barclay, R. M. R. (1992). Ecomorphology of insectivorous bats: a test of predictions using two morphologically similar species. *Ecology* **73**: 1335–1345.

Schumm, A., Krull, D. & Neuweiler, G. (1991). Echolocation in the notch-eared bat, *Myotis emarginatus*. *Behav. Ecol. Sociobiol.* **28**: 255–261.

Smithers, R. H. N. (1983). *The mammals of the southern African subregion*. University of Pretoria, Pretoria.
Speakman, J. R. (1991). The impact of predation by birds on bat populations in the British Isles. *Mammal Rev.* **21**: 123–142.
Speakman, J. R., Bullock, D. J., Eales, L. A. & Racey, P. A. (1992). A problem defining temporal pattern in animal behaviour: clustering in the emergence behaviour of bats from maternity roosts. *Anim. Behav.* **43**: 491–500.
Spofford, S. H. (1976). Roadrunner catches hummingbird in flight. *Condor* **78**: 142.
Strait, S. G. (1993). Molar morphology and food texture among small-bodied insectivorous mammals. *J. Mammal.* **74**: 391–402.
Surlykke, A. M. (1988). Interaction between echolocating bats and their prey. In *Animal sonar: processes and performance*: 531–566. (Eds Nachtigall, P. E. & Moore, P. W. B.). New York: Plenum Press. (*NATO adv. Stud. Inst. Ser., Ser. A Life Sci.* No. 156.)
Swift, S. M. & Racey, P. A. (1983). Resource partitioning in two species of vespertilionid bats (Chiroptera) occupying the same roost. *J. Zool., Lond.* **200**: 249–259.
Tuttle, M. D., Ryan, M. J. & Belwood, J. J. (1985). Acoustical resource partitioning by two species of phyllostomid bats (*Trachops cirrhosus* and *Tonatia sylvicola*). *Anim. Behav.* **33**: 1369–1371.
Vaughan, T. A. (1976). Nocturnal behavior of the African false vampire bat (*Cardioderma cor*). *J. Mammal.* **57**: 227–248.
Vaughan, T. A. (1977). Foraging behaviour of the giant leaf-nosed bat (*Hipposideros commersoni*). *E. Afr. Wildl. J.* **15**: 237–249.
Wai-Ping, V. & Fenton, M. B. (1989). Ecology of spotted bat (*Euderma maculatum*) roosting and foraging behavior. *J. Mammal.* **70**: 617–622.
Werdelin, L. (1989). Constraint and adaptation in the bone-cracking canid *Osteoborus* (Mammalia: Canidae). *Paleobiology* **15**: 387–401.
Wilkinson, G. S. (1992). Information transfer at evening bat colonies. *Anim. Behav.* **44**: 501–518.
Yager, D. D., May, M. L. & Fenton, M. B. (1990). Ultrasound-triggered, flight-gated evasive maneuvers in the praying mantis *Parasphendale agrionina*. l. Free flight. *J. exp. Biol.* **152**: 17–39.

Symp. zool. Soc. Lond. (1995) No. 67: 291–307

Street lamps and the feeding ecology of insectivorous bats

J. RYDELL
and P. A. RACEY

Department of Zoology
University of Aberdeen
Aberdeen AB9 2TN, UK

Synopsis

In southern Sweden, densities of northern bats *Eptesicus nilssonii* (Keyserling & Blasius) flying along artificially lit roads in built-up areas (villages) were assessed, by using an ultrasound detector, at 2–5 bats per kilometre. In comparison, there were 0.1–0.4 bats per kilometre of unlit road. Means of 3.2 and 3.1 common pipistrelle bats *Pipistrellus pipistrellus* (Schreber) were recorded per kilometre of lit road in England and Scotland respectively.

The density of flying insects around street lamps was determined from flash photographs of the area around the lamps. Mercury-vapour street lamps, which emit a bluish-white light, including ultraviolet (UV), were shown to attract insects. In contrast, low-pressure sodium lamps, which emit monochromatic orange light, do not attract insects. High-pressure sodium lamps, which include mercury vapour and hence emit some UV light, are intermediate in terms of insect attraction. High densities of bats (about five per kilometre) were found only near white streetlamps. The lamps attracted bats independently of the proximity of buildings and trees.

Food intake by northern bats foraging over lit roads was comparable to that over pastures rich in dung beetles and considerably higher than that in woodlands. Street lamps may thus allow the northern bat to increase its energy intake and may therefore account for the frequent occurrence of this species in built-up areas.

Street lamps do not benefit all bat species equally. They are frequently used by aerial-hawking bats adapted for echolocation away from obstacles, e.g. the vespertilionid genera *Nyctalus, Vespertilio, Eptesicus* and *Pipistrellus*. In contrast, bats whose echolocation pulses are adapted for cluttered situations, e.g. the vespertilionid genera *Myotis* and *Plecotus*, as well as the rhinolophids, do not seem to exploit insects around streetlamps. Most of the European bat species that are currently regarded as endangered belong to the latter group.

Introduction

Most modern societies illuminate roads and built-up areas for the safety of traffic and pedestrians. This may have potential effects on wildlife.

ZOOLOGICAL SYMPOSIUM No. 67
ISBN 0 19 854945 8

For instance, because the lights may attract large numbers of nocturnal insects and hence alter their spatial distribution, the availability of food for nocturnal aerial insectivores may change. By causing food to concentrate at predictable locations, street lamps may perhaps enhance the foraging success of insectivorous bats, thereby having beneficial effects on their populations. Aerial-hawking insectivorous bats frequently exploit concentrations of insects in an opportunistic fashion, including swarms that are formed naturally, as well as those that constitute aggregations around artificial light sources (Fenton & Morris 1976; Bell 1980; Vaughan 1980; Fenton this volume pp. 277–289). Moreover, streets and roads constitute 'linear landscape elements' which are important for the movements of bats in open areas (Limpens, Helmer, van Winden & Mostert 1989).

Since street lights are widely used in many parts of the world, their role as insect attractors may have implications for the understanding of current bat population trends, as well as for bat and insect conservation. However, the present review will be limited geographically to the north temperate regions (mainly Europe and North America), since this is where the relevant studies have been carried out.

Habitat selection by bats in relation to lights: case studies

Several recent studies on habitat selection by European and North American bats have been made in or near built-up areas and have specifically considered the response of bats to artificial lights.

Noctules, *Nyctalus noctula* (Schreber), monitored by radio-tracking in southern Germany (Kronwitter 1988), spent most of their foraging time (65%) either over a lake or in a town, in the latter case over open asphalt illuminated by strong lights, i.e. a car park and a road junction. Adjacent woods and farmlands were used only occasionally. Typically, the bats (about 30 individuals) fed over the lake at dusk and later moved on to feed over the lights in the town. During the second foraging period, i.e. after full darkness, as much as 75% of the foraging time was spent among the lights.

In the Bialowieza forest in Poland, in contrast, where there are several villages but no street lamps, the behaviour of *N. noctula* was different. In this case, the bats were monitored with bat detectors in five different habitats (Rachwald 1993). Activity was consistently highest over water but small forest clearings, maintained for traditional farming, were also frequently exploited. The villages were not preferred to other open areas.

A bat detector survey in Quebec, Canada (Geggie & Fenton 1985), showed that big brown bats, *Eptesicus fuscus* (Palisot de Beauvois), in urban areas concentrated their foraging efforts in habitats with houses and street lamps and, to a lesser extent, over lakes, making less use of nearby farmlands and woods. In another Canadian study, *E. fuscus*, red bats *Lasiurus borealis* (Müller), and hoary bats *L. cinereus* (Palisot de Beauvois), as well as

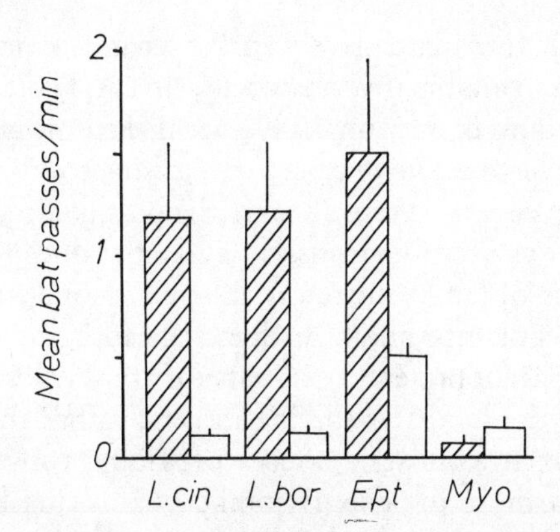

Fig. 1. Bat activity levels (mean and SE), as assessed with a bat detector at sites with lights (hatched) and away from lights in south-western Ontario, Canada. The bats are *Lasiurus cinereus (L.cin), L. borealis (L. bor), Eptesicus fuscus (Ept)* and *Myotis* spp. (*Myo*). The association with lights is significant for the first three ($P < 0.001$, $P < 0.001$ and $P < 0.025$, respectively, Mann-Whitney *U*-tests). Redrawn after Furlonger *et al.* (1987), with kind permission from the publisher and the authors.

Myotis spp., were surveyed with bat detectors in nearly 200 sites in Ontario (Furlonger, Dewar & Fenton 1987). Foraging activity of each of the first three species was positively associated with the presence of street lamps (Fig. 1). The two *Lasiurus* species mostly foraged over lit roads in rural areas or in small towns, while *E. fuscus* were also common in urban habitats. *Myotis* spp., in contrast, appeared to be negatively associated with lights.

A radio-tracking study of serotines, *Eptesicus serotinus* (Schreber), in farmland habitats in southern England (Catto 1993) showed that, second to cattle pastures which provided abundant dung beetles (*Aphodius* spp.), roads with 'white' streetlamps were the most frequently used feeding habitats.

In a Swiss study (Haffner & Stutz 1985–86), pipistrelle bats of two species were monitored with a bat detector from a moving car along 500 km of road. More than a thousand bat passes were detected. Of the common pipistrelles, *Pipistrellus pipistrellus* (Schreber), 45% were encountered near street lamps and of Kuhl's pipistrelle, *P. kuhlii* (Kuhl), the frequency was as high as 94%. The bat density varied between the various types of street lamps and was positively related to the amount of UV radiation emitted.

There is evidence, however, that some species of bats do not make use of illuminated areas, even when such habitats are available. For example, greater horseshoe bats, *Rhinolophus ferrumequinum* (Schreber), have been radio-tracked near Bristol, England, over several seasons. Although villages with 'white' streetlamps occur within the bats' home range, they are never utilized (Jones & Morton 1992; Jones, Duvergé & Ransome this volume pp. 309–324; L. Duvergé pers. comm.). This also applies to the spotted bat, *Euderma maculatum* (Allen), a well-studied long-eared vespertilionid of

North America, which feeds exclusively in woodlands even when there are areas of lighted streets available (Woodsworth, Bell & Fenton 1981; Leonard & Fenton 1983; Wai-Ping & Fenton 1989; M. B. Fenton pers. comm.).

The work briefly reviewed above raises several questions which have been further examined in Sweden and the UK over the past five years. First, to what extent do street lamps attract flying insects and how does this attractiveness vary between the types of lamps currently in use? Second, to what extent do bats forage around street lamps and how does such foraging vary among lamp types? Third, is the apparent preference of bats for areas where there are lighted streets a real effect of the lamps attracting insects, rather than an artefact of correlated variables, such as nearby houses providing roosting sites or trees providing insects? Fourth, is the energy intake of foraging bats enhanced by the presence of street lamps? Fifth, do all bat species exploit concentrations of insects around street lamps? Sixth, do street lamps have implications for population trends in bats and for conservation? These questions will be addressed in the remainder of the paper.

Spectra of the light emitted by street lamps and their attractiveness to insects

There are several types of street lamps currently in use. Most lamps belong to one of three types: mercury (Hg) vapour lamps, which give a bluish-white light and emit a considerable fraction of their energy in the ultraviolet part of the spectrum, low-pressure sodium (Na) lamps, which give a monochromatic orange light, and high-pressure sodium lamps, which give a bright orange light but, since they include mercury vapour, also emit some energy in the ultraviolet range.

In Ontario, Canada, Hickey & Fenton (1990) sampled insects, using window traps placed 2 m above the ground, and showed that moths of more than 10 mm body length, and also flying insects in general, were significantly more abundant within the light cone (<10 m from the lamp) than 50 m away from the lights. On average, the abundance of moths differed by a factor of 36.

Insect densities around street lamps were determined by taking flash photographs of an area of approximately 3 × 2 m around the lamp and subsequently counting the number of white dots, each representing a flying insect, on the projected transparency. First, in a town in southern Sweden (Rydell 1992a), insects were significantly more abundant on photographs taken close to the lamps than on those taken 25 m away from them. Furthermore, there were five times more insects around the mercury vapour lamps than around either of the sodium types. (Mercury-vapour lamps and sodium lamps that occurred adjacent to each other (50 m apart) at street crossings were directly compared, to control for the possibility that differences in the surrounding vegetation were implicated, rather than the light quality.) Likewise, high-pressure sodium lamps attracted significantly more insects than did the low-pressure type. In southern England (Blake, Hutson, Racey, Rydell & Speakman 1994), there were on average eight times more insects around mercury lamps than around sodium lamps.

There is thus good evidence that street lamps attract many flying insects and that mercury vapour lamps are much more attractive to insects than are sodium lamps of either type.

Attractiveness of street lamps to bats

Methods

Radio-tracking and use of ultrasonic detectors ('bat detectors') are both efficient methods for monitoring foraging bats. The former method has recently been used extensively on bats in order to gather information on the behaviour of a limited number of individuals (e.g. Geggie & Fenton 1985; Kronwitter 1988; Barclay 1989; Hickey & Fenton 1990; Jong & Ahlén 1991; Catto 1993). The latter method is usually unsuitable for continuous data collection from individuals. Instead, it is well suited to monitoring of foraging bats without reference to individuals. With practice, it may also allow detection and identification of several bat species simultaneously (e.g. Ahlén 1981).

In the present work, bats were monitored with a bat detector (model D-960, L. Pettersson Elektronik, Uppsala, Sweden, or a QMC S–200, QMC Instruments Ltd., London), either from a moving car or on foot. In the former case, the detector was used in the heterodyne (tuned) mode, concentrating on the frequencies emitted by a particular bat species or a small group of species that use similar frequencies. When used in this mode, these bat detectors have high sensitivity and good filtering capacity, and are able to exclude most of the noise generated by the moving car (Rydell 1991, 1992a; Blake *et al.* 1994). When bats were monitored on foot, the frequency-division (broadband) mode was used to detect all species of bats present. After detection of a bat, the detector was switched to the tuned mode for species identification (Ahlén 1981). Alternatively, tape recordings were made, through the digital time expansion system available on the D-960 model, for later analysis (Rydell 1992a).

The number of bats foraging along roads can be quantified from a moving car, if the travel speed is maintained at 40–50 km/h (Ahlén 1981). When a bat detector is used, foraging bats of most species, with the exception of some 'gleaners', like *Plecotus auritus* (Linnaeus) (Anderson & Racey 1991), can be unambiguously distinguished from commuting ones because they emit 'feeding buzzes', i.e. the sudden increase in pulse repetition rate associated with capture attempts (Griffin, Webster & Michael 1960).

Densities of bats along lit and unlit roads

Northern bats, *Eptesicus nilssonii*, were monitored once or twice each week over 14 months, along a 27-km transect in a rural area in south Sweden, by use of a bat detector from a moving car (Rydell 1991). The detector was tuned to 30 kHz. This frequency corresponds to the highest intensity of the echolocation calls of foraging *E. nilssonii* (Ahlén 1981). Altogether, bats were detected 922 times. Excluding the winter months (November–March), when no bats were encountered, the average number of bats observed along the transect was 17. The distribution of these bats along the transect was significantly affected by mercury-vapour street lamps, which occurred in some sections. In spring (April and early May) and autumn (late August–October), about 90% of the bats were detected along the 23% of the transect that was bordered by such street lamps (Fig. 2). Bat density along sections with street lamps was 2.4 per kilometre on average, compared with 0.11 per kilometre of unlit road.

In summer (late May to early August), the situation was different, with near-random distribution of bats along the transect. This may have been an effect of the light summer nights at high latitudes (in this case 57 °N), when lights attract few if any insects. Such seasonality at street lamps is not evident at lower latitudes, where the lamps seem to be used by bats throughout the summer (e.g. Hickey & Fenton 1990; Blake *et al.* 1994).

Effects of lamp type

A survey through nine small villages in another area of south Sweden during mild (>10 °C) evenings in August, by use of the same monitoring method, suggested that concentrations of bats are more likely near mercury-vapour lamps than near sodium lamps (Rydell 1992a). This is to be expected, in view of the different tendencies of the two lamp types to attract insects. A total of

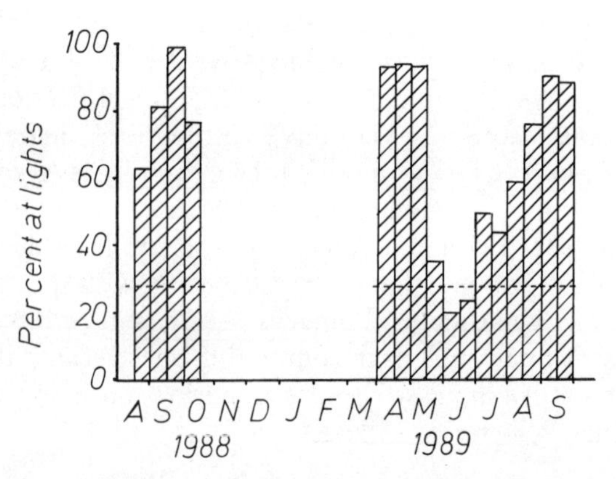

Fig. 2. Percentages of total number of observations of northern bats *Eptesicus nilssonii* that were made in illuminated sections of a 27-km transect in south Sweden. The distribution differed significantly from what would be expected if it was random (horizontal dashed line; $P < 0.001$, χ^2-test). No bats were observed between November and March. Redrawn from Rydell (1991), with kind permission from the publisher.

61 bats, mostly *E. nilssonii*, were detected. In villages lit by mercury-vapour lamps, there was a mean of 5.5 bats per kilometre of road compared with 0.52 bats per kilometre in villages lit by sodium lamps (t =1.95, $d.f.$ = 8, P = 0.088).

A 28-km transect along the Dee river valley in north-east Scotland was traversed five times during August 1993 (J. Rydell, C. M. C. Catto & P. A. Racey unpubl.), by the same method as previously used in Sweden, except that the bat detector was tuned to 50 kHz rather than 30 kHz. This tuning frequency potentially allows detection of *P. pipistrellus* as well as of *Myotis daubentonii*, both of which are common in this area (Speakman, Racey, Catto, Webb, Swift & Burnett 1991). The average density of pipistrelles was 3.1 bats per kilometre along one length of road (1.7 km) with mercury-vapour lamps, but only 0.12 bats per kilometre along roads with sodium lamps (five villages, 14.4 km) and 0.31 bat per kilometre along unlit roads in rural areas without street lamps (12.2 km). Hence, there was a significant concentration of pipistrelles in the village with mercury-vapour lamps (Fig. 3).

In summary, there are concentrations of bats over Swedish and British roads illuminated with mercury-vapour street lamps, with 2–5 bats per kilometre of road. No such concentrations are found over unlit roads or where sodium street lamps are used. In both countries, the density of foraging bats appears to be at least ten times higher over roads with mercury-vapour street lamps than over roads with sodium lamps or unlit roads.

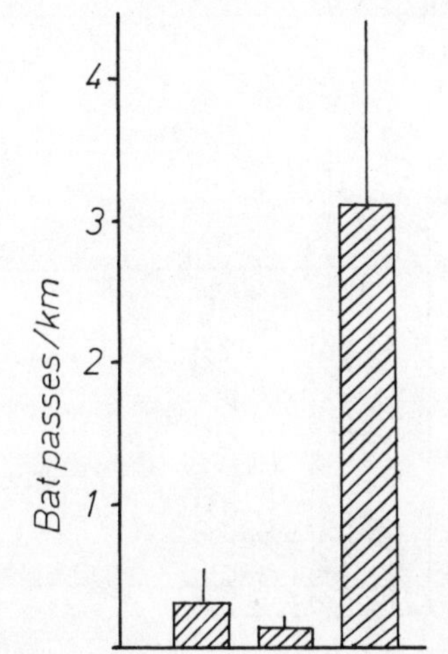

Fig. 3. Density of pipistrelles *Pipistrellus pipistrellus* (mean and SD) along a 28-km stretch of road near Aberdeen, Scotland, in (a) rural sections without street lamps, (b) villages with sodium (orange) street lamps and (c) one village with mercury (white) street lamps. The difference in bat density is significant (P<0.001, ANOVA).

Effects of other habitat variables

A 27-km long transect along lit roads through a small town and its immediate surroundings in southern Sweden was monitored for bats from a car during 13 evenings in spring and autumn (Rydell 1992a). In this case, the transect was divided into 18 sections, each characterized by lamp type (Hg or Na), the amount of woody vegetation next to the road (potential wind shelter for bats and insects), the number of buildings along the road (potential roost sites for bats) and the mean distance to the town centre (an index of noise and air pollution level). In total, 342 *E. nilssonii* and 34 particoloured bats *Vespertilio murinus* Linnaeus were detected. The mean bat density in each section was then entered into an ANOVA, using the habitat characteristics as independent variables. Lamp type and vegetation had significant positive effects on the distribution of bats, but the other variables did not. Bats were most common in rural and residential areas away from the town centre, i.e. where mercury-vapour lamps were most frequent (Fig. 4).

This result provides further evidence that foraging bats are attracted to mercury-vapour streetlamps. However, it could not be shown conclusively that the apparent effect of lamp type was independent of that of trees. Although correlations were not significant, mercury street lamps might have been slightly more frequent in areas with much vegetation, while sodium lamps may have occurred in more open situations.

In a study in southern England (Blake *et al.* 1994), roads with mercury-vapour lamps attracted about three times more bats than did roads lit by

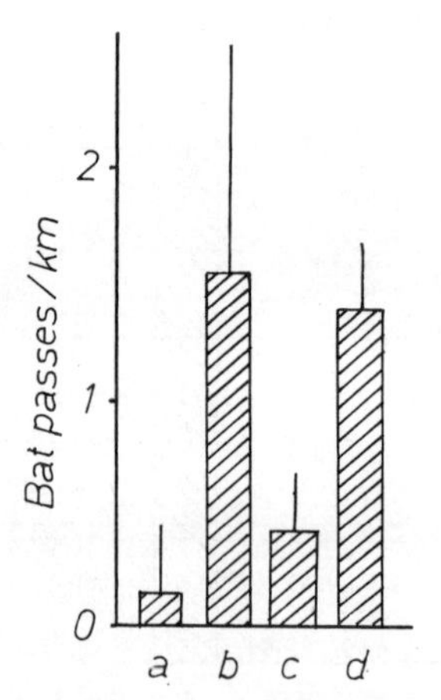

Fig. 4. Density (mean and SD) of northern bats in (a) urban, (b) residential, (c) industrial and (d) rural parts of a small town in southern Sweden. The density differed significantly between the different parts of the town and was affected by lamp type and presence of trees ($P<0.05$ and $P<0.01$, respectively). Redrawn from Rydell (1992a).

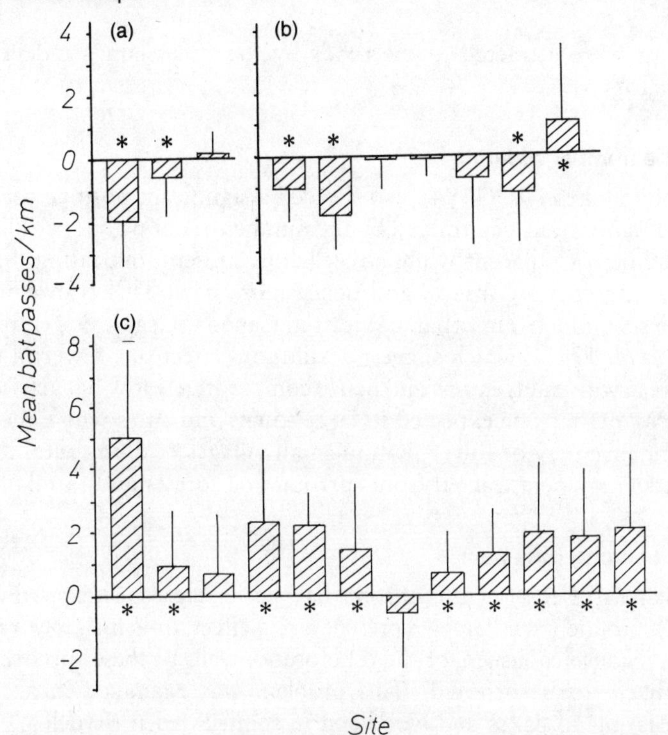

Fig. 5. Bat activity (mean and SD) along (a) unlit roads, (b) roads with sodium (orange) street lamps and (c) roads with mercury-vapour (white) street lamps in southern England. Bat activity is expressed as the difference between each of 22 sections and an adjacent unlit control, i.e. negative activity means that the activity was higher in the control section. Asterisks denote significant differences in bat activity between the section and its unlit control ($P<0.05$, t-tests). Overall, sections with mercury-vapour lamps were most attractive to bats ($P < 0.01$, ANOVA). Redrawn from Blake *et al.* (1994).

orange streetlamps or unlit roads (3.2, 1.2 and 0.7 bats per kilometre, respectively). This difference was significant, again suggesting that bats were attracted to mercury lamps. Potential effects of correlated variables were controlled for in this case. Bat activity, as measured by the mean number of bat passes per kilometre in each of 20 sections, was compared with the activity in an adjacent unlit control section by means of paired t-tests. Across all the comparisons, the bat activity in sections with mercury-vapour streetlamps, relative to their controls, was significantly higher than in sections lit by sodium lamps (Fig. 5). A direct comparison between two sections with mercury-vapour lamps and adjacent sections with sodium lamps gave the same result. In a covariance analysis on the mean bat activity in each section (relative to the activity in the adjacent control), using lamp type as the independent variable and the amount of vegetation and number of houses next to the road as covariates, only lamp type remained as a significant factor affecting mean bat activity. These results provide conclusive evidence that the bats were attracted to the roads by the street lamps independently of other factors.

Effect of the number of lights

The study by Blake *et al.* (1994) also showed a significant positive correlation between bat activity, as determined by the number of bat passes per kilometre, and the number of adjacent white street lamps present, indicating that more lamps may attract more insects and hence more bats. This is in contrast to other studies, conducted in urban habitats in Canada (Geggie & Fenton 1985; Furlonger *et al.* 1987), which suggest a 'dilution' effect (i.e. fewer insects per lamp) in areas with many lamps and hence comparatively low bat activity. Such an effect may perhaps be expected in large towns and cities where the overall insect abundance may be lower than in small villages or along rural roads, to which insects may be attracted from surrounding forests and farmland.

Differences among bat species

Does the scarcity of observations of some bat species, e.g. *Myotis* spp. (Furlonger *et al.* 1987), around street lamps represent a real effect, or is it simply a result of the relative inconspicuousness of the echolocation calls of these species, making them less likely to be detected? This problem was examined in an area of small-scale farmland, lakes and woodland in south Sweden (Rydell 1992a). At least ten species of insectivorous bats were present in this area, as confirmed by preliminary bat-detector monitoring in combination with mist-netting in front of underground mating roosts. Foraging bats were monitored with a bat detector, used in a broad-band mode, in 0.5 ha sites in representative habitats. Each site was sampled once for 30 min. Only four species were observed in sites with street lamps (n=8 sites), as compared with ten species in sites away from street lamps (n=35 sites). More importantly, the frequency of the various species observed near the lamps differed significantly form the frequency observed away from the lamps (Fig. 6). In particular, the five *Myotis* spp., as well as the brown long-eared bat *Plecotus auritus*, were only observed away from lamps.

Bats that feed regularly over lit roads, e.g. species belonging to the genera *Nyctalus, Lasiurus, Eptesicus* and *Pipistrellus*, have some characteristics in common, which are not found in those bats that apparently avoid street lamps, e.g. the genera *Myotis, Euderma, Plecotus* and *Rhinolophus*. The former fly relatively fast and usually forage in open habitats (Baagøe 1987). Their echolocation pulses have relatively slow repetition rates, include a narrow-band component (Ahlén 1981) and are adapted for target detection in the open air (Schnitzler & Henson 1980). The latter group, in contrast, fly relatively slowly and manoeuvrably. They either use broad-band echolocation pulses and fast repetition rates (the vespertilionids) or very long, high-frequency, narrow-band calls (the rhinolophids). In both cases, the pulses are suitable for target detection in cluttered situations (Schnitzler & Henson 1980;

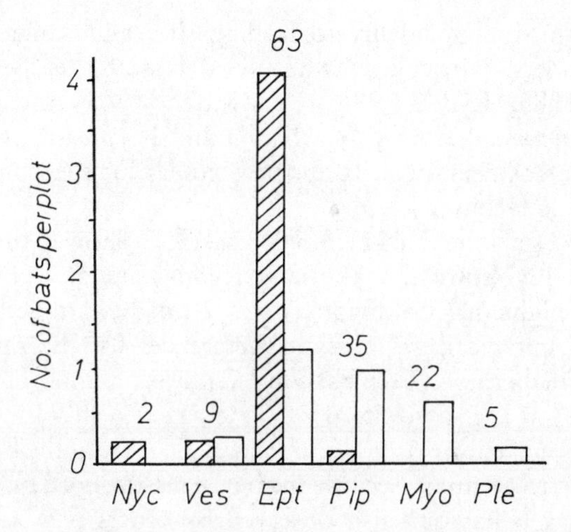

Fig. 6. The number of bats observed in 8 plots (each 0.5 ha, searched for 30 min) in villages with mercury-vapour street lamps (hatched) and in 35 plots in habitats away from street lamps (unhatched) in southern Sweden. The bats are *Nyctalus noctula* (*Nyc*); *Vespertilio murinus* (*Ves*), *Eptesicus nilssonii* (*Ept*), *Pipistrellus pipistrellus* (*Pip*), *Myotis* spp. (*Myo*) and *Plecotus auritus* (*Ple*). The distribution of bat species near street lamps differed from the distribution away from lamps ($P<0.001$, χ^2-test).

Baagøe 1987) and for a gleaning foraging technique (e.g. Swift & Racey 1983; Anderson & Racey 1991; Jones & Morton 1992).

Behaviour of bats foraging near street lamps

When feeding near street lamps, *E. nilssonii* typically fly back and forth in straight lines, at least 100 m along the road before turning back, usually staying several metres from the surrounding vegetation and 8–15 m above the ground. They occasionally make extensive vertical dives towards the ground in pursuit of insects. Hence, the bats fly above the lamps most of the time, where they usually cannot be seen. They seldom cross the light cone except when chasing insects (Rydell 1991). The behaviour of other bats that forage near street lamps appears to be similar (e.g. Belwood & Fullard 1984; Schnitzler, Kalko, Miller & Surlykke 1987; Kronwitter 1988). The habit of keeping away from the light cone, where insect density is highest (Hickey & Fenton 1990), most of the time may be a predation-avoidance strategy; alternatively, foraging on dispersed insects away from the light cone may be more efficient than foraging on dense swarms because of a 'confusion effect' (Barák & Yom-Tov 1989).

If the air space around street lamps provides profitable and predictable food sources, individual bats may be expected to show fidelity to such sites and perhaps monopolize access to them, at least under some circumstances (i.e. when the sites are economically defensible). In patchy habitats, such as open farmland with isolated trees and buildings, individual bats (*P. pipistrellus* and *E. nilssonii*) show high fidelity to feeding sites and sometimes defend them against intruders, particularly when food is scarce early in the year (Racey & Swift 1985; Rydell 1986). Kronwitter (1988) and Belwood & Fullard (1984) suggested that bats (*N. noctula* and *L. cinereus*, respectively), sometimes defend areas near street lamps in a similar manner, but conclusive evidence for this was lacking.

By means of radio-tracking, Hickey & Fenton (1990) showed that individual *L. borealis* consistently returned to the same feeding site, at a group of street lamps, night after night ($n=5$ individuals and 20 nights altogether) and the bats even showed statistically significant preferences for the same light on subsequent visits. Furthermore, each bat consistently spent most of the foraging period (about 60%) at the same site.

However, Hickey & Fenton (1990) found no evidence that the bats defended the area around the streetlamps. Because the bats eavesdropped on each other's echolocation calls, a behaviour also observed by others (e.g. Griffin 1958; Barclay 1982), the density of bats at the study site was consistently very high, with up to seven bats simultaneously exploiting the same light. Furthermore, insect-trapping near the lights showed that the resource was still not depleted significantly by the bats. Under such circumstances, territorial defence should not be expected because the cost of defending the site would be high (high invasion rate) and the benefit low (no depletion of prey).

Why do some bat species avoid street lamps?

Fenton & Morris (1976) and Bell (1980) used ultraviolet lights ('blacklights') to study the response of insectivorous bats to induced patches of prey in the North American desert. They found that the areas near the lamps were quickly exploited by several species of aerial-hawking bats and in particular by several species of *Myotis*. This suggests that the usual absence of *Myotis* spp. around streetlamps has less to do with their failure to locate or exploit concentrations of prey than with the bright light conditions prevailing over lit roads, perhaps in combination with the openness of the habitat.

As mentioned above, the echolocation pulses used by most *Myotis* and *Plecotus* spp. lack the constant frequency component believed to be necessary for long-range detection of insects in open air and they are also relatively slow fliers (Baagøe 1987). Perhaps, therefore, these bats may not be able to exploit insects efficiently in open situations. Alternatively, predation risk in bright light conditions, in combination with open situations, may be too high for slow-flying bats (Fenton & Rautenbach 1986; Speakman 1991). Evidence from high latitudes (60–65° N) in Scandinavia and Finland suggests that bats, particularly *Myotis* spp., tend to avoid open habitats, such as lakes, in the bright light conditions prevailing around midsummer, but return to such areas as nights become darker later in the year (Nyholm 1965; Rydell 1992b).

Do street lamps really enhance the foraging efficiency of bats?

Hickey & Fenton (1990) estimated that *L. borealis* that consistently foraged near a group of streetlamps captured a mean of 6.3 g of insects, mostly moths, during less than 2 h of foraging each night. This amounts to 42% of the bat's body mass. However, Hickey & Fenton (1990) did not estimate the feeding rate of bats foraging away from the lights, so no comparison with alternative habitats could be made.

Geggie & Fenton (1985) used a bat detector to count feeding buzzes emitted by *E. fuscus* that foraged in different habitats. Bats foraging along lit roads in rural settings had attack rates that were higher than those of bats foraging in parkland and over fields away from the lamps, and were comparable to the rate observed over lakes. Likewise, attack rates of bats foraging over lit roads in the rural areas were higher than those of bats foraging in urban settings. Radio-tracking also revealed that the bats in the rural areas spent less time foraging each night than their conspecifics in the urban areas.

Although the results indicate that lit roads provide more profitable feeding for *E. fuscus* than do most other habitats (except perhaps lakes), the evidence is not conclusive. This is because different habitats may provide different-sized prey items, and the infrequent occurrence of feeding buzzes in some habitats may indicate that fewer large prey items, rather than many small ones, are pursued there. Likewise, capture success, which was not controlled for, may vary considerably among different insect groups and hence among habitats (see below).

The profitability of lit roads compared with other available habitats was further examined in Sweden by using *E. nilssonii* marked with reflective tape (Rydell 1989, 1992a). Detailed observations at street lamps near the bats' maternity roost were made in August, i.e. during the bats' lactation period. In this situation, the bats fed on moths, each with a mean mass of 40 mg dry weight (moths flying in the light cone were captured with a hand net, dried and weighed). The mean attack rate on these moths was 1.6/min and the capture success rate was 36% (n=56). Hence, the bats' mean feeding rate at this site was 23 mg (dry weight) per minute. At a nearby cattle pasture, the bats captured dung beetles (mostly *Aphodius* spp; average mass 10.1 mg dry weight) which occurred in abundance. In this case the attack rate was 2.6/min and the capture success rate was 100% (n=24). Hence, in this case, the mean feeding rate was 26 mg (dry weight) per minute. Capture success rate was determined visually in the case of moths. For dung beetles, it could actually be heard when they were captured and subsequently chewed by the bats. In forest habitats, in contrast, the same bats attacked between six and ten small dipterans per minute (average mass 1.0 mg dry weight). Capture efficiency could not be determined in this case, because the dipterans were too small to be observed continuously. However, even if each feeding attempt resulted in a successful capture, the feeding rate only amounted to 6–10 mg (dry weight) per minute.

More and better estimates of the foraging efficiency of bats near street lamps and in other habitats are needed. In particular, the foraging efficiency and reproductive success of bat colonies that have access to mercury street lamps should be compared with those of colonies that do not. Meanwhile, limited evidence (Geggie & Fenton 1985; Rydell 1992a) suggests that roads with street lamps, particularly in rural areas, may enhance the food intake by *Eptesicus* species.

Implications for bat conservation

The frequent use of street lamps by foraging bats has obvious implications for conservation. In contrast to many other bat habitats, illuminated streets and roads will most probably remain widespread even in the future, to the benefit of at least some bat species. There has, however, been a recent tendency to replace mercury-vapour lamps with sodium lamps, which use less energy (usually 70 W as compared to 100 or 125 W) and do not demand processing and handling of poisonous mercury, but are of less value to bats, at least in western Europe.

Bat species are not all likely to be affected equally by the presence of street lamps. In Europe and North America, those most likely to benefit are aerial-hawking species (i.e. of the vespertilionid genera *Nyctalus, Vespertilio, Lasiurus, Eptesicus* and *Pipistrellus*). There is little evidence that any species in these genera is threatened at present (Stebbings 1988). On the other hand, there are many European and North American species of the genera *Plecotus, Myotis* (Vespertilionidae) and *Rhinolophus* (Rhinolophidae), which apparently do not take advantage of street lamps on a regular basis, that are endangered at least in some countries and have recently suffered drastic population declines (Stebbings 1988). These species, which apparently hesitate to cross illuminated areas, may perhaps even find rows of street lamps to be a problem, since their movements may be restricted by, for example, illuminated motorways (R. E. Stebbings pers. comm.).

In conclusion, it seems that street lamps cannot be expected to improve the situation for most endangered bat species, but their presence may perhaps help to explain the relative success of some other species, at least in Europe, i.e. *Vespertilio murinus, Eptesicus nilssonii, E. serotinus, Pipistrellus pipistrellus* and *P. kuhlii*, which have shown recent increases in numbers and/or range expansions (Baagøe 1986; Zukál & Gaisler 1989; Ahlén & Gerell 1989; Speakman *et al.* 1991). However, if populations of these bats have benefited from white street lamps in the past, which has yet to be shown conclusively, it will be of interest to monitor their populations as mercury-vapour lamps are exchanged for sodium lamps.

Acknowledgements

This work was funded by the Swedish Environmental Protection Agency and the Natural Science Research Council of Sweden (JR) and the University of Aberdeen (PAR). We acknowledge M. B. Fenton, J. R. Speakman and M.R. Young for constructive comments on the manuscript.

References

Ahlén, I. (1981). Identification of Scandinavian bats by their sounds. *Rapp. Sver. Lantbruksuniv. Inst. Viltekol.* **6**: 1–56.

Ahlén, I. & Gerell, R. (1989). Distribution and status of bats in Sweden. In *European bat research 1987:* 319–325. (Eds Hanák, V., Horáček, I. & Gaisler, I.). Charles University Press, Prague.

Anderson. M. E. & Racey, P. A. (1991). Feeding behaviour of captive brown long-eared bats, *Plecotus auritus*. *Anim. Behav.* **42**: 489–493.

Baagøe, H. J. (1986) [1987]. Summer occurrence of *Vespertilio murinus* Linné—1758 and *Eptesicus serotinus* (Schreber—1780) (Chiroptera, Mammalia) on Zealand, Denmark, based on records of roosts and registrations with bat detectors. *Annln naturh. Mus. Wien (B)* **88/89**: 281–291.

Baagøe, H. J. (1987). The Scandinavian bat fauna: adaptive wing morphology and free flight in the field. In *Recent advances in the study of bats*: 57–74. (Eds Fenton, M. B., Racey, P. A. & Rayner, J. M. V.). Cambridge University Press, Cambridge, U. K.

Barák, Y. & Yom-Tov, Y. (1989). The advantage of group hunting in Kuhl's bat *Pipistrellus kuhli* (Microchiroptera). *J. Zool., Lond.* **219**: 670–675.

Barclay, R. M. R. (1982). Interindividual use of echolocation calls: eavesdropping by bats. *Behav. Ecol. Sociobiol.* **10**: 271–275.

Barclay, R. M. R. (1989). The effect of reproductive condition on the foraging behaviour of female hoary bats, *Lasiurus cinereus*. *Behav. Ecol. Sociobiol.* **24**: 31–37.

Bell, G. P. (1980). Habitat use and response to patches of prey by desert insectivorous bats. *Can. J. Zool.* **58**: 1876–1883.

Belwood, J. J. & Fullard, J. H. (1984). Echolocation and foraging behaviour in the Hawaiian hoary bat, *Lasiurus cinereus semotus*. *Can. J. Zool.* **62**: 2113–2120.

Blake, D., Hutson, A. M., Racey, P. A., Rydell, J. & Speakman, J. R. (1994). Use of lamplit roads by foraging bats in southern England. *J. Zool., Lond.* **234**: 453–462.

Catto, C. M. C. (1993). *Aspects of the ecology and behaviour of the serotine bat* (Eptesicus serotinus). PhD. thesis: University of Aberdeen.

Fenton, M. B. & Morris, G. K. (1976). Opportunistic feeding by desert bats (*Myotis* spp.). *Can. J. Zool.* **54**: 526–530.

Fenton, M.B. & Rautenbach, I. L. (1986). A comparison of the roosting and foraging behaviour of three species of African insectivorous bats (Rhinolophidae, Vespertilionidae, and Molossidae). *Can. J. Zool.* **64**: 2860–2867.

Furlonger, C. L., Dewar, H. J. & Fenton, M. B. (1987). Habitat use by foraging insectivorous bats. *Can. J. Zool.* **65**: 284–288.

Geggie, J. F. & Fenton, M. B. (1985). A comparison of foraging by *Eptesicus fuscus* (Chiroptera: Vespertilionidae) in urban and rural environments. *Can. J. Zool.* **63**: 263–267.

Griffin, D. R. (1958). *Listening in the dark: the acoustic orientation of bats and men.* Yale University Press, New Haven.

Griffin, D. R., Webster, F. A. & Michael, C. R. (1960). The echolocation of flying insects by bats. *Anim. Behav.* **8**: 141–154.

Haffner, M. & Stutz, H. P. (1985–86). Abundance of *Pipistrellus pipistrellus* and *Pipistrellus kuhlii* foraging at street lamps. *Myotis* **23/24**: 167–172.

Hickey, M. B. C. & Fenton, M. B. (1990). Foraging by red bats (*Lasiurus borealis*): do intraspecific chases mean territoriality? *Can. J. Zool.* **68**: 2477–2482.

Jones, G. & Morton, M. (1992). Radio-tracking studies on habitat use by greater horseshoe bats (*Rhinolophus ferrumequinum*). In *Wildlife telemetry: remote monitoring and tracking of animals*: 521–537. (Eds Priede, I.G. & Swift, S. M.). Ellis Horwood, New York, London etc.

Jong, J. de, & Ahlén, I. (1991). Factors affecting the distribution of bats in Uppland, central Sweden. *Holarct. Ecol.* **14**: 92–96.

Kronwitter, F. (1988). Population structure, habitat use and activity patterns of the noctule bat, *Nyctalus noctula* Schreb., 1774 (Chiroptera: Vespertilionidae) revealed by radio-tracking. *Myotis* **26**: 23–85.

Leonard, M. L. & Fenton, M. B. (1983). Habitat use by spotted bats (*Euderma maculatum*, Chiroptera: Vespertilionidae): roosting and foraging behaviour. *Can. J. Zool.* **61**: 1487–1491.

Limpens, H. J. G. A., Helmer, W., van Winden, A. & Mostert, K. (1989). Bats (Chiroptera) and linear landscape elements. *Lutra* **32**: 1–20.

Nyholm, E. S. (1965). Zur Ökologie von *Myotis mystacinus* (Leisl.) und *M. daubentoni* (Leisl.) (Chiroptera). *Annls zool. fenn.* **2**: 77–123.

Racey, P. A. & Swift, S. M. (1985). Feeding ecology of *Pipistrellus pipistrellus* (Chiroptera: Vespertilionidae) during pregnancy and lactation. I. Foraging behaviour. *J. Anim. Ecol.* **54**: 205–215.

Rachwald, A. (1993). Habitat preference and activity of the noctule bat *Nyctalus noctula* in the Bialowieza Primeval Forest. *Acta theriol.* **37**: 413–422.

Rydell, J. (1986). Feeding territoriality in female northern bats, *Eptesicus nilssoni*. *Ethology* **72**: 329–337.

Rydell, J. (1989). Feeding activity of the northern bat *Eptesicus nilssoni* during pregnancy and lactation. *Oecologia* **80**: 562–565.

Rydell, J. (1991). Seasonal use of illuminated areas by foraging northern bats *Eptesicus nilssoni*. *Holarct. Ecol.* **14**: 203–207.

Rydell, J. (1992a). Exploitation of insects around street-lamps by bats in Sweden. *Funct. Ecol.* **6**: 744–750.

Rydell, J. (1992b). Occurrence of bats in northernmost Sweden (65 °N) and their feeding ecology in summer. *J. Zool., Lond.* **227**: 517–529.

Schnitzler, H. U. & Henson, O. W. (1980). Performance of airborne animal sonar systems: I. Microptera. In *Animal sonar systems*: 109–181. (Eds Busnel, R.-G. & Fish, J. F.). Plenum Press, New York & London. (*NATO adv. Stud. Inst. Ser. (A)* **28**.)

Schnitzler, H. U., Kalko, E., Miller, L. A. & Surlykke, A. (1987). The echolocation and hunting behaviour of the bat, *Pipistrellus kuhli*. *J. comp. Physiol. (A)* **161**: 267–274.

Speakman, J. R. (1991). Why do insectivorous bats in Britain not fly in daylight more frequently? *Funct. Ecol.* **5**: 518–524.

Speakman, J. R., Racey, P. A., Catto, C. M. C., Webb, P. I., Swift, S. M. & Burnett, A. M. (1991). Minimum summer populations and densities of bats in N. E. Scotland, near the northern borders of their distribution. *J. Zool., Lond.* **225**: 327–245.

Stebbings, R. E. (1988). *Conservation of European bats*. Christopher Helm, London.

Swift, S. M. & Racey, P. A. (1983). Resource partitioning in two species of vespertilionid bats (Chiroptera) occupying the same roost. *J. Zool., Lond.* **200**: 249–259.

Vaughan, T. A. (1980). Opportunistic feeding by two species of *Myotis*. *J. Mammal.* **61**: 118–119.

Wai-Ping, V. & Fenton, M. B. (1989). Ecology of spotted bat (*Euderma maculatum*) roosting and foraging behaviour. *J. Mammal.* **70**: 617–622.

Woodsworth, G. C., Bell, G. P. & Fenton, M. B. (1981). Observations of the echolocation, feeding behaviour, and habitat use of *Euderma maculatum* (Chiroptera: Vespertilionidae) in southcentral British Columbia. *Can. J. Zool.* **59**: 1099–1102.

Zukál, J. & Gaisler, J. (1989). [On the occurrence and changes in abundance of *Eptesicus nilssoni* (Keyserling et Blasius, 1839) in Czechoslovakia.] *Lynx* **25**: 83–95. [In Czech.].

Symp. zool. Soc. Lond. (1995) No. 67: 309–324

Conservation biology of an endangered species: field studies of greater horseshoe bats

Gareth JONES,
P. Laurent DUVERGÉ
and Roger D. RANSOME

*School of Biological Sciences
University of Bristol
Woodland Road
Bristol BS8 1UG, UK*

Synopsis

We overview an autecological study of the endangered greater horseshoe bat *Rhinolophus ferrumequinum* in south-west England. Bats from a woodland site generally emerged earlier than did bats from an exposed roost, except in early spring when tree leaf cover was minimal. Foliage around the roost may benefit bats by extending foraging time and reducing predation. Bats older than one year foraged between 2 and 4 km from their day roosts. Hence conservation of foraging habitats within this range is important. Ancient semi-natural deciduous woodland was used intensively by foraging bats during spring, while during late summer the bats fed mainly over pasture. Woodland was usually warmer than pasture, and the relative temperature difference between the two habitats was greatest at low temperatures. Insect abundance increased rapidly above 6–10 °C. Hence, in spring, it may be more profitable for bats to forage in woodland. The shift to feeding over pasture was associated with the dominance of *Aphodius* dung beetles in the diet during summer. Juveniles foraged independently of their mothers before weaning. Prime foraging habitat close to the maternity roost is probably important to initial and long-term juvenile survival. The hibernation requirements of *R. ferrumequinum* are briefly reviewed. Insights into the foraging needs (e.g. commuting distances) of other species may be gained from predictions based on flight morphology.

Introduction

The protection of foraging habitats is arguably the aspect of bat conservation that most urgently requires implementation. In the United Kingdom all bat species and bat roosts are protected by the Wildlife and Countryside Act (1981), but the need to protect feeding habitats has, until recently, received little attention, although legislation is being prepared to protect important feeding areas used by bats (see Hutson 1993). However, basic biological

ZOOLOGICAL SYMPOSIUM No. 67
ISBN 0 19 854945 8

information on where the bats feed, what they feed on and how far they travel is lacking for most species.

The aim of this study is to provide recommendations for the conservation of feeding grounds used by the endangered greater horseshoe bat *Rhinolophus ferrumequinum*. We give an overview of an autecological study of this species, which we hope will be of use in assessing the habitat requirements of other bats. We also aim to show how foraging ranges may be estimated for other species by using predictions based on flight morphology.

The greater horseshoe bat is considered to be endangered in Britain and Europe, with the current British population estimated at *c.* 4000 individuals in about 14 major maternity colonies (Hutson 1993). The British population of *R. ferrumequinum* has declined substantially during the past century (Stebbings & Arnold 1987), though detailed population studies show that numbers remained relatively stable from the mid 1960s in some areas until a serious decline after 1986 (Ransome 1989). A recent review of the biology of this species is given by Ransome (1991). The aim of the present paper is to answer the following questions:

1. How important are the conditions around the roost for conservation?
2. How far do the bats travel to forage, and what sort of area around a maternity roost should be conserved if feeding habitats are to be protected?
3. How is habitat use influenced by habitat microclimate and insect abundance?
4. How does diet reflect habitat use?
5. Do the habitat needs of juveniles differ from those of adults?
6. Can foraging distances of other bat species be predicted from flight morphology?

In addition we briefly review studies on the hibernation requirements of *R. ferrumequinum*.

The study populations

The work described here was performed in three areas within 50 km of Bristol, Avon. The sites are described in Ransome (1968, 1990). Most of the radio-tracking work was performed at a site on the edge of the Mendip Hills. The bats here (Site A, up to 110 adults, 30–50 young born per year) roosted in a disused mine in woodland. Breeding also occurred in a nearby (2 km distant) stable block. At least 35 bats winter in underground sites within 2 km of Site A, though many individuals fly to sites 10 km distant in the Mendips to hibernate. Site B (up to 210 adults, 70–90 young born per year) is in a coach-house in south-west Avon. Site C (currently up to 90 adults, 25–30 young born per year) is Woodchester Mansion near Stroud, Gloucestershire (see Ransome 1989, 1990 for details). Most of the work described here was performed at Site A, though smaller-scale studies were conducted at Sites B and C to investigate the generality of our findings. Sixty-six bats were radio-tracked between 1990 and 1993, giving 192 complete nights (contact maintained with bats from dusk to dawn with no gaps of longer than 1 h) of telemetric data. Bats up to three months of age are referred to as juveniles; bats which have not yet reached breeding age (usually three years) are termed subadults. Day 0 is the day of birth, and 'first-year bats' are animals in their first year of life.

The importance for conservation of conditions around the roost

Many bat species, including *R. ferrumequinum*, sometimes linger close to the roost before leaving for more distant foraging areas. Conditions around the roost may have important implications for the conservation of bats, especially in terms of protection from predators and extending potential foraging time. To illustrate the effects of cover on the emergence behaviour of the greater horseshoe bats, we documented emergence times at Site A, which is surrounded by woodland, and at the associated stable site (which has an open aspect) on the same nights (Fig. 1). Median emergence time (MET) was always earlier at Site A than at the stable block, except on the first visit of the year (6 May 1992, 21 April 1993), when bats emerged from the stable site with the same or earlier MET than at Site A. From May onwards, the woodland leaf cover caused the area around Site A to be darker than that at the stable block. Under these conditions bats emerged with a MET 5–37 min earlier at Site A. For six nights when light level data were available, bats emerged at slightly higher light intensities at Site A than at the stable block (Site A mean light level at MET = 0.83 ± 0.41 W/m^2, stable block = 0.00 W/m^2 on all six nights; paired *t*-test $t_6 = 5.00$, $P<0.01$). Thus bats at Site A emerged not only earlier, but also at higher light intensities. This may be because the perceived risk of predation was lower in the roost surrounded by trees.

Daylight flying appears to increase the risk of being captured by diurnal avian predators (Speakman 1990, 1991). Presumably such predation risks will also affect emergence time, with bats emerging only once the risks of being seen by visual predators are minimized. Predation from diurnal predators is probably important in shaping the emergence patterns of *R. ferrumequinum*. Sparrowhawk, *Accipiter nisus*, predation occurs at Site B, with at least two different hawks capturing bats in 1992–1993, and one bird capturing about 10 bats over a two-week period. The presence of tree cover around roosts may screen the bats from predatory birds. Furthermore, many aerial insects show a peak in abundance soon after dusk (Racey & Swift 1985), so the extension of feeding time into this peak may be important, especially when energy demands are high, as in lactation. Therefore, all roosts used by *R. ferrumequinum* should be at least partly surrounded by tree cover.

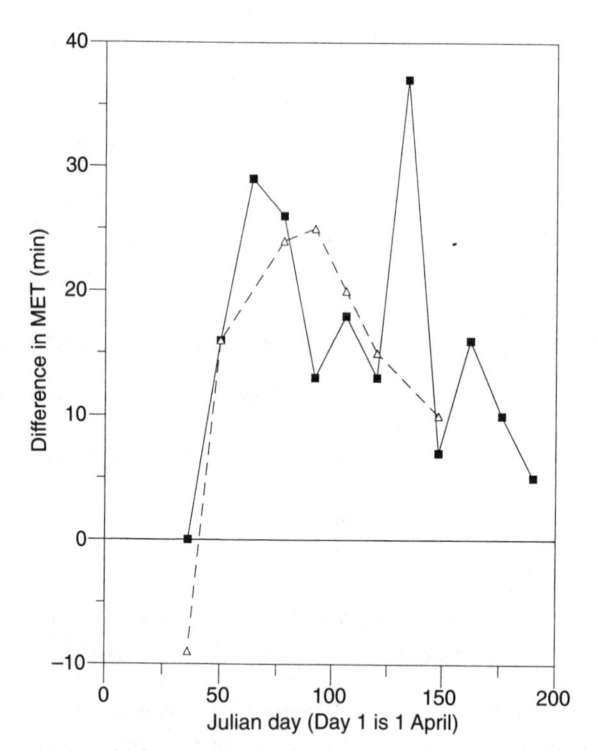

Fig. 1. The difference between median emergence time (MET) at two roost sites (Roost A and a neighbouring roost 2 km distant). Roost A is in woodland, while the neighbouring roost is exposed. The difference in MET is expressed as (Roost B minus neighbouring roost), so that positive values reflect earlier emergence at roost A. Data for 1992 are shown by squares and a solid line, 1993 by triangles and a pecked line.

Foraging distances, habitat use and diet

Foraging distances

Our data on habitat use have been obtained by radio-tracking, and the methodology is summarized by Jones & Morton (1992). All data presented here are for bats which were tracked for a complete night (defined above).

After their first year of life, *R. ferrumequinum* travel, on average, between 2 and 3 km from their day roosts to feeding areas (Table 1). These distances are representative of bats at all three study sites, so foraging range seems to be independent of colony size. Most of the data are from nursery colonies, though most bats spent at least one day in satellite roosts close to their nursery colony. Excluding the data for post-lactating females (taken from one individual only), the greatest foraging distances were found in lactating females (Table 1). Thus, lactating *R. ferrumequinum* differ from female *P.*

Table 1. The effect of sex and age on foraging distances in adult greater horseshoe bats. Data were collected in 1991 and 1992. Total distance represents the distance covered in the whole night, range is the straight line distance to the furthest point travelled.

Bat class	Bats (n)	Nights (n)	Total distance (km)	Range (km)
Females:				
Subadult	7	18	7.94±3.16	2.50±1.16
Pregnant	4	8	7.47±1.55	2.05±0.34
Lactating	3	8	12.90±2.85	2.84±0.91
Post-lactating	1	3	8.11±1.82	4.03±0.49
Males:				
Subadult	4	14	8.70±2.62	2.31±0.82

pipistrellus, whose foraging ranges remained the same or contracted between pregnancy and lactation (Racey & Swift 1985). However, the Ozark big-eared bat *Plecotus townsendii ingens* also achieves maximum foraging range during lactation (Clark, Leslie & Carter 1993). The total distance covered in a night was also greatest for lactating females. The increased flight time involved must add to the already high energetic costs of lactation (Racey & Speakman 1987) and presumably contributes to an increase in food consumption. Indeed, dawn feeds of *R. ferrumequinum* result in the production of approximately 380 ± 153 mg (n=10) dry droppings per bat in non-breeding females compared with 544 ± 118 mg (n=24) from lactating females captured on the same date in July 1993 (t_{32} = 3.38, P<0.01: see below for methodology). From our studies, it seems that to conserve foraging areas up to 3–4 km from the nursery sites is crucial for conservation of *R. ferrumequinum* habitat.

Habitat use and diet

Greater horseshoe bats at Site A forage mainly in ancient woodland during the spring, and over pasture habitats during late summer (Jones & Morton 1992; P. L. Duvergé unpubl.). A similar shift in habitat use was alluded to by Stebbings (1989), who performed preliminary radio-tracking work on this species (Stebbings 1982). This shift in habitat use was evident also from dietary studies (Jones 1990). In a study close to Site A during 1988, *R. ferrumequinum* ate mainly moths (41% of diet by volume) and beetles (33% of diet). Beetles were especially important from July to August, when scarabaeid beetles of the genus *Aphodius* dominated the diet (64% of all beetles by volume in 1988). Other beetles of the cowpat community, especially *Geotrupes* spp., were important prey in spring and autumn, as were tipulid flies. The latter, and large noctuid moths, e.g. *Noctua pronuba*, which were important from June, are often found over pasture. The shift in habitat use from woodland to pasture may be associated with *Aphodius* beetles becoming abundant during

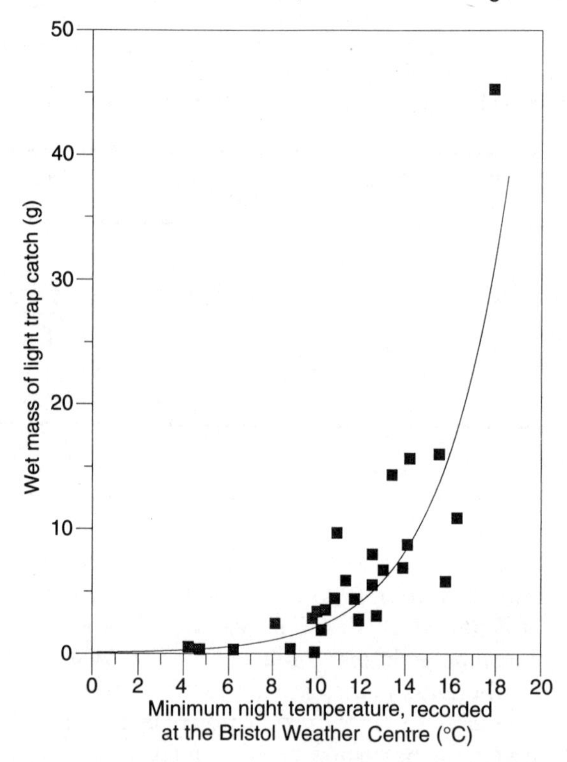

Fig. 2. The relation between light trap catch (wet mass) and minimum night temperature recorded at the Bristol Weather Centre (10 km from the study area) in 1991. The relationship is described as light trap catch (g) = $0.07 \times 10^{0.15\,(\text{temperature})}$, $r = 0.82$, $P < 0.001$.

midsummer, but microclimatic factors might also favour woodland foraging during cooler nights.

How habitat microclimate may influence insect abundance

Rydell (1989a) found a threshold for insect flight at 6–10 °C, and the temperature threshold for moth flight (as documented by light-trap catches) is similar (Fig. 2). Light trap catch biomass increases exponentially as temperature increases, and a small increase in temperature above the 6–10 °C threshold could potentially result in a large increase in the abundance of moths, which constitute most of the diet by volume in *R. ferrumequinum* (Jones 1990). It is therefore likely that habitat microclimate plays a key role in determining insect abundance and that this effect will consequently influence habitat use by bats.

We investigated the temperature differences between woodland and pasture habitats in order to determine whether woodland was consistently warmer than pasture, and whether any temperature difference between the two habitats was relatively constant. We used temperature probes accurate to

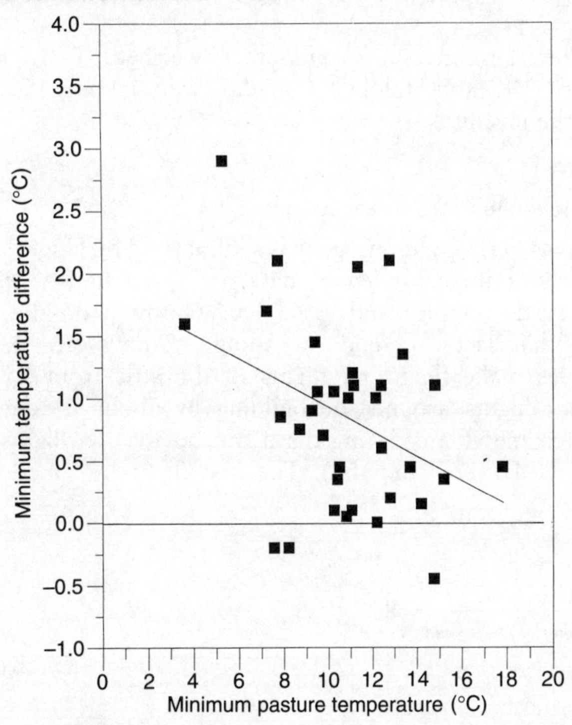

Fig. 3. Differences in minimum night-time temperatures between woodland and pasture compared with minimum temperature in pasture. See text for methods. The relationship is $y = 1.91 - 0.10x$, $r = -0.38$, $P < 0.05$.

0.2 °C on two eight-bit Squirrel Meter/Loggers. We chose three pasture and three woodland habitats in the foraging radius of bats at Site A, and measured night-time temperature in one pair of habitats chosen at random per night. Woodland was always warmer than pasture in terms of minimum overnight temperature (Fig. 3), except on three of 36 nights. Moreover, on colder nights the temperature difference between woodland and pasture was greater. Between 6 and 10 °C, woodland averaged about 1.3–0.8 °C higher minimum temperatures than did pasture. On cool nights, therefore, the temperature threshold for insect flight might be achieved in woodland, but not in pasture. The increased abundance of insects in woodland expected on these cold nights (such as occur during spring) might therefore make it profitable for bats to feed in woodland rather than over pasture.

The development of foraging in juveniles—implications for conservation

We conducted simultaneous radio-tracking of mother–young pairs to determine whether the habitat requirements of juveniles differed from those of their mothers. The transmitters used on mothers weighed 1.0–1.3 g and were made by Biotrack (UK) and Holohill (Canada). We used only 0.6 g Holohill transmitters on the juveniles.

Development in juveniles—an overview

The development of early flight ontogeny is summarized by Hughes, Ransome & Jones (1989) and further relevant data are given in Fig. 4. Evidence from the state of the nipples and rapid hair regrowth on the mammary glands suggests that lactation ends at about 45 days (R. D. Ransome unpubl.). Juveniles make their first flights in the attic from 15 days and begin exploratory flights around the building by about 24 days. At this stage juveniles become volant during the night, but restrict flights to within the building (P. L. Duvergé unpubl.). The first flights are made from the

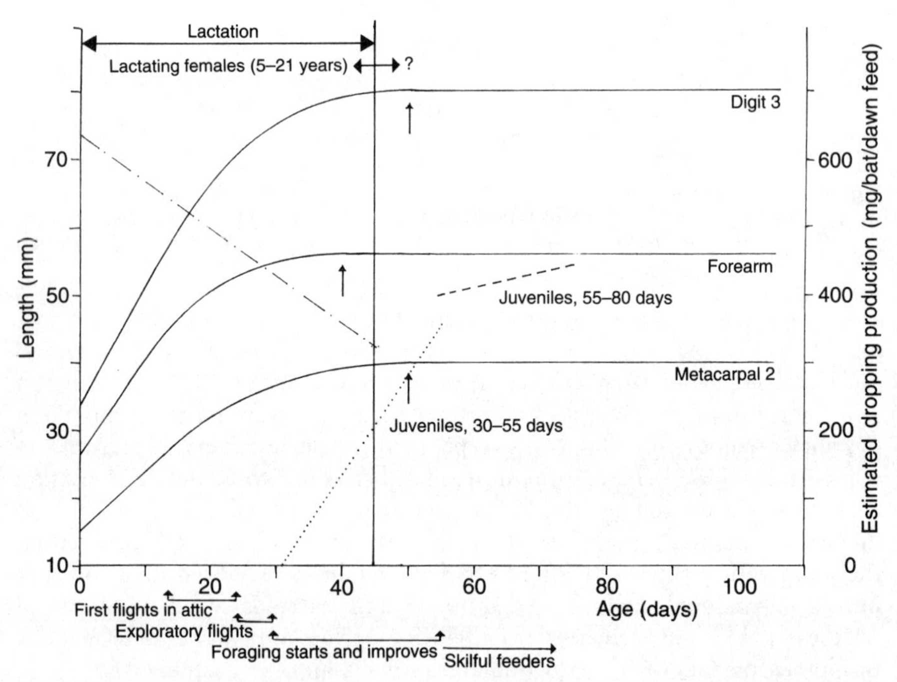

Fig. 4. Schematic representation of the development of greater horseshoe bats. Lactation ends at the vertical line, 45 days, with some individual variation. The uppermost curve is growth of digit 3, the lowest curve shows metacarpal 2. The middle curve shows forearm length. The dropping rate of juveniles is shown between days 30 and 55 on 1 September 1991, and between days 55 and 80 on 22 September 1991. The relationship for bats aged 30–55 days was described from 24 juveniles ($P < 0.001$), while that for the juveniles aged 55–80 days was not significant (range given in text). A regression line for females is drawn, based on data for females aged 5–21 years (R. D. Ransome unpubl.; $P < 0.001$).

roost at about 28–30 days (see below), and insects appear in the droppings of juveniles of these ages, suggesting that feeding begins immediately. Doppler shift compensation in echolocation first appears in a rudimentary form at about 30 days, though juveniles may still overcompensate when aged 35 days. Juveniles aged 40–45 days appear to be able to compensate for Doppler-shifts in echolocation as efficiently as can adults (Konstantinov 1989). If the bats in this study developed at the same rate as did those studied by Konstantinov, they would be flying and foraging with incomplete Doppler-shift compensation for the first 12 nights of activity outside the roost.

Dropping production as an index of food consumption

Dropping production, as an estimate of food consumption, was measured by catching bats soon after dawn and keeping them for measured periods in individual bags. Droppings were dried to constant mass and the dropping-production rate was calculated for that period of containment. Complete data were obtained from 14 individuals bagged for a total of 10 h immediately after dawn in September and moved into clean bags every 2 h. From these two-hourly data sets, mean production rates for each period were calculated. An exponential decay curve of the dry dropping rates against time since dawn was used to estimate total production for individuals on other dates when periods of containment were much shorter (R. D. Ransome unpubl.). All dropping-production estimates quoted refer to bats feeding on dung beetles, to ensure comparability. Data should be regarded as minima, since 10 h is insufficient time to permit all food eaten at dawn to be egested as faeces. Robinson & Stebbings (1993) showed that *Eptesicus serotinus* retained food items for up to 32 h after ingestion. However, since production rates fall below 5 mg/h after 10 h, and estimates for most bats lie in the range 300 to 600 mg, errors are likely to be low.

Faecal production of lactating females aged 5–21 years showed a marked decline from 640 mg/bat/dawn feed at the start of lactation to 323 mg/bat/dawn feed at the end (45 days) (Fig. 4). Younger females were much more erratic, with some showing an increase in consumption during the early part of lactation.

The dropping production of juveniles increased progressively between days 30 and 55, with an increase of about 15 mg/day up to about 350 mg/bat/dawn feed (Fig. 4). The increase in dropping production of juveniles coincides with an increase in the amount of time spent foraging (P. L. Duvergé unpubl.), distance travelled from the roost and, perhaps, with an increase in the efficiency of prey capture. Between days 55 and 100, the young may travel 2 km or more from the roost to forage (Fig. 5). At this time there is no significant relationship between dropping production and age (range = 195–744 mg/bat/dawn feed) and production falls within the range of those recorded from post-lactating females.

Range expansion by juveniles

Foraging range (straight line distance from nursery roost to furthest point visited) of juvenile *R. ferrumequinum* (Fig. 5) increases gradually with age. For two bats telemetered before first foraging flights began, one left the roost at 28 days and the other at 30 days. For the first five days of nocturnal activity outside the nursery roost juveniles remain within 1 km of the roost. Figure 5 shows how the foraging range reaches that typical of adults. Before weaning occurs at about 45 days, the juveniles return to the roost during the middle of the night, together with their mothers, presumably to suckle. By remaining close to the roost, juveniles may facilitate rapid returns for suckling, but it is also possible that flight is still handicapped by incomplete growth and a lack of ossification of the wing bones. Furthermore, they may risk disorientation and become unable to find the roost from greater distances until they have thoroughly learned the local topography. Whatever explanation, or combination of explanations, is correct, Ransome (1990) showed that the mortality rate of juveniles rises significantly between 45 and 55 days, when the foraging distance rises rapidly (Fig. 5).

After 55 to 60 days juveniles forage up to distances of 2–3 km, similar to the distances covered by adults. The range expansion described here is similar to that observed in *M. myotis* by Audet (1990). Investigations into the types of prey consumed by mothers and offspring are currently in progress, but *Aphodius* dung beetles are clearly important to juveniles for the first 2–3 weeks of foraging at Woodchester (R. D. Ransome unpubl.).

Growth of the forearm is largely complete by 40 days, but the fingers do not reach their final length until after 50 days and ossification may continue beyond 60 days (R. D. Ransome unpubl.—see Fig. 4). Hence the final skeletal size achieved by a juvenile bat depends not only on maternal quality, but also on the efficiency of its own foraging during at least the first 20 days of self-feeding. The survival of female *R. ferrumequinum* is influenced by forearm length, with large females surviving better than small females (Ransome 1989, 1990). Since the initial foraging success of this bat influences its subsequent survival potential, the provision of favourable feeding habitat (cattle-grazed permanent pasture alongside deciduous woodland) close to the maternity roost should significantly increase population levels.

Mother–young associations

In total 25 nights of data were obtained from eight mother–young pairs of bats where the location of both animals was known on the same night. Juveniles were tracked before they left the roost until 96 days of age. On only one occasion did mother and young leave the roost together, and only once were a mother and young seen foraging together. While mothers travelled rapidly to

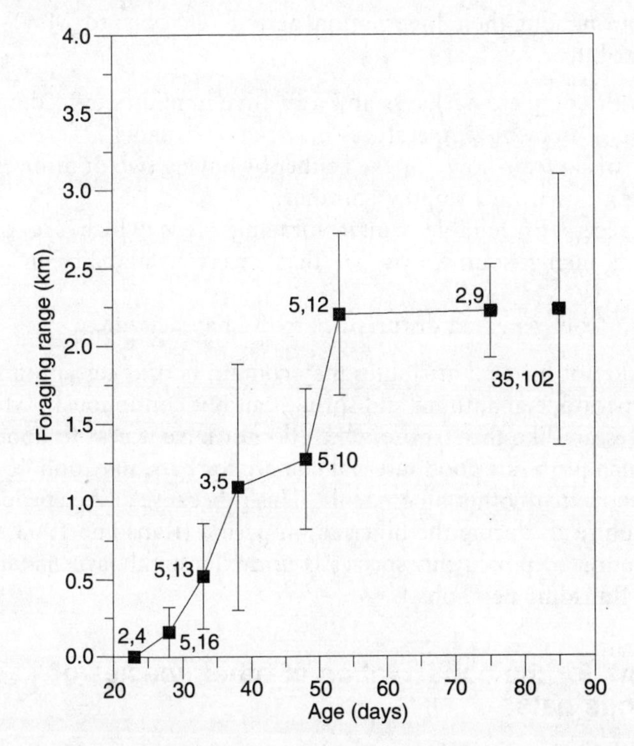

Fig. 5. The expansion of foraging range (straight line distance from maternity roost to furthest point reached) in juvenile greater horseshoe bats. Data are means for bats in five-day age classes ± SDs. The number of bats and the number of bat-nights for each data point are given, together with data for bats older than one year (point on right of graph).

foraging areas, usually 2–3 km from the roost, juveniles remained close to the roost at first. We conclude that the development of foraging behaviour in *R. ferrumequinum* involves no tuition from the mother, and is almost certainly self-taught. Some studies have suggested that maternal tuition may be important in the development of foraging in some species (*Noctilio albiventris*: Brown, Brown & Grinnel 1983; *Eptesicus fuscus*: Brigham & Brigham 1989), though *R. ferrumequinum* resembles *Myotis myotis*, in which young forage independently of their mothers (Audet 1990). Our data are valuable in that the ages of the young were known precisely, and because two bats were tagged before they started self-feeding.

Hibernation requirements relevant to conservation

Hibernation in this species has been investigated by Ransome (1968, 1971) and a more extensive coverage of hibernation, including a range of other species, is given in Ransome (1990). Suitable sites permitting greater horseshoe bats to remain throughout their hibernation period (October to May) need the following attributes:

1. Parts with complete darkness and a relative humidity exceeding 96%.
2. A range of ambient temperatures between 5 °C and 12 °C.
3. Regions of slow air-flow achieved either by having two or more entrances, or by having a downward sloping entrance.
4. Close access to suitable winter foraging areas which are preferably sheltered and facing southwards, so that insect availability in winter is maximized.
5. Freedom from repeated disturbance by human activities.

Sites which do not have all attributes may contain bats at certain times of the hibernation period, e.g. autumn and spring, but not continuously. Many male territorial sites are like this. Those which do not have access to good feeding sites, but which possess a good internal climate for bats, may only be used for short stops *en route* to other hibernacula. This is because a large proportion of the population feeds during the hibernation period (Ransome 1968). Arousal from hibernation torpor in this species is normally highly synchronized with dusk (R. D. Ransome pers. obs.).

Implications for the conservation of other species of insectivorous bats

We have identified several factors which should improve the conservation of foraging habitats for greater horseshoe bats. An important question for bat conservation is, how far do different species travel from their roosts to forage? Once this is known, effective protection or creation of suitable foraging habitats crucial to the bats at a particular colony may be attempted.

Although foraging range may be influenced by factors such as colony size, reproductive status of bats and, as shown here, by age, it should be possible to make some predictions about foraging ranges for other species. Flight performance in bats is influenced by wing morphology (Norberg & Rayner 1987). Two features of flight morphology which may be important in determining foraging range are wing loading and aspect ratio. Bats with higher wing loadings fly faster. If time constraints are important during foraging, then bats with high loadings are predicted to travel further from the roost, since they can travel further in a given time. Species with high aspect ratios (narrow, pointed wings) are predicted to range further, since unit energy costs are lower. It may therefore be possible to predict foraging ranges from wing loading and aspect ratio for species which cannot be studied directly (e.g. those too small to carry transmitters), or when rapid decisions about conservation must be made before autecological studies are possible.

A review of 18 species of microchiropteran bats showed no correlations between body mass ($r_{17} = 0.178$) or wing loading ($r_{16} = 0.313$) and foraging range. However, aspect ratio and foraging range were correlated positively (r_{16} = 0.694, P <0.01, Fig. 6). Many species have foraging ranges between 1 and 3 km, irrespective of diet or body size. High aspect ratio species commute further to forage, in line with the above predictions. One important factor which may also influence foraging range is colony size. Bats in very large

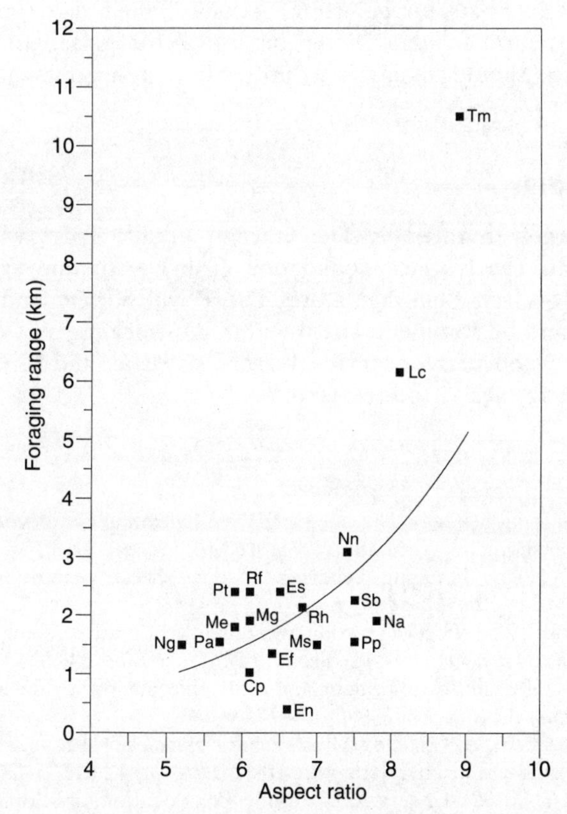

Fig. 6. The relationship between foraging range and aspect ratio in microchiropteran bats. The fitted line is described by an exponential equation: foraging radius (km) $=0.12 \times 10^{0.18(\text{aspect ratio})}$ r = 0.58, P <0.01. The line may not fit data for high aspect ratio species ideally, but fits data for low aspect ratio species better than does a linear equation. Acronyms are: Cp—*Carollia perspicillata* (Fleming & Heithaus 1986); Ef—*Eptesicus fuscus* (Brigham 1991); En—*Eptesicus nilssonii* (Rydell 1989b); Es—*Eptesicus serotinus* (Catto 1993); Lc—*Lasiurus cinereus* (Barclay 1989); Me—*Myotis emarginatus* (Krull *et al.* 1991); Mg—*Macroderma gigas* (Tidemann *et al.* 1985); Ms —*Miniopterus schreibersii* (McDonald, Rautenbach & Nel 1990); Na—*Noctilio albiventris* (Fenton, Audet *et al.* 1993); Ng—*Nycteris grandis* (Fenton, Swanepoel *et al.* 1990); Nn—*Nyctalus noctula* (Kronwitter 1988); Pa—*Plecotus auritus* (Swift & Racey 1983; Fuhrmann & Seitz (1992); Pp—*Pipistrellus pipistrellus* (Racey & Swift 1985); Pt—*Plecotus townsendii ingens* (Clark, Leslie & Carter 1993); Rf—*Rhinolophus ferrumequinum*—this study; Rh—*Rhinolophus hildebrandtii* (Fenton & Rautenbach 1986); Sb—*Scotophilus borbonicus* (Barclay 1985); Tm—*Tadarida midas* (Fenton & Rautenbach 1986). All species were radiotracked with the exception of En, Ms, Pp and Swift & Racey's study on Pa—these studies used light-tagging. Aspect ratios were taken from Norberg & Rayner (1987).

colonies, such as *Tadarida* spp., may need to travel far to reduce competition with conspecifics. The relationship between colony size and foraging range deserves more attention in both intra- and interspecific analyses.

This study has summarized some of the information necessary to formulate a conservation strategy for the endangered greater horseshoe bat. Although some data obtained are obviously species-specific, it seems that most microchiropteran bats forage within 4 km of their day roosts, and habitats within this sort of radius should be critical for conservation. High aspect ratio species should require the protection of habitats at greater distances.

Acknowledgements

This study was largely financed by The Vincent Wildlife Trust, with additional funding from The Nuffield Foundation, The Leverhulme Trust, The British Ecological Society, English Nature, The Royal Society and NERC. Matthew Morton and Ed Rimmer assisted with radio-tracking, and we thank Miriam Kelly, The Woodchester Mansion Trust, Tom Pearse and Mr and Mrs Cartwright-Hignett for access to their land.

References

Audet, D. (1990). Foraging behavior and habitat use by a gleaning bat, *Myotis myotis* (Chiroptera: Vespertilionidae). *J. Mammal.* **71**: 420–427.

Barclay, R. M. R. (1985). Foraging behavior of the African insectivorous bat, *Scotophilus leucogaster*. *Biotropica* **17**: 65–70.

Barclay, R. M. R. (1989). The effect of reproductive condition on the foraging behavior of female hoary bats, *Lasiurus cinereus*. *Behav. Ecol. Sociobiol.* **24**: 31–37.

Brigham, R. M. (1991). Flexibility in foraging and roosting behaviour by the big brown bat (*Eptesicus fuscus*). *Can. J. Zool.* **69**: 117–121.

Brigham, R. M. & Brigham, A. C. (1989). Evidence for association between a mother bat and its young during and after foraging. *Am. Midl. Nat.* **121**: 205–207.

Brown, P. E., Brown, T. W. & Grinnell, A. D. (1983). Echolocation, development, and vocal communication in the lesser bulldog bat, *Noctilio albiventris*. *Behav. Ecol. Sociobiol.* **13**: 287–298.

Catto, C. M. C. (1993). *Aspects of the ecology and behaviour of the serotine bat* (Eptesicus serotinus). PhD thesis: Aberdeen University.

Clark, B. S., Leslie, D. M. Jr & Carter, T. S. (1993). Foraging activity of adult female Ozark big-eared bats (*Plecotus townsendii ingens*) in summer. *J. Mammal.* **74**: 422–427.

Fenton, M. B., Audet, D., Dunning, D. C., Long, J., Merriman, C. B., Pearl, D., Syme, D. M., Adkins, B., Pedersen, S. & Wohlgenant, T. (1993). Activity patterns and roost selection by *Noctilio albiventris* (Chiroptera: Noctilionidae) in Costa Rica. *J. Mammal.* **74**: 607–613.

Fenton, M. B. & Rautenbach, I. L. (1986). A comparison of the roosting and foraging behaviour of three species of African insectivorous bats (Rhinolophidae, Vespertilionidae, and Molossidae). *Can. J. Zool.* **64**: 2860–2867.

Fenton, M. B., Swanepoel, C. M., Brigham, R. M., Cebek, J. & Hickey, M. B. C. (1990). Foraging behavior and prey selection by large slit-faced bats (*Nycteris grandis*; Chiroptera: Nycteridae). *Biotropica* **22**: 2–8.

Fleming, T. H. & Heithaus, E. R. (1986). Seasonal foraging behavior of the frugivorous bat *Carollia perspicillata*. *J. Mammal.* **67**: 660–671.

Fuhrmann, M. & Seitz, A. (1992). Nocturnal activity of the brown long-eared bat (*Plecotus auritus* L., 1758): data from radio-tracking in the Lenneberg forest near Mainz (Germany). In *Wildlife telemetry: remote monitoring and tracking of animals*: 538–548. (Eds Priede, I. G. & Swift, S. M.). Ellis Horwood, New York, London etc.

Hughes, P. M., Ransome, R. D. & Jones, G. (1989). Aerodynamic constraints on flight ontogeny in free-living greater horseshoe bats, *Rhinolophus ferrumequinum*. In *European bat research 1987*: 255–262. (Eds Hanák, V., Horáček, I. & Gaisler, J.). Charles University, Prague.

Hutson, A. M. (1993). *Action plan for the conservation of bats in the United Kingdom*. The Bat Conservation Trust, London.

Jones, G. (1990). Prey selection by the greater horseshoe bat (*Rhinolophus ferrumequinum*): optimal foraging by echolocation? *J. Anim. Ecol.* **59**: 587–602.

Jones, G. & Morton, M. (1992). Radio-tracking studies on habitat use by greater horseshoe bats (*Rhinolophus ferrumequinum*). In *Wildlife telemetry: remote monitoring and tracking of animals*: 521–537. (Eds Priede, I. G. & Swift, S. M.). Ellis Horwood, New York, London etc.

Konstantinov, A. I. (1989). The ontogeny of echolocation functions in horseshoe bats. In *European bat research 1987*: 271–280. (Eds Hanák, V., Horáček, I. & Gaisler, J.). Charles University, Prague.

Kronwitter, F. (1988). Population structure, habitat use and activity patterns of the noctule bat, *Nyctalus noctula* Schreb., 1774 (Chiroptera: Vespertilionidae) revealed by radio-tracking. *Myotis* **26**: 23–85.

Krull, D., Schumm, A., Metzner, W. & Neuweiler, G. (1991). Foraging areas and foraging behavior in the notch-eared bat, *Myotis emarginatus* (Vespertilionidae). *Behav. Ecol. Sociobiol.* **28**: 247–253.

McDonald, J. T., Rautenbach, I. L. & Nel, J. A. J. (1990). Foraging ecology of bats observed at De Hoop Provincial Nature Reserve, southern Cape Province. *S. Afr. J. Wildl. Res.* **20**: 133–145.

Norberg, U. M. & Rayner, J. M. V. (1987). Ecological morphology and flight in bats (Mammalia; Chiroptera): wing adaptations, flight performance, foraging strategy and echolocation. *Phil. Trans. R. Soc. (B)* **316**: 335–427.

Racey, P. A. & Speakman, J. R. (1987). The energy costs of pregnancy and lactation in heterothermic bats. *Symp. zool. Soc. Lond.* No. 57: 107–125.

Racey, P. A. & Swift, S. M. (1985). Feeding ecology of *Pipistrellus pipistrellus* (Chiroptera: Vespertilionidae) during pregnancy and lactation. I. Foraging behaviour. *J. Anim. Ecol.* **54**: 205–215.

Ransome, R. D. (1968). The distribution of the greater horse-shoe bat, *Rhinolophus ferrumequinum*, during hibernation, in relation to environmental effects. *J. Zool., Lond.* **154**: 77–112.

Ransome, R. D. (1971). The effect of ambient temperature on the arousal frequency of the hibernating greater horseshoe bat, *Rhinolophus ferrumequinum*, in relation to site selection and the hibernation state, *J. Zool., Lond.* **164**: 353–371.

Ransome, R. D. (1989). Population changes of greater horseshoe bats studied near Bristol over the past twenty-six years. *Biol. J. Linn. Soc.* **38**: 71–82.

Ransome, R. D. (1990). *The natural history of hibernating bats*. Christopher Helm, London.
Ransome, R. D. (1991). Greater horseshoe bat. In *The handbook of British mammals* (3rd edn): 88–94. (Eds Corbet, G. B. & Harris, S.). Blackwell, Oxford.
Robinson, M. F. & Stebbings, R. E. (1993). Food of the serotine bat, *Eptesicus serotinus*—is faecal analysis a valid qualitative and quantitative technique? *J. Zool., Lond.* **231**: 239–248.
Rydell, J. (1989a). Feeding activity of the northern bat *Eptesicus nilssoni* during pregnancy and lactation. *Oecologia* **80**: 562–565.
Rydell, J. (1989b). Site fidelity in the northern bat (*Eptesicus nilssoni*) during pregnancy and lactation. *J. Mammal.* **70**: 614–617.
Speakman, J. R. (1990). The function of daylight flying in British bats. *J. Zool., Lond.* **220**: 101–113.
Speakman, J. R. (1991). Why do insectivorous bats in Britain not fly in daylight more frequently? *Funct. Ecol.* **5**: 518–524.
Stebbings, R. E. (1982). Radio tracking greater horseshoe bats with preliminary observations on flight patterns. *Symp. zool. Soc. Lond.* No. 49: 161–173.
Stebbings, R. E. (1989). Conservation of the greater horseshoe bat. *Br. Wildl.* **1**: 14–19.
Stebbings, R. E. & Arnold, H. R. (1987). Assessment of trends in size and structure of a colony of the greater horseshoe bat. *Symp. zool. Soc. Lond.* No. 58: 7–24.
Swift, S. M. & Racey, P. A. (1983). Resource partitioning in two species of vespertilionid bats (Chiroptera) occupying the same roost. *J. Zool., Lond.* **200**: 249–259.
Tidemann, C. R., Priddel, D. M., Nelson, J. E. & Pettigrew, J. D. (1985). Foraging behaviour of the Australian ghost bat, *Macroderma gigas* (Microchiroptera: Megadermatidae). *Aust. J. Zool.* **33**: 705–713.

Symp. zool. Soc. Lond. (1995) No. 67: 325–344

Abundance and habitat selection of foraging vespertilionid bats in Britain: a landscape-scale approach

Allyson L. WALSH, Stephen HARRIS

School of Biological Sciences
University of Bristol
Woodland Road
Bristol BS8 1UG, UK

and A. M. HUTSON

The Bat Conservation Trust
45 Shelton Street
London WC2H 9HJ, UK

Synopsis

Large-scale surveys of the habitats required by foraging bats have seldom been attempted, despite a clear need for quantitative information in order to develop realistic conservation strategies. Accordingly, a standardized survey method was developed for a large-scale analysis of bat abundance and habitat selection in the whole of Britain. This was the first such survey of its kind undertaken anywhere. A random stratified sample of 1-km squares was surveyed by a network of volunteers walking fixed transects with bat detectors. They also collected information on the habitat features present in each 1-km square. Data from 910 squares were examined to test whether landscape and local habitat features influence bat activity. Significant regional differences were demonstrated between seven major land-class groups. Lowest bat activity levels occurred in upland, marginal upland and intensively farmed arable areas of the north. Habitat selection results are presented for two contrasting land-class groups. In both of these, woodland habitats and habitats associated with water were actively selected, whilst arable land, stone walls, scrub and parkland plus all grassland categories were avoided. Logistic regression models were used to identify habitats of critical importance within the two land-class groups. A variety of habitats were important in the pastural landscape but only one, semi-natural broadleaved woodland, was of critical importance in the arable landscape. The value of landscape-scale surveys for providing information on habitat use and a baseline from which to model the potential effects of changes in land use is discussed.

ZOOLOGICAL SYMPOSIUM No. 67
ISBN 0 19 854945 8

Introduction

A recent review of the status of British mammals suggested that eight species of bat had probably undergone no significant change in range and/or numbers in the last 30 years, whereas four species had undergone some decline, and a further three species had undergone significant declines. One other species had become extinct, and no species had shown a population increase (Harris, Morris, Wray & Yalden 1995). Since there are few quantified data on bat population changes, these assessments are, of necessity, tentative. Also, there are no quantified data on the relative importance of the various factors that may have led to bat population changes. Whilst roost destruction and the use of toxic pesticides are known to have contributed to these declines, land-use changes affecting both the quality and availability of foraging habitats may be the most significant underlying factors, at least for some species.

If previous population declines did result primarily from habitat degradation, then from a conservation perspective it is important to determine how current land-use patterns may affect populations and to predict the impact of future land-use changes. However, at present there are few quantitative data on the relationship between landscape changes and bat population changes, and there is little detailed information about bat habitat requirements in Britain. Furthermore, there is no technique to monitor future changes in land use and bat abundance in synchrony.

Quantified information on their habitat requirements was limited in the past as a result of a perceived difficulty of surveying bats away from roosts. Potential problems include large night-to-night variations in abundance at any one site, presumably due to either gross changes in weather or small changes in microclimate that affect insect abundance, and the difficulties in identifying species in flight and in obtaining comparable measures of abundance. Not surprisingly, there are only a few published studies on habitat use by bats in Britain (Swift & Racey 1983; Racey & Swift 1985; Jones & Rayner 1989; Walsh & Mayle 1991; Jones & Morton 1992). More recent studies elsewhere have attempted to quantify bat habitat use and make comparative estimates of bat abundance (McAney & Fairley 1988; Audet 1990; de Jong & Ahlén 1991; Rydell 1991; Brigham, Aldridge & Mackey 1992), but none of these is on a large enough scale to compare geographic regions. A wide variety of methods were used in these studies to examine different species in various geographical settings and so most are not directly comparable. There was therefore a need to develop a standard survey method which could be used to assess bat habitat requirements in Britain and to provide base-line abundance data for future comparisons. The requirement was for a simple survey method which would allow many volunteers to participate, thereby facilitating the collection of data on a large scale. This paper describes such a standardized approach and its use for a nationwide survey of vespertilionid bats in Britain. We discuss the advantages and limitations of such an approach in the light of the results reported here.

Methods

Stratification of the survey area

The survey area was mainland England, Scotland and Wales, plus Anglesey, the Isle of Wight and some of the Scottish Western Isles. To sample the diverse range of landscapes, we adopted a random stratified sampling system (Cochran 1963; Magurran 1988) based on the Institute of Terrestrial Ecology's land classification scheme. This assigns every 1-km square in Britain to one of 32 land classes; squares in each land class have a similar climate, physiogeography and pattern of land use (Bunce, Barr & Whittaker 1981a, b, 1983). Within each land class, a sample of 1-km squares was selected at random. This avoided observer bias in the selection of sites and ensured a standard sampling effort in different landscape types. To facilitate interpretation of the gross patterns of landscape use by bats, the 32 land classes were combined into the seven major land-class groups described in Table 1.

The survey protocol

Field work was carried out over three consecutive summers from 1990 to 1992 inclusive. The surveys were undertaken by two of us (ALW and AMH) and a large number of experienced volunteers, mainly from the existing network of bat groups in Britain. Each volunteer was allocated one or more 1-km squares and issued with several enlarged photocopies of the 1:25 000 Ordnance Survey map for that square. Detailed instruction sheets describing how to collect the bat and habitat data were issued to ensure uniformity of data collection. Each square was divided into two and a transect plotted crossing each half of the square by the shortest practical route. All habitat features at least 50 m in length or 0.5 ha in area were recorded up to approximately 20 m either side of the transect line. Forty-nine different habitat types, largely the same as those described by Cresswell, Harris & Jefferies (1990), were recognized. In practice, few 1-km squares contained more than 15 habitat categories, thus easing the recording process.

Volunteers walked their transect route on four separate occasions during defined date periods: 16 June–7 July, 8 July–28 July, 29 July–18 August and 19 August–8 September. No survey work was undertaken on nights with extreme weather conditions such as strong winds or heavy rain, when the activity of insectivorous bats is reduced (Bell 1980; Erkert 1982). Surveys began 30–45 min after sunset and tuneable bat detectors set at 45 kHz were used to maximize the range of species encountered. Volunteers rotated the start point of their walk on each of the four survey dates in order to minimize the effects of time of night on the data collected. The positions of all bat passes or feeding buzzes heard and bats seen were recorded on a separate map for each

Table 1. Descriptions of the land class composition, typical topography and land use in major land class groups I-VII.

Land-class group	Land classes included	Description of land-class group
I-Arable	2	Smooth-sloped medium to low altitude. Lowland farmland, often downland with few hedges. South England and south-west Midlands.
II-Arable	3, 4, 9, 11, 12	Flat, low-altitude alluvial plains. Intensively farmed lowland and fenland areas with large cereal fields. East Anglia, south and mid to north-east England and south-east Scotland.
III-Arable	14, 25, 26	Gently sloping, medium to low altitude, often valley floors. Intensively farmed arable lowland landscapes with fences. North England and south, central and east Scotland.
IV-Pastural	1, 5, 6, 7, 8	Medium to low altitude, often coastal areas. Lowland undulating farmland with many natural features and small field size, predominantly pasture. South England, south-west Midlands and Wales.
V-Pastural	10, 13, 15, 16, 27	Variable medium- to low-altitude valley floors with escarpments. Intricate undulating lowland landscapes with woods and many hedges plus some moorland and rough grassland. Mid to north England, north Wales and central and east Scotland.
VI-Marginal upland	17, 18, 19, 20, 28, 31	Moderate slopes, medium- to high-altitude river valley hillsides and exposed coast. Transitional farmland and grazing with walls and fences. South-west and north England, Scotland.
VII-Upland	21, 22, 23, 24, 29, 30, 32	Steep sloping ridges, high altitude. Upland/bleak moorland and mountainous landscapes with much bracken and bog. North England and Scotland including the extreme west and the Western Isles.

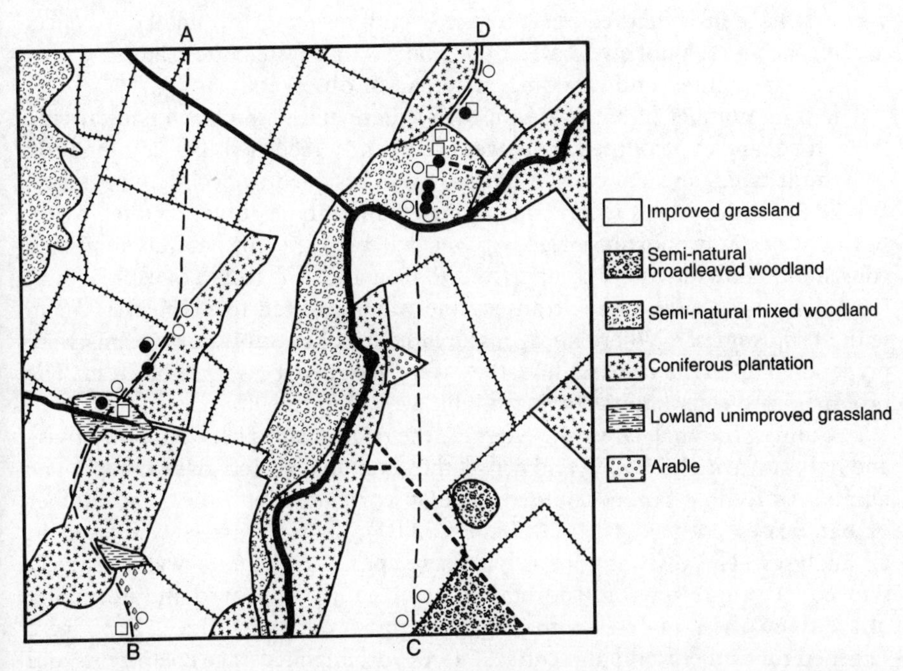

Fig. 1. Example of a square surveyed in central Scotland (land class group VII). The main habitat types are shown above. The linear features are denoted as follows: single row of trees—treelines; crossed lines—wire fences; broad broken lines—dykes with stone walls; thick solid black lines—running natural water. No roads were present; for clarity, tracks and farm buildings are not shown. Other solid lines mark habitat boundaries and not linear features. The two halves of the transect walked are marked with broken lines, and A, B, C and D mark the four start points. The bats recorded in all four walks are marked as follows: solid circles denote a pipistrelle feeding buzz, open circles a pipistrelle heard echolocating, open squares bat of an unknown species heard echolocating.

survey and given a letter code for the species, or closely related group of species, if this could be identified. The type of bat detector used, average cloud cover (on a 0–8 scale), mean night temperature (°C) and the start and finish times were also recorded. A typical 1-km square, illustrating the habitats recorded within the square, the transect routes selected and the bat data recorded, is shown in Fig. 1.

Data extraction and analysis

A potential problem with any large-scale survey conducted by different observers is ensuring the comparability of the data collected. In a previous survey of badger (*Meles meles*) habitat use, data for 2455 1-km squares were tested for potential biases and shown to be highly uniform (Cresswell *et al.*

1990). It was assumed that observer differences in the bat survey would similarly have little marked effect on the overall results. Also, small differences in approach were unlikely to affect the analysis in a systematic manner, owing to the large number and randomized spread of observers.

The total number of bat passes in each square and within each habitat type in each square was counted. Bats were frequently recorded in a combination of habitat types. In such cases a bat pass was assigned to both habitat types. Where the bat pass was recorded in a single habitat type, it was counted twice. In this way the proportion of activity recorded in different habitats remained consistent, allowing direct comparisons to be made. The length of each habitat type along both sides of the transect line was measured to the nearest 50 m with an opisometer. Where linear features such as hedgerows or streams were perpendicular to the transect line, they were assigned a distance of 50 m. The proportion of each habitat type available was then calculated.

The aim of the analysis was to examine the relationships between bat activity and habitat variables within and across the seven major land-class groups. The number of feeding buzzes counted strongly correlated with the total number of bat passes counted (r_s = 0.56, $P<0.001$), so bat passes were used in all analyses. The distribution of bat passes per 1-km square was positively skewed; a square-root transformation significantly improved the normality of the data when analyses with this assumption were performed. Data were screened for outliers, and influential squares with high leverage coefficients and large standardized residuals were discarded (Hosmer & Lemeshow 1989).

The length of the transect walked and time spent monitoring were used to calculate a variable to allow for different walking speeds in each square. The effects of this and other non-habitat factors on the number of bat passes recorded per 1-km square for each land-class group were examined by analysis of variance using SPSS (Norusis 1990). Avoidance or selection of habitat types in each date period and over all four date periods was examined by constructing Bonferroni confidence intervals around the observed use of each habitat type (Neu, Byers & Peek 1974). Multiple logistic regression analysis was employed to evaluate habitats of critical importance in determining high bat activity in each land-class group. Models used habitat variables to discriminate between squares in the bottom third of the data-set ('low' activity squares) and in the top third ('high' activity squares). In all analyses, habitat variables were truncated to 26 discrete types and only those occurring in more than ten squares and with five or more bat passes were used (Hayes & Winkler 1970).

Results

Of the 1030 1-km squares surveyed, 120 (11.7%) were not included in the analysis because the sheets were either incorrectly or inadequately completed, or because of high leverage coefficients. Thus the analysis is based on data from

910 squares which were of a high standard and uniform approach. Surveying these squares involved 2700 h of search effort and nearly 30 000 bat passes were counted in the 9000 km walked. Only 6% of squares were negative for bats and these were mostly in Scotland. The proportion of bat passes identified to a particular species or species group was 24.4%. Of these, 71.0% were *Pipistrellus pipistrellus*, 17.0% *Myotis* spp., 7.6% *Nyctalus noctula*, 2.7% *Plecotus* spp. and 1.7% *Eptesicus serotinus*. It is highly likely that a similar proportion of the unidentified bat passes were also *Pipistrellus pipistrellus*. Thus whilst the results refer to the total number of bat passes counted for all species, the predominant species sampled is most probably *Pipistrellus pipistrellus*.

Figure 2 shows the effects of increasing sample size (i. e. the number of squares surveyed) on mean bat activity levels recorded in a single land class and land-class group. A fairly stable mean was obtained after surveying 25–30 squares in most land classes/land-class groups.

Non-habitat factors influencing bat activity

A simple one-way ANOVA indicated that the type of bat detector significantly affected the level of activity recorded (F=5.8, *d.f.* 3, P<0.01). However, multiple comparison tests revealed that only two out of the four types of detector tested differed to a significant extent (Tukey's test, P<0.05). These were the two least frequently used detectors, contributing only 10% of the records (Walsh, Hutson & Harris 1993). Furthermore, the majority of volunteers (70%) used only one type of detector (QMC Mini-2), and the use of different detectors was randomly distributed amongst the various land classes and land-class groups. Thus we assumed that the observed differences between bat detectors had no marked effect on the survey results.

Activity did not appear to vary between date periods once the significant effects of mean night temperature and walking rate (P<0.001) were factored out (Table 2). However, land-class group proved to be a significant factor in influencing the abundance of bat passes counted (P<0.001).

Relative abundance of bats

Mean bat activity (passes counted per 1-km square) was calculated for each land-class group once activity had been pooled across date periods (Fig. 3). The upland group VII and the northern arable group III exhibited significantly lower activity levels than did any of the other groups (Tukey's test, P<0.05). Marginal upland (group VI) showed intermediate levels of bat activity, and this also differed significantly from any of the other groups (Tukey's test, P<0.05). Pastural groups IV and V and arable groups I and II had very similar high levels of bat activity. The more southerly distributed land-class groups (arable group I and pastural group IV) showed significantly elevated

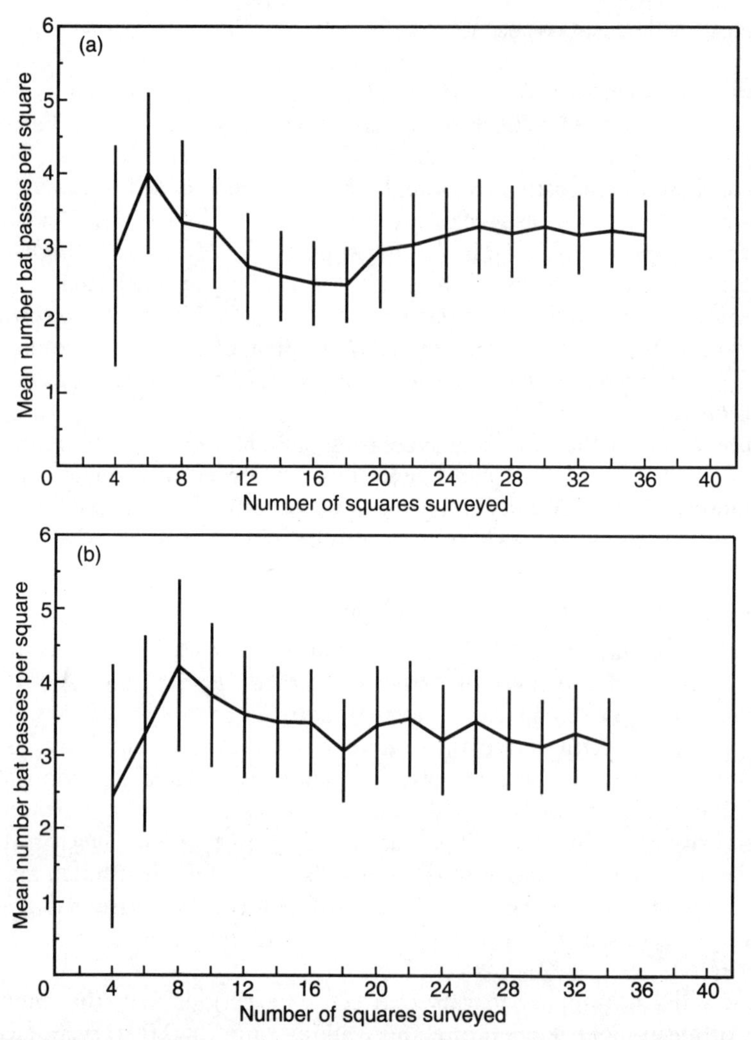

Fig. 2. The effect of increasing sample size (number of squares surveyed) on mean activity measured as the number of passes (±SE) per 1-km square. The two examples are (a) land class 6 and (b) land-class group III.

bat activity levels in comparison with arable group II and pastural group V (Tukey's test, $P<0.05$).

These regional differences suggested that there may be a gradient of decreasing levels of bat activity from the south to the north of Britain. To test this hypothesis, the relationship between bat activity and the east and north grid co-ordinate for each square was investigated by means of a multiple linear regression analysis. The model used controlled for any differences due to habitat, mean night temperature and walking rate and identified a significant and negative relationship between bat activity and distance north (Fig. 4), but no obvious relationship with distance east.

Table 2. Analysis of variance summary for the variation in mean bat activity between four date periods in the seven major land-class groups I–VII. Nuisance co-variables walking rate and night temperature were controlled for by forcing them into the model first.

Source of variation	*d.f.*	*F*	Significance
Covariates	2	70.4	***
Mean nightly temperature	1	81.4	***
Walking rate	1	68.7	***
Main effects	9	21.7	***
Land-class group	6	32.5	***
Date period	3	0.3	n.s.
Interaction:			
Land-class group × date period	18	0.5	n.s.
Explained	29	11.9	***
Residual	3550		
Total	3579		

***$P < 0.001$; n.s., not significant

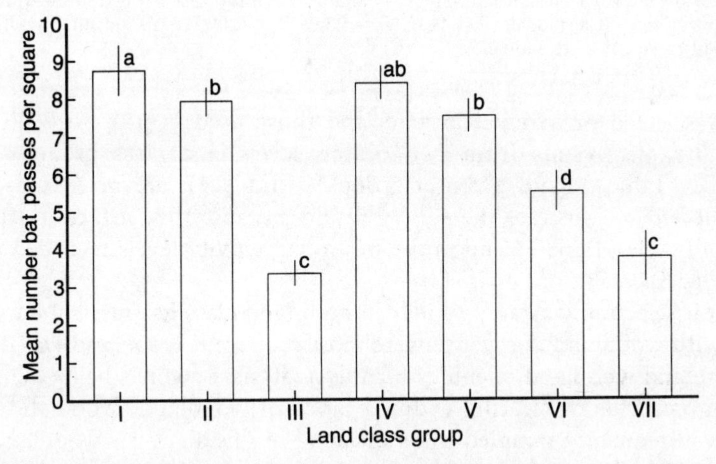

Fig. 3. Mean bat activity measured as the mean number (±SE) of passes per 1-km square for the seven major land class groups. I = arable I, n = 68; II = arable II, n = 184; III = arable III, n = 93; IV = pastural IV, n = 222; V = pastural V, n = 202; VI = marginal upland, n = 86; VII = upland, n = 55, where n is the number of squares surveyed. The supercripts indicate significantly different land-class groups ($P < 0.05$, Tukey's test).

Habitat selection

Temporal shifts in habitat use appeared not to occur, since very few differences in habitat selection were found between the four date periods. Hence the results were pooled across date periods. To demonstrate the types of

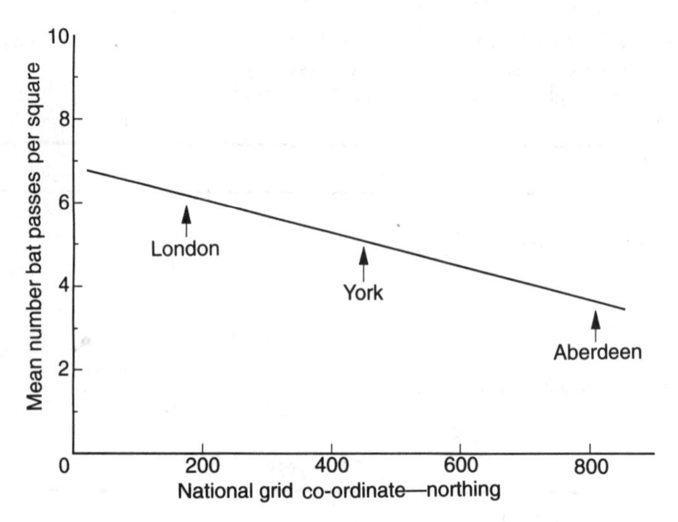

Fig. 4. Partial linear relationship between bat activity (mean number of passes per 1-km square square root transformed) and distance north as described by a multiple linear regression analysis. The regression included nuisance variables—walking rate and night temperature—and a suite of habitat variables (full model, $r^2 = 0.32$, $F = 23.2$, $P < 0.0001$). To aid interpretation, the positions of three cities are shown.

habitats selected or avoided by bats and those used in proportion to their availability, the results from two of the seven land class groups, arable group III (Table 3) and pastural group V (Table 4), are presented. These were selected to illustrate the survey results because they are contrasting in their landscape (Table 1) and in the mean bat activity levels recorded within them (Fig. 3).

Habitat selection was very similar in both land-class groups; habitats associated with woodland and water were most preferred. Woodland was divided into edge and woodland openings for this analysis, openings being defined as an open path/ride or clearing within a block of woodland. Woodland centre was too infrequently sampled to be included in this analysis. Woodland edge categories (conifer, mixed and broadleaved), broadleaved woodland openings and all habitats associated with water were strongly selected as foraging sites, whilst arable land, scrub, stone walls without hedgerows or treelines and parkland plus all grassland categories were avoided by bats. Some minor habitat selection differences occurred between these two particular land-class groups. Thus, ditches were only avoided in arable group III (Table 3) and hedgerows and urban areas were only selected in pastural group V (Table 4). Overall, in pastural group V, 26.9% of the area available consisted of selected habitats compared with only 13.8% in arable group III. Mean bat activity levels in the same land-class groups were 7.51 and 3.34 passes per 1-km square respectively, which suggests that lack of suitable foraging habitats may be limiting bat abundance.

Table 3. Ninety per cent Bonferroni confidence intervals for the proportion ($\hat{P}_i$) of bat pass locations within 26 different habitat types in arable land-class group III.'+' and '−' indicate where the habitat was used significantly more or less than expected from its availability. 'p/a' indicates if a habitat was used in proportion to its availability (n =93 squares).

Habitat type	No. of bat pass locations	$\hat{P}_i^a$	Bonferroni confidence interval[b]	Selection
Hedgerows	53	0.0443	$0.0402 < P < 0.0870$	p/a
Treelines	45	0.0308	$0.0318 < P < 0.0750$	+
Open ditches	<5			(−)[c]
Covered ditches	<5			(−)[c]
Open stone walls	17	0.0368	$0.0070 < P < 0.0344$	−
Streams	15	0.0113	$0.0051 < P < 0.0303$	p/a
Broadleaved woodland edges	86	0.0275	$0.0740 < P < 0.1324$	+
Mixed woodland edges	44	0.0120	$0.0309 < P < 0.0736$	+
Coniferous woodland edges	46	0.0138	$0.0328 < P < 0.0764$	+
Broadleaved woodland centre	54	0.0197	$0.0412 < P < 0.0884$	+
Mixed woodland centre	59	0.0345	$0.0465 < P < 0.0958$	+
Coniferous woodland centre	19	0.0138	$0.0085 < P < 0.0371$	p/a
Scrub	31	0.0566	$0.0190 < P < 0.0554$	−
Parkland/amenity grasslands	22	0.0450	$0.0108 < P < 0.0414$	−
Young/recently felled plantations		*		
Rivers/canals	28	0.0063	$0.0161 < P < 0.0505$	+
Ponds		*		
Lake/reservoir margins		*		
Water (no riparian edge) margins		*		
Heather moorlands/heathlands/bogs		*		
Improved grasslands	9	0.0454	$0.0007 < P < 0.0203$	−
Arable land	168	0.3687	$0.1632 < P < 0.2402$	−
Upland unimproved grasslands		*		
Lowland unimproved grasslands	11	0.0245	$0.0225 < P < 0.0242$	−
Semi-improved grasslands	18	0.0673	$0.0075 < P < 0.0352$	−
Urban areas	66	0.0824	$0.0536 < P < 0.1055$	p/a

[a]* indicates when the extreme scarcity or absence of a particular habitat precluded a test.
[b]P is the proportion of habitats available.
[c]Where <5 bat passes occurred, no test was conducted and avoidance was assumed (−).

Table 4. Ninety per cent Bonferroni confidence intervals for the proportion ($\hat{P}_i$) of bat pass locations within 26 different habitat types in pastural land-class group V. '+' and '−' indicate where the habitat was used significantly more or less than expected from its availability. 'p/a' indicates if a habitat was used in proportion to its availability (n=202 squares).

Habitat type	No. of bat pass locations	$\hat{P}_i^a$	Bonferroni confidence interval[b]	Selection
Hedgerows	770	0.1544	$0.1994 < P < 0.2397$	+
Treelines	245	0.0410	$0.0575 < P < 0.0823$	+
Open ditches	31	0.0090	$0.0042 < P < 0.0132$	p/a
Covered ditches	33	0.0084	$0.0047 < P < 0.0141$	p/a
Open stone walls	42	0.0187	$0.0067 < P < 0.0173$	−
Streams	106	0.0157	$0.0219 < P < 0.0386$	+
Broadleaved woodland edges	366	0.0210	$0.0895 < P < 0.1192$	+
Mixed woodland edges	58	0.0069	$0.0102 < P < 0.0226$	+
Coniferous woodland edges	55	0.0069	$0.0096 < P < 0.0217$	+
Broadleaved woodland centre	100	0.0158	$0.0204 < P < 0.0366$	+
Mixed woodland centre	46	0.0119	$0.0076 < P < 0.0187$	p/a
Coniferous woodland centre	37	0.0132	$0.0056 < P < 0.0155$	p/a
Scrub	55	0.0268	$0.0096 < P < 0.0217$	−
Parkland/amenity grasslands	30	0.0136	$0.0041 < P < 0.0130$	−
Young/recently felled plantations	31	0.0080	$0.0043 < P < 0.0134$	p/a
Rivers/canals	91	0.0068	$0.0182 < P < 0.0336$	+
Ponds	25	0.0023	$0.0030 < P < 0.0112$	+
Lake/reservoir margins	44	0.0028	$0.0072 < P < 0.0187$	+
Water (no riparian edge) margins	46	0.0033	$0.0076 < P < 0.0187$	+
Heather moorlands/heathlands/bogs		*		
Improved grasslands	238	0.1108	$0.0556 < P < 0.0800$	−
Arable land	168	0.1938	$0.0375 < P < 0.0583$	−
Upland unimproved grasslands	38	0.0176	$0.0058 < P < 0.0159$	−
Lowland unimproved grasslands	104	0.0449	$0.0214 < P < 0.0379$	−
Semi-improved grassland	268	0.1447	$0.0635 < P < 0.0893$	−
Urban areas	382	0.0923	$0.0938 < P < 0.1241$	+

[a] * indicates when the extreme scarcity or absence of a particular habitat precluded a test.
[b] P is the proportion of habitats available.

Table 5. Logistic regression analysis of habitat factors affecting high bat activity within squares in arable land class group III (n=93 squares). Nuisance variables walking rate and night temperature were forced into the model, then a forward stepwise procedure was followed. The regression correctly classified 84.4% of squares, model χ^2=42.2, $d.f.$=3, P<0.0001.

Variable	Coefficient	Wald χ^2	R	P	−2 log likelihood ratio	Significance of −2 log likelihood ratio
Walking rate	−0.0521	1.72	0.0000	0.190		
Mean temperature	0.4844	2.92	0.1049	0.088		
Broadleaved woodlands	0.0291	3.59	0.1383	0.048	25.3	0.0000
Coniferous woodlands	0.0025	1.85	0.0000	0.173	5.5	0.0194
Mixed woodlands	0.0046	1.51	0.0000	0.218	8.6	0.0033
Constant	−5.0219	3.83		0.050		

The likelihood-ratio test was used to determine variable removal from the model, hence the associated significance is shown in addition to the usual coefficients and Wald statistics.

Predictive critical habitat models

One of the aims of this study was to establish a functional relationship between bat activity and habitat type and availability. The logistic regression equations constructed (Tables 5 and 6) provide such a relationship. They have high predictive power once the effects of walking rate and night temperature have been controlled for (correct classification rate: 84% for arable III, 80% for pastural V). Chi-squared tests, comparing the observed occurrences of high bat abundance to those predicted by the models, showed a high goodness of fit (P<0.94 for arable III, P<0.30 for pastural V). The most successful combination of habitat variables for defining a 'high' bat activity square in pastural group V (Table 6) was a higher proportion of hedgerows, treelines, broadleaved woodland (edge plus openings) and habitats associated with water plus a lower proportion of scrub and arable land. In contrast, a less diverse combination of a higher proportion of all woodland categories (mixed, conifer and broadleaved edge plus openings) best defined those squares with 'high' bat activity in the arable III land class group (Table 5). Broadleaved woodland was a particularly critical resource in this landscape.

The functional relationships described by these equations may also be used to predict the effects of change in land use. Given habitat information for any 1-km square in Britain, the probability of the square being of high activity can be estimated. Then, for a measured increase or decrease in a particular habitat type, a resultant change in the probability that that square will be of high bat activity can similarly be estimated, thereby measuring the potential effects of landscape changes.

Table 6. Logistic regression analysis of habitat factors affecting high bat activity within squares in pastural land-class group V (n=202 squares). Nuisance variables walking rate and night temperature were forced into the model, then a forward stepwise procedure was followed. The regression correctly classified 80.4% of squares, model χ^2=66.8, $d.f.$=8, P<0.0001.

Variable	Coefficient	Wald χ^2	R	P	−2 log likelihood ratio	Significance of −2 log likelihood ratio
Walking rate	−0.0720	12.07	−0.2350	0.000		
Mean temperature	0.1436	1.23	0.0000	0.267		
Hedgerows	0.0011	11.76	0.2316	0.001	14.4	0.0001
Treelines	0.0015	4.15	0.1086	0.042	4.6	0.0326
Broadleaved woodlands	0.0019	7.97	0.1811	0.005	9.8	0.0180
Streams	0.0054	7.30	0.1708	0.007	10.9	0.0010
Scrub	−0.0018	4.59	−0.0119	0.032	5.5	0.0191
Arable	−0.0008	13.95	−0.0256	0.000	18.0	0.0000
Rivers/canals	0.0032	4.64	0.1210	0.031	6.2	0.0127
Lakes	0.0042	4.95	0.1270	0.026	5.7	0.0168
Constant	0.0485	0.06		0.808		

The likelihood-ratio test was used to determine variable removal from the model, hence the associated significance is shown in addition to the usual coefficients and Wald statistics.

Discussion

A large-scale survey method for bats

In this study, we did not expect to overcome all of the potential problems associated with surveying foraging bats, but by careful design of the survey protocol we aimed to minimize them. The survey method relied entirely on the ultrasonic detection of bats, as have many smaller-scale studies (Fenton 1970, 1982; Kunz & Brock 1975; Bell 1980; Fenton & Thomas 1980; Ahlén 1980/81). For a country-wide survey, techniques that rely on visual counts of roosting or foraging bats, direct capture or expensive automated equipment (Kunz & Brock 1975; Gaisler 1979; Kunz 1988) are logistically impractical to apply and may introduce many more biases. Thus we used a standard line transect system to collect data on bat activity and habitat use; by walking transects it was possible to generate large sample sizes for many habitat types. Walking transects rather than driving avoids biasing the results towards linear features associated with roadsides (Ahlén 1980/81; Rydell 1986; Jüdes 1989), and hence ensures that all available habitats are randomly sampled.

As with any survey technique, the use of bat detectors has strengths and limitations. Portable bat detectors provide data in the form of 'bat passes' per unit time or area, thereby permitting comparable quantitative estimations of bat activity within different regions and habitats. However, because detectors do not differentiate between several passes by the same bat and single passes by several bats (Thomas & West 1989), the counts of bat passes represent an index of relative abundance and cannot be interpreted as population density figures. In this project, species identification was limited by the demand for continuous steady progression and by having the detector tuned to 45 kHz. Despite a range of vespertilionid bat species registering on the detector, differences in maximum detection distance of each species prevent quantitative comparisons (Fenton & Bell 1981) and also bias the survey towards species with the majority of their call component at 45 kHz, in this case *Pipistrellus pipistrellus*.

Differences in the sensitivity of different types of detector are also an important consideration. Although the type of bat detector used did not appear to affect the overall results from this study, significant differences were identified between two types of detector which would potentially bias the results of a smaller-scale study. Waters & Walsh (1994) report sensitivity differences between several types of bat detector and suggest that, to improve the quantification of the results, efforts should be made to standardize bat detector type and to minimize the variability caused by individual user effects by following a strict survey protocol. However, despite the practical restrictions imposed on such a broad-scale survey, we believe there are many advantages with the approach adopted. In particular, the quantitative information gained provides a valuable new insight into the habitat requirements of vespertilionid bats, in particular our most widespread species *Pipistrellus pipistrellus*.

Habitat selection

The types of habitat preferred and avoided in the two land-class groups used as examples are representative of the general pattern observed in all land-class groups. Bats were observed in a diverse range of habitats and appeared to be using habitats in accordance with what is known from past observations and the literature on habitat selection by those bat species occurring in Britain. Thus the main foraging habitats of vespertilionid bats in Britain are associated with woodland and water. Where available, bats were shown strongly to select riparian habitats, from small streams and ponds to larger rivers, canals, lakes and reservoirs. Fenton (1970), Bell (1980) and Walsh & Mayle (1991) also recorded high bat activity over ponds. In Scotland, Racey & Swift (1985) reported that *Pipistrellus pipistrellus* favoured riverine habitats in addition to ponds, and in Poland Rachwald (1992) found that *Nyctalus noctula* most often fed at riversides. All woodland edge and opening habitats, including broadleaved, mixed and coniferous woodlands, were similarly highly favoured foraging habitats. An exception was openings within coniferous woodland in both land-class groups. The selection of woodland sites as foraging areas is well documented (Racey & Swift 1985; Furlonger, Dewar & Fenton 1987; Rachwald 1992; Clark, Leslie & Carter 1993), and is attributed to the high insect densities associated with woodlands. Coniferous woodlands in this study were mainly plantations, which are known to have fewer species of insect than woodlands containing broadleaved trees (Winter 1983). This could explain why fewer bats were recorded within this habitat.

Linear landscape features such as hedgerows and treelines appeared to be an important part of the ecological infrastructure available to bats, linking roosts and suitable feeding habitats. Their pattern of selection implied that they were used, not only as commuting routes, but also as linear foraging sites. Limpens & Kapteyn (1991) proposed that bats, particularly *Myotis* spp., preferred commuting along linear landscape elements because of the availability of insects, shelter from wind and predation and a reliance on these features for orientation within the landscape. Kalko & Schnitzler (1993) demonstrate that bats are not reliant on these features for orientation. In the present study, an avoidance of stone walls without hedgerows or treelines associated with them, plus an avoidance of hedgerows in arable land-class group III, where they are rarer and more likely to be fragmented, suggests that the association of cover and/or high insect densities are important factors influencing the suitability of different linear landscape elements as flight routes.

A very strong and previously unquantified relationship exhibited in both land-class groups was an avoidance of arable land by bats. In arable land-class group III, this amounted to nearly 40% of the land area sampled, and in pastural land-class group V nearly 20%. In contrast, all woodland habitats combined make up only 12% and 8% of the arable and pastural group respectively, highlighting the differences in the availability of suitable foraging habitat between the two land-class groups. All grassland categories, from unimproved grassland to improved pasture, were avoided in the two land-class groups illustrated in this paper. In other land-class groups, unimproved pasture was often used in proportion to availability. Other authors have similarly recorded comparatively low activity over pasture (Fenton 1970; Lunde & Harestad 1986; Walsh & Mayle 1991), but the use of pasture is likely to differ between species. Larger species such as *Eptesicus serotinus, Eptesicus nilssonii* and *Rhinolophus ferrumequinum* readily exploit pasture, foraging for beetles and moths (Robinson & Stebbings 1993; Rydell 1986; Jones & Morton 1992), whereas species such as *Pipistrellus pipistrellus* tend to avoid open fields (Racey & Swift 1985).

Influence of region on bat abundance

Bat activity and/or abundance decreases with latitude northwards, but does not appear to change on an east–west axis. Only eight breeding species of bat extend their range as far north as the Scottish border, of which only four are widespread in Scotland. Hence the observed decrease in bat activity may be connected with the decrease in species diversity. However, since the majority of our records were for *Pipistrellus pipistrellus*, a species which is widespread in Scotland, we propose that the observed trend is likely to represent a real decrease in abundance. This further implies that population numbers may be lower as the northern edge of the range of many bat species is approached. This relationship is contradictory to the findings of Speakman *et al.* (1991), who found no evidence of a latitudinal effect on *Pipistrellus pipistrellus* populations. However, they compared only two sites, one in an area in north-east Scotland and the other in northern England (near York). Possible causes for this observed trend include a less favourable climate and the availability of suitable foraging sites (Caughley, Grice, Barker & Brown 1988).

Monitoring, prediction and conservation strategies

The primary aim of this study was to provide a means of assessing the significance of habitat changes for some bat populations. To do this we secured a means of detecting change, establishing its direction and measuring its magnitude, by developing a standard and repeatable protocol for measuring habitat and bat activity changes synchronously. This provides a viable method to complement monitoring bat population trends by colony counts, a technique with several drawbacks. Basing population monitoring solely on colony counts may lead to invalid comparisons between different areas because of unequal sampling effort and because shifts in colony size may not necessarily correlate with changes in population numbers if changes in roost location and emigration and/or immigration rates are high and variable between years. Thus the further development of the technique outlined in this paper and its repetition in the future should both test the predictions about the effects of changing land-use patterns on bat populations and aid the verification of estimates of bat population trends.

The logistic regression models described represent the first step towards predicting the most significant effects of habitat change in a quantified manner, and further, more refined, modelling will extend this process. Despite vespertilionid bats utilizing a diverse array of habitats, the models identify a few (riparian, woodland and linear corridor habitats) as being of pivotal importance (Tables 5 and 6). Riparian and woodland habitats are patchily distributed, and habitat fragmentation may lead to further dispersion of these habitats. It may similarly disrupt the continuity of the landscape infrastructure formed by linear habitat corridors such as hedgerows. It can thus be predicted that bats are likely to be significantly affected by habitat fragmentation, a view supported by the analysis of Bright (1993). The advantage of the data collected in the present survey is that they can be used to predict the relative magnitude of any such effect. They should, however, be utilized with caution at present, since the results reflect the habitat preferences of *Pipistrellus pipistrellus* on the whole and other species may have more specific habitat needs.

The data from this study can be used to identify specific preferred habitats in different landscapes and thus help to develop the conservation strategy for British bats. One of the major points of the European Bats Agreement 1991 is to protect key habitats. The effectiveness of this agreement, and also of a new EC Habitats and Species Directive brought into force in 1994, depends on the quality of the information available on the types of habitats preferred by bats. This survey has enabled the construction of a quantified database on bat habitat use in all areas of the country, thereby providing some of the basic information required for the implementation of these agreements. A report on the state of the British countryside which details land-use changes from 1984 to 1990 (Barr *et al.* 1993) highlights the rapid rate of loss of many of the habitats this study has shown to be favoured by bats; for example, the net loss of hedgerows in the period was 23%. Such a rate of change reinforces the need to monitor the effects of landscape change on bat abundance, and further to extend this base-line survey so as to quantify the habitat requirements of individual species.

Acknowledgements

A large number of people helped with the collection of the field data, often under very difficult conditions. We are extremely grateful for all their help. The survey was funded by the Joint Nature Conservation Committee, the National Westminster Bank plc through WWF, and the World Wide Fund for Nature; their generous support is gratefully acknowledged. SH would like to thank the Dulverton Trust for financial support.

References

Ahlén, I. (1980/81). Field identification of bats and survey methods based on sounds. *Myotis* **18–19**: 128–136.

Audet, D. (1990). Foraging behavior and habitat use by a gleaning bat, *Myotis myotis* (Chiroptera: Vespertilionidae). *J. Mammal.* **71**: 420–427.

Barr, C. J., Bunce, R. G. H., Clarke, R. T., Fuller, R. M., Furse, M. T., Gillespie, M. K., Groom, G. B., Hallam, C. J., Hornung, M., Howard, D. C. & Ness, M. J. (1993). *Countryside survey 1990—main report.* Department of the Environment, London.

Bell, G. P. (1980). Habitat use and response to patches of prey by desert insectivorous bats. *Can. J. Zool.* **58**: 1876–1883.

Brigham, R. M., Aldridge, H. D. J. N. & Mackey, R. L. (1992). Variation in habitat use and prey selection by Yuma bats, *Myotis yumanensis. J. Mammal.* **73**: 640–645.

Bright, P. W. (1993). Habitat fragmentation—problems and predictions for British mammals. *Mammal Rev.* **23**: 101–111.

Bunce, R. G. H., Barr, C. J. & Whittaker, H. A. (1981a). *An integrated system of land classification.* Annual report for 1980. Institute of Terrestrial Ecology, Cambridge.

Bunce, R. G. H., Barr, C. J. & Whittaker, H. A. (1981b). *Land classes in Great Britain: preliminary descriptions for users of the Merlewood method of land classification.* Merlewood Research and Development Paper No. 86. Institute of Terrestrial Ecology, Grange-over-Sands, Cumbria.

Bunce, R. G. H., Barr, C. J. & Whittaker, H. A. (1983). A stratified system for ecological sampling. In *Ecological mapping from ground, air and space*: 39–46. (Ed. Fuller, R. M.). Institute of Terrestrial Ecology, Abbots Ripton.

Caughley, G., Grice, D., Barker, R. & Brown, B. (1988). The edge of the range. *J. Anim. Ecol.* 57: 771–785.

Clark, B. S., Leslie, D. M. & Carter, T. S. (1993). Foraging activity of adult female Ozark big-eared bats (*Plecotus townsendii ingens*) in summer. *J. Mammal.* 74: 422–427.

Cochran, W. G. (1963). *Sampling techniques*. John Wiley & Sons, New York.

Cresswell, P., Harris, S. & Jefferies, D. J. (1990). *The history, distribution, status and habitat requirements of the badger in Britain*. Nature Conservancy Council, Peterborough.

de Jong, J. & Ahlén, I. (1991). Factors affecting the distribution pattern of bats in Uppland, central Sweden. *Holarct. Ecol.* **14**: 92–96.

Erkert, H. G. (1982). Ecological aspects of bat activity rhythms. In *Ecology of bats*: 201–242. (Ed. Kunz, T. H.). Plenum Press, New York.

Fenton, M. B. (1970). A technique for monitoring bat activity with results obtained from different environments in southern Ontario. *Can. J. Zool.* **48**: 847–851.

Fenton, M. B. (1982). Echolocation, insect hearing and feeding ecology of insectivorous bats. In *Ecology of bats*: 261–285. (Ed. Kunz, T. H.). Plenum Press, New York.

Fenton, M. B. & Bell, G. P. (1981). Recognition of species of insectivorous bats by their echolocation calls. *J. Mammal.* **62**: 233–243.

Fenton, M. B. & Thomas, D. W. (1980). Dry-season overlap in activity patterns, habitat use, and prey selection by sympatric African insectivorous bats. *Biotropica* **12**: 81–90.

Furlonger, C. L., Dewar, H. J. & Fenton, M. B. (1987). Habitat use by foraging insectivorous bats. *Can. J. Zool.* **65**: 284–288.

Gaisler, J. (1979). Results of bat census in a town (Mammalia: Chiroptera). *Věstník čsl. spol. Zool.* **43**: 7–21.

Harris, S., Morris, P., Wray, S. & Yalden, D. (1995). *A review of British mammals: population estimates and conservation status of British mammals other than cetaceans.*. Joint Nature Conservation Committee, Peterborough.

Hayes, W. L. & Winkler, R. L. (1970). *Statistics, probability, inference and decision* **1**. Holt, Rinehart & Winston, New York.

Hosmer, D. W. & Lemeshow, S. (1989). *Applied logistic regression*. John Wiley & Sons, New York.

Jones, G. & Morton, M. (1992). Radio–tracking studies on habitat use by greater horseshoe bats (*Rhinolophus ferrumequinum*). In *Wildlife telemetry: remote monitoring and tracking of animals*: 521–537. (Eds Priede, I. G. & Swift, S. M.). Ellis Horwood, Chichester.

Jones, G. & Rayner, J. M. V. (1989). Foraging behaviour and echolocation of wild horseshoe bats *Rhinolophus ferrumequinum* and *R. hipposideros* (Chiroptera, Rhinolophidae). *Behav. Ecol. Sociobiol.* **25**: 183–191.

Jüdes, U. (1989). Analysis of the distribution of flying bats along line-transects. In *European bat research 1987*: 311–318. (Eds Hanák, V., Horácek, I. & Gaisler, J.). Charles University Press, Praha.

Kalko, E. K. V. & Schnitzler, H.-U. (1993). Plasticity in echolocation signals of European pipistrelle bats in search flight: implications for habitat use and prey detection. *Behav. Ecol. Sociobiol.* **33**: 415–428.

Kunz, T. H. (Ed.) (1988). *Ecological and behavioral methods for the study of bats*. Smithsonian Institution Press, Washington, D. C.

Kunz, T. H. & Brock, C. E. (1975). A comparison of mist nets and ultrasonic detectors for monitoring flight activity of bats. *J. Mammal.* **56**: 907–911.

Limpens, H. J. G. A. & Kapteyn, K. (1991). Bats, their behaviour and linear landscape elements. *Myotis* **29**: 39–47.

Lunde, R. E. & Harestad, A. S. (1986). Activity of little brown bats in coastal forests. *NW. Sci.* **60**: 206–209.

Magurran, A. E. (1988). *Ecological diversity and its measurement.* Croom Helm, London.

McAney, C. M. & Fairley, J. S. (1988). Habitat preference and overnight seasonal variation in the foraging activity of lesser horseshoe bats. *Acta theriol.* **33**: 393–402.

Neu, C. W., Byers, C. R. & Peek, J. M. (1974). A technique for analysis of utilization-availability data. *J. Wildl. Mgmt* **38**: 541–545.

Norusis, M. J. (1990). *SPSS base system users guide.* SPSS Inc., Chicago.

Racey, P. A. & Swift, S. M. (1985). Feeding ecology of *Pipistrellus pipistrellus* (Chiroptera: Vespertilionidae) during pregnancy and lactation. I. Foraging behaviour. *J. Anim. Ecol.* **54**: 205–215.

Rachwald, A. (1992). Habitat preference and activity of the noctule bat *Nyctalus noctula* in the Bialowieza Primeval Forest. *Acta theriol.* **37**: 413–422.

Robinson, M. F. & Stebbings, R. E. (1993). Food of the serotine bat, *Eptesicus serotinus*—is faecal analysis a valid qualitative and quantitative technique? *J. Zool., Lond.* **231**: 239–248.

Rydell, J. (1986). Foraging and diet of the northern bat *Eptesicus nilssoni* in Sweden. *Holarct. Ecol.* **9**: 272–276.

Rydell, J. (1991). Seasonal use of illuminated areas by foraging northern bats *Eptesicus nilssoni. Holarct. Ecol.* **14**: 203–207.

Speakman, J. R., Racey, P. A., Catto, C. M. C., Webb, P. I., Swift, S. M. & Burnett, A. M. (1991). Minimum summer populations and densities of bats in N. E. Scotland, near the northern borders of their distributions. *J. Zool., Lond.* **225**: 327–345.

Swift, S. M. & Racey, P. A. (1983). Resource partitioning in two species of vespertilionid bats (Chiroptera) occupying the same roost. *J. Zool., Lond.* **200**: 249–259.

Thomas, D. W. & West, S. D. (1989). Wildlife–habitat relationships: sampling procedures for Pacific northwest vertebrates. Sampling methods for bats. *U. S. For. Serv. gen. tech. Rep. PNW* No. 243: 1–20.

Walsh, A. L., Hutson, T. M. & Harris, S. (1993). UK volunteer bat groups and the British bats and habitats survey. In *Proceedings of the first European bat detector workshop*: 113–123. (Ed. Kapteyn, K.). Netherlands Bat Research Foundation, Amsterdam.

Walsh, A. L. & Mayle, B. A. (1991). Bat activity in different habitats in a mixed lowland woodland. *Myotis* **29**: 97–104.

Waters, D. A. & Walsh, A. L. (1994). The influence of bat detector brand on the quantitative estimation of bat activity. *Bioacoustics* **5**: 205–221.

Winter, T. G. (1983). *A catalogue of phytophagous insects and mites on trees in Great Britain.* Forestry Commission Booklet No. 53. Her Majesty's Stationery Office, London.

Symp. zool. Soc. Lond. (1995) No. 67: 345–360

Information transfer in bats

Gerald S. WILKINSON

Department of Zoology
University of Maryland
College Park
Maryland 20742, USA

Synopsis

Here I review, and present new, evidence indicating that bats gain information about their environment, i.e. the location or quality of food, roosting sites, predators or mates, from other bats. Four mechanisms—local enhancement, social facilitation, imitative learning, and intentional signalling—of acquiring or transmitting information are described. Local enhancement or inadvertent direction to a resource has been demonstrated for two situations—eavesdropping and following. Playback experiments with several species show that insectivorous bats approach echolocation calls. Non-random departures from nursery colonies have been described for several species; in *Nycticeius humeralis* individuals improve their foraging success by following previously successful foragers to feeding sites. Social facilitation may occur in at least one group-foraging species where *per capita* feeding attempts increase with group size. Imitative learning has been reported for several species learning to feed under novel situations in the laboratory, but the data are consistent with local enhancement and individual learning. Matching of echolocation call frequency in one species provides the best evidence for imitative learning, while matching of isolation calls to directive calls may simply reflect heritable ontogenetic change.

Intentional or voluntary signalling occurs among bats during courtship, attempted predation, resource defence and resource advertisement. For each of these situations evidence from one or more species indicates that the type or frequency of vocalization changes depending on the identity of potential receivers of the signal. In a few cases, such as defensive calls while feeding from horses in *Desmodus rotundus* and group foraging calls in *Phyllostomus hastatus*, individuals selectively give or withhold calls depending on whether or not individuals are from the same social group. Signalling situations relevant to recent theory regarding honest advertisement are identified as exciting opportunities for further study.

Introduction

Although many bats are notable for their ability to extract information from the environment by using echolocation, in this paper I evaluate evidence indicating that bats also obtain information about the location or quality of

ZOOLOGICAL SYMPOSIUM No. 67
ISBN 0 19 854945 8

food, roosting sites or mates from conspecifics. I focus on auditory signals because this sensory modality has received the most study. To clarify function I distinguish four mechanisms for information transfer—local enhancement, social facilitation, imitative learning and intentional signalling. Although some of these terms were developed to describe social learning, they differ from each other with respect to the potential costs and benefits associated with information transfer. Therefore, by categorizing bat behaviours with these terms I hope to illuminate the ecological and evolutionary conditions that favour information transfer.

Local enhancement refers to one organism inadvertently directing the behaviour of another to some region of the environment. While local enhancement is often used for situations where animals can observe each other feed, inadvertent direction to a resource can occur away from feeding sites if animals forage from a central place. For this reason, cases in which animals follow conspecifics from a colony to feeding or roosting sites represent examples of local enhancement. Local enhancement is not advantageous to the animal which is followed unless group foraging is beneficial or information about the location of unpredictable resources is exchanged over time. As I have argued elsewhere (Wilkinson 1992b), information exchange can result from conditional following without reciprocity.

Social facilitation operates when individual feeding rate is improved in the presence of a conspecific. Thus, in contrast to local enhancement, all members of a group benefit immediately from sharing information about the location of a foraging site. While many forms of social facilitation are possible, here I emphasize cases of group foraging involving transfer of information about feeding sites or individual identity.

Imitative learning permits more rapid acquisition of novel behaviours than does trial-and-error learning, and could allow group-living individuals to share feeding skills (Giraldeau 1984). Imitation can be inadvertent if a naive individual observes a knowledgeable animal, or directed, i.e. involving teaching, if a knowledgeable animal repeats a behaviour more often in the presence than in the absence of a naive conspecific. Genetic models predict that imitative learning should be favoured by natural selection when the environment is predictable and individual learning is either inaccurate or costly to make accurate (Boyd & Richerson 1988). Distinguishing imitation from local enhancement and trial-and-error learning is, however, difficult. The best evidence for imitation is association of a particular motor pattern with a novel behaviour (Galef 1988). Because unnecessary repetitions of a behaviour consume energy and may increase predation risk, teaching should be very uncommon.

Intentional signalling requires that signals should be produced only in association with particular aspects of the environment and should be under voluntary control. For example, alarm and food calls alert conspecifics about particular features of an individual's immediate environment. Whether or not intentional signals provide honest information is controversial (Hauser & Nelson 1991). Until recently, behavioural ecologists thought that signals rarely provide accurate information and function to manipulate the behaviour of recipients for individual gain (Krebs & Dawkins 1984). Honest signalling was assumed to be limited to close relatives or long-term associates, thereby enhancing inclusive (Hamilton 1964) or direct fitness through reciprocity (Axelrod & Hamilton 1981), respectively. However, recent theory suggests that signals should provide honest information because only costly signals that still provide a net benefit to the signaller represent evolutionarily stable strategies (Grafen 1990; Maynard Smith 1991). Because this controversy is unlikely to be settled without more evidence, I identify possible cases of intentional signalling where deception could occur.

Because few studies on bats have been designed to assess information transfer, below I review cases of bat social interactions involving vocalizations which could represent examples of local enhancement, social facilitation, imitative learning or intentional signalling. Where function is uncertain, I provide possible adaptive explanations. My objective is two-fold: to illustrate the diversity of vocal signals used by bats to mediate social behaviour and to stimulate further work on communication call function in bats.

Local enhancement

Eavesdropping

Because most insectivorous bats use an active detection system for capturing prey, echolocation pulses, especially those used for capturing prey, can provide information about the location of feeding sites to other bats. Aggregations of bats at rich feeding sites could form more quickly if individuals responded to each other's vocalizations than if they independently discovered prey patches by searching. Accordingly, groups of several temperate insectivorous bat species have been observed forming rapidly around ephemeral insect patches, e.g. *Myotis* spp. (Fenton & Morris 1976; Bell 1980), *Pipistrellus pipistrellus* (Racey & Swift 1985), *P. kuhlii* (Barak & Yom-Tov 1989). That eavesdropping leads to group formation has been demonstrated with playback experiments for adult and recently volant *Myotis lucifugus* which approached speakers broadcasting echolocation calls emitted by both *M. lucifugus* and *Eptesicus fuscus* (Barclay 1982). Playbacks to *Lasiurus borealis* indicate that this species responds only to conspecifics (Balcombe & Fenton 1988).

Eavesdropping as a method for acquiring information about foraging sites should differ between species, depending on frequency and amplitude of echolocation pulses, as well as on the presence of obstructions between foragers, because atmospheric attenuation has larger effects on high- than on low-frequency sounds (Griffin 1971). Atmospheric attenuation can vary from 0.2 to 15 dB/m depending on frequency and humidity, and adds to the 6 dB loss per distance doubling caused by spherical spreading (Griffin 1971;

Lawrence & Simmons 1982). Many insectivorous bats partially overcome attenuation by producing echolocation calls with amplitudes in excess of 110 dB at 10 cm (Griffin & Novick 1955). In contrast, phyllostomid bats produce much quieter echolocation calls of 75 dB at 10 cm (Griffin & Novick 1955). To illustrate the consequences of frequency and amplitude variation on eavesdropping, I have estimated the distance a 110 dB or a 75 dB sound would travel before reaching 0 dB under several temperature and humidity conditions experienced by temperate and tropical bats (Fig. 1), using ANSI predictive equations (Acoustical Society of America 1978). These calculations predict that an *Eptesicus fuscus* emitting 110 dB, 40 kHz calls (Masters, Jacobs & Simmons 1991) on a 30 °C evening with 100% relative humidity can detect conspecifics calling at 60 m. In contrast, a *Leptonycteris curasoae* with 75 dB, 35 kHz calls (Howell 1974) should detect conspecifics within 20 m on a 30 °C, 25% relative humidity night, while *Glossophaga soricina* with 75 dB, 100 kHz calls (Griffin & Novick 1955) would hear conspecifics no more than 9 m away on a 30 °C, 100% relative humidity night. Actual detection distances may be substantially less if obstructions, such as trees or bushes, are present in the foraging area or if there is noise from wind or insects that would mask the echolocation signal and raise the detectability threshold above 0 dB. Conversely, longer distances may be possible for bats foraging over water because the water surface reflects sound and can act as a wave guide (Wiley & Richards 1978).

If foraging success declines in the presence of conspecifics, then bats should modify their echolocation calls in the presence of conspecifics to reduce eavesdropping. One method for decreasing transmission distance is to increase call frequency. Reduction of eavesdropping could explain cases where more energy

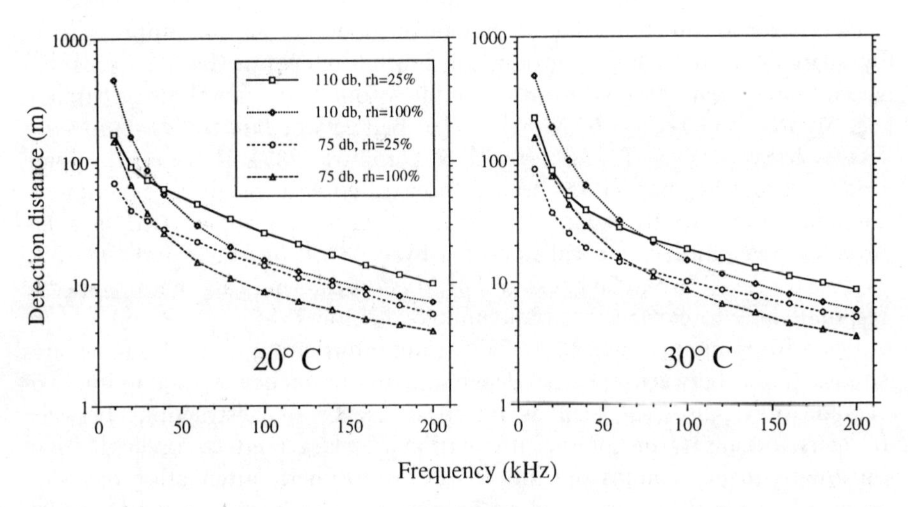

Fig. 1. Distance at which a 110 dB or 75 dB signal of a given frequency will decrease to 0 dB. Attenuation occurs through spherical spreading and atmospheric absorption.

is put into high-frequency components of a call, e.g. by foraging *Rhinolophus rouxi* (Neuweiler, Metzner, Heilmann, Rubsamen, Eckrich & Costa 1986). Alternatively, echolocation may be used less frequently. Observations of *Antrozous pallidus* (Bell 1982) and *Macrotus californicus* (Bell 1985) not using echolocation while capturing prey may, therefore, represent behavioural adaptations to avoid alerting competitors to the presence of large food items rather than to avoid alerting prey to imminent capture. In support of this proposition, ultrasonic hearing organs have never been described (D. Yager pers. comm.) for two favourite prey items of *Antrozous pallidus*, scorpions and Jerusalem crickets (Hatt 1923).

Following

An alternative, but not exclusive, mechanism for locating feeding sites is to follow conspecifics. Following behaviour is expected when resources are unpredictable but persist long enough for multiple foraging trips to be profitable. Although many, if not most, temperate insectivorous bats may experience such food dispersion, only individuals which make multiple foraging trips per night and gather at a common site between trips can benefit from following. These conditions occur in some species during lactation when young are left at a communal nursery roost and females take several foraging trips per night to meet the energetic demands of producing milk. Recent efforts to monitor departure intervals from communal roosts have demonstrated that at least three temperate insectivores, *Eptesicus fuscus* (Brigham & Fenton 1986), *Pipistrellus pipistrellus* (Speakman, Bullock, Eales & Racey 1992) and *Nycticeius humeralis* (Wilkinson 1992b), depart in small clusters more often than would be expected by chance. Although Speakman *et al.* (1992) suggest that clustered departures may be due to a bottleneck at the roost exit, clustered departures of *N. humeralis* occurred after the initial evening exodus between 2200 and 0500, at a time when departures were sufficiently infrequent to preclude a bottleneck (Wilkinson 1992b).

Evidence from variation in prey density, location of foraging radio-tagged individuals, and individual weights before and after departing within 10 s of a previously successful or unsuccessful forager provide strong evidence that *N. humeralis* follow each other in order to improve foraging success (Wilkinson 1992b). At least 20% of departing animals did not gain weight while out of the roost, indicating that they failed to locate ephemeral, rich prey patches. On average, 16% of second or later departures occurred within 10 s of the departure of another bat. Because radio-telemetry data indicate that females sometimes change feeding sites within and between nights as well as forage with other bats, we postulate that following occurs only after independent searching has failed. Following apparently improves foraging success because the weights of bats that followed previously successful foragers were greater than the weights of bats that did not follow (Fig. 2).

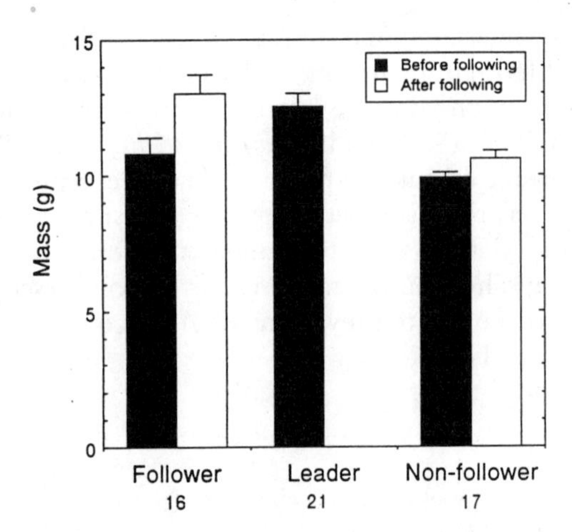

Fig. 2. Average (±SE) weight of *Nycticeius humeralis* after a foraging trip (before following) and, on the same night, after a subsequent trip on which they followed another bat (after following).

Because followers benefit at the expense of searchers, following would provide higher inclusive fitness if exchanged among relatives than if exchanged among unrelated individuals. However, the average relatedness among leader-follower pairs was −0.12 (SE = 0.07, n = 41 pairs), estimated by genotypic correlation (Queller & Goodnight 1989) using four polymorphic allozymes (Wilkinson 1992a). Furthermore, only one of ten leader–follower pairs shared the same mitochondrial DNA sequence haplotype (Wilkinson 1992a)—a result to be expected if bats arbitrarily choose colony mates to follow.

These results are consistent with previously unsuccessful foragers simply following a colony member chosen at random on a subsequent trip. The weight data indicate that such a strategy would allow a follower to avoid paying search costs on 80% of subsequent departures. Why do not some females forego independent searching and simply wait to follow another bat? I suspect the energetic demands of lactation are so great in this species that a female which did not attempt to hunt at dusk would not be able to capture enough prey during the night to meet her energy requirements and those of her two or three young. At peak lactation female *N. humeralis* produce half their body weight in milk per day (Steele 1991) and make an average of four foraging trips per night (Wilkinson 1992b).

Social facilitation

Group foraging is commonly observed among bats which forage either close to water or on nectar and pollen. Although some cases of flocking may represent opportunistic aggregations formed by local enhancement, several species forage in persistent groups and exhibit co-ordinated flight.

For example, *Rhynchonycteris naso* (Bradbury & Vehrencamp 1976), *Myotis adversus* (Dwyer 1970), *Pipistrellus kuhlii* (Barak & Yom-Tov 1989), *Noctilio leporinus* (A. Brooke pers. comm.), *Leptonycteris curasoae* (Howell 1979), and *Phyllostomus discolor* (Sazima & Sazima 1977) have all been observed flying one bat after another while hawking insects, gaffing fish or visiting flowers, respectively. Whether such co-ordinated tandem flight provides similar information to conspecifics in each of these situations is currently unknown. Tandem flight may enable bats to monitor the foraging activity of others in the group in order to avoid recently visited feeding sites. Alternatively, fish or insects fleeing from one bat may be more easily captured or detected by other bats in the group. In support of this assertion, *per capita* feeding buzz rate correlated with group size for *P. kuhlii* foraging around street lights (Barak & Yom-Tov 1989).

If foraging success improves as group size increases, then signals that reliably attract conspecifics would be favoured by natural selection. One method for advertising prey availability is to lower the frequency of the terminal buzz in an attack sequence so as to increase detection distance, as has been observed in *P. kuhlii* (Schnitzler, Kalko, Miller & Surlykke 1987) and *Myotis daubentonii* (Kalko & Schnitzler 1989). Whether such calls represent a physiological constraint (Kalko & Schnitzler 1989) or a mechanism for increasing foraging-group size requires further study. Low-frequency calls could also be used to advertise location. For example, the directive calls of *Antrozous pallidus* (O'Shea & Vaughan 1977) appear to recruit conspecifics to roost sites, possibly to improve thermoregulation (Trune & Slobodchikoff 1976).

Co-ordinated tandem flight requires that bats monitor the location of conspecifics, presumably by eavesdropping on echolocation pulses. When foraging groups contain the same individuals over time, e.g. *P. pipistrellus* (Racey & Swift 1985), *N. leporinus* (A. Brooke pers. comm.), *P. discolor* (Wilkinson 1987), and *P. hastatus* (G. S. Wilkinson & J. Boughman unpubl.), individual recognition must also occur. Individual recognition of echolocation cries has been demonstrated in *Eptesicus fuscus* (Masters & Jacobs 1989) and could occur in bats where echolocation pulses differ between individuals, e.g. *P. pipistrellus* (Miller & Degn 1981), *Rhinopoma hardwickii* (Habersetzer 1981), *E. fuscus* (Masters *et al.* 1991), *Lasiurus borealis* (Brigham, Cebek & Hickey 1989) and *Pteronotus parnellii* (Suga, Niwa, Taniguchi & Margoliash 1987).

Imitative learning

Foraging technique

Imitative learning has been reported for *Eptesicus fuscus, Myotis lucifugus* and *Antrozous pallidus* (Gaudet & Fenton 1984). Naive bats allowed to search for mealworms suspended from a wall learned to locate prey more quickly when a knowledgeable bat was present, whereas naive bats allowed to search alone did not discover food during the testing period. Similarly, naive *Phyllostomus discolor* searching for one accessible food cup among an array of 16 cups in the presence of a knowledgeable conspecific found food more rapidly than did naive bats searching alone (Wilkinson 1987).

Neither of these examples requires observational learning. If solitary bats search less than bats hunting in a group, i.e. demonstrate a 'fear effect' (Galef 1988), then naive bats hunting in a group should locate food more rapidly by individual learning than do solitary bats. Alternatively, if naive bats followed knowledgeable bats, then the area to search would be reduced. Food could then be located by individual learning more quickly as a consequence of local enhancement.

Offspring might also be expected to learn how to feed from observing their mothers or other adults. Although the young of some bat species learn to fly and feed independently of their mothers (Buchler 1980), anecdotal reports on *Eptesicus fuscus* (Brigham & Brigham 1989), *Noctilio albiventris* (Brown, Brown & Grinnell 1983) and *Lavia frons* (Vaughan & Vaughan 1987) indicate that females occasionally associate with recently volant young outside the roost. Video records of *Nycticeius humeralis* departing from an attic roost on three nights reveal that six of 25 recently volant (average age 30 days) pups departed after 2200 within 10 s of an adult female. This represents more paired departures than would be expected ($P < 0.001$, 1000 randomizations) if adults and juveniles were departing at random. However, only two of the six juveniles followed their mothers: the remaining four followed unrelated adult females. In contrast, female *Desmodus rotundus* often feed in the same pastures and sometimes at the same bite with daughters during their first year (Wilkinson 1985). Species with difficult feeding techniques and protracted periods of association between mothers and young, such as *D. rotundus* or *Vampyrum spectrum*, deserve further study to document imitative learning.

Vocalizations

One method for advertising group membership is to adopt a common vocalization. This requires imitation of a call—an uncommon ability in mammals. Observations that *Phyllostomus discolor* infant isolation calls converge on maternal directive calls during a 45-day period after birth have been promoted as evidence for imitative learning (Esser & Schmidt 1989). Since isolation calls closely resemble female directive calls in several species, such as *Tadarida brasiliensis* (Balcombe & McCracken 1992) and *P. hastatus* (D.L. O'Reilly & G.S. Wilkinson unpubl.), directive call imitation could be widespread. However, infant isolation calls should also eventually resemble maternal directive calls if these two call types are ontogenetically related (Brown 1976) and heritable. In support of this suggestion we have estimated significant heritabilities for spectral variables measured from isolation calls of 39 *Nycticeius humeralis* sibling pairs (Scherrer & Wilkinson 1993) and from 22 groups of *Phyllostomus hastatus* (D. L. O'Reilly & G.S. Wilkinson unpubl.). Because harem male *P. hastatus* typically father all of the young in a female group (McCracken & Bradbury 1977), the young within a group are paternal half-siblings. The recent claim (Rasmuson & Barclay 1992) that isolation calls of *Eptesicus fuscus* are not heritable should be viewed with caution as only two pairs of siblings were measured.

Recent observations of extreme similarity in the constant frequency portion of the echolocation pulse between mother and young *Rhinolophus ferrumequinum* suggest that imitative learning can influence echolocation calls (Jones & Ransome 1993). Deafening experiments have demonstrated that *R. rouxi* need to hear to produce appropriately tuned echolocation pulses (Rubsamen & Schafer 1990). However, the ability to modify echolocation call resting frequency may be a consequence of a matched signal-receiver system and have little functional significance for information transfer except to provide a mechanism for generating individual differences.

Intentional signalling

Mating calls

The most spectacular mating calls in bats occur among epomophorine bats (Wickler & Seibt 1976; Bradbury 1977a) which employ lek or exploded lek mating systems (Bradbury 1981). In addition, a diverse array of microchiropteran species, including *Saccopteryx bilineata* (Bradbury & Emmons 1974), *Cardioderma cor* (Vaughan 1976), *Nyctalus noctula* (Miller & Degn 1981) and *Pipistrellus pipistrellus* (Miller & Degn 1981), use audible calls during courtship. At least for *Hypsignathus monstrosus*, courtship calls represent intentional signals because call rate increases when females approach (Bradbury 1977a). Individual males may also have distinctive courtship calls, e.g. *Epomophorus wahlbergi* (Wickler & Seibt 1976) and *P. pipistrellus* (G. Jones pers. comm.). Further study is needed to determine if variation in some aspect of these calls influences female visitation rates by honestly or deceptively signalling male quality.

Distinct vocalizations associated with copulation have been described for *Myotis lucifugus* (Barclay & Thomas 1979) and *Hypsignathus monstrosus* (Bradbury 1977a). Barclay & Thomas (1979) argue that copulation calls signal to the female being mated that a male intends to copulate, not fight, with her. Such an interpretation seems unlikely for *H. monstrosus*, since these calls occur at the end of a copulation in this species (Bradbury 1977a). An intriguing alternative hypothesis is that males are signalling their mating success to other females. In many lekking birds, females visit male arenas and observe copulations for several days before selecting a mate (Bradbury & Gibson 1983). Because bat copulations usually occur in darkness, females may need to rely on sound to monitor mating activity. If copulation calls are costly for males to produce, then they could provide reliable indicators of male mating activity. This explanation is consistent with male *M. lucifugus* not calling during mating when females are in torpor and with call differences among males (Barclay & Thomas 1979).

Alarm calls

Some of the best evidence indicating that non-human animals can associate signals with environmental referents involve species that use different alarm calls for different types of predators (Seyfarth, Cheney & Marler 1980). Among bats, alarm calls have been reported only for *Pteropus poliocephalus* (Nelson 1964). Distinct calls associated with different classes of predators have not been described. However, I suspect that alarm calls may be more common among bats. During my work with nursery colonies of *N. humeralis* and *P. hastatus* I have noticed that once animals have been captured in a roost, subsequent approach with a headlamp causes an increase in the intensity and frequency of audible vocalizations.

Many bats emit loud, low-frequency vocalizations when captured. These distress calls might startle or acoustically impair potential predators. However, they also often attract conspecifics (August 1979). Given the prevalence of this behaviour, playback studies would be useful in determining whether distress calls elicit mobbing behaviour.

Territorial calls

Observations of chases and vocalizations among marked or recognizable individuals indicate that resource defence is particularly common among tropical bats, quite possibly because resource dispersion in these areas is more stable over time than in temperate areas. For example, several emballonurid bats, *Rhynchonycteris naso*, *Saccopteryx leptura* and *S. bilineata*, maintain group territories by chasing and vocalizing at intruders (Bradbury & Emmons 1974; Bradbury & Vehrencamp 1976). A variety of species that utilize a sit-and-wait foraging strategy defend territories by chasing intruders or vocalizing. Bats which maintain solitary feeding territories include *Hipposideros commersoni* (Vaughan 1977) and *Rhinolophus rouxii* (Neuweiler *et al.* 1986). The megadermatids, *Cardioderma cor* (McWilliam 1987) and *Lavia frons* (Vaughan & Vaughan 1986), form reproductive pairs in which one or both members of the pair defends the territory by chasing and vocalizing at intruders. Among vespertilionids, *Myotis daubentonii* (Wallin 1961) and *M. adversus* (Dwyer 1970) have been observed chasing conspecifics at foraging sites, while *P. pipistrellus* foraging on a regular beat will chase and vocalize at conspecifics when food density is low (Racey & Swift 1985). *D. rotundus* maintain exclusive group feeding areas and will attack, chase and vocalize (Sailler & Schmidt 1978) at non-roost group members that attempt to feed from the same prey animal (Wilkinson 1985). Because some individuals, including relatives, are allowed to feed from the same wound without aggression (Wilkinson 1985), this example demonstrates intentional signalling.

Food calls

At least two species of bats, *Macroderma gigas* and *Phyllostomus hastatus*, give audible vocalizations at feeding sites without evidence of territorial behaviour. *M. gigas*, like other megadermatids, use a sit-and-wait hunting strategy but produce loud chirps while hunting (Tidemann, Priddel, Nelson & Pettigrew 1985). Observations and radio-tracking provide no evidence that these calls advertise feeding territories. In contrast, up to 20 individuals have been observed foraging in a common area and calling at a rate that correlated with hunting activity (Tidemann *et al.* 1985). Further study is needed to determine if these calls advertise food and if social facilitation occurs.

For the last two years my students and I have been studying *Phyllostomus hastatus* to determine the function of loud, low-frequency screech vocalizations (Fig. 3) given by flying bats when departing from a roost and at feeding sites (Greenhall 1965). These 75–100 g bats feed on fruit, pollen and insects (Gardner 1977) and have been observed foraging in groups on termites (Bloedel 1955), sapacaia nut fruits (Greenhall 1965), and *Hymenaea courbaril* pollen (McCracken & Bradbury 1981). By monitoring bats flying into and out of a cave roost on video using infra-red illumination, we have discovered that bat departures and screech calls are clustered throughout the night, while

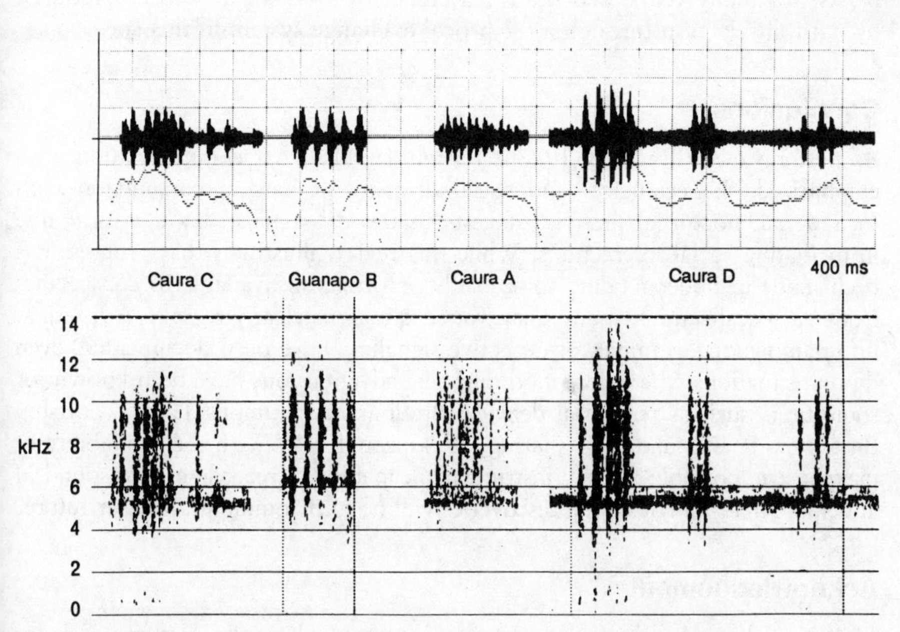

Fig. 3. *Phyllostomus hastatus* screech calls from four individuals recorded at two caves in Trinidad, W. I. The grid interval in the upper portion of the figure is 400 ms.

arrivals are random. By broadcasting screech calls from speakers we have demonstrated that these calls attract conspecifics and elicit additional calling both at the roost site and at flowering balsa trees, *Ochroma lagopus*, where we have observed groups of bats to feed and produce screech calls. We have heard screech calls from groups, but rarely from solitary individuals, flying to feeding sites. By attaching light-emitting diodes of a single colour to all adult females in a roosting group, we have observed group mates departing in pairs and giving screech calls more often than expected. Bats that return to the cave after a feeding trip and fly directly into the cave almost never give screech calls, while those that return, circle and subsequently depart give calls and are accompanied by a group mate more often than expected.

Because screech calls are not confined to either the lactation or the mating period, they do not involve parental care or courtship. Furthermore, an alarm call function seems dubious because screech calls occur frequently at the cave as well as in transit to feeding sites in the absence of any aerial predators. Current evidence suggests that these calls function to recruit conspecific bats into foraging groups and are given at times when group members will hear them. We are currently investigating whether these calls carry information about individual identity and are given by animals which have either located or failed to discover feeding sites. Because adult females within roost groups are unrelated to each other (McCracken 1987) but form stable associations that persist for many years, active advertisement of food sites could be favoured by natural selection through a reciprocal exchange system in this species.

Conclusions

As past reviews (Bradbury 1977b; Fenton 1985) of bat communication have emphasized, the ecological and social diversity of bats in conjunction with their aerial, nocturnal lifestyle suggests that vocal signals play a critical role in mediating social interactions. While this review illustrates that progress has been made in understanding some of the information available to conspecifics from bat vocalizations, many unanswered questions clearly remain. At present, no unambiguous examples of deceptive signalling have been documented, even though situations where deception might be advantageous have been known for some time, such as territorial defence (Bradbury & Emmons 1974), courtship (Bradbury 1977a) and food sharing (Wilkinson 1984). With the availability of inexpensive portable digital instruments capable of recording ultrasound, I anticipate that many exciting discoveries will be forthcoming in the near future.

Acknowledgements

My research on bat behaviour has been supported by the National Science Foundation and a Searle Scholar Award from the Chicago Community Trust. I thank J. Boughman and S. Swift for comments on the manuscript.

References

Acoustical Society of America (1978). American national standard method for the calculation of the absorbtion of sound by the atmosphere. *ANSI* S1.26: 1–12.

August, P. V. (1979). Distress calls in *Artibeus jamaicensis*: ecology and evolutionary implications. In *Vertebrate ecology in the northern Neotropics*: 151–159. (Ed. Eisenberg, J. F.). Smithsonian Institution Press, Washington.

Axelrod, R. & Hamilton, W. D. (1981). The evolution of cooperation. *Science* **211**: 1390–1396.

Balcombe, J. P. & Fenton, M. B. (1988). The communication role of echolocation calls in vespertilionid bats. In *Animal sonar*: 625–628. (Eds Nachtigall, P. E. & Moore, P. W. B.). Plenum Press, New York.

Balcombe, J. P. & McCracken, G. F. (1992). Vocal recognition in Mexican free-tailed bats: do pups recognize mothers? *Anim. Behav.* **43**: 79–87.

Barak, Y. & Yom-Tov, Y. (1989). The advantage of group hunting in Kuhl's bat *Pipistrellus kuhli* (Microchiroptera). *J. Zool., Lond.* **219**: 670–675.

Barclay, R. M. R. (1982). Interindividual use of echolocation calls: eavesdropping by bats. *Behav. Ecol. Sociobiol.* **10**: 271–275.

Barclay, R. M. R. & Thomas, D. W. (1979). Copulation call of *Myotis lucifugus*: a discrete situation-specific communication signal. *J. Mammal.* **60**: 632–634.

Bell, G. P. (1980). Habitat use and responses to patches of prey by desert insectivorous bats. *Can. J. Zool.* **58**: 1876–1883.

Bell, G. P. (1982). Behavioral and ecological aspects of gleaning by a desert insectivorous bat *Antrozous pallidus* (Chiroptera: Vespertilionidae). *Behav. Ecol. Sociobiol.* **10**: 217–223.

Bell, G. P. (1985). The sensory basis of prey location by the California leaf-nosed bat *Macrotus californicus* (Chiroptera: Phyllostomidae). *Behav. Ecol. Sociobiol.* **16**: 343–347.

Bloedel, P. (1955). Observations on the life histories of Panama bats. *J. Mammal.* **36**: 232–235.

Boyd, R. & Richerson, P. J. (1988). An evolutionary model of social learning: the effects of spatial and temporal variation. In *Social learning: psychological and biological perspectives*: 29–48. (Eds Zentall, T. R. & Galef, B. G. Jr.). Lawrence Erlbaum Associates, Hillsdale, N. J., Hove & London.

Bradbury, J. W. (1977a). Lek mating behavior in the hammer-headed bat. *Z. Tierpsychol.* **45**: 225–255.

Bradbury, J. W. (1977b). Social organization and communication. In *Biology of bats* **3**: 1–72. (Ed. Wimsatt, W.). Academic Press, New York.

Bradbury, J. W. (1981). The evolution of leks. In *Natural selection and social behavior: recent research and new theory*: 138–169. (Eds Alexander, R. D. & Tinkle, D. W.). Chiron, New York & Concord.

Bradbury, J. W. & Emmons, L. H. (1974). Social organisation of some Trinidad bats. I. Emballonuridae. *Z. Tierpsychol.* **36**: 137–183.

Bradbury, J. W. & Gibson, R. M. (1983). Leks and mate choice. In *Mate choice*: 109–138. (Ed. Bateson, P.). Cambridge University Press, Cambridge.

Bradbury, J. W. & Vehrencamp, S. L. (1976). Social organization and foraging in emballonurid bats. I. Field studies. *Behav. Ecol. Sociobiol.* **1**: 337–381.

Brigham, R. M. & Brigham, A. C. (1989). Evidence for association between a mother bat and its young during and after foraging. *Am. Midl. Nat.* **121**: 205–207.

Brigham, R. M., Cebek, J. E. & Hickey, M. B. C. (1989). Intraspecific variation in the echolocation calls of two species of insectivorous bats. *J. Mammal.* **70**: 426–428.

Brigham, R. M. & Fenton, M. B. (1986). The influence of roost closure on the roosting and foraging behaviour of *Eptesicus fuscus* (Chiroptera: Vespertilionidae). *Can. J. Zool.* **64**: 1128–1133.

Brown, P. (1976). Vocal communication in the pallid bat, *Antrozous pallidus. Z. Tierpsychol.* **41**: 34–54.

Brown, P. E., Brown, T. W. & Grinnell, A. D. (1983). Echolocation, development, and vocal communication in the lesser bulldog bat, *Noctilio albiventris. Behav. Ecol. Sociobiol.* **13**: 287–298.

Buchler, E. R. (1980). The development of flight, foraging, and echolocation in the little brown bat (*Myotis lucifugus*). *Behav. Ecol. Sociobiol.* **6**: 211–218.

Dwyer, P. D. (1970). Foraging behaviour of the Australian large-footed *Myotis* (Chiroptera). *Mammalia* **34**: 76–80.

Esser, K.-H. & Schmidt, U. (1989). Mother-infant communication in the lesser spearnosed bat *Phyllostomus discolor* (Chiroptera, Phyllostomidae): evidence for acoustic learning. *Ethology* **82**: 156–168.

Fenton, M. B. (1985). *Communication in the Chiroptera.* Indiana University Press, Bloomington.

Fenton, M. B. & Morris, G. K. (1976). Opportunistic feeding by desert bats (*Myotis* spp.). *Can. J. Zool.* **54**: 526–530.

Galef, B. G., Jr. (1988). Imitation in animals: history, definition, and interpretation of data from the psychological laboratory. In *Social learning: psychological and biological perspectives*: 3–28. (Eds Zentall, T. R. & Galef, B. G. Jr.). Lawrence Erlbaum Associates, Hillsdale, N. J., Hove & London.

Gardner, A. L. (1977). Biology of bats of the New World family Phyllostomatidae. Part 2. Feeding habits. *Spec. Publs Texas tech. Univ.* No. 13: 293–350.

Gaudet, C. L. & Fenton, M. B. (1984). Observational learning in three species of insectivorous bats (Chiroptera). *Anim. Behav.* **32**: 385–388.

Giraldeau, L.-A. (1984). Group foraging: the skill pool effect and frequency-dependent learning. *Am. Nat.* **124**: 72–79.

Grafen, A. (1990). Biological signals as handicaps. *J. theoret. Biol.* **144**: 517–546.

Greenhall, A. M. (1965). Sapucaia nut dispersal by greater spear-nosed bats in Trinidad. *Caribb. J. Sci.* **5**: 167–171.

Griffin, D. R. (1971). The importance of atmospheric attenuation for the echolocation of bats (Chiroptera). *Anim. Behav.* **19**: 55–61.

Griffin, D. R. & Novick, A. (1955). Acoustic orientation of neotropical bats. *J. exp. Zool.* **130**: 251–299.

Habersetzer, J. (1981). Adaptive echolocation sounds in the bat *Rhinopoma hardwickei. J. comp. Physiol.* **144**: 559–566.

Hamilton, W. D. (1964). The genetical evolution of social behavior. *J theoret. Biol.* **7**: 1–51.

Hatt, R. T. (1923). Food habits of the Pacific pallid bat. *J. Mammal.* **4**: 260–261.

Hauser, M. D. & Nelson, D. A. (1991). 'Intentional' signaling in animal communication. *Trends Ecol. Evol.* **6**: 186–189.

Howell, D. J. (1974). Acoustic behavior and feeding in glossophagine bats. *J. Mammal.* **55**: 293–308.

Howell, D. J. (1979). Flock foraging in nectar-feeding bats: advantages to the bats and to the host plants. *Am. Nat.* **114**: 23–49.

Jones, G. & Ransome, R. D. (1993). Echolocation calls of bats are influenced by maternal effects and change over a lifetime. *Proc. R. Soc. (B)* **252**: 125–128.

Kalko, E. K. V. & Schnitzler, H.-U. (1989). The echolocation and hunting behaviour of Daubenton's bat, *Myotis daubentoni*. *Behav. Ecol. Sociobiol.* **24**: 225–238.

Krebs, J. R. & Dawkins, R. (1984). Animal signals: mind-reading and manipulation. In *Behavioural ecology: an evolutionary approach* (2nd edn): 380–402. (Eds Krebs, J. R. & Davies, N. B.). Blackwell Scientific Publications, Oxford.

Lawrence, B. D. & Simmons, J. A. (1982). Measurements of atmospheric attenuation at ultrasonic frequencies and the significance for echolocation by bats. *J. acoust. Soc. Am.* **71**: 585–590.

Masters, W. M. & Jacobs, S. C. (1989). Target detection and range resolution by the big brown bat (*Eptesicus fuscus*) using normal and time-reversed model echoes. *J. comp. Physiol. (A)* **166**: 65–73.

Masters, W. M., Jacobs, S. C. & Simmons, J. A. (1991). The structure of echolocation sounds used by the big brown bat *Eptesicus fuscus*: some consequences for echoprocessing. *J. acoust. Soc. Am.* **89**: 1402–1413.

Maynard Smith, J. (1991). Honest signalling: the Philip Sydney game. *Anim. Behav.* **46**: 1034–1035.

McCracken, G. F. (1987). Genetic structure of bat social groups. In *Recent advances in the study of bats*: 281–298. (Eds Racey, P. A., Fenton, M. B. & Rayner, J. M. V.). Cambridge University Press, Cambridge.

McCracken, G. F. & Bradbury, J. W. (1977). Paternity and genetic heterogeneity in the polygynous bat, *Phyllostomus hastatus*. *Science* **198**: 303–306.

McCracken, G. F. & Bradbury, J. W. (1981). Social organization and kinship in the polygynous bat *Phyllostomus hastatus*. *Behav. Ecol. Sociobiol.* **8**: 11–34.

McWilliam, A. N. (1987). Territorial and pair behaviour of the African false vampire bat, *Cardioderma cor* (Chiroptera: Megadermatidae), in coastal Kenya. *J. Zool., Lond.* **213**: 243–252.

Miller, L. A. & Degn, H.-J. (1981). The acoustic behaviour of four species of vespertilionid bats studied in the field. *J. comp. Physiol.* 142: 67–74.

Nelson, J. E. (1964). Vocal communication in Australian flying foxes (Pteropodidae; Megachiroptera). *Z. Tierpsychol.* **21**: 857–570.

Neuweiler, G., Metzner, W., Heilmann, U., Rubsamen, R., Eckrich, M. & Costa, H. H. (1986). Foraging behaviour and echolocation in the rufous horseshoe bat (*Rhinolophus rouxi*) of Sri Lanka. *Behav. Ecol. Sociobiol.* **20**: 53–67.

O'Shea, T. J. & Vaughan, T. A. (1977). Nocturnal and seasonal activities of the pallid bat, *Antrozous pallidus*. *J. Mammal.* **58**: 269–284.

Queller, D. C. & Goodnight, K. F. (1989). Estimating relatedness using genetic markers. *Evolution* **43**: 258–275.

Racey, P. A. & Swift, S. M. (1985). Feeding ecology of *Pipistrellus pipistrellus* (Chiroptera: Vespertilionidae) during pregnancy and lactation. I. Foraging behaviour. *J. Anim. Ecol.* **54**: 205–215.

Rasmuson, T. M. & Barclay, R. M. R. (1992). Individual variation in the isolation calls of newborn big brown bats (*Eptesicus fuscus*): is variation genetic? *Can. J. Zool.* 70: 698–702.

Rubsamen, R. & Schafer, M. (1990). Audiovocal interactions during development? Vocalisation in deafened young horseshoe bats vs. audition in vocalisation-impaired bats. *J. comp. Physiol. (A)* **167**: 771–784.

Sailler, H. & Schmidt, U. (1978). Die sozialen Laute der Gemeinen Vampirfledermaus *Desmodus rotundus* bei Konfrontation am Futterplatz unter experimentellen Bedingungen. *Z. Säugetierk.* **43**: 249–261.

Sazima, I. & Sazima, M. (1977). Solitary and group foraging: two flower-visiting patterns of the lesser spear-nosed bat *Phyllostomus discolor*. *Biotropica* **9**: 213–215.
Scherrer, J. A. & Wilkinson, G. S. (1993). Evening bat isolation calls provide evidence for heritable signatures. *Anim. Behav.* **46**: 847–860.
Schnitzler, H.-U., Kalko, E., Miller, L. & Surlykke, A. (1987). The echolocation and hunting behavior of the bat, *Pipistrellus kuhli*. *J. comp. Physiol. (A)* **161**: 267–274.
Seyfarth, R. M., Cheney, D. L. & Marler, P. R. (1980). Vervet monkey alarm calls: semantic communication in a free-ranging primate. *Anim. Behav.* **28**: 1070–1094.
Speakman, J. R., Bullock, D. J., Eales, L. A. & Racey, P. A. (1992). A problem defining temporal pattern in animal behaviour: clustering in the emergence behaviour of bats from maternity roosts. *Anim. Behav.* **43**: 491–500.
Steele, S. R. E. (1991). *The energetics of reproduction in the evening bat,* Nycticieus humeralis. MS thesis, University of Maryland.
Suga, N., Niwa, H., Taniguchi, I. & Margoliash, D. (1987). The personalized auditory cortex of the mustached bat: adaptation for echolocation. *J. Neurophysiol.* **58**: 643–654.
Tidemann, C. R., Priddel, D. M., Nelson, J. E. & Pettigrew, J. D. (1985). Foraging behaviour of the Australian ghost bat, *Macroderma gigas* (Microchiroptera: Megadermatidae). *Aust. J. Zool.* **33**: 705–713.
Trune, D. R. & Slobodchikoff, C. N. (1976). Social effects of roosting on the metabolism of the pallid bat (*Antrozous pallidus*). *J. Mammal.* **57**: 656–663.
Vaughan, T. A. (1976). Nocturnal behavior of the African false vampire bat (*Cardioderma cor*). *J. Mammal.* **57**: 227–248.
Vaughan, T. A. (1977). Foraging behaviour of the giant leaf-nosed bat (*Hipposideros commersoni*). *E. Afr. Wildl. J.* **15**: 237–249.
Vaughan, T. A. & Vaughan, R. P. (1986). Seasonality and the behavior of the African yellow-winged bat. *J. Mammal.* **67**: 91–102.
Vaughan, T. A. & Vaughan, R. P. (1987). Parental behavior in the African yellow-winged bat (*Lavia frons*). *J. Mammal.* **68**: 217–223.
Wallin, L. (1961). Territorialism on the hunting ground of *Myotis daubentoni*. *Säugetierk. Mitt.* **9**: 156–159.
Wickler, W. & Seibt, U. (1976). Field studies on the African fruit bat *Epomophorus wahlbergi* (Sundevall), with special reference to male calling. *Z. Tierpsychol.* **40**: 345–376.
Wiley, R. H. & Richards, D. G. (1978). Physical constraints on acoustic communication in the atmosphere: implications for the evolution of animal vocalizations. *Behav. Ecol. Sociobiol.* **3**: 69–94.
Wilkinson, G. S. (1984). Reciprocal food sharing in the vampire bat. *Nature, Lond.* **308**: 181–184.
Wilkinson, G. S. (1985). The social organization of the common vampire bat. I. Pattern and cause of association. *Behav. Ecol. Sociobiol.* **17**: 111–121.
Wilkinson, G. S. (1987). Altruism and co-operation in bats. In *Recent advances in the study of bats*: 299–323. (Eds Racey, P. A., Fenton, M. B. & Rayner, J. M. V.). Cambridge University Press, Cambridge.
Wilkinson, G. S. (1992a). Communal nursing in evening bats. *Behav. Ecol. Sociobiol.* **31**: 225–235.
Wilkinson, G. S. (1992b). Information transfer at evening bat colonies. *Anim. Behav.* **44**: 501–518.

Symp. zool. Soc. Lond. (1995) No. 67: 361–376

The trophic niches of sympatric sibling *Myotis myotis* and *M. blythii*: do mouse-eared bats select prey?

Raphaël ARLETTAZ
Institute of Zoology and Animal Ecology
University of Lausanne
CH-1015 Lausanne, Switzerland

and Nicolas PERRIN
Ethological Station Hasli
University of Bern
CH-3032 Hinterkappelen, Switzerland

Synopsis

The two sibling species *Myotis myotis* and *M. blythii*, while sympatric in the Swiss Alps, exhibit highly distinct and narrow trophic niches. Does this differentiation arise purely from the use of distinct feeding habitats or, alternatively, do the species exert active prey selection within their feeding habitats? To answer this question, we compared the distribution of prey in the species' diets with that in their respective feeding habitats, looking for discrepancies with respect to taxon and/or size. Although the general correlation is fairly good, some local discrepancies arise: (1) *M. myotis* underexploited small prey items (<0.05 g dry weight), and (2) *M. blythii* overexploited cockchafers (*Melolontha melolontha*) and underexploited ground arthropods. We argue, however, that these discrepancies do not result from active prey selection, but from (1) the low availability of some categories of prey (due to either low detectability or low accessibility) and (2) the opportunistic exploitation of atypical feeding habitats. In fact, our results do not falsify the parsimonious hypothesis that the two species exert no active prey selection within their feeding habitats. We are furthermore not aware of any paper from the bat literature that provides definite support for active prey selection among insectivorous vespertilionids. Their echolocation system may actually preclude it.

Introduction

The greater and the lesser mouse-eared bats, *Myotis myotis* and *Myotis blythii* respectively, are two morphologically and genetically closely related species of vespertilionid bats which occur in sympatry over wide areas (Strelkov 1972; Felten, Spitzenberger & Storch 1977; Bogan, Setzer, Findley & Wilson 1978).

ZOOLOGICAL SYMPOSIUM No. 67
ISBN 0 19 854945 8

These sibling species often coexist in their nursery roosts, frequently building up mixed reproductive clusters (Constant 1960; Ariagno 1973; Ruedi, Arlettaz & Maddalena 1990). Recently, they also have been found together within the same mating roost, but there was no evidence for mixed mating pairs (R. Arlettaz & M. Lutz unpubl.). Despite similar karyotypes (Ruedi *et al.* 1990), the two species do not seem to hybridize, since no hybrid was found among more than 400 individuals biochemically identified (Arlettaz, Ruedi & Hausser 1993).

According to the principles of competitive exclusion (or Volterra-Gause principle; Hutchinson 1957) and of limiting similarity (MacArthur & Levins 1967), we would expect such an intimate coexistence of cryptic but genetically distinct bats to occur only if they had evolved ways of partitioning resources. Arlettaz *et al.* (1993) have recently shown that sympatric mouse-eared bats from the Swiss Alps exploit different prey spectra. Carabid beetles (Carabidae) are by far the most frequent prey in the diet of *M. myotis*, whereas *M. blythii* feeds mainly on bush crickets (Tettigoniidae). As these insects live on the ground or close to it, *M. myotis* and *M. blythii* must both be considered as ground-gleaning bats. However, both species may occasionally switch from their normal foraging method, especially when large concentrations of prey suddenly become available, e.g. cockchafers (*Melolontha melolontha*) in April–June (Kolb 1958). These insects are caught on the wing around trees or gleaned from the canopy foliage (R. Arlettaz & R. Güttinger unpubl.).

Previous radiotracking of sympatric mouse-eared bats in the study area has shown distinct interspecific habitat partitioning (R. Arlettaz unpubl.): *M. myotis* forages mostly in forest and orchards, whereas *M. blythii* exploits mountain slopes covered by steppe, a typical climactic grassland found in the driest valleys of the Central Alps. The aim of this study was to test whether the trophic niche separation of the two species of mouse-eared bats results entirely from habitat segregation, or if they also feed selectively on certain arthropod taxa and/or sizes within their foraging habitats.

A non-selective forager feeds on prey according to availability in the foraging environment. Conversely, a selective feeder captures only certain categories or sizes of prey among those it encounters within its feeding habitat: some prey taxa or sizes are neglected or avoided (underexploited), whereas others are represented in the diet in a higher proportion than would be expected from their availability (overexploited). According to Wittenberger (1981: 211), resource availability refers to the amount of resource an animal can actually capture, utilize or otherwise exploit. In bats, food choice may be heavily restricted by the physical characteristics of their specific echolocation calls (Gould 1955; Simmons, Fenton & O'Farrell 1979; Neuweiler 1984, 1990). Prey availability is difficult or impossible to measure accurately (Faure & Barclay 1992). Instead, we compared bat diets with food abundance, looking for discrepancies between prey abundance in the environment and in the diets, with respect to taxon and/or size.

Material and methods

Species identification and faecal analysis

This study took place in the Alps of Valais (southern Switzerland, 46°15' N, 7°30' E) from April to August 1992. In order to assess the diet of mouse-eared bats, droppings were collected at a mixed nursery roost located inside the attic of a church. Individuals flying back to the colony late at night or early in the morning were caught at the attic's entrance with the help of a specially designed harp trap (Arlettaz 1987). Bats were identified according to Arlettaz, Ruedi & Hausser (1991) or on the basis of an electrophoretical analysis of blood samples in the laboratory (Ruedi *et al.* 1990). They were kept in linen bags until defecation was completed. A total of 120 (70 *M. myotis* and 50 *M. blythii*) samples were thus gathered and stored in 70% ethanol. Samples were collected between early May and mid-August. Faeces were dissected under a binocular microscope. Remains were identified, usually to family level. The relative volume (to the nearest 5–10%) of the different prey categories within each sample (4–15 pellets each) was estimated. Volume proportions indirectly provide information about the relative 'biomass' of the different prey items; this could not be achieved by frequency analyses (Kunz & Whitaker 1983). The overall relative proportion of each category of prey in the diet was estimated for each species separately.

Food abundance

Food abundance was investigated through pitfall trapping and hand netting of the ground and/or grass arthropod fauna from early April to late July 1992. The different habitats were chosen among previously (1989–1991) radio-delimited hunting grounds of the bats from the colony where faeces were collected. The three sampling sites (woodland, orchards and steppe) were located within a radius of 2 to 15 km from the colony. Each site was set up with three separate (150–600 m distant) groups of five pitfall traps. The distance between two successive pitfall traps was 5–10 m. The trapped arthropods were collected every 10 days from the beginning to the end of the experiment. In steppe (but not in forest and orchards because of the absence of grass on the ground), hand netting was performed on the same dates as collection from pitfall traps, on a 15–30 m long transect along rows of traps. Samples were stored in 70% ethanol. The content of samples was sorted and the frequency of the different category items was estimated for each sample and each sampling period separately. Insects were dried in an oven for 72 h at 65 °C. The dry weight of each category from a single sample and a single sampling period was measured to the nearest 0.001 g. Items from pitfall traps and hand netting were pooled for analysis.

The arthropods sampled were divided into two groups: larger items (>5–7 mm body length) and smaller ones (<5–7 mm). Bauerová (1978) and Pont & Moulin (1985) concluded that 12–15 mm represents the minimum body length of prey captured by the greater mouse-eared bat. Only items belonging to the larger class (>5–7 mm) have therefore been considered in the subsequent analyses on prey selection. A lower threshold than that proposed by these authors was chosen, since the lesser mouse-eared bat can be expected to feed on smaller prey than does its relative, owing to its slightly smaller body size (21.8 vs 25. 1 g: Arlettaz *et al.* 1991). Finally, the biomass of the different categories was converted into a proportion of the total biomass of food abundance.

Prey selection

Taxa

Since the distributions of prey categories were strongly skewed, and even remained clumped after logarithmic transformation, we performed non-parametric and contingency table analyses. The relationships between the relative frequencies of prey categories in diet and habitats were estimated with Spearman's rank correlation. The distributions of prey categories in the bats' diet and in food supply were tested through χ^2 contingency table analysis. Analyses were restricted to the period May–July, when both faeces and insect samples were obtained. Average category proportions from woodland and orchard were used for the comparison in *M. myotis*, since their arthropod faunas are rather similar (comparison of proportions of the prey categories: $r_s = 0.398$, $n = 28$, $P = 0.042$).

Size

In order to test for selection of prey size, the frequency distribution of the body mass of the arthropods captured in the field was compared with the estimated body mass of those found in faeces. Only the predominant prey groups were considered in this analysis, namely Carabidae for *M. myotis* and Tettigoniidae for *M. blythii*, owing to the scarcity of comparative material obtained for other taxa.

The commonest fragment that enabled body size to be estimated from prey in faeces was the last segment of the tarsus. In order to achieve data independency, we considered only the mean length of all the last tarsus segments found within each faecal sample. Regression of the item's dry mass on tarsus last segment length was calculated for insects trapped and/or hand-netted. For Carabidae, the mean length of the last tarsus segment was calculated for each beetle as the average of the three measurements taken on the three types of legs (foreleg, midleg and backleg). In bush crickets, only the fore- and midleg were taken into account, since previous experiments with captive mouse-eared bats have shown that *M. blythii* always discards the weakly attached backlegs of bush crickets, whilst this never happens with carabid beetles taken by *M. myotis*.

Results

Dietary niches

Twenty-one prey categories were distinguished within the 120 individual samples analysed (May–August): 16 in *M. myotis* and 11 in *M. blythii* (Fig. 1). Carabidae were by far the most abundant prey items in the diet of *M. myotis* (58.5% by volume), followed by Lepidoptera larvae (23.6%) and Gryllotalpidae (10.6%). The bulk of the diet of *M. blythii* consisted of Tettigoniidae (65.1%), Lepidoptera larvae (18.7%) and cockchafers *Melolontha melolontha* (7.7%) (Fig. 1). Altogether, these taxa made up 92.7% and 91.5% of the biomass consumed by *M. myotis* and *M. blythii*, respectively. All the remaining prey taxa were present in the diets of both species at less than 5% by volume. Seasonal variation in the diets only showed discrepancies from the general pattern (Fig. 1) during May, when Gryllotalpidae predominated in the diet of *M. myotis* (45.6% vs 31.9% for Carabidae, the second most important prey), whereas *Melolontha melolontha* was the most important prey of *M. blythii* (48.1%, against 20.6% for Lepidoptera larvae and 18.8% for

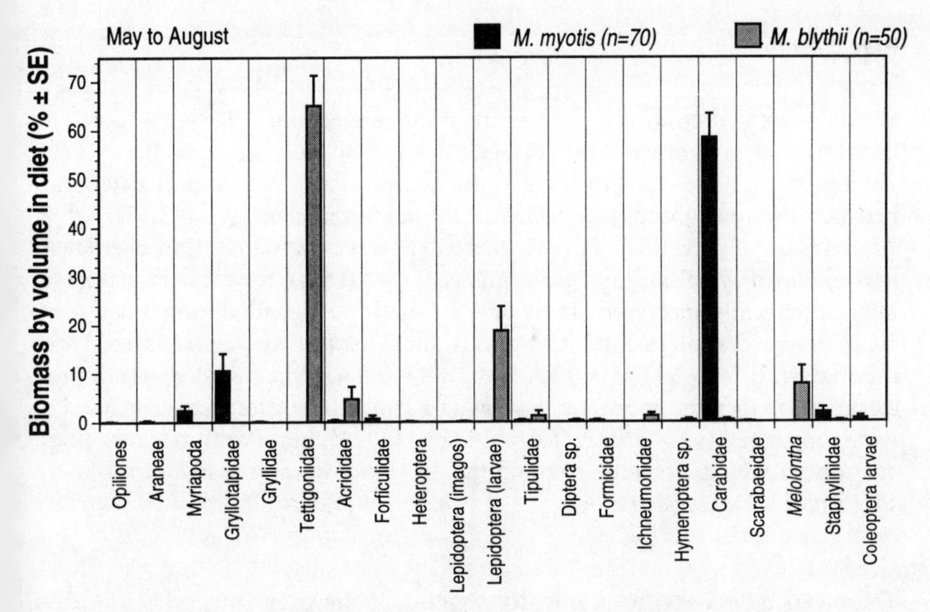

Fig. 1. Percentage (biomass by volume) of the 21 prey categories found in the diets of *M. myotis* and *M. blythii* from May to August. Error bars show the intraspecific variation in diet composition (SE = standard error of the mean). Taxa are arranged according to systematic order.

bush crickets). Overall, the trophic niches of the greater and lesser mouse-eared bats were narrow, with a few prey categories dominating the diet.

Relative biomass and phenology of prey

A total of 13 666 arthropod items were identified. The habitats of *M. myotis* and *M. blythii* contained 28 and 25 prey categories, respectively. Larger arthropod items (>5–7 mm body length) made up 77.4% of the total biomass (51.9% of the total frequency) of arthropods in woodland, 85.7% (60.5%) in orchards, and 80.9% (40.1%) in steppe (Fig. 2). However, since the frequency of occurrence of the different prey categories was highly correlated with their respective biomass within all three habitats (woodland, $r_s = 0.956$, $n = 20$, $P = 0.0001$; orchards, $r_s = 0.962$, $n = 14$, $P = 0.0005$; steppe, $r_s = 0.918$, $n = 25$, $P = 0.0001$), data on relative biomass also provide overall information about the relative frequency of the prey categories. Carabidae were by far the commonest prey in forested areas and orchards throughout the season (Fig. 2a,b). In orchards, *Gryllotalpa* was also common (Fig. 2b). Arachnida (essentially Aranaeidea) were predominant in orchards only early in the season, when no bat faeces were collected (April, Fig. 2b). Bush crickets (Tettigoniidae) comprised most of the biomass in steppe from June onwards. Their delayed larval development makes them a rare prey item early in the season (Fig. 2c).

Prey selection

Taxa

For *M. myotis*, there was a good correlation between the relative proportions of arthropod categories in the diet and in the trap samples ($r_s = 0.619$, $n = 28$, $P = 0.001$) and the difference in the distributions of these prey categories between the two groups was statistically not significant ($\chi^2 = 8.67$, $d.f. = 27$, $P = 0.998$; Fig. 3a). For *M. blythii*, a very low correlation coefficient was obtained when all taxa were pooled ($r_s = 0.046$, $n = 25$, $P = 0.814$), but it drastically increased when only the arthropods inhabiting grass were considered, i.e. if all ground arthropods and *Melolontha* were removed from the analysis ($r_s = 0.643$, $n = 13$, $P = 0.02$). Similarly, the distributions of prey categories in diet and food supply showed a significant difference when all taxa were considered ($\chi^2 = 39.27$, $d.f. = 24$, $P = 0.026$), but if only grass-dwelling arthropods were taken into account, the difference became insignificant ($\chi^2 = 15.73$, $d.f. = 12$, $P = 0.204$).

Size

The mean biomasses (in grams dry weight) of the prey categories found in the three habitats are shown in Fig. 4. The main prey eaten by either bat species was not only the most abundant, but also the largest among European arthropod fauna.

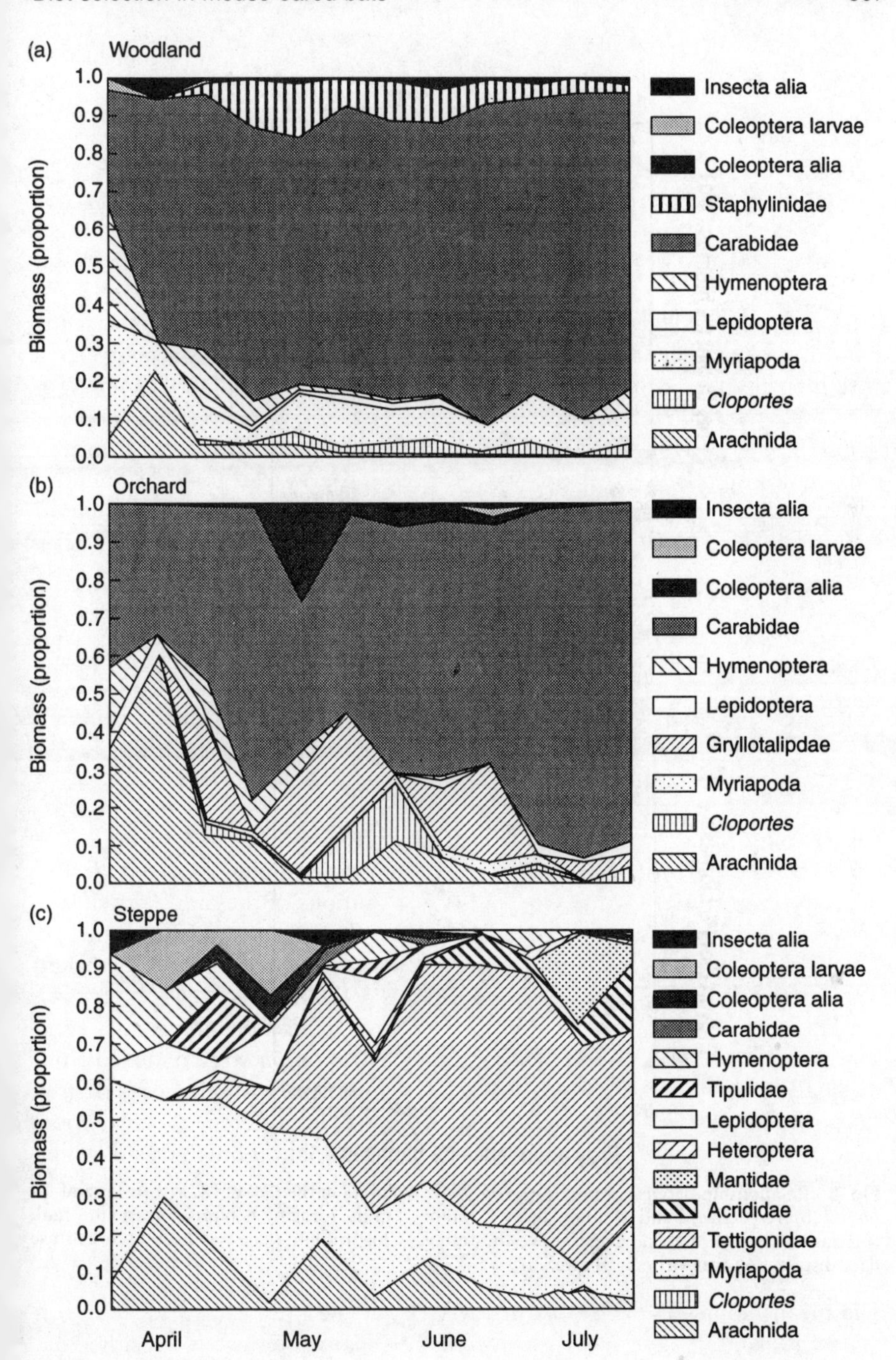

Fig. 2. Relative proportion and phenology of the main arthropod categories sampled in forest (a), in orchards (b) and in steppe (c). Taxa are arranged according to systematic order, from bottom to top.

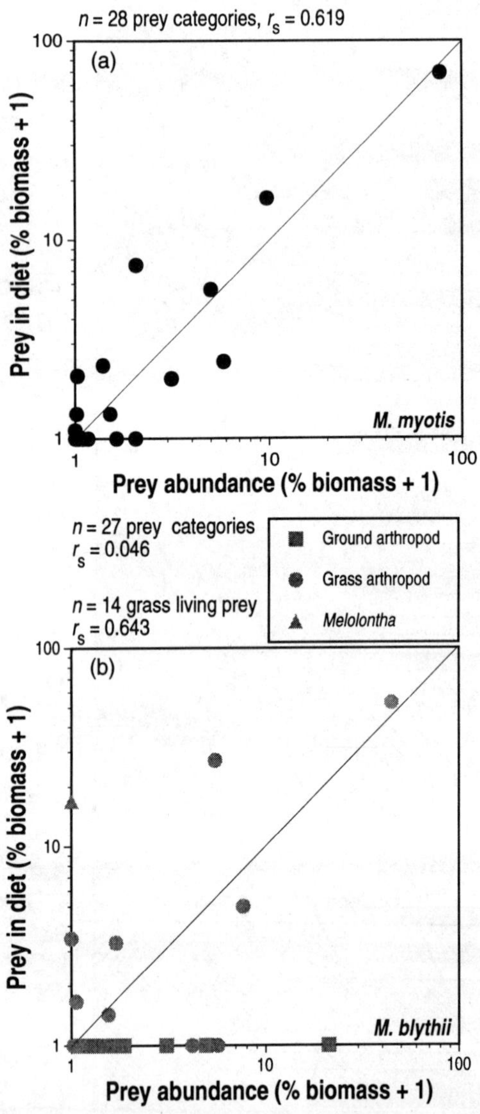

Fig. 3. Relationship between a prey category (% by volume) in the diet of *M. myotis* (a) and *M. blythii* (b), respectively, and its abundance (% biomass) in the foraging habitats. Spearman's rank correlation coefficient is indicated above the frame. A logarithmic scale has been used because data distribution was strongly skewed.

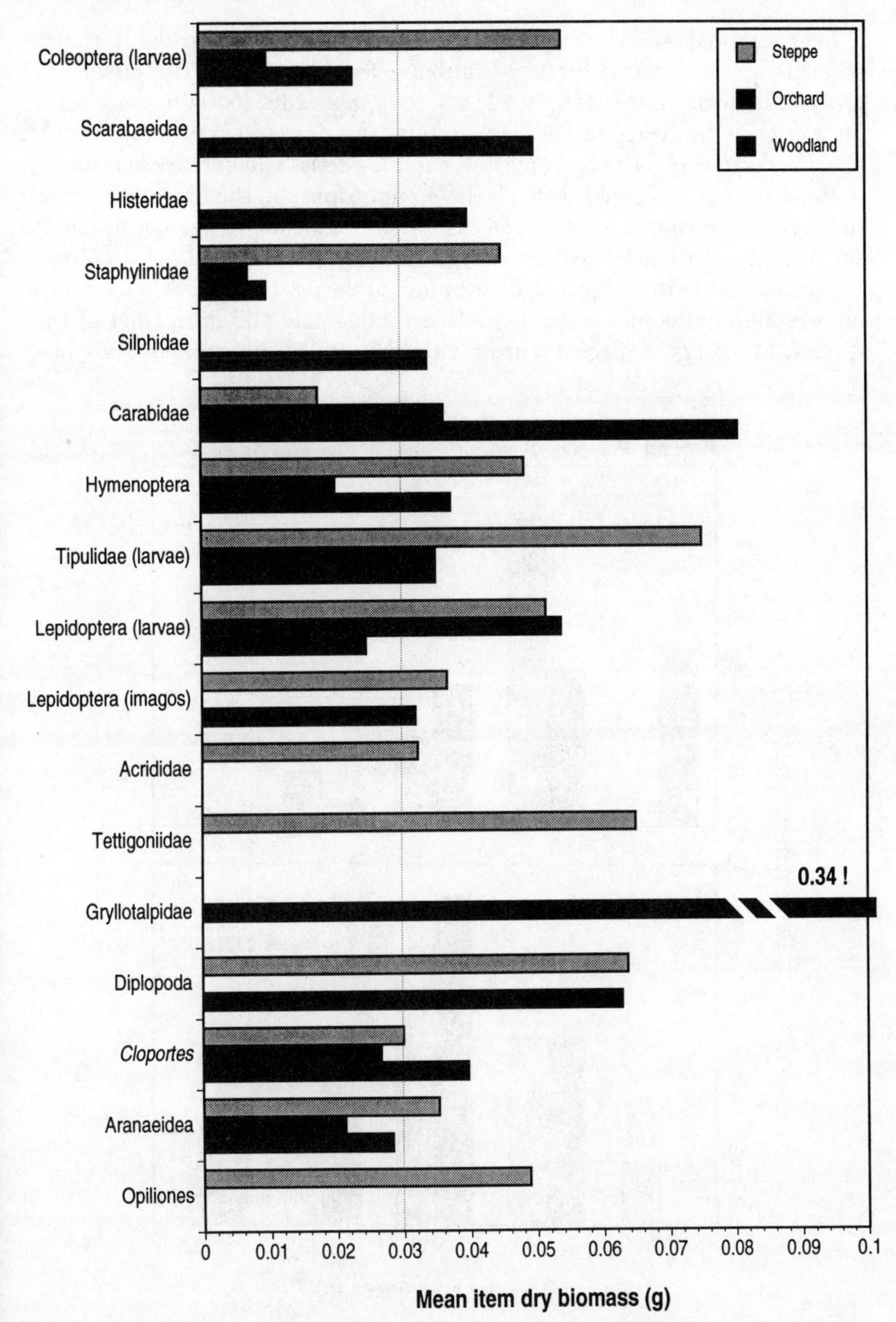

Fig. 4. Mean item biomass (g) of the main arthropod categories sampled in the field. Prey items whose average biomass was less than 0.03 g (line) in each of the three habitats are not shown. Taxa are arranged according to systematic order, from bottom to top.

Regression equations of the dry body mass vs the length of the last tarsus segment were calculated for 189 Carabidae from woodland (comprising 522 tarsus fragments; $y = 0.15x-0.42$, $r = 0.88$, $n = 189$, $P<0.001$) and for 35 bush crickets from steppe (64 tarsus fragments; $y = 0.337x-0.2$, $r = 0.503$, $n = 35$, $P = 0.002$). The body mass of carabid beetles and bushcrickets present in the diets was estimated through these regressions, on the basis of 29 faecal samples (comprising a total of 154 last tarsus segments) in *M. myotis* and 23 samples (comprising 50 last tarsus segments) in *M. blythii*.

Figure 5 shows the frequency distribution of the dry body mass of Carabidae (above) and Tettigoniidae (below) in the food supply and in the diet of each species. *M. myotis* neglected items lighter than 0.05 g (mean item body mass

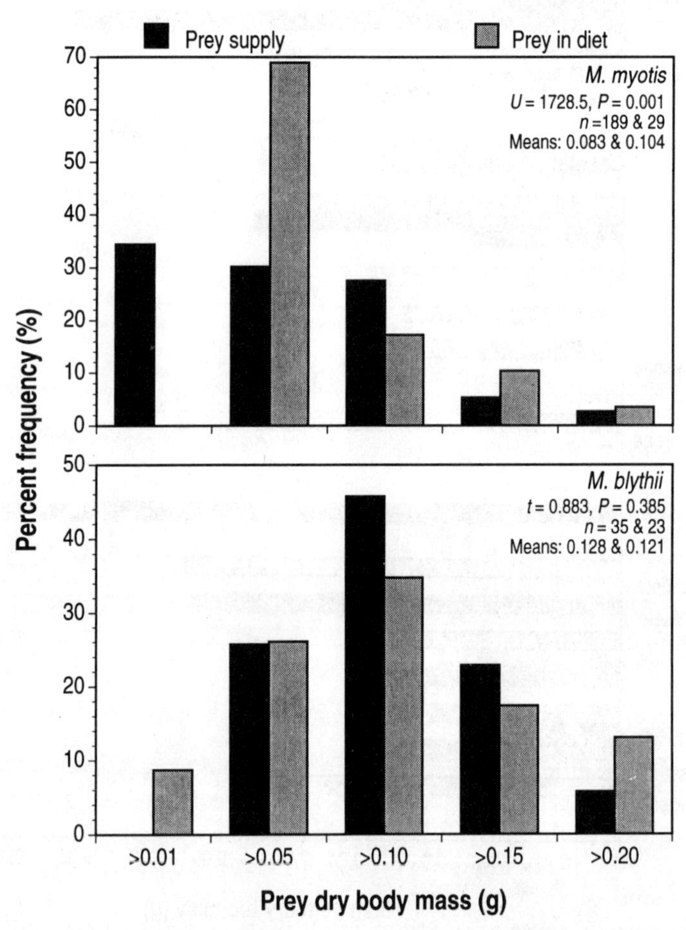

Fig. 5. Relative frequency distribution (%) of the different body mass classes of prey in the food supply and in the diet for *M. myotis* (above) and *M. blythii* (below). Results of statistical tests on means are indicated.

of carabid beetles from supply: = 0.83 g; from diet: = 0.104 g; $U = 1728.5$, $n = 189$ & 29, $P = 0.001$). On the other hand, *M. blythii* did not show any size preference for larger bush crickets (mean item body mass of bush crickets from supply: = 0.128 g; from diet: = 0.121 g; $t = 0.883$, $n = 35$ & 23, $P = 0.385$). On average, bush cricket prey were larger than carabid beetle prey. However, the lightest prey body mass from *M. myotis*' faecal samples was 0.063 g, which corresponds to a body length of 14.8 mm (regression of body length vs dry body mass of carabid beetles: $y = 70.888x+10.314$, $r = 0.96$, $n = 36$, $P < 0.001$), whereas the overall smallest estimated prey body mass of *M. blythii* was 0.043 g, which represents a body length of 13.5 mm (regression of body length against dry body mass of bush crickets: $y = 51.387x+11.324$, $r = 0.913$, $n = 27$, $P < 0.001$). The minimum size of the prey items captured by the lesser mouse-eared bat *M. blythii* would hence be slightly smaller than the size of prey eaten by its larger sibling species.

Discussion

Although the overall correlation between food abundance and diet was fairly good, some discrepancies arose: *M. myotis* underexploited small prey items (<0.05 g dry weight, or <14.8 mm body length), whereas *M. blythii* underexploited ground arthropods and overexploited cockchafers (*M. melolontha*). This may suggest selective feeding. However, these discrepancies may also stem from violations of two assumptions: first, food abundance may differ from actual food availability and second, foraging may have taken place in habitats other than those in which arthropods were sampled.

Because of their special acoustic sensory systems involved in predation, investigating prey availability in insectivorous bats is difficult (Faure & Barclay 1992). The absence of smaller prey items in the diet of *M. myotis* may be an effect of low detectability. Moreover, the use of passive acoustical cues for locating prey by *M. myotis* (Deutschmann 1991) probably further reduces the availability of smaller items, much more so than for echolocating foragers (Barclay 1985–1986; Barclay & Brigham 1991). Nevertheless, active food selection in terms of prey profitability cannot be definitely excluded in *M. myotis*. We have seen that *M. myotis* is able to catch molecrickets (Gryllotalpidae) which are, on average, four times heavier than carabid beetles (Fig. 4); handling costs thus probably do not differ substantially between carabid beetles of different sizes, and the capture of larger items would be energetically advantageous. *M. blythii* clearly neglected arthropods on the soil surface and preferred the grass-dwelling ones; this suggests that this species has evolved species-specific adaptations to detect prey in grass vegetation instead of on the ground.

Mouse-eared bats sometimes switch from their ground- or grass-gleaning behaviour to aerial feeding (R. Arlettaz unpubl.). Some apparent selection may be explained by such changes in foraging strategy. The most striking example is the cockchafers, which were overexploited by *M. blythii*. In May, these beetles were found in large numbers in bats' diet but never in the pitfall traps or during hand netting. In fact, cockchafers do not occur on slopes covered by steppe because the soil is so shallow that their underground larvae do not find suitable conditions; in the study area, this beetle occurs exclusively in cultivated landscapes (A. Schmidt pers. comm.). Every three or four years, cockchafer populations explode, attracting mouse-eared bats, which then switch from their normal feeding behaviour to forage by hawking around tree crops (R. Arlettaz & R. Guttinger unpubl.). The relative discrepancy between the abundance of the second most important prey type, caterpillars, and their proportions in the diet may also indicate foraging activity which took place away from the traditionally used habitats. Flexibility in feeding behaviour of *M. myotis* was suspected by Kolb (1958), who also observed seasonal peaks in given categories of prey such as *M. melolontha, Tortrix viridana* or *Geotrupes*.

Our results do not falsify the hypothesis that the two species exert no active prey selection within their foraging habitats. Which peculiarities, if any, may then explain the narrow trophic niches observed in mouse-eared bats? *M. myotis* and *M. blythii* are among the largest vespertilionid bat species and they are by far the largest representatives of their genus, which includes 93 species (Nowak 1991). They are ground- and grass-gleaning bats, which may restrict the types of prey they can detect. They select species-specific foraging habitats where they use passive acoustical cues for prey detection (Deutschmann 1991). Hence, mouse-eared bats are specialized predators because they have evolved particular morphological, behavioural and physiological adaptations which restrict the resources they could use. Faure & Barclay (1992) termed such bats 'passive specialists'.

Bauerová (1978) compared the diet of *M. myotis* with prey abundance. Unfortunately, she set up pitfalls in half-open habitats, in the immediate surroundings of the colony, where she believed foraging took place, whereas recent radiotracking experiments showed that the main foraging grounds are located between 1.5 and 25 km away from the nursery roost but never in the immediate vicinity of the colony (Rudolph 1989; Audet 1990; R. Arlettaz unpubl.). Bauerová (1978) limited her analysis of ground-dwelling arthropods to carabid beetles and compared the proportions of the different species with their presence in the diet. Proportions of taxa were similar, but bats seemed to neglect the smaller items. Although there is a probable bias in her data on food supply, since smaller carabid beetles are much more frequent in open habitats than in forest (Thiele 1977), the apparent selection of carabid beetles of larger body size by *M. myotis* agrees with the present study.

The contrasted pattern observed in mouse-eared bats (narrow niches but absence of evidence for actively selective feeding) is not unique within insectivorous bat communities. At least one similar case is reported by

Faure & Barclay (1992), who considered *Myotis evotis* as a non-selective feeder despite the prevalence of moths in its diet. As pointed out by Fenton (1982), 'although the data from a number of studies have been used to support the suggestion that some insectivorous bats [actively] specialize to some level on particular types of insects, the evidence does not justify these suggestions'. Most microchiropteran species whose diet is documented exhibit broader niches than do mouse-eared bats or long-eared *Myotis*. Doubt must thus be expressed about reports of selective feeding in insectivorous bats unless reliable data on food abundance, or better food availability, can support it (Kunz 1988; Faure & Barclay 1992).

In a sample of 80 papers dealing with trophic ecology in insectivorous bats, 61 articles described diet composition but not food supply. Nevertheless, among them, nine addressed the question of dietary selectivity: four pleaded for a selective diet and five against. Although the 19 remaining papers all provided further information about food supply, eight did not address precisely the question of dietary selection. Six of the other 11 papers clearly concerned non-selective feeders (Swift & Racey 1983; Swift, Racey & Avery 1985; Barclay 1985–1986; Hoare 1991; Barclay & Brigham 1991; Fenton, Rautenbach, Chipese, Cumming, Musgrave, Taylor & Volpers 1993). Four further papers do not present definite evidence for trophic selectivity: (1) Brigham & Fenton (1991) hesitated between active and passive selection; (2) Anthony & Kunz (1977) demonstrated an apparent temporary prey selectivity in *M. lucifugus*; (3) Brigham & Saunders (1990) suggested that *Eptesicus fuscus* is probably a selective forager, but they did not make a clear distinction between food abundance and food availability. As they themselves hypothesized, small dipterans were thus overrepresented in their data on food 'availability', yet may well not be detectable by the bats, i.e. not actually available *sensu stricto*. Furthermore, Coleoptera, which made up the bulk of the diet, were scarce in traps, but, as hypothesized by Brigham & Saunders (1990) and stated by Jones (1990), there is no reliable technique for sampling correctly the abundance of flying beetles. (4) Sample & Whitmore (1993) reported an overexploitation of Lepidoptera but an avoidance of Coleoptera by *Plecotus townsendii*. However, as this species is probably a forest foliage-gleaner (Dalton, Brack & McTeer 1986; Krull 1992; and Sample & Whitmore 1993; *contra* Bell in Kunz & Martin 1982), passive specialization *sensu* Faure & Barclay (1992) may thus also act in this species. In fact, there is only a single paper that, in our opinion, seems to demonstrate feeding selectivity in an insectivorous bat species. Jones (1990) found that *Rhinolophus ferrumequinum* actually prefers preying on moths when abundant, avoiding other types of insects during moth population peaks in midsummer. The sophisticated echolocation systems of rhinolophids (mainly finely tuned constant-frequency calls) may allow much better prey discrimination than do other acoustic systems evolved in other bat families. Is this the reason why most, if not all, vespertilionids seem to feed unselectively?

Acknowledgements

We thank Dr Bert Jenkins for his help in designing the arthropod sampling, René Güttinger for numerous fruitful discussions on the ecological and behavioural aspects of mouse-eared bats and Dr Augustin Schmidt for information about cockchafers. We also thank Dr Ivan Horacek, Thomas Kokurewicz, Felix Matt and Jean-Michel Serveau, as well as all colleagues from the University of Lausanne who took part either in the field work or in stimulating discussions. We are particularly indebted to Prof. Jacques Hausser, Prof. Paul Racey, Dr Sue Swift and Dr Jens Rydell who made a thorough appraisal of the manuscript and greatly improved the English. Financial support was provided by grants of the 'Conseil de la culture de l'Etat du Valais' and the 'Fondation Dr Ignace Mariétan'.

References

Anthony, E. L. P. & Kunz, T. H. (1977). Feeding strategies of the little brown bat, *Myotis lucifugus*, in southern New Hampshire. *Ecology* **58**: 775–786.

Ariagno, D. (1973). Observations sur une colonie de petits et de grands murins (*Myotis oxygnathus* et *Myotis myotis*). *Annls Spéléol.* **28**: 125–130.

Arlettaz, R. (1987). Le molosse: première capture au gîte en Suisse. *Rhinolophe* No. 3: 10–14.

Arlettaz, R., Ruedi, M. & Hausser, J. (1991). Field morphological identification of *Myotis myotis* and *Myotis blythi* (Chiroptera, Vespertilionidae): a multivariate approach. *Myotis* **29**: 7–16.

Arlettaz, R., Ruedi, M. & Hausser, J. (1993). Ecologie trophique de deux espèces jumelles et sympatriques de chauves-souris: *Myotis myotis et Myotis blythi* (Chiroptera: Vespertilionidae). Premiers résultats. *Mammalia* **57**: 519–531.

Audet, D. (1990). Foraging behavior and habitat use by a gleaning bat, *Myotis myotis* (Chiroptera: Vespertilionidae). *J. Mammal.* **71**: 420–427.

Barclay, R. M. (1985–1986). Foraging strategies of silver haired (*Lasionycteris noctivagans*) and hoary (*Lasiurus cinereus*) bats. *Myotis* **23–24**: 161–166.

Barclay, R. M. R. & Brigham, R. M. (1991). Prey detection, dietary niche breadth, and body size in bats: why are aerial insectivorous bats so small? *Am. Nat.* **137**: 693–703.

Bauerová, Z. (1978). Contribution to the trophic ecology of *Myotis myotis*. *Folia zool.* **27**: 305–316.

Bogan, M. A., Setzer, H. W., Findley, J. S. & Wilson, D. E. (1978). Phenetics of *Myotis blythi* in Morocco. In *Proceedings of the fourth international bat research conference*: 217–230. (Eds Olembo, R. J., Castelino, J. B. & Mutere, F. A.). Kenya Literature Bureau, Nairobi.

Brigham, R. M. & Fenton, M. B. (1991). Convergence in foraging strategies by two morphologically and phylogenetically distinct nocturnal aerial insectivores. *J. Zool., Lond.* **223**: 475–489.

Brigham, R. M. & Saunders, M. B. (1990). The diet of big brown bats (*Eptesicus fuscus*) in relation to insect availability in southern Alberta, Canada. *NW.* Sci. **64**: 7–10.

Constant, P. (1960). Contribution à l'étude de *Myotis myotis* et *Myotis blythi oxygnathus*. *Sous le Plancher* **2–3**: 32–34.

Dalton, V. M. T., Brack, V., Jr. & McTeer, P. M. (1986). Food habits of the big-eared bat, *Plecotus townsendii virginianus*, in Virginia. *Va J. Sci.* **37**: 248–254.
Deutschmann, K. (1991). *Verhalten von* Myotis myotis *(Borkh., 1797) beim Fang fliegender Insekten und der Lokalisation von Beute am Boden.* Unpubl. Master's thesis: University of Tübingen, Germany.
Faure, P. A. & Barclay, R. B. (1992). The sensory basis of prey detection by the long-eared bat, *Myotis evotis*, and the consequences for prey selection. *Anim. Behav.* **44**: 31–39.
Felten, H., Spitzenberger, F. & Storch, G. (1977). Zur Kleinsäugerfauna West-Anatoliens. Teil IIIa. *Senckenberg. biol.* **58**: 1–44.
Fenton, M. B. (1982). Echolocation, insect hearing, and feeding ecology of insectivorous bats. In *Ecology of bats*: 261–285. (Ed. Kunz, T. H.). Plenum Press, New York and London.
Fenton, M. B., Rautenbach, I. L., Chipese, D., Cumming, M. B., Musgrave, M. K., Taylor, J. S. & Volpers, T. (1993). Variation in foraging behaviour, habitat use, and diet of large slit-faced bats (*Nycteris grandis*). *Z. Säugetierk.* **58**: 65–74.
Gould, E. (1955). The feeding efficiency of insectivorous bats. *J. Mammal.* **36**: 390–407.
Hoare, L. R. (1991). The diet of *Pipistrellus pipistrellus* during the pre-hibernal period. *J. Zool., Lond.* **225**: 665–670.
Hutchinson, G. E. (1957). Concluding remarks. *Cold Spring Harb. Symp. quant. Biol.* **22**: 415–427.
Jones, G. (1990). Prey selection by the greater horseshoe bat (*Rhinolophus ferrum-equinum*): optimal foraging by echolocation? *J. Anim. Ecol.* **59**: 587–602.
Kolb, A. (1958). Nahrung und Nahrungsaufnahme bei Fledermäusen. *Z. Säugetierk.* **23**: 84–95.
Krull, D. (1992). *Jagdverhalten und Echoortung bei* Antrozous pallidus (*Chiroptera: Vespertilionidae*). PhD thesis: University of München, Germany.
Kunz, T. H. (1988). Methods of assessing the availability of prey to insectivorous bats. In *Ecological and behavioral methods for the study of bats*: 191–210. (Ed. Kunz, T. H.). Smithsonian Institution Press, Washington and London.
Kunz, T. H. & Martin, R. A. (1982). *Plecotus townsendii. Mammalian Spec.* No. **175**: 1–6.
Kunz, T. H. & Whitaker, J. O. Jr. (1983). An evaluation of fecal analysis for determining food habits of insectivorous bats. *Can. J. Zool.* **61**: 1317–1321.
MacArthur, R. & Levins, R. (1967). The limiting similarity, convergence, and divergence of coexisting species. *Am. Nat.* **101**: 377–385.
Neuweiler, G. (1984). Foraging, echolocation and audition in bats. *Naturwissenschaften* **71**: 446–455.
Neuweiler, G. (1990). Echoortende Fledermäuse. Jagdbiotope, Jagdstrategien und Anpassungen des Echohörens. *Biol. unserer Zeit* **20**: 169–176.
Nowak, R. M. (1991). *Walker's Mammals of the world* **1** (5th edn). Johns Hopkins University Press, Baltimore & London.
Pont, B. & Moulin, J. (1985). Etude du régime alimentaire de *Myotis myotis.* Méthodologie—premiers résultats. In *Actes du 9° colloque francophone de mammalogie, les chiroptères, Rouen* 1985: 23–33.
Rudolph, B. U. (1989). *Habitatwahl und Verbreitung des Mausohrs* (Myotis myotis) *in Nordbayern.* Unpubl. Master's thesis: University of Erlangen, Germany.
Ruedi, M., Arlettaz, R. & Maddalena, T. (1990). Distinction morphologique et biochimique de deux espèces jumelles de chauves-souris: *Myotis myotis* (Bork.) et *Myotis blythi* (Tomes) (Mammalia; Vespertilionidae). *Mammalia* **54**: 415–429.

Sample, B. E. & Whitmore, R. C. (1993). Food habits of the endangered Virginia big-eared bat in West Virginia. *J. Mammal.* **74**: 428–435.

Simmons, J. A., Fenton, M. B. & O'Farrell, M. J. (1979). Echolocation and pursuit of prey by bats. *Science* **203**: 16–21.

Strelkov, P. P. (1972). *Myotis blythi* (Tomes, 1857): distribution, geographical variability and differences from *Myotis myotis* (Borkhausen, 1797). *Acta theriol.* **17**: 355–380.

Swift, S. M. & Racey, P. A. (1983). Resource partitioning in two species of vespertilionid bats (Chiroptera) occupying the same roost. *J. Zool., Lond.* **200**: 249–259.

Swift, S. M., Racey, P. A. & Avery, M. J. (1985). Feeding ecology of *Pipistrellus pipistrellus* (Chiroptera: Vespertilionidae) during pregnancy and lactation. II. Diet. *J. Anim. Ecol.* **54**: 217–225.

Thiele, H. U. (1977). Carabid beetles in their environments. A study on habitat selection by adaptations in physiology and behaviour. *Zoophysiol. Ecol.* No. 10: 1–369.

Wittenberger, J. F. (1981). *Animal social behaviour.* Duxbury Press, Boston.

Symp. zool. Soc. Lond. (1995) No. 67: 377–386

Characterization of mitochondrial DNA variability within the microchiropteran genus *Pipistrellus*: approaches and applications

BARRATT, E. M.[1],
BRUFORD, M. W.[1],
BURLAND, T. M.[1],
JONES, G.[2],

RACEY, P. A.[3]

and WAYNE, R. K.[1,4]

[1]*Conservation Genetics Group*
Institute of Zoology
Regent's Park, London NW1, 4RY, UK

[2]*School of Biological Sciences*
University of Bristol
Bristol BS8 1UG, UK

[3]*Department of Zoology*
University of Aberdeen
Aberdeen AB9 2TN, UK

[4]*Department of Biology*
University of California
Los Angeles, CA 90024, USA

Synopsis

Past systematic studies of the genus *Pipistrellus* have used either morphological or karyological characters to reconstruct phylogenetic relationships. We provide a new molecular perspective on the relationships of *Pipistrellus* species through a phylogenetic analysis of mitochondrial cytochrome *b* DNA sequence data. We have used the polymerase chain reaction to amplify DNA of individuals from both living populations and museum collections. Our results show that *Pipistrellus* is a diphyletic grouping, containing two distinct clades that differ by about 12% in mitochondrial DNA sequence. Surprisingly, we find that genotypes from the UK population of *P. pipistrellus* are found in both clades. Three possible explanations for this observation are (1) that hybridization has occurred between species from both clades, (2) that two distinct ancestral lineages have been retained in the species, or (3) that the lineages comprise separate species.

Introduction

The genus *Pipistrellus* comprises 47 (Corbet & Hill 1991) to 50 (Koopman 1993) species, depending upon the classification followed. Species within this

ZOOLOGICAL SYMPOSIUM No. 67
ISBN 0 19 854945 8

genus are found worldwide in temperate and tropical zones except South America. The species diversity is highest in Asia and Africa, and only a few species occur in North America and Australia (Honacki, Kinman & Koeppl 1982). The taxonomic affinities of *Pipistrellus* to other genera and the relationships within the genus are controversial (Ellerman & Morrison-Scott 1951; Hill 1966; Kitchener & Halse 1978; Heller & Volleth 1984): for example *Pipistrellus* has been described as paraphyletic or, alternatively, as a subgenus of *Eptesicus* (Ellerman & Morrison-Scott 1951; Kitchener, Caputi & Jones 1986). Australian pipistrelles have been reclassified as *Falsistrellus* on the basis of a suite of morphological characteristics and allozyme electrophoretic data (Kitchener *et al.* 1986; Adams, Baverstock, Watts & Reardon 1987). Furthermore, *P. savii* has been reassigned to the genus *Hypsugo*, on the basis of cranial, dental and baculum morphology and electrophoretic data (Horáček & Hanák 1985–1986; Ruedi & Arlettaz 1991). Finally, the taxonomic validity of many species and subspecies has been questioned (Horáček & Hanák 1985–1986) and identification of some pipistrelle species using external morphological characteristics is difficult. For instance the three currently recognized European species (*P. pipistrellus, P. kuhlii* and *P. nathusii*) overlap in geographic range and can be difficult to distinguish when sympatric.

Classification and phylogenetic relationships of the Chiroptera have, to date, been based primarily on morphological characters such as forearm dimensions, measurements of skulls and dentition, bacula morphology or karyotypes (e.g. Baker & Patton 1967; Bickham 1979; Baker & Bickham 1980; Hill & Harrison 1987). However, the analysis of certain morphological characters (e.g. dentition; Tate 1942) can be difficult, since such features may not necessarily reflect the true phylogenetic relationships of the bat species examined. Furthermore, the use of karyological analysis to determine phylogenetic relationships requires a knowledge of the magnitude and rates of chromosomal change and the degree to which such changes correlate with the evolution of species-specific characters (Bush, Case, Wilson & Patton 1977; Lande 1979). Although differences in chromosomal structure are often correlated with taxonomic differentiation, the actual role of such changes in speciation is controversial (Patton & Sherwood 1983). In contrast, some organisms show little or no cytogenetic variation but have extensive changes in morphology and biochemistry, suggesting that chromosomal morphology may be constrained.

Recent advances in molecular biology techniques, specifically the advent of the polymerase chain reaction (PCR) and the discovery of universal PCR primers for mitochondrial DNA sequences (Kocher, Thomas, Meyer, Edwards, Pääbo, Villablanca & Wilson 1989), have led to the extensive use of mtDNA sequences to reconstruct evolutionary relationships. For example, mtDNA sequence analyses have been used to test theories of monophyletic or diphyletic origin of the Chiroptera (Mindell, Dick & Baker 1991), to reconstruct the phylogenetic relationships of Neotropical bats (Phillips, Pumo, Genoways, Ray

& Briskey 1991; van den Bussche 1992: van den Bussche & Baker 1993) and to demonstrate alloparenting in free-tailed bats (McCracken 1984). Because the overall rate of mtDNA sequence evolution is rapid and some regions evolve more rapidly than others (Brown 1986; Shields & Wilson 1987), analysis of different mtDNA regions can reveal variation at the population, subspecific or specific level (Avise 1994). Moreover, molecular techniques have been developed to extract DNA from museum specimens, potentially allowing historical populations to be studied and adding an important temporal perspective to patterns of genetic diversity (Higuchi, Wrischnik, Oakes, George, Tong & Wilson 1987; Pääbo 1989; Thomas, Schaffner, Wilson & Pääbo 1989; Patton & Smith 1992; Stanley, Kadwell & Wheeler 1994).

In this study we analysed the DNA sequence from the mitochondrial cytochrome *b* gene of *Pipistrellus* species. Using samples from extant populations and museum collections, we resolve phylogenetic relationships of species within the genus and relate our finding to past morphological and karyological studies (Zima 1982; Menu 1984; Heller & Volleth 1984; Yin, Xingfu, Yijun & Caihua 1985).

Materials and methods

Samples

Tissue samples from wing membrane or flight muscle were collected from 11 *Pipistrellus* species (Table 1). Fresh samples were stored at −20 °C and museum samples were stored in 70% ethanol.

DNA extraction

DNA was extracted from wing membrane punches by a standard proteinase K digestion method (as described in Stanley *et al.* 1994) followed by a modified

Table 1. *Pipistrellus* species examined in this study

Species	Location	Museum reference	Storage medium	Sample age in years	Number of samples
P. pipistrellus	U.K.	Extant	Frozen/dried	> 1	190
P.p. mediterraneus[a]	Malta	HZM[b]	Ethanol	20	4
P. rueppelli	Egypt	HZM	Ethanol	11	2
P. nanus	Tanzania	HZM	Ethanol	34	3
P. somalius	Kenya	HZM	Ethanol	22	1
P. tenuipinnis	Ethiopia	HZM	Ethanol	4	1
P. stenopterus	Indonesia	HZM	Ethanol	18	1
P. javanicus	Indonesia	HZM	Ethanol	16	1
P. mimus	India	HSM	Ethanol	> 1	1
P. ariel	Israel	Extant	Ethanol	>10	1
P. bodenheimeri	Israel	Extant	Ethanol	>10	3

[a]Cabrera Latorre (1904: 273).
[b]Harrison Zoological Museum.

Table 2. Details of tissue and DNA obtained from different species of *Pipistrellus*

Species	Tissue	DNA quality	PCR product
P. pipistrellus	Flight muscle and wing	High MW	<5 ng/μl
P.p. mediterraneus	Wing	Degraded	>5 ng/μl
P. rueppelli	Wing	Not visible	>5 ng/μl
P. nanus	Wing	Not visible	>5 ng/μl
P. somalicus	Wing	Not visible	>5 ng/μl
P. tenuipinnis	Wing	Degraded	<5 ng/μl
P. stenopterus	Wing	Degraded	>5 ng/μl
P. javanicus	Wing	Degraded	>5 ng/μl
P. mimus	Wing	High MW	>5 ng/μl
P. ariel	Wing	Degraded	<5 ng/μl
P. bodenheimeri	Wing	Degraded	<5 ng/μl

salting-out procedure (B. Petri pers. comm.). Both the quality and quantity of the DNA extracted were dependent upon the integrity of the original sample and the DNA isolated from museum specimens was often degraded (Table 2).

Amplification and sequencing of mitochondrial DNA

Two universal primers, H15149 and L14841 (Kocher *et al.* 1989), were used to amplify a 308 bp segment of the mitochondrial cytochrome *b* gene (Kocher *et al.* 1989). In addition, a pair of species-specific primers, H15063 and L14911, were designed to amplify a region 152 bp long in DNA extracted from museum specimens of *P. pipistrellus* (E.M. Barratt, T.M. Burland, G. Jones, P.A. Racey & R.K. Wayne unpubl.)

The typical PCR conditions utilized 100 ng of template DNA and 6.26 pmol of primer DNA at a denaturation temperature of 94 °C for 30 s followed by annealing at 45 °C for 45 s and extension at 72 °C for 45 s. Thirty-five cycles of amplification were performed in a Perkin-Elmer 9600 DNA thermal cycler. Double-stranded PCR products (50 μl) were prepared for sequencing with geneclean kit (Flowgen) and sequenced directly with a Sequenase kit (US Biomedical). Owing to the degraded condition of the DNA extracted from museum specimens, only the 152 bp fragment was amplified. Because the possibility of DNA contamination is increased with museum specimens, negative (no DNA) extraction and amplification controls were used (Pääbo 1989). In addition, sequences were matched to those in sequence data bases compiled by the National Centre for Biotechnology Information (Entrez, NCBI, National Institute of Health, Bethesda, USA).

The DNA sequences were aligned by means of the programme CLUSTAL V for Apple Macintosh (Higgins & Sharp 1989). Most-parsimonious trees were constructed from sequence data using the heuristic search algorithm of PAUP (phylogenetic analysis using parsimony) version 3.0 for the Apple Macintosh (Swofford 1989). Members of the family Vespertilionidae, the noctule bat, *Nyctalus noctula*, and the barbastelle bat, *Barbastella barbastellus*, were used as outgroups to define the polarity of character-state changes.

Results

In general, approximately 250 bp and 150 bp of sequence data were obtained from the amplified 308 bp and 152 bp fragments respectively. A consensus of the 100 most parsimonious trees revealed two significantly distinct clades for the *Pipistrellus* species that differed by approximately 12% in cytochrome *b* sequence (Fig. 1). The sequence divergence between the pipistrelle clades and the outgroups was approximately 19% (*N. noctula*) and 22% (*B. barbastellus*). One clade, designated group A, contained *P. pipistrellus*, *P. ariel*, *P. bodenheimeri* and *P. rueppelli*. Group B contained *P. pipistrellus*, *P. p. mediterraneus*, *P. somalicus*, *P. tenuipinnis*, *P. javanicus*, *P. stenopterus*, *P. mimus* and *P. nanus*. The sequence divergence between the species in group A ranged between 1.6 and 4.2%, whilst that for the species in group B was only 0–2.5%. A geographic pattern to the clustering of the species within each clade was only obvious in group A, where three of the four species were from the Middle East (*P. ariel*, *P. bodenheimeri* and *P. rueppelli*). Interestingly, *P. pipistrellus* occurred in both of the species clusters, suggesting that the species is a paraphyletic grouping. This observation is supported by additional cytochrome *b* sequence of 208 individuals from across the geographic range of *P. pipistrellus* (E.M. Barratt *et al.* unpubl.).

Discussion

Our preliminary results support the hypothesis that the genus *Pipistrellus* may be a diphyletic group (Kitchener *et al.* 1986). No geographical partitioning of the two clades is obvious, although *P. ariel*, *P. bodenheimeri* and *P. rueppelli* are found in the Middle East. Such highly resolved diphyletic groupings have not been previously proposed but the bootstrap resampling values for the two clades (100 and 99% respectively) are highly significant, especially given the amount of sequence data used.

Our data may support the classification of *P. tenuipinnis* and *P. somalicus* as *Pipistrellus* species (Hill & Harrison 1987) rather than *Eptesicus* (Koopman 1993). However, no correlation has so far been observed between karyological analysis and morphological analysis (Hill & Harrison 1987), nor between the data presented here and the species groupings proposed following classification of baculum structure (Hill & Harrison 1987). Although further sequence evidence could result in a convergence between the two phylogenies, this

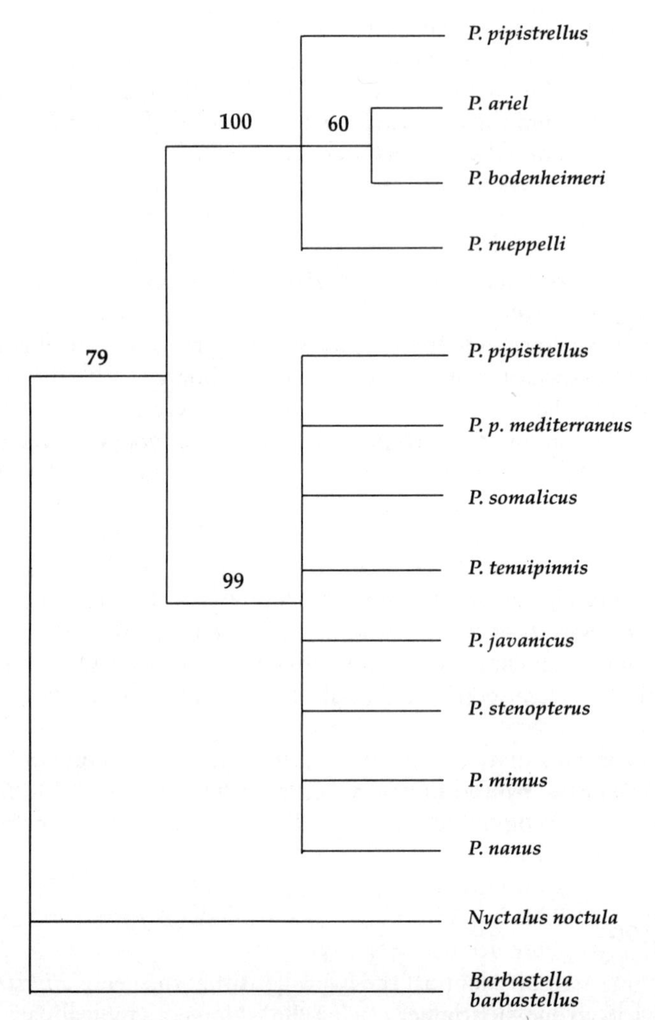

Fig. 1. Phylogenetic relationships of the genus *Pipistrellus* based on parsimony analysis of cytochrome *b* gene sequences. A consensus tree from 100 trees was generated by using the heuristic search option contained in PAUP, with *Nyctalus noctula* and *Barbastella barbastellus* as the outgroups. Numbers on the branches represent bootstrap node confidence values from 100 replications and *n* values per species are given in Table 1.

seems unlikely given the strength of support for clades within the cytochrome *b* sequence. On the basis of bacula morphology (Hill & Harrison 1987) five species are in the same subgenus *Pipistrellus (Pipistrellus): P. pipistrellus, P. p. mediterraneus, P. javanicus, P. mimus* and *P. rueppelli*, (although *P. rueppelli* is placed in a distinct subgroup). *P. ariel, P. nanus* and *P. bodenheimeri* are placed in the same subgenus *Pipistrellus (Hypsugo savii)*; however, *P. nanus* is placed in a different subgroup. *P. stenopterus* is placed in the same subgenus

Hypsugo but in a distinct group. *P. somalicus* and *P. tenuipinnis* belong to another *Pipistrellus* subgenus (*Neoromicia*) but are placed within different groups. The bacula data have, therefore, resulted in more partitioning of the *Pipistrellus* species between subgenera, groups and subgroups than is indicated by the molecular genetic data available so far.

Interestingly, *P. pipistrellus* was resolved as a paraphyletic species that includes genotypes from the two distinct clades. This raises the possibility that hybridization has occurred between taxa in both clades, transferring the mtDNA genotype of one species to species in another clade (e.g. Lehman, Clarkson, Mech, Meier & Wayne 1992). However, this seems unlikely, since hybridization would need to be very extensive as both clades are represented in samples from throughout Britain and there is evidence that the clades overlap in some parts of the species range.

Alternatively, but less probably, there may not have been sufficient time for the lineage sorting to result in reciprocal monophyly of *Pipistrellus* genotypes (Neigel & Avise 1986; Avise & Nelson 1989). Lineages within two isolated populations might be expected to coalesce to separate ancestors (reciprocal monophyly) after about $4N$ generations, where N is the female effective population size. If female population sizes are typically a few thousand individuals, and if generation times are about one year, genotypes within a lineage would be expected to coalesce after a few thousand years. The chiropteran fossil record is poor, especially for non-cave-dwelling bats such as *Pipistrellus*. The first known fossil record of the genus *Pipistrellus* is from the Late Pliocene of Africa about 2–3 million years ago, with earlier Pleistocene records in the North American Irvingtonian, about 1 million years ago (Savage & Russell 1983). *P. pipistrellus* first appeared in European cave faunas about 250 000 years ago (Kurtén 1968) but it is not known if representatives of both clades were present. If both clades were present, and given the early timing of the appearance of North American and European species, we might expect reciprocal monophyly to have developed between reproductively isolated lineages.

The most likely explanation of our results is that an unrecognized species of *Pipistrellus* exists in Europe. Data on echolocation frequency have revealed the occurrence of two distinct phonic groups of *P. pipistrellus*. This observation, combined with further behavioural data, has already led to the suggestion that cryptic species of *P. pipistrellus* are present (Jones & Van Parijs 1993). Preliminary molecular data indicate that the two different genotype groups correspond to the different echolocation frequencies, suggesting that they are reproductively isolated (E. M. Barratt *et al.* unpubl.).

The data presented here demonstrate the potential of modern molecular techniques to address taxonomic issues that have not been resolved by traditional methods. The use of museum samples adds a temporal aspect to such phylogenetic studies and permits the inclusion of extinct species/subspecies/races and the analysis of type specimens. There is now great potential for the identification of new species by a combination of behavioural, morphological and molecular data. This promises an exciting future for the field of molecular phylogenetics.

Acknowledgements

The museum specimens used in this study have been drawn from the collections of the Harrison Zoological Museum, Sevenoaks, Kent (HZM): *P. p. mediterraneus* HZM101.7221, 102.7222, 103.7223, 104. 7224; *P. rueppelli* HZM5.6297, 12.12.114; *P. nanus* HZM185.6298, 186.6374; *P. somalicus* HZM2.9113; *P. tenuipinnis* HZM3.24721; *P. stenopterus* HZM3.12859; *P. javanicus* HZM2.12862; *P. mimus* HZM MM15.

References

Adams, M., Baverstock, P. R., Watts., C. H. S. & Reardon, T. (1987). Electrophoretic resolution of species boundaries in Australian Microchiroptera. II. The *Pipistrellus* group (Chiroptera: Vespertilionidae). *Aust. J. biol. Sci.* **40**: 163–170.

Avise, J. C. (1994). *Molecular markers, natural history and evolution.* Chapman & Hall, New York.

Avise, J. C. & Nelson, W. S. (1989). Molecular genetic relationships of the extinct dusky seaside sparrow. *Science* **243**: 646–648.

Baker, R. J. & Bickham, J. W. (1980). Karyotypic evolution in bats: evidence of extensive and conservative chromosomal evolution in closely related taxa. *Syst. Zool.* **29**: 239–253.

Baker, R. J. & Patton, J. L. (1967). Karyotypes and karyotypic variation of North American vespertilionid bats. *J. Mammal.* **48**: 270–286.

Bickham, J. W. (1979). Chromosomal variation and evolutionary relationships of vespertilionid bats. *J. Mammal.* **60**: 350–363.

Brown, W. M. (1986). The mitochondrial genome of animals. In *Molecular evolutionary biology*: 95–128. (Ed. MacIntyre, R.). Cornell University Press, New York.

Bush, G. L., Case, S. M., Wilson, A. C. & Patton, J. L. (1977). Rapid speciation and chromosomal evolution in mammals. *Proc. natn. Acad. Sci. USA* **74**: 3942–3946.

Cabrera Latorre, D. A. (1904). Ensayo monográfico sobre las quirópteros de España. *Mems R. Soc. esp. Hist. nat.* **2**: 249–287.

Corbet, G. B. & Hill, J. E. (1991). *A world list of mammalian species.* (3rd edn). British Museum (Natural History), London & Oxford University Press, Oxford.

Ellerman, J. R. & Morrison-Scott, T. C. S. (1951). *Checklist of Palaearctic and Indian mammals 1758 to 1946.* British Museum (Natural History), London.

Heller, K.-G. & Volleth, M. (1984). Taxonomic position of 'Pipistrellus societatis' Hill, 1972 and the karyological characteristics of the genus *Eptesicus* (Chiroptera: Vespertilionidae). *Z. zool. Syst. EvolForsch.* **22**: 65–77.

Higgins, D. G. & Sharp, P. (1989). Fast and sensitive multiple sequence alignments on a microcomputer. *Cabios* **5**: 151–153.

Higuchi, R. G., Wrischnik, L. A., Oakes, E., George, M., Tong, B. & Wilson, A. C. (1987). Mitochondrial DNA of the extinct quagga: relatedness and extent of postmortem change. *J. molec. Evol.* **25**: 283–287.

Hill, J. E. (1966). The status of *Pipistrellus regulus* Thomas (Chiroptera, Vespertilionidae). *Mammalia* **30**: 302–307.

Hill, J. E. & Harrison, D. L. (1987). The baculum in the Vespertilioninae (Chiroptera: Vespertilionidae) with a systematic review, a synopsis of *Pipistrellus* and *Eptesicus*, and the descriptions of a new genus and subgenus. *Bull. Br. Mus. nat. Hist. (Zool.)* **52**: 225–305.

Honacki, J. H., Kinman, K. E. & Koeppl, J. W. (1982). *Mammal species of the world: a taxonomic and geographic reference.* Allen Press & Association of Systematics Collections, Lawrence, Kansas.

Horáček, I. & Hanák, V. (1985–1986) Generic status of *Pipistrellus savii* and comments on classification of the genus *Pipistrellus* (Chiroptera: Vespertilionidae). *Myotis* **23–24**: 9–16.

Jones, G. & van Parijs, S. M. (1993). Bimodal echolocation in pipistrelle bats: are cryptic species present? *Proc. R. Soc.* (*B*) **251**: 119–125.

Kitchener, D. J., Caputi, N. & Jones, B. (1986). Revision of Australo-Papuan *Pipistrellus* and of *Falsistrellus* (Microchiroptera: Vespertilionidae) *Rec. W. Aust. Mus.* **12**: 435–495.

Kitchener, D. J. & Halse, S. A. (1978). Reproduction in female *Eptesicus regulus* (Thomas) (Vespertilionidae) in south-western Australia. *Aust. J. Zool.* **26**: 257–267.

Kocher, T. D., Thomas, W. K., Meyer, A., Edwards, S. V., Pääbo, S., Villablanca, F. X. & Wilson, A. C. (1989). Dynamics of mitochondrial DNA evolution in animals: amplification and sequencing with conserved primers. *Proc. natn. Acad. Sci. USA* **86**: 6196–6200.

Koopman, K. F. (1993). Order Chiroptera. In *Mammal species of the world: a taxonomic and geographic reference* (2nd edn): 137–241. (Eds Wilson, D. E. & Reeder, D. M.). Smithsonian Institution Press, Washington & London.

Kurtén, B. (1968). *Pleistocene mammals of Europe.* The World Naturalist Series. Weidenfeld & Nicolson, London.

Lande, R. (1979). Effective deme sizes during long-term evolution estimated from rates of chromosomal rearrangement. *Evolution* **33**: 234–251.

Lehman, N., Clarkson, P., Mech, L. D., Meier, T. J. & Wayne, R. K. (1992). A study of the genetic relationships within and among wolf packs using DNA fingerprinting and mitochondrial DNA. *Behav. Ecol. Sociobiol.* **30**: 83–94.

McCracken, G.F. (1984) Communal nursing in Mexican free-tailed bat maternity colonies. *Science* **223**: 1090–1091.

Menu, H. (1984). Revision du statut de *Pipistrellus subflavus* (F. Cuvier, 1832). Proposition d'un taxon générique nouveau: *Perimyotis* nov. gen. *Mammalia* **48**: 409–416.

Mindell, D. P., Dick, C. W. & Baker, R. J. (1991). Phylogenetic relationships among megabats, microbats, and primates. *Proc. natn. Acad. Sci. USA* **88**: 10322–10326.

Neigel, J. E. & Avise, J. C. (1986). Phylogenetic relationships of mitochondrial DNA under various demographic models of speciation. In *Evolutionary processes and theory*: 515–534. (Eds Nevo, E. & Karlin, S.), Academic Press, Orlando etc.

Pääbo, S. (1989). Molecular genetics in archaeology: a prospect. *Anthrop. Anz.* **45**: 9–17.

Patton, J. L. & Sherwood, S. W. (1983). Chromosome evolution and speciation in rodents. *A. Rev. Ecol. Syst.* **14**: 139–158.

Patton, J. L. & Smith, M. F. (1992). mtDNA phylogeny of Andean mice: a test of diversification across ecological gradients. *Evolution* **46**: 174–183.

Philips, C. J., Pumo, D. E., Genoways, H. H., Ray, P. E. & Briskey, C. A. (1991). Mitochondrial DNA evolution and phylogeography in two Neotropical fruit bats, *Artibeus jamaicensis* and *Artibeus lituratus*. In *Latin American mammology: history, biodiversity, and conservation*: 97–123. (Eds Mares, M. A. & Schmidly, D. J.). University of Oklahoma Press, Norman & London.

Ruedi, M. & Arlettaz, R. (1991). Biochemical systematics of the Savi's bat (*Hypsugo savii*) (Chiroptera: Vespertilionidae). *Z. zool. Syst. EvolForsch.* **29**: 115–122.

Savage, D. E. & Russell, D. E. (1983). *Mammalian paleofaunas of the world.* Addison-Wesley Publishing Company, Reading, Mass.

Shields, G. F. & Wilson, A. C. (1987). Calibration of mitochondrial DNA evolution in geese. *J. molec. Evol.* **24**: 212–217.

Stanley, H. F., Kadwell, M. & Wheeler, J. C. (1994). Molecular evolution of the family Camelidae—a mitochondrial DNA study. *Proc. R. Soc. (B)* **256**: 1–6.

Swofford, D. L. (1989). *PAUP: phylogenetic analysis using parsimony*. Version 3.0. Illinois Natural History Surv., Illinois.

Tate, G. H. H. (1942). Results of the Archbold Expeditions. No. 47. Review of the vespertilionine bats, with special attention to genera and species of the Archbold collections. *Bull. Am. Mus. nat. Hist.* **80**: 221–297.

Thomas, R. H., Schaffner, W., Wilson, A. C. & Pääbo, S. (1989). DNA phylogeny of the extinct marsupial wolf. *Nature, Lond.* **340**: 465–467.

van den Bussche, R. A. (1992). Restriction-site variation and molecular systematics of New World leaf-nosed bats. *J. Mammal.* **73**: 29–42.

van den Bussche, R. & Baker, R. J. (1993). Molecular phylogenetics of the New World bat genus *Phyllostomus* based on cytochrome b DNA sequence variation. *J. Mammal.* **74**: 793–802.

Yin, L., Xingfu, X., Yijun, S. & Caihua, S. (1985). Chromosomes of house bat *Pipistrellus abramus* Temminck. *Acta zool. sin.* **31**: 296–298.

Zima, J. (1982). Chromosomal homology in the complements of bats of the family Vespertilionidae. II. G-band karyotypes of some *Myotis, Eptesicus* and *Pipistrellus* species. *Folia zool.* **31**: 31–36.

Symp. zool. Soc. Lond. (1995) No. 67: 387–396

Genetic population structure of the noctule bat *Nyctalus noctula*: a molecular approach and first results

Frieder MAYER

Institut für Zoologie II
Universität Erlangen-Nürnberg
Staudtstrasse 5
D-91058 Erlangen, Germany

Synopsis

The noctule *Nyctalus noctula* is widespread throughout northern Europe and migrates several hundred or even more than a thousand kilometres south for hibernation.

Two observations lead to a hypothesis about the genetic population structure. First, banding experiments showed that females breed in the area where they were born and are therefore expected to be philopatric. Second, in autumn, during migration, the females mate with territorial males in tree holes, and thereafter spermatozoa, probably from several males, are stored in the uterus until ovulation in spring. Because of the maternal inheritance of mitochondria, these observations would suggest that genetic subpopulations might occur in the breeding area that could be identified by using a mitochondrial genetic marker. In contrast to mitochondrial genes, nuclear genes are recombined during mating. At the nuclear level, therefore, fewer or no substructures of populations are expected if the females mate with unrelated males from other geographic regions. This is more likely to be the case in a migratory species than in non-migratory species like horseshoe bats or *Myotis* species, for example *Myotis bechsteinii*.

Polymorphic genetic markers can be used to test these hypotheses. The following three approaches are described and first results are given. (1) Multilocus DNA fingerprinting showed very high levels of variability of unrelated individuals. (2) Single-locus DNA typing using microsatellites allowed a distinction to be made between maternal and paternal alleles of juveniles within a nursery colony of unknown relationships. Therefore, comparisons between populations can be made on the basis of maternal or paternal alleles, even though the fathers of the juveniles are not known. (3) Sequence variation in the mitochondrial control region (D-loop) was very high, even within nursery colonies. Juveniles could be matched to their mother according to their mitochondrial DNA. These results agreed with the data obtained from the microsatellite typing. Mitochondrial sequences might also be useful for the analysis of relatedness within and between local male or female associations.

ZOOLOGICAL SYMPOSIUM No. 67
ISBN 0 19 854945 8

Introduction

The noctule bat is a common European species. The main breeding areas are located in northern parts of Europe and Asia. Females usually give birth to two offspring each year (Gaisler, Hanák & Dungel 1979; Heise 1989). In August, after weaning of the young, these bats start to migrate south for more than 1000 km (Strelkov 1969). Along migratory routes males occupy tree holes during summer and try to attract females by courtship calls. Such tree holes are expected to be mating roosts (Sluiter & van Heerdt 1966). After mating, spermatozoa are stored in the uterus until ovulation in spring (Racey 1975). Both sexes hibernate together. In spring, only females are known to migrate back to the area where they were born, while males are believed to stay in the mating area (Heise 1985, 1989). These data were collected from banded individuals and during field observations.

These observations allow hypotheses to be formed about the genetic population structure of this bat. If the females do return every year to their native area, one would expect to be able to identify genetic differences between populations by using a genetic marker which is inherited exclusively by the females, i.e. the mitochondrial DNA (mtDNA). If the females mate with an unrelated male or several unrelated males (as we expect from field observations), less or no genetic substructuring should be found at the nuclear genomic level.

The aim of this paper is to show how these hypotheses can be tested quantitatively by using highly polymorphic genetic markers. I used three different molecular approaches: multilocus DNA fingerprinting, single-locus microsatellite typing and sequence analysis of the control region of the mitochondrial genome.

Results and discussion

Multilocus DNA fingerprinting

DNA fingerprinting in general is based upon repetitive sequences which are widely dispersed in all eukaryotic species studied so far. Two major classes of repetitive DNA have been used for DNA typing: minisatellites, which are DNA sequences comprising multiple copies of a sequence usually less than 65 base pairs in length, and microsatellites, which are sequences with tandem repeat units less than 10 base pairs long. Loci comprising such sequences are known to have high mutation rates, which result in alleles with different copy numbers of the repeat unit. Such variability in repeat units can be detected by restriction fragment length analysis (Fig. 1).

A multilocus DNA fingerprint is a complicated, bar-code-like banding pattern (Fig. 2). It results from many fragments, each containing a stretch of the simple repetitive sequences at many different loci scattered throughout the genome. In the noctule bat, all three synthetic oligonucleotides tested

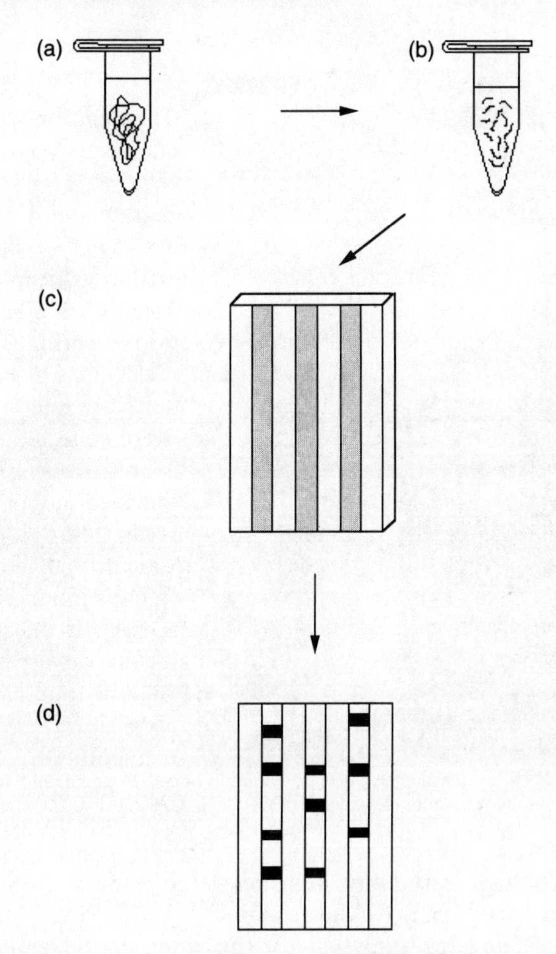

Fig. 1. Multilocus DNA fingerprinting: DNA of an individual is extracted from, e.g., a blood sample (150 μl) or a small piece of the wing membrane (0.2 cm^2) (a). The DNA is digested with a restriction enzyme which cuts the DNA at a specific recognition site (b). The resulting DNA fragments are separated on an agarose gel according to their molecular weight (c). Fragments containing repetitive sequences can be visualized by hybridization with, for example, a radioactive labelled DNA probe which is in its sequence complementary to the repetitive sequence (d).

($(GATA)_4$, $(GACA)_4$ and $(GTG)_5$) produced a highly polymorphic individual-specific banding pattern among unrelated individuals. Family analysis, of both parents and at least one offspring, revealed inheritance of the bands according to Mendelian rules. These genetic markers could therefore be used for relatedness analysis. For that purpose, the proportion of bands that are identical in two individuals is determined (Lynch 1991). The band-sharing of unrelated noctule bats was between 13 and 16%. The probability of two unrelated noctules having an identical genetic fingerprint is less than 3.8

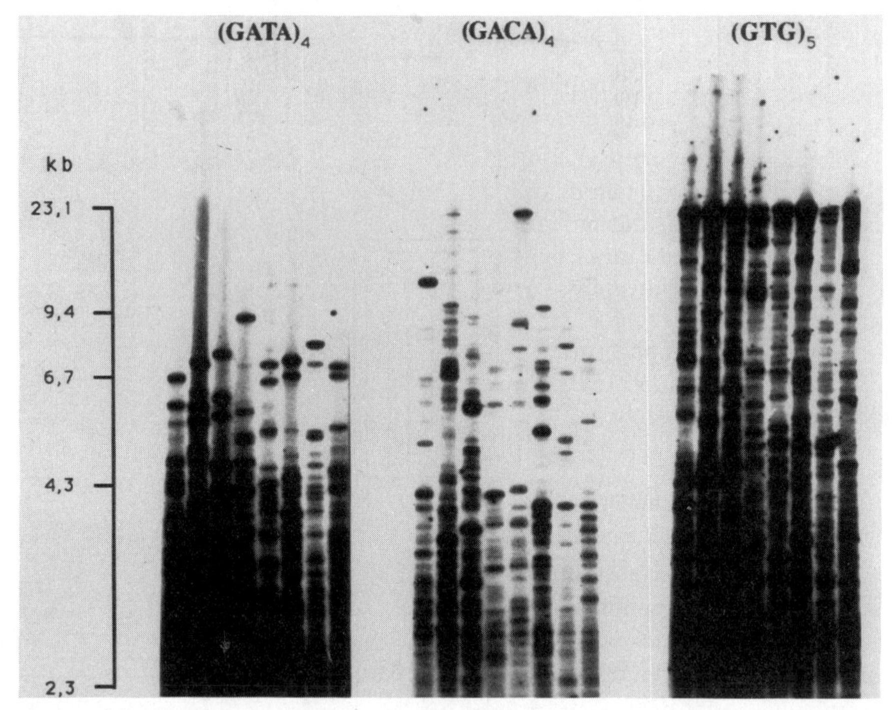

Fig. 2. Multilocus DNA fingerprints of unrelated noctule bats. The same gel was subsequently hybridized with the synthetic oligonucleotides $(GATA)_4$, $(GACA)_4$ and $(GTG)_5$.

$\times 10^{-11}$. Such a high variability indicates large population sizes and no inbreeding in the noctule bat (F. Mayer unpubl.).

Multilocus DNA fingerprinting was used to study the relatedness of twins in the noctule bat. In general, twins may be related in three different ways: monozygotic twins, full sibs and half sibs. Each family investigated consisted of a mother and her two offspring. Fathers were not known because males do not participate in raising the young and mating occurs several months and hundreds of kilometres away from the breeding area. In the case of monozygotic twins, both sibs have an identical fingerprint. Full sibs, having the same father, share about half of their paternal bands, while half sibs have no ancestral paternal bands in common. In analysis of several families, one case of monozygotic twins was found and full and half sibs were detected at about the same frequency (Mayer, von Helversen & Epplen 1991; F. Mayer & O. von Helversen unpubl.).

Multilocus DNA fingerprinting has been widely used for assigning paternity in many vertebrates, leading to a new understanding of animal mating systems, especially in birds (Burke 1989). In the horseshoe bat *Rhinolophus sedulus*, the method was used to confirm that a male which was found together with a female and its offspring, was in fact the father, which suggested a monogamous mating system in this species (Heller, Achmann & Witt 1993). In another study, selective nursing of the individual's own offspring was shown in the pipistrelle bat *Pipistrellus pipistrellus* by Bishop, Jones, Lazarus & Racey (1992).

There are two major disadvantages of the multilocus approach in bat research. First, several micrograms of DNA are needed to generate a multilocus fingerprint. Second, the multilocus banding pattern is rather complex and alleles or genotypes cannot be revealed from a multilocus DNA fingerprint. The analysis of relatedness beyond first-order relatives therefore becomes difficult, if not impossible. Investigations of population substructures based on multilocus DNA fingerprinting are rare and are limited to small and inbred populations (Lynch 1991). The analysis of hypervariable microsatellites via the polymerase chain reaction (PCR) overcomes both problems.

Single-locus microsatellite typing

The amplification of DNA by the PCR method allows small sample sizes to be used. A piece of wing tissue of about 0.2 cm^2 is sufficient for several hundred amplification reactions. Specific primers allow the amplification of one locus out of the whole genomic DNA. To generate such locus-specific primers entails many steps of molecular cloning (Tautz 1989; Rassmann, Schlötterer & Tautz 1991). A partial genomic library has to be established, clones must be screened for microsatellites and positive clones must be sequenced. Finally, primers can be designed that will be complementary to the flanking regions of the microsatellites. Using these primers in an amplification reaction and separating the amplification products on a high-resolution sequencing gel produces one or two bands, depending on whether the individual is homo- or heterozygotic at a specific locus (Figs 3 and 4).

In the noctule bat, five loci have been investigated so far and they showed a high degree of polymorphism. Ten to 15 alleles were detected and rates of heterozygosity were between 0.75 and 0.93 (F. Mayer unpubl.).

By comparing the adult female's genotypes with those of the juveniles, many females could be excluded as the mother of a juvenile. By doing this for several loci, it was possible to match the mothers to their offspring (Fig. 5). This example of a nursing colony, sampled in mid July, shows that mothers and offspring do not necessarily roost together in late lactation.

Knowing the mother of a juvenile allows maternal, and hence paternal, alleles of this offspring to be distinguished. Therefore, the frequencies of maternal or paternal alleles of the juveniles within nursing colonies can be analysed separately, even though the fathers are not known owing to the mating behaviour.

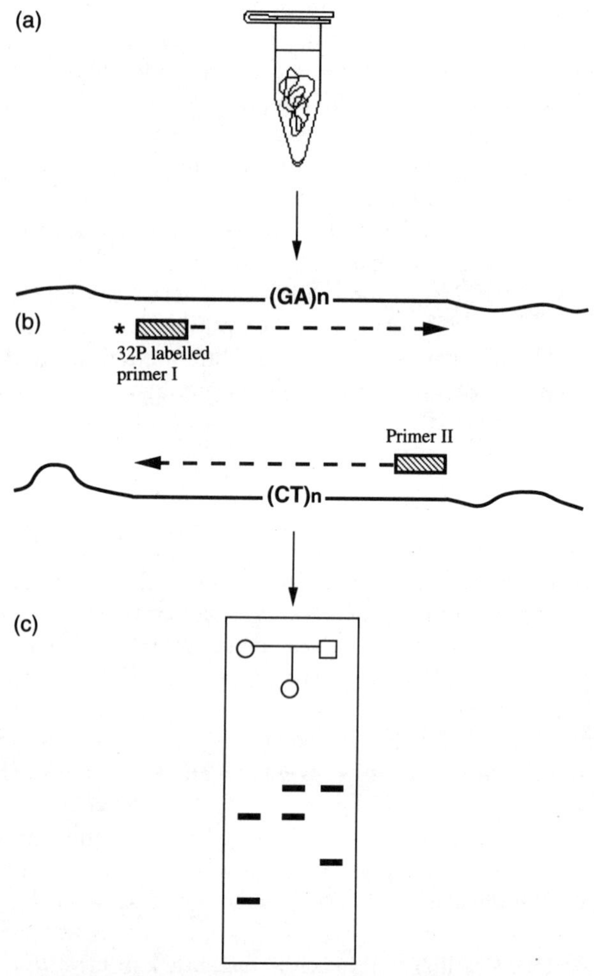

Fig. 3. Microsatellite typing using the polymerase chain reaction (PCR). Whole genomic DNA is isolated (a) and used in a PCR reaction (b) to amplify the two alleles of an individual. The amplification products, for example of a family consisting of the mother (○), the father (□) and a daughter (○), are separated on a high-resolution sequencing gel (c).

The large number and high variability of microsatellite loci should allow them to be used to determine effective population sizes, inbreeding and gene flow between populations and to estimate relatedness, even of more distantly related individuals or groups (Queller, Strassmann & Hughes 1993).

The major problem of this approach is the elaborate work of molecular cloning and the high costs of DNA polymerases and radioactivity. Primers are often species- or genus-specific and they must be established anew for almost every species.

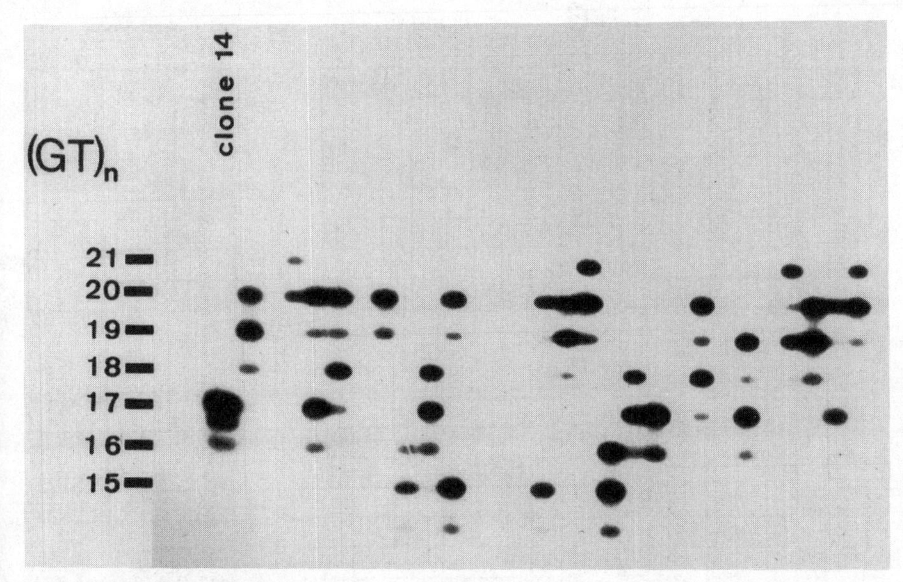

Fig. 4. Single-locus DNA fingerprints of unrelated noctules. Each column represents an individual. The alleles are named according to the copy number of the GT/CA-repeats. Clone 14 was sequenced, and 17 repeats were detected.

Sequencing of mitochondrial DNA

The mitochondrial genome is expected to be primarily inherited from the female and a growing set of primers is available to study various parts of the mtDNA (Kocher *et al.* 1989; Wilkinson & Chapman 1991; Barratt *et al.* this volume pp. 377–386; Petri, Neuweiler & Pääbo this volume pp. 397–403).

In the noctule bat a part of the control region (D-loop) of the mitochondrial genome was amplified by PCR. The amplification products were visualized on an agarose gel and were sequenced (Fig. 6). Two levels of polymorphisms were detected. First, the amplified part of the control region contained a variable number (four to nine) of repeats of an 81 base pair sequence. Second, the sequence was different within the repeats. In addition, many individuals were heteroplasmatic, i.e. had more than one mitochondrial lineage. With such a variable repeat number and differences in the sequence level, the control region of the noctule bat seems to be almost individual-specific. All 14 females of two nursery assemblages could be distinguished by their control region sequence (F. Mayer unpubl.). By sequencing adult females and juveniles of nursery colonies, it was possible to match the offspring to their mothers. This agreed with the results of microsatellite typing.

These findings show stable inheritance of the mother's mitochondrial genotype by her offspring. In the noctule, the control region of the mitochondrial

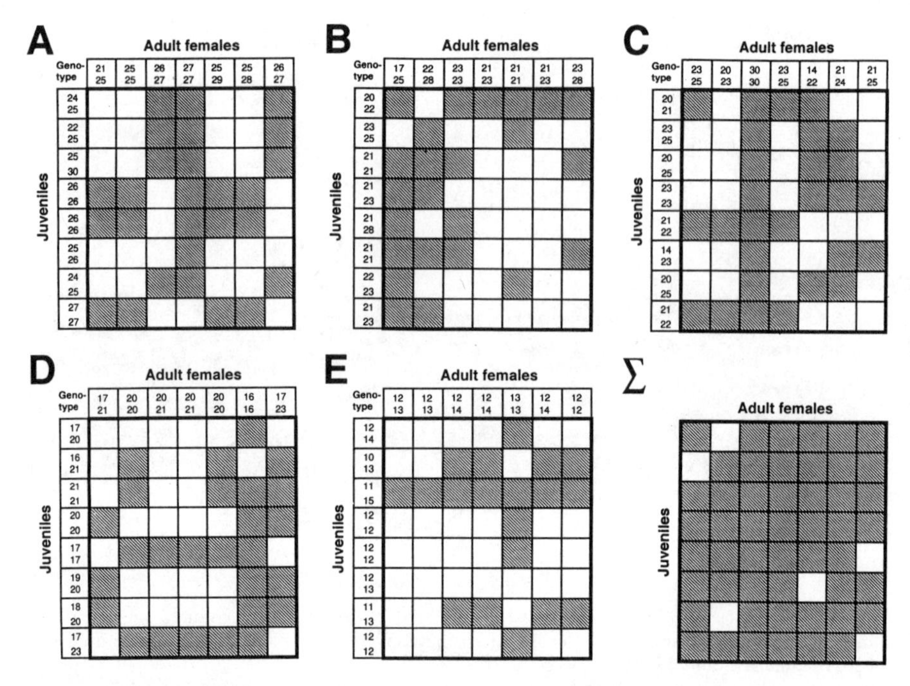

Fig. 5. Genotypes from five loci (A–E) of seven adult females and eight juveniles from a nursing colony. Locus D corresponds to locus 14 shown in Fig. 4. Maternity exclusions are indicated by shaded squares. All exclusions are summarized in Σ. This matrix shows that three females had no offspring within the roost and two juveniles did not roost together with their mother. All bats were sampled in a bat box during the day in late lactation.

genome can therefore be used as a genetic marker (1) to distinguish matrilineages within nursing colonies, (2) to assign even weaned juveniles to their mothers and (3) to analyse the maternal relatedness of female or male associations. Less variable mitochondrial markers might be useful to study genetic structures more on the level of populations, for example the relatedness between nursing colonies.

Acknowledgements

This work was supported by the Deutsche Forschungsgemeinschaft and the Konrad-Adenauer-Stiftung e.V. For advice and many interesting discussions I would like to thank Roland Achmann, Jörg Epplen, Otto von Helversen, Svante Pääbo, Barbara Petri and Diethard Tautz.

References

Bishop, C. M., Jones, G., Lazarus, C. M. & Racey, P. A. (1992). Discriminate suckling in pipistrelle bats is supported by DNA fingerprinting. *Molec. Ecol.* **1**: 255–258.

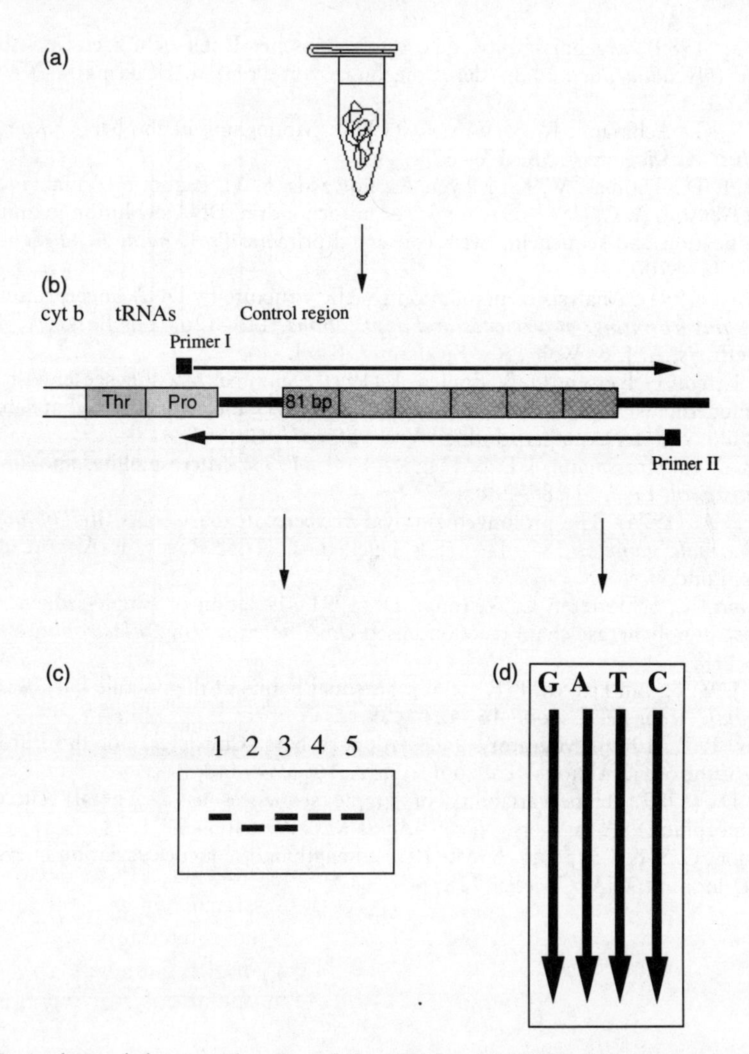

Fig. 6. Analysis of the control region of the mitochondrial DNA. Whole genomic DNA is isolated (a) and the variable region is amplified with two PCR primers specific for flanking regions (b). Amplification products are separated on an agarose gel to determine the copy number of the 81 bp-repeat (c) and sequenced to detect point mutations within the repeats (d). Individual 3 of the schematic example of five individuals (c) is heteroplasmatic.

Burke, T. (1989). DNA fingerprinting and other methods for the study of mating success. *Trends Ecol. Evol.* **4**: 139–144.

Gaisler, J., Hanák, V. & Dungel, J. (1979). A contribution to the population ecology of *Nyctalus noctula* (Mammalia: Chiroptera). *Přírod. Pr. Cesk. Akad. Věd.* (N. S.) **13**: 1–38.

Heise, G. (1985). Zu Vorkommen, Phänologie, Ökologie und Altersstruktur des Abendseglers (*Nyctalus noctula*) in der Umgebung von Prenzlau/Uckermark. *Nyctalus* (N. F.) **2**: 133–146.

Heise, G. (1989). Ergebnisse reproduktionsbiologischer Untersuchungen am Abendsegler (*Nyctalus noctula*) in der Umgebung von Prenzlau/Uckermark. *Nyctalus* (N. F.) **3**: 17–32.

Heller, K.-G., Achmann, R. & Witt, K. (1993). Monogamy in the bat *Rhinolophus sedulus*? *Z. Säugetierk.* **58**: 376–377.

Kocher, T. D., Thomas, W. K., Meyer, A., Edwards, S. V., Pääbo, S., Villablanca, F. X. & Wilson, A. C. (1989). Dynamics of mitochondrial DNA evolution in animals: amplification and sequencing with conserved primers. *Proc. natn. Acad. Sci. USA* **86**: 6196–6200.

Lynch, M. (1991). Analysis of population genetic structure by DNA fingerprinting. In *DNA fingerprinting: approaches and applications*: 113–126. (Eds Burke, T., Dolf, G., Jeffreys, A. J. & Wolff, R.). Birkhäuser, Basel.

Mayer, F., von Helversen, O. & Epplen, J. (1991). Stammen Zwillingsgeschwister bei der Fledermaus *Nyctalus noctula* vom selben Vater? Eine Verwandtschaftsanalyse mit Hilfe von 'DNA fingerprinting'. *Verh. dt. zool. Ges.* **84**: 318–319.

Queller, D. C., Strassmann, J. E. & Hughes, C. R. (1993). Microsatellites and kinship. *Trends Ecol. Evol.* **8**: 285–288.

Racey, P. A. (1975). The prolonged survival of spermatozoa in bats. In *The biology of the male gamete*: 385–416. (Eds Duckett, J. G. & Racey, P. A.). Academic Press, London.

Rassmann, K., Schlötterer, C. & Tautz, D. (1991). Isolation of simple-sequence loci for use in polymerase chain reaction-based DNA fingerprinting. *Electrophoresis* **12**: 113–118.

Sluiter, J. W. & van Heerdt, P. F. (1966). Seasonal habits of the noctule bat (*Nyctalus noctula*). *Archs néerl. Zool.* **16**: 423–439.

Strelkov, P. P. (1969). Migratory and stationary bats (Chiroptera) of the European part of the Soviet Union. *Acta zool. cracov.* **16**: 393–440.

Tautz, D. (1989). Hypervariability of simple sequences as a general source for polymorphic DNA markers. *Nucleic Acids Res.* **17**: 6463–6471.

Wilkinson, G. S. & Chapman, A. M. (1991). Length and sequence variation in evening bat D-loop mtDNA. *Genetics* **128**: 607–617.

Symp. zool. Soc. Lond. (1995) No. 67: 397–403

Mitochondrial diversity and heteroplasmy in two European populations of the large mouse-eared bat, *Myotis myotis*

Barbara PETRI,
Gerhard NEUWEILER
and Svante PÄÄBO

*Zoologisches Institut
der Universität München
Luisenstr. 14
80333 Munich, Germany*

Synopsis

A total of 101 animals were sampled from three nursery colonies of *Myotis myotis* in Portugal and in Bavaria, southern Germany. A segment of the triple-stranded part of the mitochondrial control region ranging from 400 bp to 800 bp was amplified via PCR and analysed by gel electrophoresis. Substantial length variation was found among different animals and 33% of the bats were heteroplasmic for up to six different size classes. When the DNA sequence of the heteroplasmic segment was determined, the length variation was found to be due to a variable number of tandemly repeated motifs of 82 bp, as has also been described for the American evening bat, *Nycticeius humeralis*. Heteroplasmy, as well as homoplasmy, were stably inherited in 25 mother–pup pairs. Mechanisms for the generation of length and sequence heteroplasmy are thought to be slippage mutations caused by the D-loop replication procedures and to a lesser degree by biparental inheritance of mitochondrial DNA.

Introduction

Mammalian mitochondrial DNA (mtDNA) is a highly conserved genome carrying genes for 13 proteins, 22 tRNAs and two rRNAs (Anderson *et al.* 1981). It is believed to be strictly maternally inherited. Its rate of sequence evolution is higher than that of the nuclear genome. A unique feature of the otherwise extremely economic organization of mtDNA is the control region or D-loop, the only large non-coding region of the molecule, where sequence divergence of up to 10% can be observed within species. A part of this contains a short third strand (7S DNA) which, during replication, may displace the two parental strands. The D-loop varies in length among species, often owing to repetition of distinct sequences. Size variations are found not only among and within species, but also within one individual, leading to heteroplasmy, the existence of two or more types of mitochondrial genome. Recently length

ZOOLOGICAL SYMPOSIUM No. 67
ISBN 0 19 854945 8

heteroplasmy has been reported in a number of mammalian species, including cow (Laipis, Van de Walle & Hauswirth 1988), rabbit (Mignotte *et al.* 1990; Biju-Duval *et al.* 1991), American evening bat (Wilkinson & Chapman 1991), pig (Ghivizzani *et al.* 1993) and elephant seal (Hoelzel, Hancock & Dover 1993). Heteroplasmy due to sequence variation is more rare; when observed, the differences are usually not greater than 1%. Buroker *et al.* (1990) presented a model of duplication and deletion of repeat units which would generate heteroplasmy. Another explanation of heteroplasmy could be 'paternal leakage' of mitochondrial DNA. Hybridization experiments in mice (Gyllensten *et al.* 1991) and observation of high-sequence divergence in heteroplasmic mussels (Hoeh, Blakley & Brown 1991) and anchovy (Magoulas & Zouros 1993) presented examples of biparental inheritance of mitochondria.

Material and methods

Collecting and DNA isolation

Tissue samples (3 mm^2) were taken from the tail membranes of about 30% of *M. myotis* individuals from two nursery colonies (Au and Beyharting) in Bavaria, and of 50 individuals from one nursery colony (Mina Lousal) in Portugal. Mother–pup pairs of the two Bavarian colonies were sampled when pups were still attached to their mothers' nipples so that their relatedness was unambiguous. DNA was either phenol- or salt-extracted by standard methods. The yield of genomic DNA was approximately 15 μg.

MtDNA, PCR and sequencing

First amplifications of the total control region of the mitochondrial genome of five bat species (*M. myotis, M. emarginatus, M. nattereri, Nyctalus noctula* and *Rhinolophus hipposideros*) were performed using tRNAthr and tRNAphe universal primers (Kocher *et al.* 1989). Direct sequencing (dideoxy chain termination method after Sanger) was used to select the region that showed the most sequence variation intra- as well as interspecifically. For this segment, two specific *M. myotis* primers were designed, L, 5′– GAACTTATGCAAAGCTTCCA-3′ and H, 5′-GGGTTGGTTTCACGGAGGTA-3′, yielding amplification products of 400–800 bp (Fig. 1). PCR products were electrophoretically separated on agarose gels to determine size classes (Fig. 2). Double-strand amplification products were sequenced directly with the primers used for amplifications.

Results

Sequence analysis of bat mitochondrial DNA

Sequence comparison was performed of the complete non-coding control region of the mitochondrial genome of three individuals of *M. myotis* and of one individual each of the species *M. emarginatus* and *M. nattereri*, closely

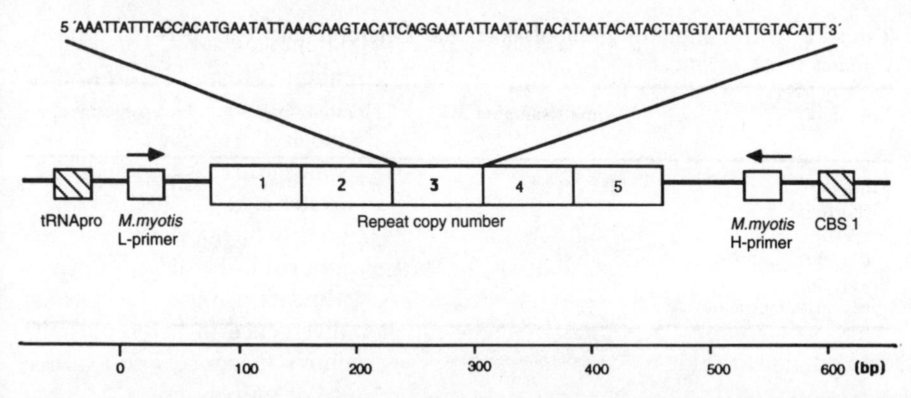

Fig. 1. Schematic diagram of *Myotis myotis* D-loop region containing five copies of an 82-bp repeat unit. Inset shows 5′ to 3′ light strand sequence of the most frequent repeat type within the Portuguese and the German populations of *M. myotis*.

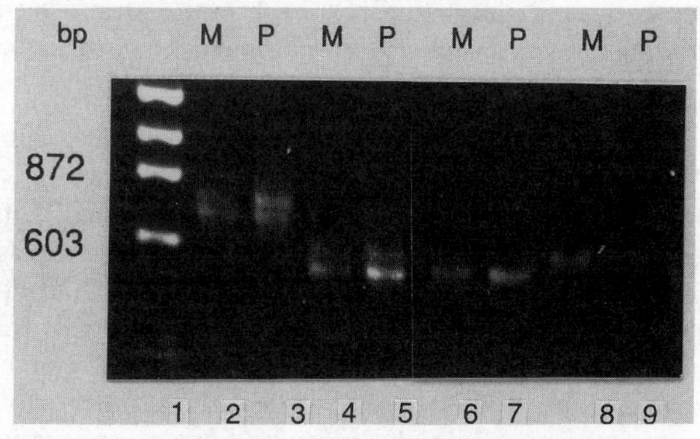

Fig. 2. PCR amplification products using *M. myotis* specific primers that flank the repeat region. Lanes show heteroplasmic and homoplasmic mother–pup pairs, where M stands for mother and P stands for pup. Lanes 2 and 3, 4 and 5 show two heteroplasmic pairs with size classes 8 and 7, and 6 and 5, respectively. The two homoplasmic mother–pup pairs have size class 5 (lanes 6 and 7) and size class 6 (lanes 8 and 9). The picture demonstrates stable inheritance of size variants within one generation.

related to *M. myotis*, the less closely related *Nyctalus noctula* and one species from a different family, *Rhinolophus hipposideros*. This comparison showed the highest sequence divergence between the 3′ end of the tRNA gene for proline and 5′ end of CSB1 (conserved sequence block), a region of high homology within mammalian mitochondrial genomes.

Detection of mitochondrial DNA size variants

Specific *M. myotis* primers located within the hypervariable D-loop segment spanning the region between the gene for tRNA proline and CSB1 (Fig. 1) amplified PCR products that varied in length between 400 and 800 bp

Table 1. Distribution of homoplasmic and heteroplasmic bats between nursery colonies of *M. myotis*

Colonies	Animals sampled (n)	Homoplasmy (%)	Heteroplasmy (%)
Mina Lousal (Portugal)	21	67	33
Au (Bavaria, Germany)	50	72	28
Beyharting (Bavaria, Germany)	30	63	37

among individuals and within individuals (Fig. 2). Direct sequencing of the amplification products showed that the different size variants were due to a variable copy number of an 82-bp repeat unit that was tandemly repeated between three and eight times (Fig. 1). Each single strand 82-bp repeat is capable of folding into a secondary self-complementary structure similar to that of tRNAs.

MtDNA heteroplasmy

Of the 101 bats analysed, 33% were heteroplasmic. Among 21 individuals of the Portuguese colony, Mina Lousal, seven (33%) were heteroplasmic; among 80 animals of the two Bavarian colonies 25 (Au 28%, Beyharting 37%) were heteroplasmic (Table 1). Analysis of 25 mother–pup pairs of the Au colony showed that all pups inherited their mothers' homoplasmy or heteroplasmy respectively (Fig. 2). The smallest observed size variant consisted of three repeat copies. This size class was never observed in a homoplasmic animal, nor was the four-copy size class. The largest size class contained eight repeat copies. The distributions of the different size classes (Beyharting data, not shown) clearly reveal that the sizes of five and six repeat copies are the most frequent among homoplasmic (66% and 79% respectively), as well as heteroplasmic (45%), bats in the three colonies (Fig. 3). The two size variants in each of the heteroplasmic individuals from Mina Lousal differed by one repeat copy only. Heteroplasmic animals from Au and Beyharting colonies showed size combinations in individuals that differed by up to three repeat copies.

Sequence analysis of repeats

The complete sequence of 74 copies of the 82-bp repeat from heteroplasmic Au bats and 49 repeat copies from heteroplasmic Mina Lousal bats revealed 27 different repeat types in the Au colony and 24 different types in the Portuguese colony. Among the two colonies, seven types were identical, with one repeat type being shared by all bats sequenced. In 6068 bp of 74 repeat copies (Au) and 4018 bp of 49 repeat copies (Mina Lousal) 177 and 148

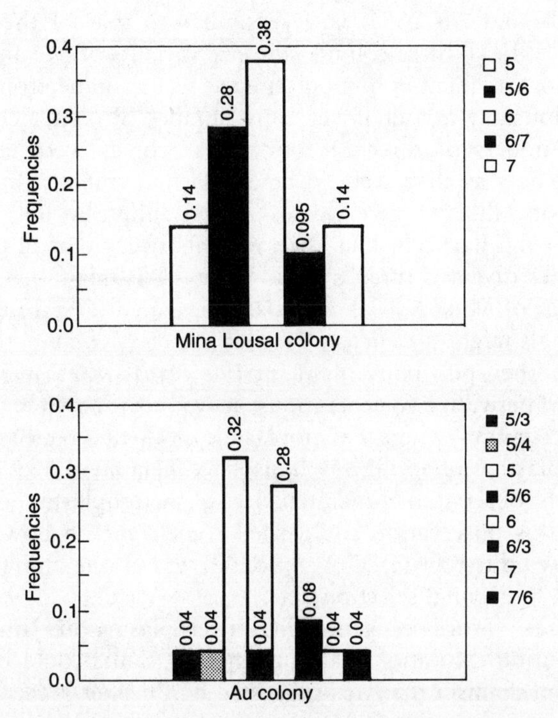

Fig. 3. Frequency distribution of length variants in homoplasmic (white bars) and heteroplasmic (black bars) *M. myotis* from Mina Lousal and Au colonies. Size classes are shown left of the diagram.

substitutions respectively were found, with a percentage difference of 2.9% (Au) and 3.6% (Mina Lousal) respectively. A hundred and sixty-eight of the 177 substitutions in the Au colony were transitions, nine were transversions. Of the 148 substitutions in the Portuguese colony, 118 were transitions and 30 transversions, yielding a transition: transversion ratio of 18:1 for the German colony and 4:1 for the Portuguese colony.

Discussion

Heteroplasmy (molecular heterogenity) is believed to be present among the hundreds of mitochondrial DNA molecules in a cell. Such heteroplasmy may be transient if mtDNA molecules are sorted out through repeated stochastic sampling at each cell division. Nevertheless, heteroplasmy is observed in a number of animal species. Wilkinson & Chapman (1991) reported 27.7% heteroplasmy in *Nycticeius humeralis* from seven colonies in the USA. The current study shows that in 101 individuals of *Myotis myotis* from three colonies, 33% of the bats were heteroplasmic for up to seven different size classes (Fig. 3). The level of heteroplasmy in the two German populations (Beyharting 37% and Au 28%) is very similar to that of the Portuguese population (33%). In all three colonies the size variants of five and six repeat copies are the most frequent in homoplasmic as well as in heteroplasmic bats. Although selection may favour lower copy number (Brown, Beckenbach & Smith 1992), numbers of three or four repeat copies were not observed in homoplasmic bats as they were in heteroplasmic animals. In Portuguese heteroplasmic bats, the two size classes always differ by only one repeat copy, whereas in the Bavarian bats, size variants occur that differ by up to three repeat copies (five and three, six and three). This might be explained by the lower number of Mina Lousal animals investigated (20). Testing of more Portuguese animals might also detect varying size class combinations.

Analysis of mother–pup pairs of *M. myotis* reveals that heteroplasmy is stably maintained between two generations, as was also shown for the evening bat—with one exception, where a homoplasmic mother had one homoplasmic and one heteroplasmic offspring (Wilkinson & Chapman 1991). It would be interesting to observe the maintenance of heteroplasmy over several generations, as this observation could shed some light on how long equal distribution of two different mtDNA molecules within one individual will be maintained until 'purifying' selection acts.

The evolutionary processes generating heteroplasmy are mutation and paternal contribution. Mutations causing length variants could be explained by a molecular mechanism that was suggested by Buroker *et al.* (1990). This model takes advantage of the triple-stranded part of the control region with an 82-bp repeat unit in sturgeon D-loop, of which loss or duplication during replication causes length polymorphism and heteroplasmy. This model does not explain heteroplasmy due to extensive sequence differences, suggesting that one molecule must have derived from the other. Presence of two independently diverged lineages might therefore be attributed to paternal contribution of mtDNA, although the stable inheritance would argue against biparental inheritance.

Acknowledgements

We thank Luisa Rodrigues and Jorge Palmeirim for help and sampling permission of Mina Lousal bats in Portugal. This work was supported by Grant Ne 146/14–1 from the DFG, Germany.

References

Anderson, S., Bankier, A. T., Barrell, B. G., de Bruijn, M. H. L., Coulson, A. R., Drouin, J., Eperon, I. C., Nierlich, D. P., Roe, B. A., Sanger, F., Schreier, P. H., Smith, A. J. H., Staden, R. & Young, I. G. (1981). Sequence and organization of the human mitochondrial genome. *Nature, Lond.* **290**: 457–465.

Biju-Duval, C., Ennafaa, H., Dennebouy, N., Monnerot, M., Mignotte, F., Soriguer, R. C., El Gaaied, A., El Hili, A. & Mounolou, J. C. (1991). Mitochondrial DNA evolution in lagomorphs: origin of systemic heteroplasmy and organization of diversity in European rabbits. *J. molec. Evol.* **33**: 92–102.

Brown, J. R, Beckenbach, A. T. & Smith, M. J. (1992). Mitochondrial DNA length variation and heteroplasmy in populations of white sturgeon (*Acipenser transmontanus*). *Genetics* **132**: 221–228.

Buroker, N. E., Brown, J. R., Gilbert, T. A., O'Hara, P. J., Beckenbach, A. T., Thomas, W. K. & Smith, M. J. (1990). Length heteroplasmy of sturgeon mitochondrial DNA: an illegitimate elongation model. *Genetics* **124**: 157–163.

Ghivizzani, S. C., Makay, S. L. D., Madsen, C. S., Laipis, P. J. & Hauswirth, W. W. (1993). Transcribed heteroplasmic repeated sequences in the porcine mitochondrial. *J. molec. Evol.* **37**: 36–47.

Gyllensten, U., Wharton, D., Josefsson, A. & Wilson, A. C. (1991). Paternal inheritance of mitochondrial DNA in mice. *Nature, Lond.* **352**: 255–257.

Hoeh, W. R., Blakley, K. H. & Brown, W. M. (1991). Heteroplasmy suggests limited biparental inheritance of *Mytilus* mitochondrial DNA. *Science* **251**: 1488–1490.

Hoelzel, A. R., Hancock, J. M. & Dover, G. A. (1993). Generation of VNTR variation and heteroplasmy in the control region of two elephant seal species. *J. molec. Evol.* **37**: 190–197.

Kocher, T. D., Thomas, W. K., Meyer, A., Edwards, S. V., Pääbo, S., Villablanca, F. X. & Wilson, A. C. (1989). Dynamics of mitochondrial DNA evolution in animals: amplification and sequencing with conserved primers. *Proc. natn. Acad. Sci. USA* **86**: 6196–6200.

Laipis, P. L., Van de Walle, M. J. & Hauswirth, W. W. (1988). Unequal partitioning of bovine mitochondrial genotypes among siblings. *Proc. natn. Acad. Sci. USA* **85**: 8107–8110.

Magoulas, A. & Zouros, E. (1993). Restriction-site heteroplasmy in anchovy (*Engraulis encrasicolus*) indicates incidental biparental inheritance of mitochondrial DNA. *Molec. Biol. Evol.* **10**: 319–325.

Mignotte, F., Gueride, M., Champagne, A. M. & Mounolou, J. C. (1990). Direct repeats in the non-coding region of rabbit mitochondrial DNA. Involvement in the generation of intra- and inter-individual heterogeneity. *Eur. J. Biochem.* **194**: 561–571.

Wilkinson, G. S. & Chapman, A. M. (1991). Length and sequence variation in evening bat D-loop mtDNA. *Genetics* **128**: 607–617.

Index

Note: page numbers in *italics* refer to figures and tables.